W9-BUW-408

MATHEMATICS

MATHEMATICS

MODELS AND APPLICATIONS

Lawrence C. Eggan
Charles L. Vanden Eynden

Illinois State University

D. C. HEATH AND COMPANY Lexington, Massachusetts Toronto

Published simultaneously in Canada.

Printed in the United States of America.

International Standard Book Number: 0-669-01051-0

Library of Congress Catalog Card Number: 78-19918

To Chris and Joan

PREFACE

Students who are not required to take any specific college mathematics course quite often enroll in one of several introductory courses. One type of course might be called "mathematics appreciation"; it includes diverse topics such as sets, groups, number theory, and graphs, chosen to give an idea of what modern mathematics is all about. Another type is specifically aimed at elementary education majors. "Finite mathematics" is the label most often given to a third type of course. Although the content of our book substantially overlaps with material usually found in finite mathematics texts, the emphasis is different.

Designed for those students who will probably take only one college mathematics course, this text presupposes only one year of high school algebra. Our purpose is to provide a direct and explicit answer to the question, What good is mathematics? Since students are not satisfied by simply being told that the material they study will be needed in a later course, we have only included material that can be applied immediately to real problems. In particular the chapters on sets and logic that begin most finite mathematics texts will not be found here.

The great majority of exercises in this book are word problems. The ability to manipulate symbols is useless unless it can be applied to life in some way. The stage in the application of mathematics to real-world problems that comes *before* the manipulation of symbols is slighted throughout the education of most students; here it is emphasized.

The problems we have provided are tied to the real world, and the numbers involved are realistic. Sometimes students become so accustomed to working exercises that have been prearranged to come out in terms of pleasant, round numbers that they conclude they have made a mistake when an "unpleasant" number appears. In an age of computers and pocket calculators there is no excuse for this. For most calculations in this book three-digit accuracy is ample (except for finance problems, where each penny is considered sacred).

Pocket calculators today are relatively common in all fields and cannot be ignored, least of all in a mathematics course: A student can now buy a calculator that makes unnecessary most of the tables at the back of mathematics texts. We believe any student owning a calculator should be encouraged to use it in this course as much as possible. However, all appropriate tables *are* provided, and of course possession of a calculator is not at all necessary. Frequently we point out how a calculator can handle problems that would be impossible or excessively tedious with pencil and paper.

The metric system is clearly entering everyone's life. We feel that it is important for present-day students to take the metric system seriously and be able to convert between metric and British units as the need arises. To this end we have used both types of units throughout our examples and exercises and included many cases where conversion is necessary.

Many examples are given for each mathematical technique presented since at this level most learning is done by studying examples rather than by following and applying abstract mathematical explanations.

Chapter 1 begins with a review of how to translate word problems into algebraic expressions. This leads naturally to solving and graphing linear equations, graphing linear inequalities, and an introduction to linear programming.

The chapter on the computer is placed second for two reasons. Besides being of interest in its own right, computer knowledge allows those who have access to a computing facility an opportunity to have this powerful tool available for problems that arise in subsequent chapters. Many special exercises have been included that are too long and tedious to solve without the aid of a computer, or at least an electronic calculator. We have given enough explanation that students can actually learn to write simple programs in either BASIC or WATFIV (a variation of FORTRAN IV). We suggest that the students learn one language or the other but not both. The Appendix contains some additional information about each language, but, as with other chapters, our intention is to give a brief introduction rather than to be definitive in any sense. It is not necessary to have a computer facility available for students to get some insight into the computer from this chapter, nor is this chapter necessary for any of the subsequent chapters.

In Chapter 3 probability questions lead naturally to problems of counting. Many of our examples on probability use games and gambling because students seem to find these more interesting than pulling marbles from an urn. The use of mathematical expectation in decision-making is also emphasized.

The subject of statistics is introduced in Chapter 4 from the standpoint of trying to make sense in human terms of a set of measurements. The normal curve provides the main focus of the chapter.

Chapter 5 is aimed toward making sensible decisions as consumers. The consumer activities involving the most mathematics, borrowing, installment buying, investing, and annuities, get an especially thorough treatment here.

One of the rapidly expanding and increasingly important areas of mathematics furnishes the material for Chapter 6. Barely 200 years old, graph theory is finding its way into the solution of an ever-increasing variety of real-world problems. Our main intent is to get students to recognize when a graph is an appropriate model for a situation rather than to present technical theorems of graph theory. The chapter ends with the PERT method for scheduling complex projects.

The final chapter introduces matrices, one of the most practical tools of modern mathematics. We give applications to inventory matrices, solving systems of linear equations, and Markov chains.

Traditional mathematics is not neglected, for example, graphing (in linear programming and statistics), exponents (in probability and interest problems), and solving equations (in linear programming and Markov chains). Yet the emphasis is never mathematics for the sake of mathematics, but always solving some problem of the outside world. For example, a straight line is first graphed not to picture an arbitrarily presented equation but to show how a couple's federal income tax depends on their taxable income.

We have found that there is ample material in this book for a one-semester course. Thus the instructor may wish to skip some chapters or even rearrange them. The chapters are for the most part independent, as the chart indicates.

Chapter 2 is optional. However, if it *is* covered, it is preferable to

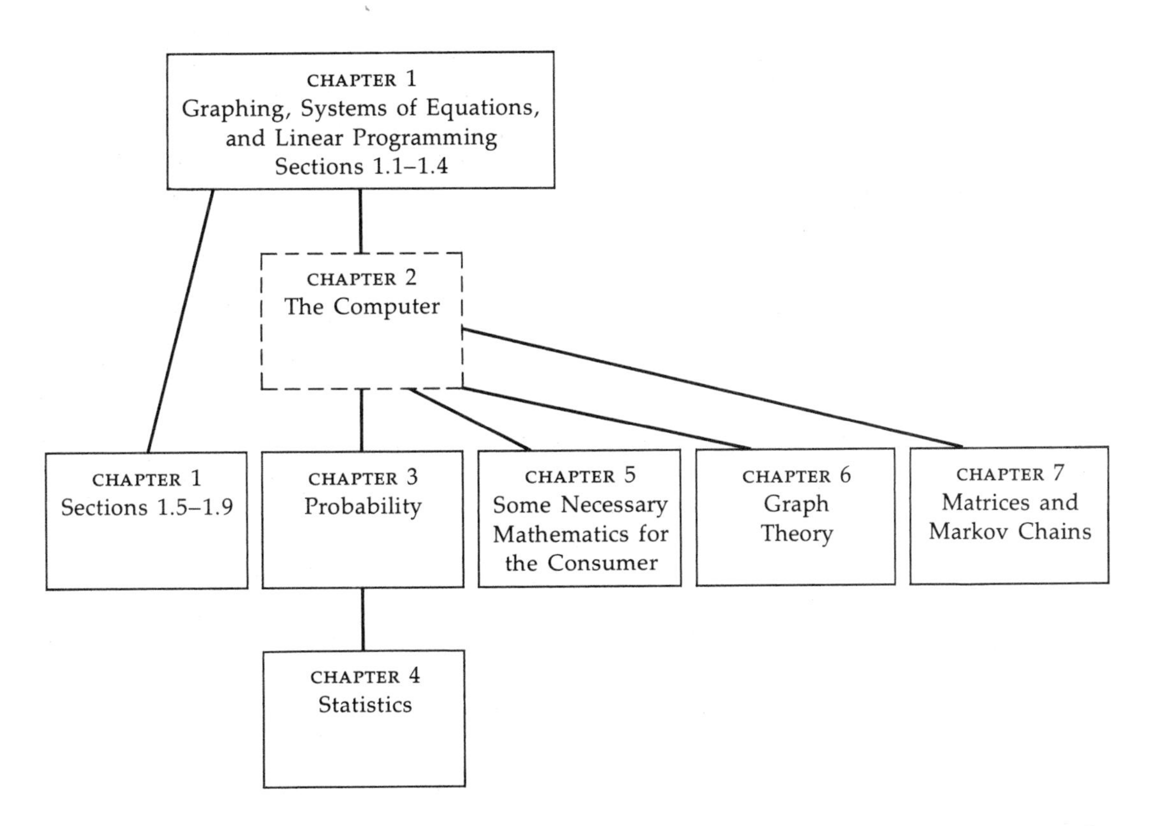

do so directly after Chapter 1 so that the computer may be used as a tool in later chapters.

Throughout this book the mathematics background assumed is kept to a minimal level; Chapters 2 and 6 in particular might form a good introduction to the course.

It is a paradox that at a time when the influence of mathematics in all phases of life has grown and is growing enormously, the understanding of how mathematics is *applied* seems to have decreased, even among college-educated people. We hope that this book can play a part in remedying this situation.

L. C. E.
C. L. V. E.

CONTENTS

CHAPTER ONE GRAPHING, SYSTEMS OF EQUATIONS, AND LINEAR PROGRAMMING 2

1.1 Mathematical Statements 2
1.2 More-Complicated Mathematical Statements 7
1.3 Solving Equations and Inequalities 14
1.4 Graphing 19
1.5 Graphing Inequalities 29
1.6 Systems of Linear Equations 37
1.7 Systems of Linear Inequalities 48
1.8 Linear Programming 60
1.9 More Linear Programming 69

CHAPTER TWO THE COMPUTER 82

2.1 A Brief History and Introduction 82
2.2 Flow Charts 94
2.3 Flow Charts with Looping 105
2.4 An Introduction to BASIC 111
2.5 Summary of Some BASIC Commands 121
2.6 Running a BASIC Program 128
2.7 Transfers and Looping 132
2.8 A New Loop Command and a Last Example 138
2.9 An Introduction to FORTRAN (WATFIV) 145
2.10 Summary of Some FORTRAN Commands 152
2.11 Addressing, Transfers, and Looping 159
2.12 A Last Example and DO-Loops 165

CHAPTER THREE PROBABILITY 174

3.1 Probability 174
3.2 Expectation 179
3.3 The Multiplication Principle 185
3.4 A Multiplication Rule for Probabilities 191
3.5 An Addition Rule for Probabilities 196

3.6 Permutations and Combinations 204
3.7 More Probability Problems 214

CHAPTER FOUR STATISTICS 222

4.1 The Mean and Median 222
4.2 Standard Deviation 228
4.3 The Standard Normal Distribution 235
4.4 The General Normal Distribution 244
4.5 Repeated Trials 252
4.6 The Binomial Distribution 258

CHAPTER FIVE SOME NECESSARY MATHEMATICS FOR THE CONSUMER 268

5.1 What Is Interest? 268
5.2 Savings and Retirement Plans 280
5.3 Variations and Extending Table V 287
5.4 Annuities 291
5.5 Installment Buying and Retirement Planning 295
5.6 On Being a Consumer 299

CHAPTER SIX GRAPH THEORY 326

6.1 Scheduling and Coloring 326
6.2 The Four-Color Theorem and The Salesman Problem 336
6.3 The Mailman Problem 344
6.4 Directed Graphs 351
6.5 Transitive Directed Graphs 359
6.6 PERT 366

CHAPTER SEVEN MATRICES AND MARKOV CHAINS 380

7.1 Introduction and Problem Examples 380
7.2 What Is a Matrix? 383
7.3 Operations on Matrices 389
7.4 Further Properties 400
7.5 Markov Chains 406

7.6 Result after n Transitions 410
7.7 The Main Theorem: A 2×2 Example 414
7.8 The Main Theorem: The General Case 419
7.9 Solving Systems of Equations 429
7.10 Two Extensions 436

APPENDIX 443

Table I Squares and Square Roots 468
Table II Areas Under the Normal Distribution 469
Table III Compound Interest 470
Table IV Present Value of a Dollar 471
Table V Accumulated Amounts 472
Table VI Present Value of an Annuity 473

Answers to Odd-Numbered Exercises 474
Index 521

MATHEMATICS

GRAPHING, SYSTEMS OF EQUATIONS, AND LINEAR PROGRAMMING

CHAPTER ONE

1.1 MATHEMATICAL STATEMENTS

We live in a mathematical world. If this is not generally recognized, it is because many people are not sensitive to the mathematical relationships around them.

A large newspaper may easily contain as many quantitative statements as a mathematics text. Consider the first two paragraphs of an article by Gene Salorio, headlined "Pumpkin Prices Climb," appearing in the *New York Times* of November 11, 1973*:

> Like turkeys, the pumpkin business reaches its peak at this time of year, with millions of Thanksgiving and Christmas turkeys followed by pumpkin pies. Those pies will, however, be a bit more expensive this year.
>
> Pumpkin farmers are receiving about $12 a ton, up 20 percent from last year's price and 50 percent from that of 1971. During the preceding two decades the price had remained stable at $8 per ton.

A number of precise mathematical statements are made in these two paragraphs, but because words rather than symbols are used to express them, the reader may not realize this. In the first paragraph, for example, we learn that pumpkin pies will cost more in 1973 than previously. Let us denote by C_3 the average cost of a pumpkin pie in 1973, and let C_2 denote the corresponding cost in 1972. Then we could express the fact of rising pie cost by

SUBSCRIPTS
The 3 in C_3 is a *subscript*, and C_3 is read "C-sub-three," or, more briefly, "C-three."

$$C_3 > C_2.$$

Here $>$ is the usual mathematical symbol for "is greater than."

In the same way we translate "Pumpkin farmers are receiving about $12 a ton, up 20 percent from last year's prices and 50 percent from that of 1971" into mathematical symbolism by letting P_3, P_2, and P_1 be the price per ton of pumpkins to farmers in 1973, 1972, and 1971, respectively, and writing

PERCENT
"Percent" means "hundredths," so 20% of the 1972 pumpkin price P_2 is $\frac{20}{100}$ of P_2, or $0.20P_2$. If prices are *up* 20% from 1972, the new price must be $P_2 + 0.20P_2$.

$$P_3 = \$12,$$

$$P_3 = P_2 + 0.20P_2,$$

and

$$P_3 = P_1 + 0.50P_1.$$

We do not claim, of course, that our symbolic formulations are better than the ones we started with. It depends on how they are to be

*An article by Gene Salorio headlined "Pumpkin Prices Climb," appearing in the *New York Times* of November 11, 1973. (© 1973 By the New York Times Company. Reprinted by permission.)

used. For a newspaper article the first paragraph of Mr. Salorio's account is certainly better than

> PUMPKIN PRICES CLIMB
> Let C_2 and C_3 denote the average price of a pumpkin pie in 1972 and 1973, respectively. Then $C_2 < C_3$.

For the purpose of further mathematical analysis, however, the latter formulation may be better.

There are various advantages mathematical language may have over plain English. One is that the *conciseness* of mathematics allows the relationships expressed to be understood at a glance. A second is that mathematics has a *precision* that ordinary language often lacks. A third is that mathematical expressions can be *manipulated* (by algebra, for example) to derive facts not at first obvious.

We need to be able to translate ordinary sentences into mathematical symbolism. Let us continue to do this with the pumpkin article.

The statement that pumpkin prices to farmers held steady at $8 per ton during the two decades preceding 1971 might be expressed as follows. Let $P(n)$ be the price per ton farmers got for pumpkins in year n. Then $P(n) = \$8$ for $n = 1951, 1952, \ldots, 1970$.

The article continues:

> "All my expenses are up, and tonnage per acre is down to 10 or 12 where it usually runs between 15 and 18," declared Robert Robson, a pumpkin farmer in Geneva, N.Y., a major growing area.

Let Y be the 1973 pumpkin yield of Robert Robson, in tons per acre, and let U be his usual yield. Then

$$10 \leq Y \leq 12$$

and

$$15 \leq U \leq 18.$$

> Despite higher prices, growers are finding a brisk market. The pumpkins, harvested during the past month, go principally to bakeries and canneries, although Mr. Robson estimated that 10 percent of his crop became Halloween jack-o'-lanterns.

Let B be the part of the 1973 pumpkin crop going to bakeries and canneries, let J be the part becoming jack-o'-lanterns, and let T be the total crop. Then

$$B > \tfrac{1}{2}T,$$

and

$$J = 0.10T.$$

It should be noted that not all the mathematical statements we found above were equations. Although the idea is common that mathematics consists chiefly in manipulating equations, this is not the case. Other symbols that find frequent use are $<$ ("is less than"), $\leq$ ("is less than or equal to"), $>$ ("is greater than"), and $\geq$ ("is greater than or equal to"). For example, the following statements are true:

$$3 < 5,$$

$$3 \leq 5,$$

$$3 \leq 3;$$

while the following are false:

$$3 < 3,$$

$$5 < 3,$$

$$5 \leq 3.$$

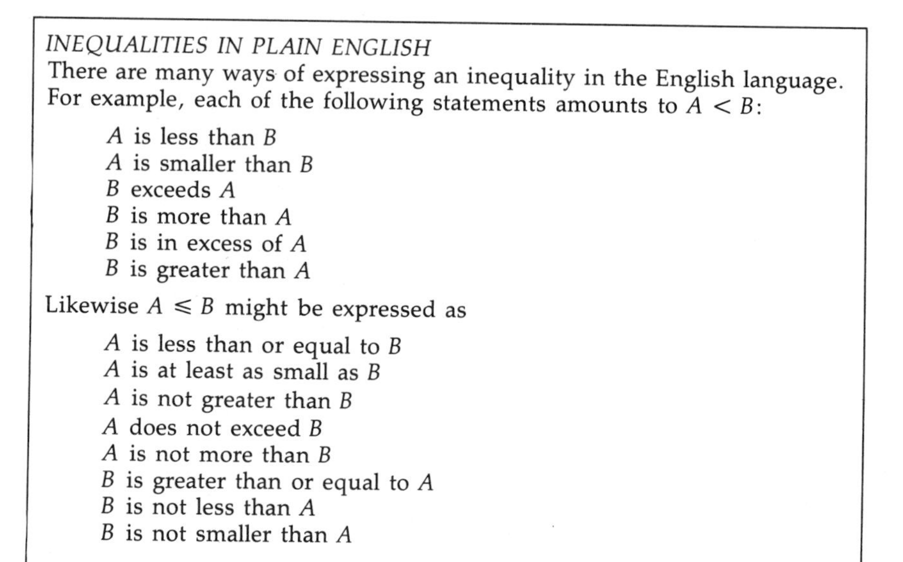

INEQUALITIES IN PLAIN ENGLISH

There are many ways of expressing an inequality in the English language. For example, each of the following statements amounts to $A < B$:

A is less than B
A is smaller than B
B exceeds A
B is more than A
B is in excess of A
B is greater than A

Likewise $A \leq B$ might be expressed as

A is less than or equal to B
A is at least as small as B
A is not greater than B
A does not exceed B
A is not more than B
B is greater than or equal to A
B is not less than A
B is not smaller than A

A statement like

$$J = 0.10T$$

is meaningless in itself, that is, without any explanation of what J and T represent. In our examples we tried to choose letters suggesting the quantities they represented, for example, T for the total pumpkin crop and J for the part of the crop going for jack-o'-lanterns. Although this is not necessary, it is often helpful for remembering the meaning of the symbols.

Exercises 1.1

Express symbolically the quantitative information given in each problem. In the first eight problems use the letters given to denote the various quantities. In the remaining problems invent (and explain) your own notation.

1. "At Filasky's [Brookville, L.I.] roadside stand pumpkins are selling for 12 cents a pound, the same price as last year." [From the same article.] Let F_3 be Filasky's 1973 price per pound for pumpkins, and let F_2 be their corresponding 1972 price.

2. In 1960 hurricane Donna swept the entire U.S. Atlantic coast. High wind speeds of 121 and 130 miles per hour were recorded in Ft. Myers, Florida, and Block Island, R.I., respectively. The storm caused 50 deaths in the United States and an estimated $150 million damage.

Let Y be the year hurricane Donna hit. Let FMW and BIW denote the high wind speeds recorded from it in Ft. Myers and Block Island, and let d and D represent the number of deaths and value of damage due to the hurricane in the United States. (Here FMW and BIW are to be considered as single symbols, not as F times M times W, for example. Although such multiletter symbols are not common in traditional mathematics (except as subscripts), they are often used in computer programming.)

3. In July 1967 there were 97,945,000 men and 101,173,000 women in the United States. The median age of the men was 26.4, of the women, 29.0.

Let M and W be the number of men and women in the United States in July 1967 and let A_M and A_W be the median ages of these two groups.

4. In England and Wales there are over 800,000 births every year, of which 500,000 occur in hospitals. In the United States, where an even greater proportion of births take place in hospitals, there are nearly 4 million births a year, and more than 100,000 women in maternity wards on any given day.

Let B and I denote the number of births and births in hospitals in England and Wales, and let B' and I' denote the corresponding numbers in the United States. Let n denote the number of women in maternity wards in the United States on an arbitrary day.

5. The diameter of the planet Jupiter is 11 times the diameter of Earth, while the mass of Jupiter is more than 300 times Earth's. Let D_J and D_E be the diameters of Jupiter and Earth and let M_J and M_E be the masses of these planets.

6. In 1974 the suicide rate in the United States (per 100,000) was 12.2. This was more than twice that of Italy but less than one-third the rate for Hungary. Let the suicide rates for the United States, Italy, and Hungary be denoted by SUS, SI, and SH.

7. In 1969 the U.S. county with more than 50,000 population with the highest median family income was Montgomery County in Maryland. The median family income there, $16,710, was more than $1000 more than that of the runner-up, Fairfax County, Virginia. Let M and F be the 1969 median family incomes in Montgomery and Fairfax Counties.

8. The total number of people in the regular military forces of the United Kingdom is more than four times the figure for Canada, which is 80,000. The forces of the United Kingdom and Canada combined are still less than those of France, 502,100. Let UK, C, and F represent the number of people in the forces of the United Kingdom, Canada, and France, respectively.

9. The largest city and capital of Paraguay is Asunción, with a population of 305,160, according to a 1962 census. The population of the whole country was estimated to be 2,161,000 in 1967.

10. Roger Bannister of Britain became the first man to run the mile in less than 4 minutes in 1954, when he ran it in 3:59.4. The previous record was 4:01.4, run by Gunder Haegg of Sweden in 1945. The record time in 1864 was 4:56, held by Charles Lawes of Britain.

11. The average (adult) human eyeball weighs about $\frac{1}{4}$ ounce and is about 1 inch in diameter. The average male eye is about $\frac{1}{50}$ inch larger than the average female eye. About $\frac{1}{4000}$ of the weight of an adult is eyeball; in a baby the ratio is $\frac{1}{400}$.

12. In September 1969 production workers in manufacturing in the United States earned an average of $3.24 per hour. The average price of a pound of American Cheddar cheese at that time was 95.4¢, for which an average worker would have to work about 18 minutes. In about 11 minutes he could earn enough to buy a package of 48 tea bags.

13. Write five sentences expressing in English the idea $A > B$. (See the box "Inequalities in Plain English" above.)

14. Write five sentences expressing the idea $A \geqslant B$.

1.2 MORE-COMPLICATED MATHEMATICAL STATEMENTS

We will now consider some examples of statements whose symbolic expression is slightly more complicated. The purpose of this section is simply to get more practice in converting ideas expressed in English into mathematical language.

EXAMPLE 1

Income Tax Table 1.1 reproduces a portion of the federal income tax schedule applying to married taxpayers filing a joint return for 1977. This table actually amounts to 18 different sentences, the applicable one to be chosen by the taxpayer. For example, one of the sentences is

TABLE 1.1

If the amount on Schedule TC, Part I, line 3, is:		Enter on Schedule TC, Part I, line 4:	
Not over $3,200		—0—	
Over—	*But not over—*		*of the amount over—*
$3,200	$4,200	14%	$3,200
$4,200	$5,200	$140+15%	$4,200
$5,200	$6,200	$290+16%	$5,200
$6,200	$7,200	$450+17%	$6,200
$7,200	$11,200	$620+19%	$7,200
$11,200	$15,200	$1,380+22%	$11,200
$15,200	$19,200	$2,260+25%	$15,200
$19,200	$23,200	$3,260+28%	$19,200
$23,200	$27,200	$4,380+32%	$23,200
$27,200	$31,200	$5,660+36%	$27,200
$31,200	$35,200	$7,100+39%	$31,200
$35,200	$39,200	$8,660+42%	$35,200
$39,200	$43,200	$10,340+45%	$39,200
$43,200	$47,200	$12,140+48%	$43,200
$47,200	$55,200	$14,060+50%	$47,200
$55,200	$67,200	$18,060+53%	$55,200
$67,200	$79,200	$24,420+55%	$67,200

Form **1040** Department of the Treasury—Internal Revenue Service **U.S. Individual Income Tax Return 1977**

For the year January 1–December 31, 1977, or other taxable year beginning , 1977 ending , 19 .

Use IRS label. Otherwise, print or type.			
First name and initial (if joint return, give first names and initials of both)	Last name		Your social security number
Present home address (Number and street, including apartment number, or rural route)	For Privacy Act Notice, see page 3 of Instructions.		Spouse's social security no.
City, town or post office, State and ZIP code	Occupation	Yours ▶	
		Spouse's ▶	

If the amount on Schedule TC, Part I, line 3 is
over \$4,200 but not over \$5,200
enter on Schedule TC, Part I, line 4
\$140 + 15% of the amount over \$4,200

If the amount on Schedule TC, Part I, line 3, was \$5000, then the "amount over \$4200" would be \$5000 − \$4200 = \$800.

The amount on Schedule TC, Part I, line 3, referred to is called the taxable income and represents total income less exemptions and deductions. The amount calculated from the table is the tax (except possibly for a tax credit to be subtracted later, which we will ignore).

Let F be the tax due from a couple having a taxable income of \$10,000. Then, since F is between \$7,200 and \$11,200,

$$\begin{aligned} F &= \$620 + 19\% \text{ of the amount over } \$7200 \\ &= 620 + .19(10{,}000 - 7200) \\ &= 620 + .19(2800) \\ &= 620 + 532 = \$1152. \end{aligned}$$

As another example we compute the tax G on a taxable income of \$25,000. Since this is between \$23,200 and \$27,200,

$$\begin{aligned} G &= \$4380 + 32\% \text{ of the amount over } \$23{,}200 \\ &= 4380 + .32(25{,}000 - 23{,}200) \\ &= 4380 + .32(1800) \\ &= 4380 + 576 = \$4956. \end{aligned}$$

EXAMPLE 2

More Taxes The tax schedule given in Table 1.1 is arranged so that the greater the taxable income, the greater the tax. A couple using the schedule tells us that their taxable income is more than \$15,000. Letting their tax be T, we conclude that

THE TAX SCHEDULE

The tax schedule is designed to be without jumps; that is, if a couple's taxable income changes by a small amount, then so does their tax. This is clear enough if the change in taxable income does not alter which line of the schedule applies. It is not so clear if the change entails going from one line to the next.

For example, line 1 applies to a taxable income of \$4200, while line 2 applies to a taxable income slightly over \$4200. The tax on the latter would be \$140 plus 15% of a small amount, or slightly over \$140. But from the first line the tax on \$4200 is 14% of (\$4200 − \$3200), or 14% of \$1000, which is \$140.

In the same way the tax on \$5200 is $\$140 + .15(5200 - 4200) = \$290$ from line 2, and this is where the line 3 taxes start. The same principle holds throughout the schedule.

$$\text{(their tax)} > \text{(the tax on \$15,000)},$$

or

$$T > 1380 + .22(15{,}000 - 11{,}200),$$

and calculate that their tax is more than \$2216.

EXAMPLE 3

Income Unknown A couple used the tax schedule in Table 1.1 and came up with \$1776 as their tax. What can be said about their taxable income?

Let us denote their taxable income by I. Since the tax is \$1380 on a taxable income of \$11,200 and \$2260 on a taxable income of \$15,200, we can conclude that their taxable income must lie between \$11,200 and \$15,200; and so line 6 of the schedule applies. Then

$$\$1776 = \$1380 + \text{(22\% of the amount over \$11,200)},$$

or

$$1776 = 1380 + .22(I - 11{,}200).$$

Of course this equation could be solved for I, but such problems will be deferred until the next section.

EXAMPLE 4

The Cost of Tea In 1969 the average production worker earned \$3.24 per hour and had to work 11 minutes to buy a package of 48 tea bags. We might ask what the price T of a package of tea bags was. Since the worker worked 11 minutes, or $\frac{11}{60}$ hours, at \$3.24 per hour to earn T, we have

$$\text{(price of 48 tea bags)} = \text{(earnings per hour) (hours to pay for tea bags)},$$

or

$$T = (3.24)(\tfrac{11}{60}).$$

We calculate $T = 59.4¢$.

UNITS

The important thing to remember about units is that one must be consistent with them in any given mathematical expression. Thus in the tea bag example time was measured in two different ways; hours (the wage was $3.24 per hour), and minutes (it took 11 minutes to pay for a package of tea bags). Minutes were converted to hours so that these data could be combined in a single equation. One trick useful for such conversion is that of multiplying by 1. Since 1 hour = 60 minutes, to convert 11 minutes to hours we multiply by 1 = 1 hour/60 minutes as follows:

$$11 \text{ minutes} = 11 \cancel{\text{minutes}} \left(\frac{1 \text{ hour}}{60 \cancel{\text{minutes}}}\right) = \frac{11}{60} \text{ hours.}$$

The same trick may be used to convert between British and metric units. For example, knowing that 1 kilogram is 2.2 pounds, we have

$$\begin{aligned} 40 \text{ pounds} &= 40 \text{ pounds} \left(\frac{1 \text{ kilogram}}{2.2 \text{ pounds}}\right) \\ &= \frac{40}{2.2} \text{ kilograms} = 18 \text{ kilograms.} \end{aligned}$$

EXAMPLE 5

The Size of Alaska Even if California and Texas were combined into a single state, its area would not exceed that of Alaska, 586,412 square miles.

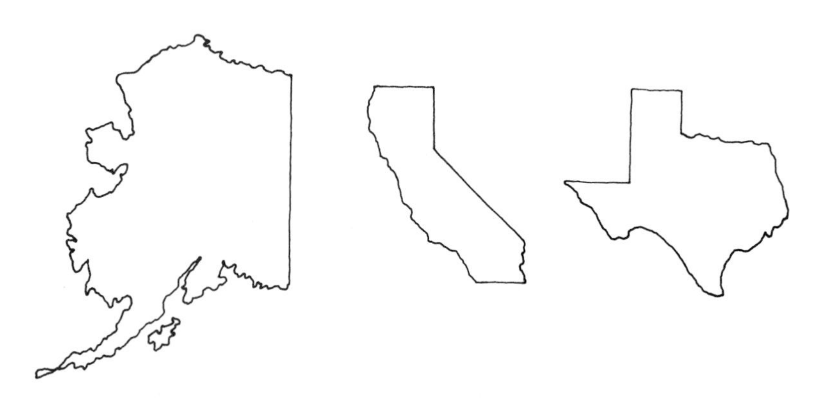

If we let A_T and A_C be the areas of Texas and California, the above statement could be written

$$\text{(area of Texas)} + \text{(area of California)} \leqslant \text{(area of Alaska)},$$

or $$A_T + A_C \leqslant 586{,}412.$$

EXAMPLE 6

The Foreign Car Maria just bought a new foreign car. The owner's manual says not to drive it faster than 100 kilometers per hour for the first 10,000 kilometers. She has $2\frac{1}{2}$ hours to get to a train station 160 miles away. Can she make it? One mile is 1.609 kilometers.

Suppose D is the distance the car goes in 2.5 hours. Then

$$\begin{aligned} D &\leq (2.5 \text{ hours}) \left(100 \frac{\text{kilometers}}{\text{hour}}\right) \\ &= (2.5)\,(100) \text{ kilometers} \left(\frac{1 \text{ mile}}{1.609 \text{ kilometers}}\right) \\ &= 155 \text{ miles.} \end{aligned}$$

She can't make it.

EXAMPLE 7

Pushing it Driver Maria of Example 6 decides she must ignore the owner's manual and drive fast enough to catch her train 160 miles away in $2\frac{1}{2}$ hours. What can we say about the speed she must maintain on her speedometer, which reads in kilometers per hour?

If we let S be her speed in kilometers per hour, then she will go a distance

$$\begin{aligned} (2.5 \text{ hours}) \left(\frac{S \text{ kilometers}}{\text{hour}}\right) &= (2.5S \text{ kilometers}) \left(\frac{1 \text{ mile}}{1.609 \text{ kilometers}}\right) \\ &= \frac{2.5}{1.609} S \text{ miles.} \end{aligned}$$

We must have $(2.5/1.609)\, S \geq 160$.

EXAMPLE 8

Hauling Fruit A trucker will load his truck with crates of pears and apples. A crate of pears weighs 30 pounds and a crate of apples weighs 25 pounds. His truck can carry no more than 14,000 pounds.

Suppose we let P be the number of crates of pears he takes and let A be the number of crates of apples. What can we say about P and A?

The weight of the P crates of pears, weighing 30 pounds each, will be $30P$. Likewise the weight of the A crates of apples, weighing 25 pounds each, will be $25A$. The total weight of the fruit will thus be

$$30P + 25A.$$

Thus we have

$$30P + 25A \leq 14{,}000.$$

Exercises 1.2

In each problem write one or more mathematical statements (equations or inequalities) expressing the information given. Reduce purely numerical expressions to a single number. For example, if the problem said "Kurt's weight W is more than twice the 120 pounds Jane weighs," the answer would be $W > 2(120)$, or $W > 240$.

1. A couple had a 1977 taxable income of $3500 and used the schedule given in Example 1 to compute their tax T.
2. A couple had a taxable income of less than $5000 and a tax T.
3. A couple with a taxable income I pays $1000 tax. [*Hint:* A taxable income of $7200 implies a tax of $620 and a taxable income of $11,200 implies a tax of $1380. Thus I is between $7200 and $11,200.]
4. The worker earning $3.24 per hour mentioned in Example 4 had to work 14 minutes to pay C dollars for a can of coffee.
5. The worker of problem 4 had to work t minutes to pay for a pound of butter, costing 85¢.
6. In September 1969 the average price P in dollars of a medium grade wool suit was such that the worker of problem 4 had to work more than 24 hours to pay for it.
7. The Brooklyn Bridge spans 1595 feet, but the Golden Gate Bridge spans G feet, which is more than two and one-half times this distance.
8. The span of the Verrazano–Narrows Bridge between Staten Island and Brooklyn exceeds by more than 50 feet the span of the Golden Gate Bridge. Denote by V and G the spans of the two bridges, in feet.
9. The last major league baseball player to bat over .400 was Ted Williams, who batted .406 for Boston in 1941. The batting average Y of Carl Yastrzemski, also of Boston, was more than 100 points less when he won the American League batting championship in 1968.
10. In 1970 there were an estimated 14,467,000 people in Czechoslovakia, 77% of them Roman Catholics. Let C be the number of Roman Catholics in Czechoslovakia.
11. The 1970 revenue of General Motors in millions of dollars was 18,752, and this was 22.8% less than its 1969 revenue. Let its 1969 revenue in millions of dollars be R.

"PERCENTAGE" IN BASEBALL
Batting (and other) averages in baseball are always given as a three-digit decimal. Thus a hitter batting .300 is one who gets a hit $\frac{300}{1000}$ or $\frac{3}{10}$ of his times at bat. He is nevertheless said to have a "percentage" of 300, even though this means 300 thousandths, not hundredths. An average of .315 would be called 15 "percentage points" or 15 "points" higher than one of .300.

12. The 1970 revenues of IBM and Mobil Oil were $7504 million and $7261 million, respectively. The total of the two companies was still less than the revenue of AT&T, however. Let the latter be R', in millions of dollars.

13. In 1947 a blue whale was caught which was more than 90 feet long. Its heart weighed more than 1500 pounds. Let its length in feet be L and let H be the weight of its heart, in tons.

14. The largest known dog (a St. Bernard) weighed more than 294 pounds more than the smallest known fully grown dog (a Chihuahua). Let S be the weight of the St. Bernard and let C be the weight of the Chihuahua, both in pounds.

15. The senior class at Wilson High School has $120 from a paper drive. They decide to use it to buy records for the school library. Some records cost $5.98 and some cost $6.98. Let x be the number of $5.98 records and y be the number of $6.98 records they buy.

16. A sales representative must sell at least $2000 worth of encyclopedias each month to keep her job. Suppose she sells S of the standard sets, at $450 each, and D of the deluxe sets, at $600 each. She keeps her job.

17. A girl buys some candy bars for 20¢ each and some comic books for 30¢ each, receiving change from a $2 bill. Denote by CAB the number of candy bars she buys and by COB the number of comic books she buys.

18. A packer is putting boxes of candy of two different sizes into a carton. The small box takes up 22 cubic inches and the large box takes up 35 cubic inches. The carton is 355 cubic inches. Let S denote the number of small boxes and L the number of large boxes packed.

19. A fisherman is only allowed to take a total of 15 pounds of fish from a certain lake. Since there are only two varieties of fish in the lake, for simplicity all fish of the smaller variety are assumed to weigh 6 ounces and all fish of the larger variety are assumed to weigh 9 ounces by the game wardens. Let F_s and F_l denote the number of small and large fish a law-abiding fisherman catches one day.

20. A certain country allows a resident to bring into the country no more than 3 gallons of liquor. A traveler brings in some fifth and quart bottles, F of the former and Q of the latter. (A fifth bottle holds $\frac{1}{5}$ of a gallon.)

21. Three children need to take at least 100 orders for boxes of greeting cards in order to win a tape recorder. They get the recorder after

Amos brings in A orders, Betty brings B orders, and Carlos brings C orders.

22. A house-plant supplier has a contract to supply at least 1000 plants to a store. The contract is fulfilled with S sheffleras, B Boston ferns, and A African violets.

23. A department store buyer puts in an order for d dresses, b blouses, and v vests. Each dress costs \$30, each blouse \$12, and each vest \$14. She gets a discount because the order exceeds \$500.

24. A trucker loads his truck with 60-pound television sets, 50-pound stereos, and 10-pound radios. He takes x televisions, y stereos, and z radios, and the truck can carry at most 24,000 pounds.

25. A grower needs at least 500 flower seedlings. She plants S snapdragon seeds (germination 55%) and C columbine seeds (germination 85%).

26. A shopper buys x oranges at 12¢ each and y apples at 15¢ each and spends \$5.67.

27. A railroad car carries F full-sized cars weighing 3400 pounds each and C compact cars weighing 2200 pounds each for a total weight of 28,000 pounds.

28. A family buys B_1 boxes of Christmas cards containing 12 cards each and B_2 boxes of cards containing 10 cards each. This gives them a total of 78 cards.

29. A grocer buys A cartons of Chockobars and B cartons of Milkimellos. A carton of Chockobars contains 120 candy bars, which sell for 15¢ each. A carton of Milkimellos contains 90 candy bars, which sell for 20¢ each. The candy eventually all gets sold for more than \$400.

30. A man buys x 6-packs of 12-ounce bottles of beer and y 8-packs of 11-ounce bottles, for a total of more than 3 gallons.

1.3 SOLVING EQUATIONS AND INEQUALITIES

EXAMPLE 1

Figuring a Friend's Income At a party a friend mentions that he paid \$900 federal income tax for 1977. Let us try to compute his taxable income. Suppose we know that he filed a joint return. Referring back to the table at the beginning of Section 1.2, we see that \$620 tax is paid on a taxable income of \$7200 and \$1380 tax is paid on a taxable income of \$11,200. Since the greater the taxable income, the greater the tax, we conclude his taxable income I lies between \$7200 and \$11,200. Then, according to the table,

$$(\text{his tax}) = 620 + .19(\text{excess of his taxable income over } \$7200),$$

$$\text{or} \quad 900 = 620 + .19(I - 7200).$$

We subtract 620 from both sides of this equation.

$$900 - 620 = .19(I - 7200),$$
$$280 = .19(I - 7200).$$

We divide both sides by .19.

$$\frac{280}{.19} = I - 7200,$$
$$1474 = I - 7200$$

(rounding off to the nearest dollar). We add 7200 to both sides.

$$1474 + 7200 = I,$$
$$8674 = I.$$

Our conclusion is that our friend's taxable income was $8674.

The General Method

The method we used in Example 1 was to make various transformations on the equation we started with until the desired unknown quantity, I, was isolated. Each equation was changed to an *equivalent* equation, that is, one having the same solution.

An equation is changed to an equivalent equation when any of the following transformations are performed on it:

1. *Adding the same number to both sides*
2. *Subtracting the same number from both sides*
3. *Multiplying both sides by the same nonzero number*
4. *Dividing both sides by the same nonzero number.*

EXAMPLE 2 **Ford's 1969 Revenues** The Ford Motor Company had 1970 revenues of $14,980 million, 1.5% above the year before. What were the company's 1969 revenues?

Call the 1969 revenues R. Then (in millions of dollars)

$$(1970 \text{ revenues}) = (1969 \text{ revenues}) + (1.5\% \text{ of } 1969 \text{ revenues}),$$

or $$14{,}980 = R + .015R = 1(R) + .015R = (1 + .015)R = 1.015R.$$

Thus $$\frac{14{,}980}{1.015} = R \qquad \text{(using rule 4).}$$

We calculate R to be \$14,759 million.

Solving Inequalities

Inequalities can be solved in a similar way, but the rules are slightly different.

An inequality is changed to an equivalent inequality when any of the following transformations are performed on it:

1′. *Adding the same number to both sides*
2′. *Subtracting the same number from both sides*
3′. *Multiplying both sides by the same* ***positive*** *number*
4′. *Dividing both sides by the same* ***positive*** *number.*

EXAMPLE 3

Another Tax Deduction A couple says that they are in the 25% bracket and paid more than \$3000 tax for 1977. What can we deduce about their taxable income I?

The 25% bracket statement means they figure their tax from the seventh line of the tax table in Section 1.2, according to which extra income is taxed at a 25% rate. Thus we have

$$\$2260 + \left(\begin{array}{c}25\% \text{ of their excess}\\ \text{taxable income over } \$15{,}200\end{array}\right) > \$3000,$$

or $$2260 + .25(I - 15{,}200) > 3000.$$

Then
$$.25(I - 15{,}200) > 3000 - 2260 \qquad \text{(using rule 2′),}$$
$$.25(I - 15{,}200) > 740,$$
$$I - 15{,}200 > \frac{740}{0.25} \qquad \text{(using rule 4′),}$$
$$I - 15{,}200 > 2960,$$
$$I > 2960 + 15{,}200 \qquad \text{(using rule 1′),}$$
$$I > 18{,}160.$$

Since the upper limit on taxable income for the 25% bracket is \$19,200 (refer to the tax table again), we have

$$18{,}160 < I \leq 19{,}200.$$

EXAMPLE 4 **Breaking Dishes** A dish manufacturer makes a 10¢ profit on each dish he sells; but every defective dish returned to the factory costs him 40¢. A profit of more than \$5000 was made on a recent batch of 55,000 dishes. What can we say about the number d of defective dishes?

We have (in dollars)

$$5000 < (\text{profit in sales}) - (\text{loss due to returns}),$$

or

$$5000 < (\text{number sold})(.10) - (\text{number returned})(.40),$$

or

$$5000 < (55{,}000 - d)(.10) - d(.40).$$

Thus

$$\begin{aligned}
5000 &< 55{,}000(.10) - d(.10) - d(.40),\\
5000 &< 5500 - d(.10 + .40),\\
5000 &< 5500 - .50d,\\
0 &< 5500 - .50d - 5000 \qquad (\text{using rule } 2'),\\
0 &< 500 - .50d,\\
.50d &< 500 \qquad (\text{using rule } 1'),\\
d &< \frac{500}{.50} \qquad (\text{using rule } 4'),\\
d &< 1000.
\end{aligned}$$

Notice that we avoided dividing by the negative number $-.50$ above by adding $.50d$ to both sides of the equation. Rule 4′ only allows division by *positive* numbers. Rule 3′ is restricted in a similar way.

Exercises 1.3

In each problem write an equation or inequality involving the unknown quantity, then solve it by using allowable transformations.

1. A man buys R records for \$5.88 each and a book for \$8.95, paying a total of \$32.47.
2. Claire bought C candy bars for 15¢ each, receiving 5¢ change from a \$2 bill.

3. A truck is loaded with 12 washers weighing 130 pounds each and 9 dryers weighing D pounds each, for a total of 2100 pounds.

4. In a 4-hour shift a worker can bolt on P metal plates, taking 12 minutes for each one, and have 24 minutes left over for a coffee break.

5. Sally's car can go 300 miles on a tankful of gas. It gets M miles to the gallon and the tank holds 20 gallons.

6. Angelo can take a 200-mile trip in his car and have 2 gallons of gas left in the tank. The car gets 25 miles to the gallon and the tank holds T gallons.

7. A company buys 10,000 bolts for B cents each, paying \$850.

8. The 10,000 bolts in problem 7 weigh W ounces each, for a total weight of 2500 pounds.

9. A coffee mug holds 2 ounces more than a cup, which holds C ounces. Four mugs and six cups hold a total of 88 ounces.

10. An eight-track tape costs \$1 more than a record, which costs R dollars. Three records and two tapes cost \$21.90.

11. The tax of a couple using the schedule given in Section 1.2 is \$500. Find their taxable income I.

12. A couple in the 22% bracket paid a tax of more than \$2000. What can you conclude about their taxable income I?

13. A worker who was paid \$4.38 per hour in September 1978 worked t minutes to pay for a 73¢ half-gallon of milk.

14. Ten New Hampshires would fit into the 262,134 square miles of Texas. What can be said about N, the area of New Hampshire?

15. A woman has invented a game. She has found a manufacturer who will produce the games for 50¢ each, plus \$10,000 to set up the machinery. The woman can sell the games to retailers for \$1.50 each. She wants to have N games made and make a profit.

16. The woman in problem 5 has only \$20,000 to invest and the manufacturer requires payment in advance. What does this say about N?

17. A trucker is making up a load of apples and peaches. A box of apples weighs 20 pounds; a box of peaches weighs 25 pounds. He has contracted to carry exactly 200 boxes of apples; and his truck can carry no more than 10,000 pounds. Suppose he carries P boxes of peaches.

18. A manufacturer has a budget of \$1,000,000 to advertise his razor blades. He has already committed \$400,000 for TV ads, and plans to spend the rest on N newspaper ads, at \$1500 per ad.

19. An apple contains 76 calories and an ounce of peanut brittle has 125. An apple and two raw carrots do not total as many calories as an ounce of peanut brittle. Let C denote the number of calories in a carrot.

20. Three doughnuts have more calories than 5 apples. (See problem 19.) Let D denote the number of calories in a doughnut.

1.4 GRAPHING

Pictures of the Tax Table

A couple filing a joint federal tax return for 1977 having a taxable income I between \$7200 and \$11,200 pays a tax T, where, according to Table 1.1,

$$T = 620 + .19(I - 7200).$$

This formula allows T to be calculated for any given value of I in the range from \$7200 to \$11,200. The following table shows a few pairs I and T.

I	$T = 620 + .19(I - 7200)$
7,200	620
8,200	810
9,200	1,000
10,200	1,190
11,200	1,380

We might make a bar graph to illustrate our calculations.

Figure 1.1 lists the taxable incomes along a horizontal axis; the length of the bar above each income indicates the corresponding tax. Since only the length of each bar is important, it would suffice to plot only its top point, as shown in Figure 1.2 (see page 20).

Notice that our picture involves both a horizontal and a vertical scale. Taxable incomes I are measured along the horizontal scale; we will call the horizontal line in our picture the *I-axis*. The vertical line we will call the *T-axis* since the tax T is measured along it.

Figure 1.1

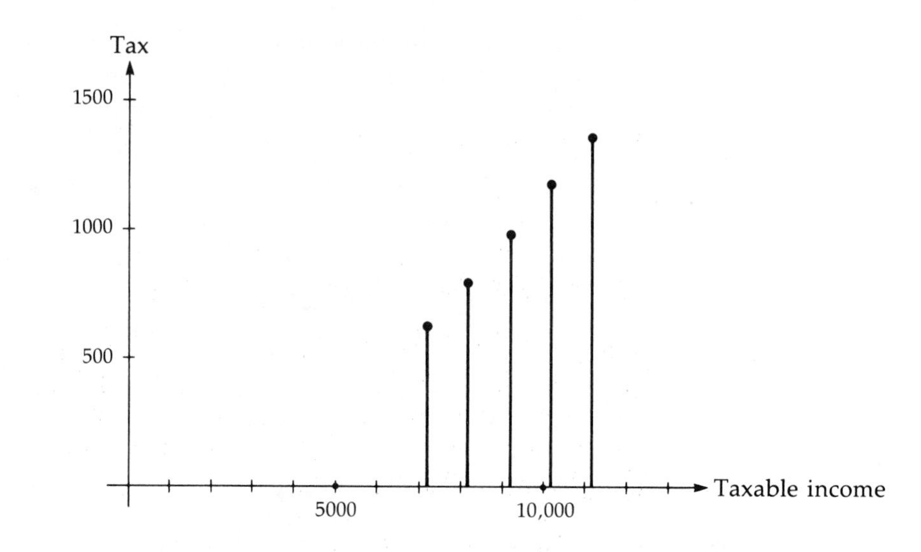

The leftmost point in our picture corresponds to the pair $I = 7200$, $T = 620$. This is indicated graphically by placing it above the 7200 unit point on the I-axis and level with the 620 unit point on the T-axis.

We can plot as many points as we want. Setting $I = 7700$, for example, we compute $T = 715$. We indicate this by placing a point

Figure 1.2

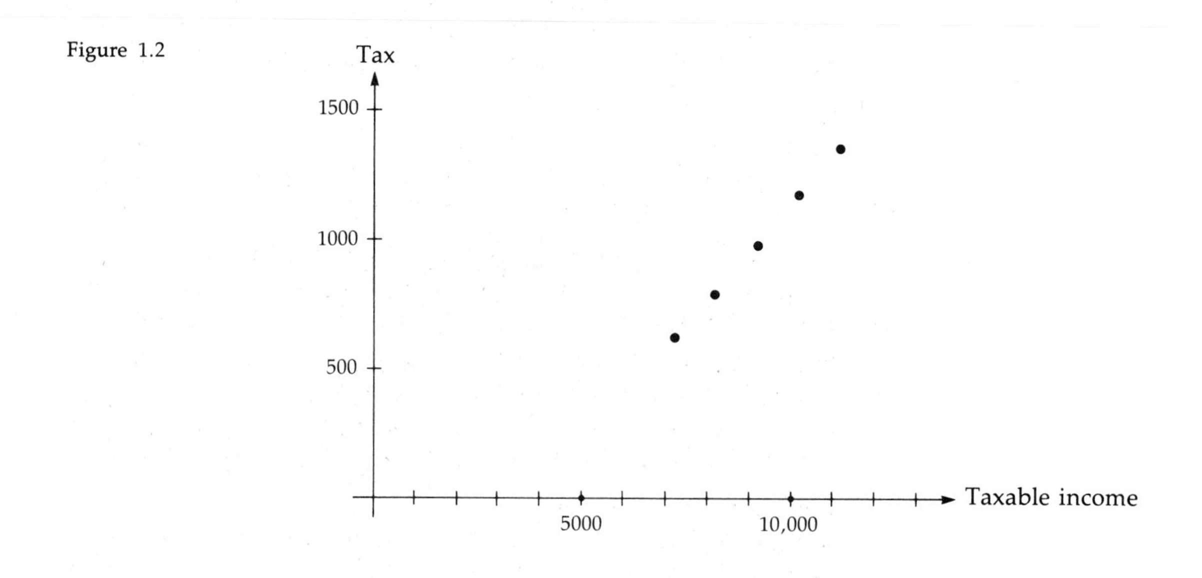

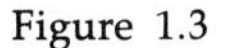
Figure 1.3

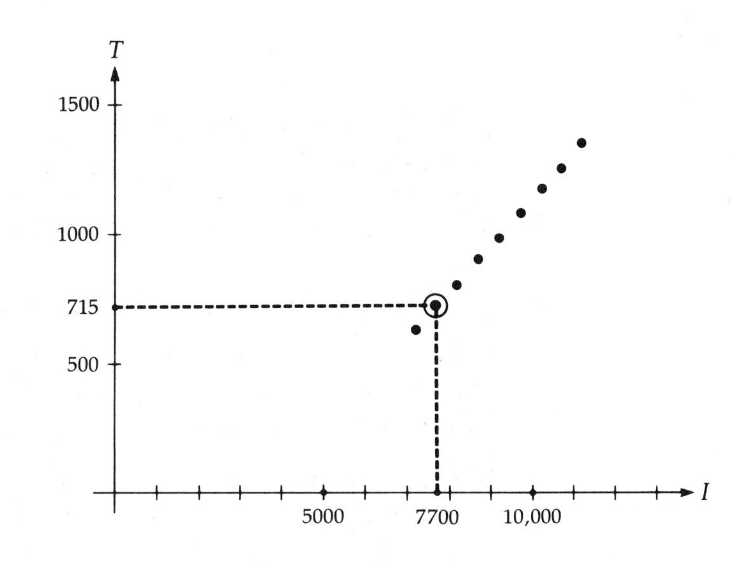

above the 7700 unit point on the I-axis and level with the 715 unit point on the T-axis. This point has been circled in Figure 1.3, in which a number of additional points have been added.

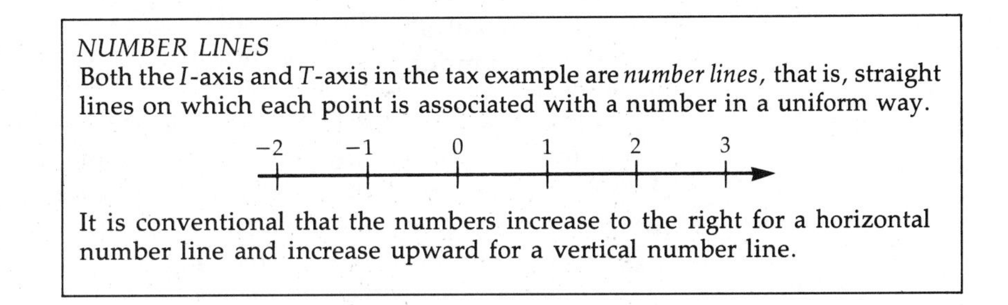

NUMBER LINES

Both the I-axis and T-axis in the tax example are *number lines,* that is, straight lines on which each point is associated with a number in a uniform way.

It is conventional that the numbers increase to the right for a horizontal number line and increase upward for a vertical number line.

There are infinitely many numbers between 7200 and 11,200 that we could choose for I. It seems plausible that if we plotted a point for each of them we would produce a picture such as that shown in Figure 1.4 at the top of page 22.

Our picture appears to be a straight line segment. (It is.) It is called the *graph* of the equation

$$T = 620 + .19(I - 7200)$$

for $7200 \leqslant I \leqslant 11,200$.

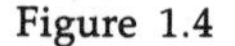
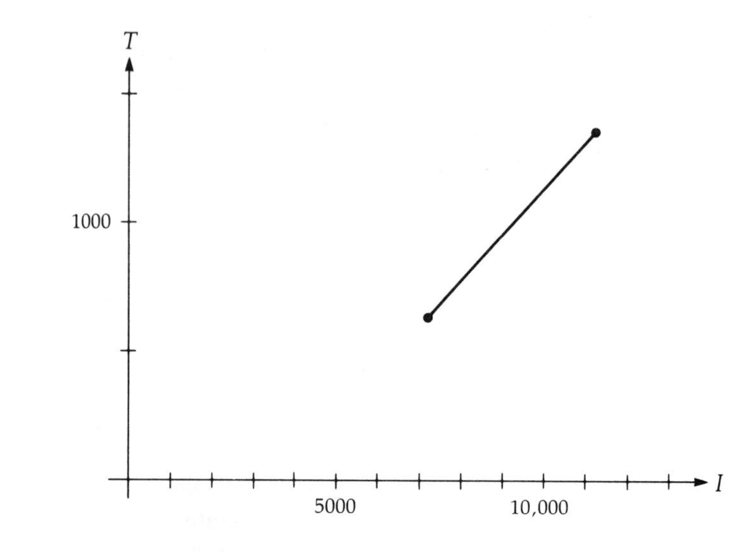

Figure 1.4

Graphs in General

We will now give a general definition of the word "graph." First, we note that although in the example above we had the variables I and T and the corresponding I-axis and T-axis, it is traditional in mathematics to use the letters x and y, calling the horizontal axis the x-axis and the vertical axis the y-axis. (This convention causes no problem, since we could just as easily label taxable income by x as call it I; likewise we could let y denote the tax.)

Suppose a horizontal number line (the x-axis) and a vertical number line (the y-axis) are drawn, intersecting at the zero point of each line. We associate with each pair of numbers x, y a point directly above the number x on the x-axis and level with the number y on the y-axis. This point is called (x, y) (see Figure 1.5).

Conversely, suppose P is any point in the plane. Assume P is directly above the number x on the x-axis and level with the number y on the y-axis. Then $P = (x, y)$. We call x the *x-coordinate* of P and y the *y-coordinate* of P. For example, the x-coordinate of (3, 1) is 3 (see Figure 1.6).

HOW MANY POINTS MAKE A LINE?
It might be argued that since the penny is the smallest unit of American money, the set of values that the variable I might assume is not infinite, but rather limited to 7200.00, 7200.01, 7200.02, . . ., 11,199.99, 11,200.00. Our interpretation of the graph as a straight line segment is correct *as a picture of the purely mathematical statement*

$$T = 620 + .19(I - 7200), \qquad 7200 \leqslant I \leqslant 11{,}200,$$

however.

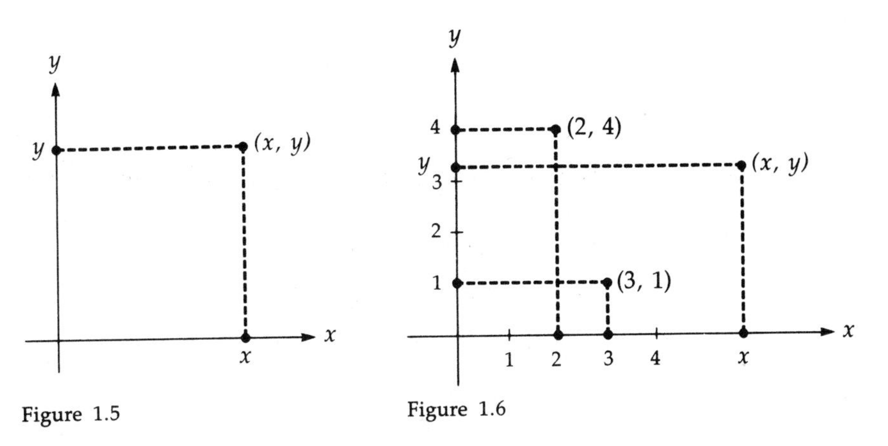

Figure 1.5

Figure 1.6

Now, suppose S is any statement about x and y such that for any pair of numbers x and y the statement is either true or false. (S might be, for example, an equation or inequality involving x and y.) Then the *graph* of S is defined to be the set of all points (x, y) such that S is true about x and y.

The statement

$$\text{“}y = 620 + .19(x - 7200) \quad \text{and} \quad 7200 \leqslant x \leqslant 11{,}200\text{”}$$

has the graph shown in Figure 1.7.

Figure 1.7

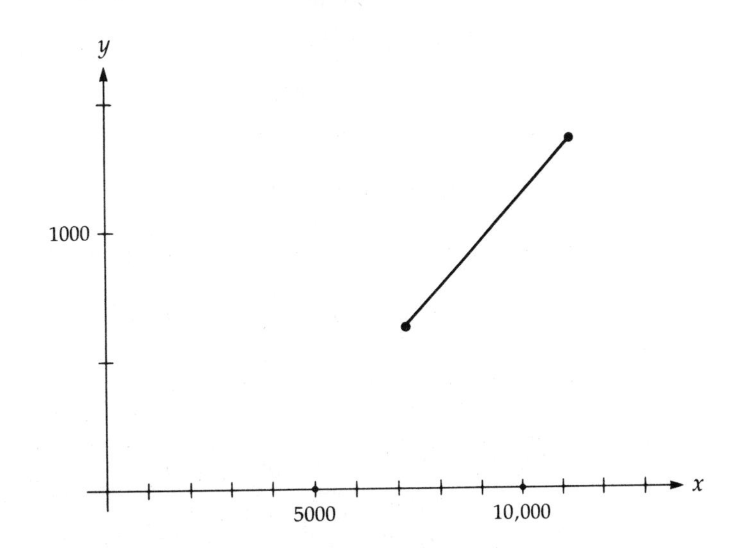

EXAMPLE 1

The Economical Jailer A jailer usually serves hamburger and bread to his prisoners for lunch. He does not want to be accused of underfeeding them. He reads that the recommended daily calorie intake for males 18 to 35 years old is 2900 calories. Dividing by 3 and allowing for a lack of

THE NONCOMMITTAL x AND y

Of course, there is a clear disadvantage to using the letters x and y to represent the variables in a problem rather than letters, such as initials, more suggestive of the real-world quantities for which they stand. It is easy to remember that I is the income and T the tax, and not so easy to remember what x and y stand for. Also, in another problem x and y may stand for entirely new quantities, and this sudden switching of meaning takes some getting used to.

Although there is no prohibition against using names other than x and y when graphing, these letters do carry one advantage, namely, one need never wonder which is measured along the horizontal and which along the vertical axis. It is one of the strongest conventions of mathematics that x is always measured horizontally and y vertically.

exercise, he decides 900 calories should be sufficient for lunch. He plans to serve *exactly* 900 calories to save money.

Hamburger contains 105 calories per ounce and a slice of white bread has 63 calories. If he serves x ounces of hamburger and y slices of bread, we have

(calories in x ounces hamburger) + (calories in y slices bread) = 900,

or

$$105x + 63y = 900.$$

Solving for y, we have

$$y = \frac{900 - 105x}{63}.$$

We make a table of corresponding values of x and y and plot the points (x, y) in Figure 1.8.

x	0	1	4	6	8
y	14.3	12.6	7.6	4.3	1.0

The graph is shown in Figure 1.9.

CHOOSING UNITS FOR GRAPHING

It is not necessary to use the same unit length (that is, the distance between 0 and 1) on the two axes when making a graph. Since we may measure entirely different quantities along the axes, there usually would be no point to this. For example, the x-axis might correspond to hours worked in a week and the y-axis to dollars earned. Rather units should be chosen on the axes so as to allow all the important information to be graphed, while filling up as much of the space available, and thus providing the most information, as possible.

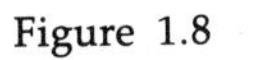
Figure 1.8

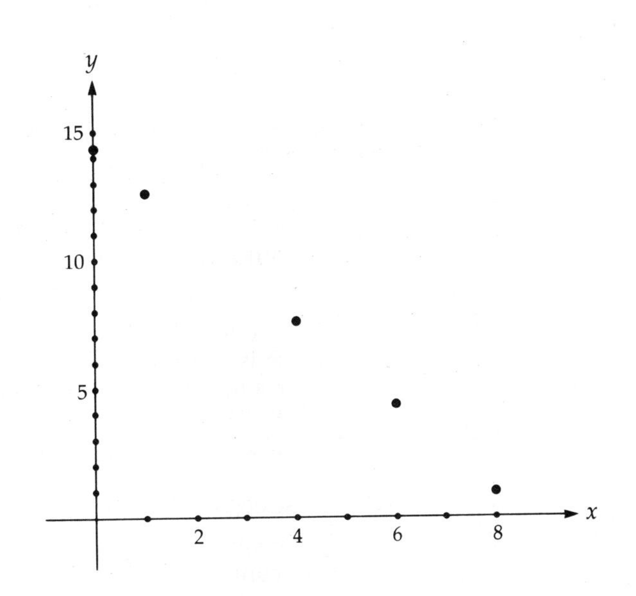

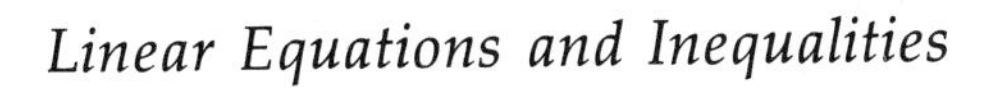

Linear Equations and Inequalities

The equation

$$105x + 63y = 900$$

graphed in Figure 1.9 is an example of a *linear equation,* an equation equivalent to one of the form

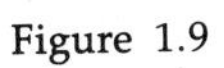
Figure 1.9

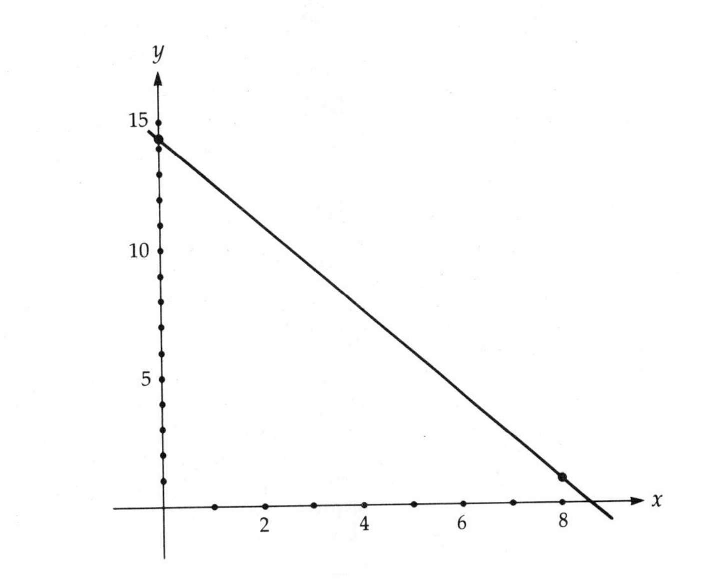

$$ax + by = c,$$

where a, b, and c are constants. (In the above equation $a = 105$, $b = 63$, and $c = 900$.) It can be shown that *the graph of any linear equation is a straight line.*

Since a straight line is determined by any two points on it, once we notice that an equation is linear, it is easily graphed. We need only to find two pairs (x, y) satisfying the equation, plot them, and then draw the straight line through the two points.

The point where $x = 0$ and the point where $y = 0$ are often conveniently calculated. For the equation

$$105x + 63y = 900,$$

for example, we find that if $x = 0$, then $105 \cdot 0 + 63y = 900$, or

$$63y = 900,$$

and $$y = \frac{900}{63} = 14.3.$$

Likewise, if $y = 0$, then $105x + 63 \cdot 0 = 900$, or

$$105x = 900,$$

and $$x = \frac{900}{105} = 8.6.$$

We see that the points (0, 14.3) and (8.6, 0) are on the graph. (In practice it is wise to plot three points. If they are not in a straight line and the equation is linear, then some mistake must have been made.)

Similarly, a *linear inequality* is defined to be an inequality equivalent to one of the form

$$ax + by < c,$$

$$ax + by \leqslant c,$$

$$ax + by > c,$$

or $$ax + by \geqslant c,$$

where a, b, and c are constants. An example is

$$3x + 50 < y,$$

since by using the transformations listed in Section 1.3 this can be changed to

$$3x + (-1)y < -50.$$

EXAMPLE 2

Buying Furniture A buyer for a furniture store has \$10,000 to be spent on an order of coffee tables, which wholesale for \$150 each, and end tables, which wholesale for \$80 each. Suppose she orders x coffee tables and y end tables. Then

$$(\text{cost of coffee tables}) + (\text{cost of end tables}) = 10{,}000$$

or

$$150x + 80y = 10{,}000.$$

This is a linear equation and so its graph is a straight line. Setting $x = 0$ gives $80y = 10{,}000$, and

$$y = \frac{10{,}000}{80} = 125.$$

Thus (0, 125) is on the graph.

Similarly, setting $y = 0$ gives $150x = 10{,}000$, and

$$x = \frac{10{,}000}{150} = 66.7,$$

and we see that (66.7, 0) is on the graph as seen in Figure 1.10.

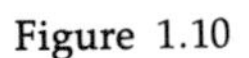

Figure 1.10

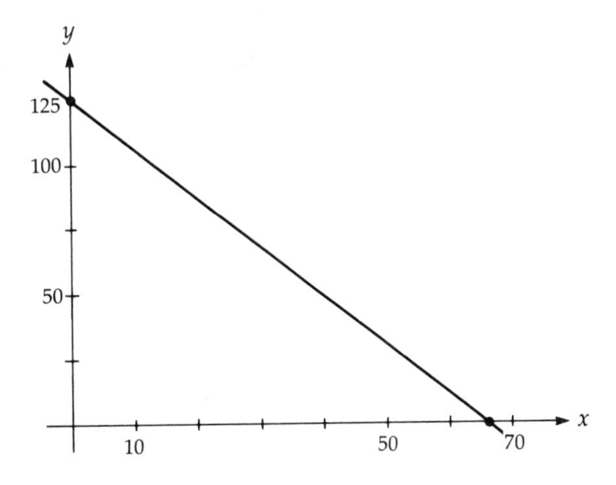

Exercises 1.4

1. Suppose the jailer in Example 1 decides to serve exactly 4 ounces of hamburger. How many slices of bread should he serve? Suppose he serves 3 slices of bread. How many ounces of hamburger should he serve? Graph the points corresponding to these assumptions.

2. The jailer switches to bread and frankfurters. The latter contain 124 calories each. He still wishes to serve 900 calories. Let x be the number of frankfurters and y the number of slices of bread. Write an equation linking x and y. Make a table of x and y for $x = 0, 2, 4, 6$. Plot the corresponding points. Graph your equation.

3. It costs a manufacturer $5000 to buy a machine for making steak knives. After the initial purchase he can make a knife for 50¢ and sell it for $1. He manufactures x knives before the machine wears out, making a profit of y dollars. Write an equation involving x and y. Make a table of x and y for $x = 10{,}000, 20{,}000, 30{,}000$, and $40{,}000$. Plot the corresponding points. Graph your equation.

4. A man spends $1 to buy x pounds of bananas at 19¢ a pound and y pounds of tomatoes at 29¢ a pound. Write an equation relating x and y. Plot at least four points and graph your equation.

5. A person selling sets of children's books is paid $100 per week plus $15 for each set she sells. One week she sells x sets and makes y dollars. Write an equation relating x and y and graph it.

6. If a weight of x pounds is hung on a certain spring it will stretch $y = 3x$ inches. Graph this equation. Circle the point corresponding to a stretch of 5 inches.

7. A man spends 63¢ to buy x pounds of peaches at y cents per pound. Write an equation for x and y. Graph it.

8. One week a worker works x hours for a total pay of y; she makes $3.50 an hour for the first 40 hours and $7 an hour for overtime. Assuming $x > 40$, write an equation for x and y. Graph it.

9. The first sentence of problem 8 is a statement relating x and y. Graph it for $0 \leqslant x \leqslant 80$.

10. A mail order company has a postage and handling charge that is 10% of the amount of the order, with a $1 minimum charge. Let x be the amount of the order and y the postage and handling charge. Graph the relationship between x and y for $0 \leqslant x \leqslant \$30$.

11. A paper carrier always starts out when collecting with \$10 in change, all in nickels and dimes. Suppose x is the number of nickels and y the number of dimes started with. Write an equation linking x and y. Graph it.

12. A civic group is allowed to take 40 adults into the baseball game free on a certain night. They may substitute two children for each one adult. They use their free passes to the utmost, taking x adults and y children. Write an equation involving x and y. Graph it.

13. A boy may spend up to \$5 on balloons and noisemakers for his birthday party. He buys x packages of balloons at 25¢ each and y noisemakers at 10¢ each. Write a linear inequality for x and y.

14. An insurance salesperson must sell at least \$25,000 worth of insurance each month to meet her quota. She sells x \$1000 policies and y \$2500 policies. Write an inequality involving x and y, assuming she meets her quota.

15. Safety rules say a boat may not carry more than 1000 pounds of passengers. The operator is allowed to assume each adult weighs 150 pounds and each child weighs 60 pounds for reasons of simplicity. He allows x adults and y children to board the boat. Write a linear inequality linking x and y.

16. Two types of pens are sold by mail, one type costing 20¢ and the other 30¢. There is a \$1.50 handling charge on all orders, and the total order, including this charge, must be at least \$5. An order is filled for x of the first type and y of the second. Write a linear inequality for x and y.

1.5 GRAPHING INEQUALITIES

The dietitian at a school cafeteria is planning a lunch of macaroni and cheese. Each cup of cooked macaroni contains 7.5 grams of protein; each ounce of cheese contains 7.1 grams of protein. The dietician wants each serving to contain at least 20 grams of protein. Let a serving contain x cups of macaroni and y ounces of cheese. Then

(grams protein per cup macaroni) (no. cups macaroni)
+ (grams protein per ounce cheese) (no. ounces cheese) $\geqslant 20$,

or $$7.4x + 7.1y \geqslant 20.$$

We would like to find the graph of this inequality.

We already know how to graph the corresponding *equality*,

$$7.4x + 7.1y = 20.$$

The graph is the straight line shown in Figure 1.11.

Figure 1.11

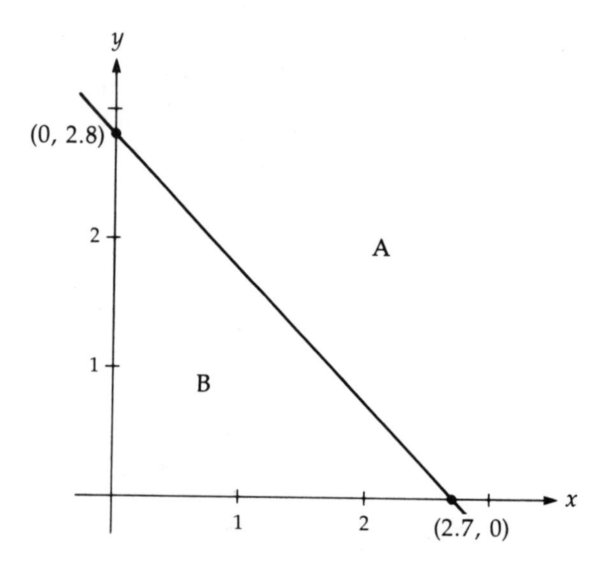

It turns out that the graph of a linear *inequality* always consists of all points on one side or the other of the straight line that is the graph of the corresponding *equality*. Thus the graph of

$$7.4x + 7.1y \geqslant 20$$

is in either region A, which consists of all points on or above and to the right of the line graphed in Figure 1.11, or else region B, which consists of all points on or below and to the left of the line. To find out which is the case, it suffices to choose a single point on one side of the line or the other and to test whether it satisfies the inequality.

In our example a simple point to use for the test is (0, 0). Substituting $x = 0$ and $y = 0$ into the inequality

$$7.4x + 7.1y \geqslant 20$$

yields $$(7.4)(0) + (7.1)(0) \geqslant 20,$$

or $$0 \geqslant 20.$$

CROSS OUT WHAT YOU DON'T WANT
It is common in other books to hatch the region of points *satisfying* the inequality, contrary to our convention. There is a definite advantage to hatching the region of nonsolutions, however, which will become clear when systems of inequalities are treated.

Since this is *false*, and since (0, 0) clearly lies in region B of the plane, below and to the left of the line, we conclude that *the graph of*

$$7.4x + 7.1y \geqslant 20$$

consists of all points on and above and to the right of the line

$$7.4x + 7.1y = 20.$$

We indicate this graphically by drawing a short arrow from this line into the solution region A. For added emphasis we will also crosshatch the region B of nonsolutions. Thus the graph of

$$7.4x + 7.1y \geqslant 20$$

is the region that is *not* crosshatched in Figure 1.12, along with its boundary line.

Figure 1.12

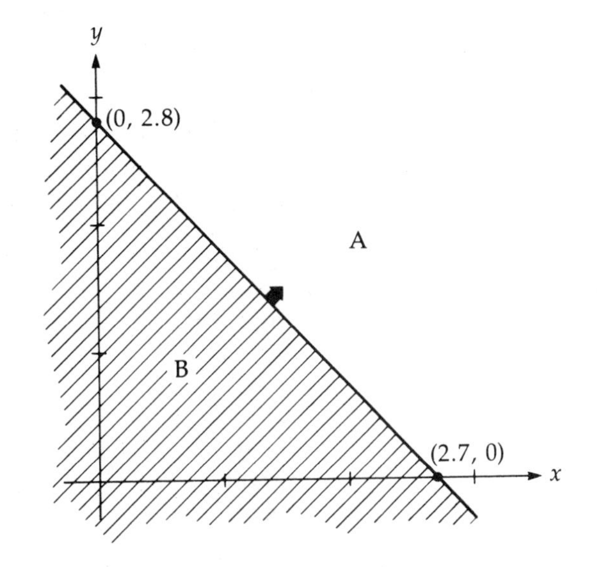

EXAMPLE 1

The Cost of Macaroni and Cheese Let us now suppose that cooked macaroni costs 8¢ per cup and that cheese costs 10¢ per ounce. If the dietitian can spend no more than 50¢ per lunch, then, letting x and y be as above,

(cost of macaroni per cup)(no. cups macaroni)
+ (cost of cheese per ounce)(no. ounces cheese) $\leqslant$ 50,

Figure 1.13

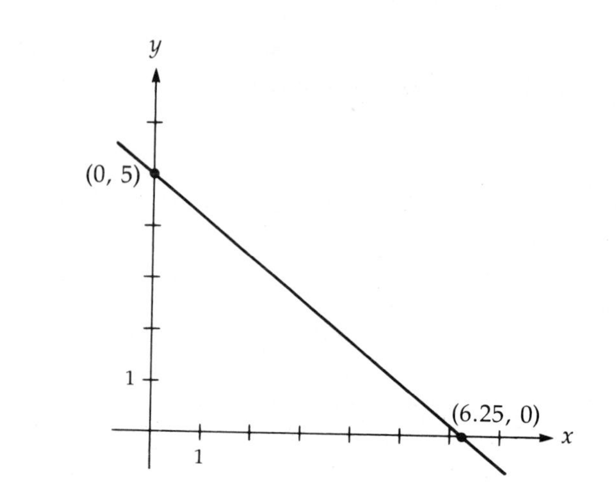

or $$8x + 10y \leqslant 50.$$

To graph this inequality, we start by graphing the corresponding equality

$$8x + 10y = 50.$$

(See Figure 1.13.) Again, an easy point to try in the inequality is (0, 0). Letting $x = 0$ and $y = 0$ in

$$8x + 10y \leqslant 50$$

yields $$0 \leqslant 50,$$

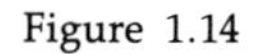

Figure 1.14

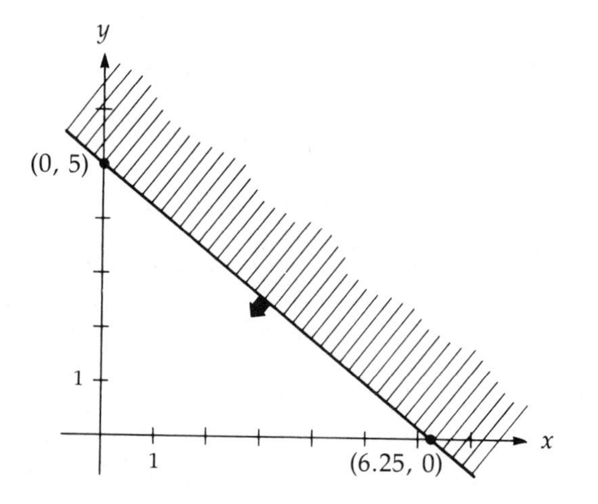

a true statement. We see that the graph of

$$8x + 10y \leq 50$$

is the region on and below the line

$$8x + 10y = 50,$$

and indicate this as shown in Figure 1.14.

EXAMPLE 2

Graphing Two Inequalities Can the dietitian prepare a macaroni and cheese lunch that costs no more than 50¢ and still provides at least 20 grams of protein? We can answer this question by graphing the *system* of inequalities

$$7.4x + 7.1y \geq 20, \quad (1)$$

$$8x + 10y \leq 50. \quad (2)$$

We simply graph each inequality separately on the same set of axes (Figure 1.15).

Figure 1.15

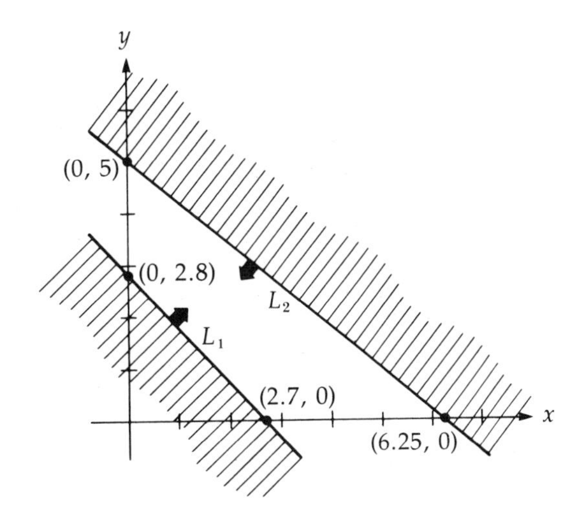

Any point in the unhatched region in Figure 1.15 is a solution to the system. For example, the point (3, 1) is in the unhatched region, and $x = 3$, $y = 1$ can be seen to satisfy both inequalities:

$$(1) \qquad 7.4(3) + 7.1(1) = 22.2 + 7.1 = 29.3 \geq 20,$$

$$(2) \qquad 8(3) + 10(1) = 24 + 10 = 34 \leq 50.$$

We have labeled the lines corresponding to inequalities 1 and 2 in our graph as L_1 and L_2.

EXAMPLE 3

The Frame Up A home craftsman has a 12-foot piece of wood molding from which he wishes to make a rectangular picture frame (Figure 1.16).

Figure 1.16

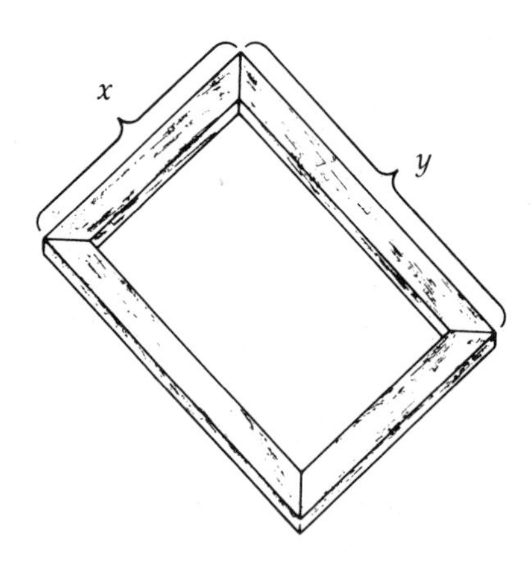

If the frame is to be x inches by y inches, what are the choices for x and y?

Since 12 feet make 144 inches, we have

$$2x + 2y \leq 144.$$

We start by graphing the equation

$$2x + 2y = 144.$$

When $x = 0$ we have $2y = 144$ and $y = 72$. Likewise, when $y = 0$, then $x = 72$. Thus (0, 72) and (72, 0) are on the graph of the equation.

Finally, the point $(x, y) = (0, 0)$ satisfies our inequality, since

$$2(0) + 2(0) = 0 \leq 144.$$

Thus the graph of the inequality is the side of the line through (0, 72) and (72, 0) containing the point (0, 0). We indicate this by hatching the other side of this line in Figure 1.17.

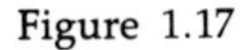
Figure 1.17

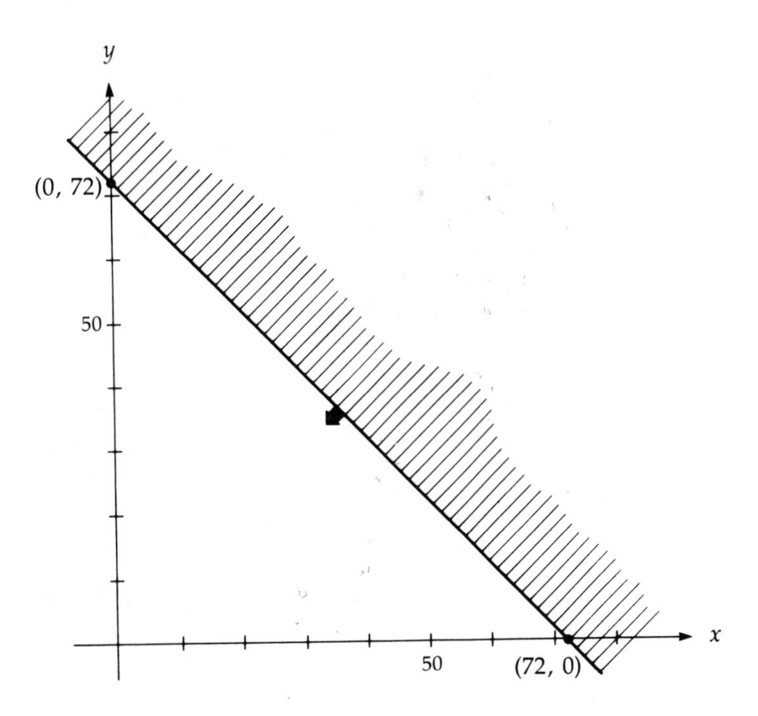

Exercises 1.5

In each problem write an inequality expressing the facts given and graph it. If more than one inequality is implied, graph all the inequalities on the same axes.

1. A cup of cooked macaroni contains 218 calories; an ounce of cheese contains 113 calories. At least 1000 calories is to be provided in each lunch of x cups of macaroni and y ounces of cheese.

2. A dairy buys eggs from farmers to put in cartons of a dozen and sell. They buy x eggs and y cartons a week, buying at least 13 eggs per carton to allow for breakage.

3. The dairy in the previous problem pays $3\frac{1}{2}$¢ per egg and 2¢ per carton. They spend no more than $100 per week on this part of their business.

4. A trucker is making up a load of x crates of apples, weighing 25 pounds each, and y crates of peaches, weighing 30 pounds each. His truck can carry at most 5000 pounds.

5. A library can spend at most $1000 per year on records. These cost either $4.98 or $5.98 each. It buys x of the former and y of the latter in a certain year.

6. A corporation allows at most $1,000,000 per year to advertise a certain product. One year it buys x minutes of television time at $5000 per minute and y pages of magazine ads at $15,000 per page.

7. A door-to-door salesperson sells x children's encyclopedias at $79 each and y medical dictionaries at $24 each. He must sell at least $500 worth of books per month to keep his job.

8. The salesperson in the previous problem receives a commission of $15 per encyclopedia and $5 per dictionary. He will quit his job unless he makes at least his expenses of $200 per month.

9. A gas station operator can obtain only a daily total of at most 4000 gallons of gasoline, regular and premium, from her supplier. She makes 3¢ per gallon of regular and 4¢ per gallon of premium, and must make a daily profit of at least $150 to meet her expenses. Suppose she sells x gallons of regular and y gallons of premium per day.

10. A radio disc jockey gets x favorable and y unfavorable letters per week from listeners. He will be fired if there are more unfavorable letters than favorable. He will also be fired if he does not get at least 100 favorable letters per week. He is not fired.

11. A certain television program will be canceled if the number x of favorable letters the network gets about it per week is not at least four times the number of unfavorable letters, y. It will also be canceled if it does not get at least 3000 favorable letters per week or if it gets more than 1000 unfavorable letters per week. It is not canceled.

12. A factory makes plaster mice and rabbits. It takes a worker 12 minutes to paint a mouse and 10 minutes to paint a rabbit. A worker paints x mice and y rabbits in an eight-hour day.

13. A grower provides 40-pound crates of apples and 30-pound crates of cherries to a supermarket chain. A contract provides that he must supply at least 2000 pounds of these fruits per week. One week he delivers x crates of apples and y crates of cherries.

14. In order to collect her allowance, Terry must earn at least 100 points from her parents in a week. She gets 10 points for making her bed and 8 points for taking out the garbage. One week she got her allowance by making her bed x times and taking out the garbage y times.

15. The psychology department budgets up to $200 per year for typing paper and carbon paper. Typing paper is $6 per ream and carbon paper is $11 per ream. No more than one ream of carbon paper is

bought for each two reams of typing paper. One year x reams of typing paper and y reams of carbon paper are bought.

16. Agriculturalists find that to grow a certain strain of watermelon successfully there must be at least 70 sunny days. Also there must be at least 1 rainy day for each 8 sunny days. These melons were successful one summer with x sunny days and y rainy days.

1.6 SYSTEMS OF LINEAR EQUATIONS

A trucker buys crates of apples and pears from farmers to sell in the city. A crate of apples costs him \$4 and weighs 25 pounds. A crate of pears costs \$3.50 and also weighs 25 pounds. He has \$780 to spend, his truck carries 5000 pounds, and he decides that by spending the whole amount and loading his truck to its capacity he will make the most money. (We will see later in this chapter how such a decision might be justified.) Let us decide how many crates of each fruit he should buy.

Suppose he buys x crates of apples and y crates of pears. Then

$$(\text{cost of apples}) + (\text{cost of pears}) = \$780,$$

or

$$4x + 3.5y = 780.$$

Also

$$(\text{weight of apples}) + (\text{weight of pears}) = 5000,$$

or

$$25x + 25y = 5000.$$

We arrive at the *system* of linear equations

$$4x + 3.5y = 780, \quad (1)$$

$$25x + 25y = 5000. \quad (2)$$

These equations are graphed in Figure 1.18 (page 38).

We are interested in a pair x, y satisfying *both* equations. The corresponding point (x, y) would be on both lines. This is the point P in the figure.

We will calculate the coordinates of P algebraically by the method of *substitution*. This method consists of four steps.

1. *Solve one of the equations for one of the unknowns in terms of the other unknown.*
2. *Substitute the expression equaling the solved-for unknown in the other equation, using the result of step 1.*
3. *Solve the resulting equation for the second unknown.*
4. *Determine the value of the first unknown by substituting the value obtained in step 3 into the formula obtained in step 1.* We illus-

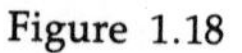
Figure 1.18

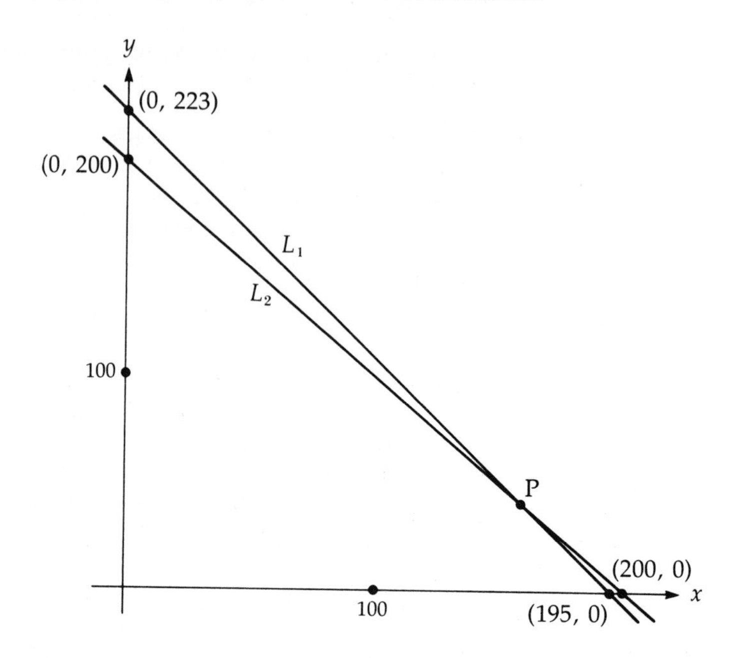

We illustrate this method with the system

$$4x + 3.5y = 780,$$
$$25x + 25y = 5000.$$

1. Solving the first equation for x, we find

$$4x = 780 - 3.5y,$$
$$x = \tfrac{1}{4}(780 - 3.5y) = 195 - .875y.$$

2. Substituting for x in the second equation, we get

$$25(195 - .875y) + 25y = 5000.$$

3. We solve this for y. Dividing by 25 first simplifies the calculation.

$$195 - .875y + y = 200,$$
$$195 + (-.875 + 1)y = 200,$$
$$195 + .125y = 200,$$
$$.125y = 200 - 195 = 5,$$
$$y = \frac{5}{.125} = 40.$$

4. Then from the equation of part 1,

$$\begin{aligned} x &= 195 - .875y \\ &= 195 - .875(40) \\ &= 195 - 35 = 160. \end{aligned}$$

The trucker should buy 160 crates of apples and 40 crates of pears.

EXAMPLE 1

The Jail Lunch As another example consider the jailer who serves his prisoners bread and hamburger meat for lunch. Hamburger costs \$1.28 per pound and contains 105 calories per ounce; a slice of bread costs 1¢ and contains 63 calories. State law requires that the lunch contain at least 900 calories and cost at least 40¢. The jailer decides to spend exactly 40¢ and serve exactly 900 calories.

Let the lunch consist of x ounces of hamburger and y slices of bread. Then

$$(\text{calories in hamburger}) + (\text{calories in bread}) = 900,$$

or

$$105x + 63y = 900.$$

Since hamburger costs \$1.28 per pound, or

$$\frac{128¢}{\text{pound}} \; \frac{1 \text{ pound}}{16 \text{ ounces}} = 8¢ \text{ per ounce},$$

we have

$$(\text{cost of hamburger}) + (\text{cost of bread}) = 40,$$

or

$$8x + y = 40.$$

We proceed to solve the system

$$\begin{cases} 105x + 63y = 900, \\ 8x + y = 40 \end{cases}$$

by substitution, starting (for convenience) by solving the second equation for y.

1. $8x + y = 40,$

$$y = 40 - 8x,$$

2. $105x + 63(40 - 8x) = 900,$

3. $105x + 2520 - 504x = 900,$

$$2520 - 900 = 504x - 105x,$$

$$1620 = 399x,$$

$$x = \frac{1620}{399} = 4.06,$$

4. $y = 40 - 8x = 40 - 8(4.06)$

$= 40 - 32.48 = 7.52.$

The jailer should serve about 4 ounces of meat and $7\frac{1}{2}$ slices of bread. The graph is shown in Figure 1.19.

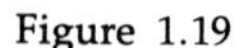

Figure 1.19

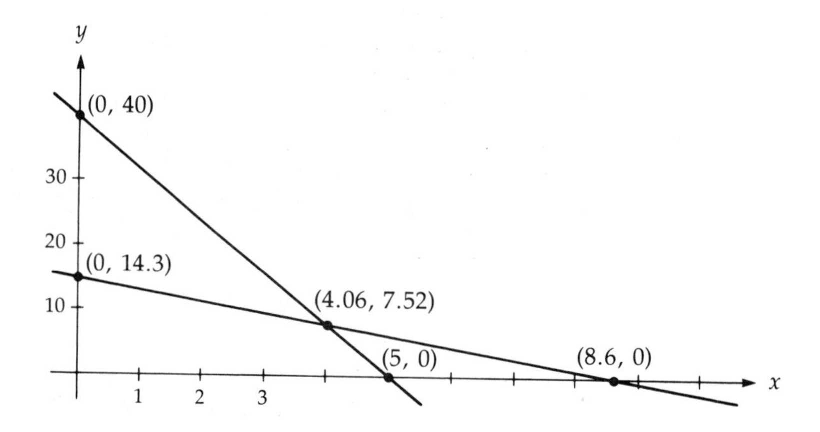

EXAMPLE 2

The Peas A farmer plans to spend \$300 to plant peas. The seed costs \$2 per pound. For each 50 pounds of seeds he must buy a pound of nitrogen-fixing-bacteria inoculant for \$1.50. How many pounds of each should he buy?

Let him buy x pounds of seeds and y pounds of inoculant. Since he is to spend \$300, we have

$$2x + 1.5y = 300.$$

Since he needs 50 times as much seed as inoculant, we have

$$x = 50y.$$

Our system is

$$\begin{cases} 2x + 1.5y = 300, \\ x = 50y. \end{cases}$$

The easiest way to proceed is to substitute the value of x given by the second equation into the first. We get

$$\begin{aligned} 2(50y) + 1.5y &= 300, \\ 100y + 1.5y &= 300, \\ 101.5y &= 300, \\ y &= \frac{300}{101.5} = 2.96. \end{aligned}$$

Then $x = 50y = 50(2.96) = 148$. The farmer should buy 148 pounds of seed and 3 pounds of inoculant.

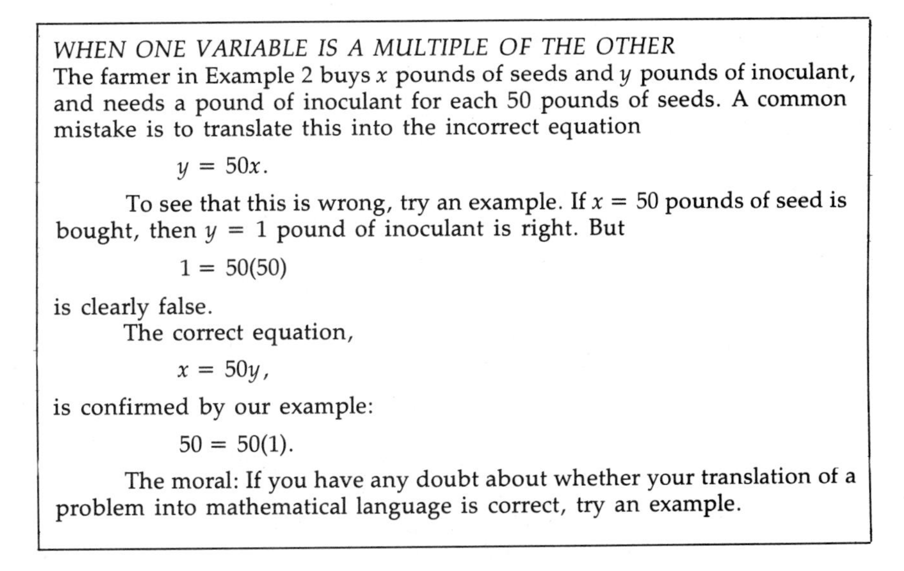

WHEN ONE VARIABLE IS A MULTIPLE OF THE OTHER
The farmer in Example 2 buys x pounds of seeds and y pounds of inoculant, and needs a pound of inoculant for each 50 pounds of seeds. A common mistake is to translate this into the incorrect equation

$$y = 50x.$$

To see that this is wrong, try an example. If $x = 50$ pounds of seed is bought, then $y = 1$ pound of inoculant is right. But

$$1 = 50(50)$$

is clearly false.

The correct equation,

$$x = 50y,$$

is confirmed by our example:

$$50 = 50(1).$$

The moral: If you have any doubt about whether your translation of a problem into mathematical language is correct, try an example.

EXAMPLE 3 **Skirts and Blouses** A dress factory has 1860 yards of material to be made into skirts and blouses. Each skirt uses 3 yards of material and each blouse 2 yards. Each skirt also requires 3 feet of trim; a blouse uses 4 feet of trim. There are 2700 feet of trim available. Is it possible for the factory to use up all the material on hand?

Suppose x skirts and y blouses are made. The requirements for material and trim imply

$$\begin{cases} 3x + 2y = 1860 & \text{(material)}, \\ 3x + 4y = 2700 & \text{(trim)}. \end{cases}$$

We start by solving the first equation for x.

$$3x = 1860 - 2y,$$
$$x = \tfrac{1}{3}(1860 - 2y).$$

Substituting this in the second equation gives

$$\begin{aligned} 3[\tfrac{1}{3}(1860 - 2y)] + 4y &= 2700, \\ 1860 - 2y + 4y &= 2700, \\ 1860 + 2y &= 2700, \\ 2y &= 2700 - 1860 = 840, \\ y &= \frac{840}{2} = 420. \end{aligned}$$

Then

$$\begin{aligned} x &= \tfrac{1}{3}(1860 - 2y) = \tfrac{1}{3}[1860 - 2(420)] \\ &= \tfrac{1}{3}(1860 - 840) = \tfrac{1}{3}(1020) = 340. \end{aligned}$$

Making 340 skirts and 420 blouses will use up all the material and trim.

Adding and Subtracting Equations

A method of solving a system of two linear equations that is often simpler than substitution involves adding or subtracting the equations. In Example 3 the system

$$\begin{cases} 3x + 2y = 1860, \\ 3x + 4y = 2700 \end{cases}$$

could have been solved by subtracting both sides of the first equation from the corresponding sides of the second:

$$\begin{array}{rrcl} & 3x + 4y & = & 2700 \\ (-) & 3x + 2y & = & 1860 \\ \hline & 2y & = & 840 \\ & y & = & 420. \end{array}$$

Then $y = 420$ could have been substituted into either of the original equations to find x. Using the first equation, for example,

$$\begin{aligned} 3x + 2(420) &= 1860, \\ 3x + 840 &= 1860, \\ 3x &= 1860 - 840 = 1020, \\ x &= 340. \end{aligned}$$

The reason this method works so well for the system

$$\begin{aligned} 3x + 2y &= 1860, \\ 3x + 4y &= 2700 \end{aligned}$$

is that $3x$ appears in both equations, so subtracting them leaves an equation involving only y.

Likewise, given

$$\begin{cases} 2x - 5y = 15, \\ 3x + 5y = 5, \end{cases}$$

we add the equations.

$$\begin{array}{rrcl} & 2x - 5y & = & 15 \\ (+) & 3x + 5y & = & 5 \\ \hline & 5x & = & 20 \end{array}$$

We see $x = 4$.

Putting $x = 4$ in the first equation then gives

$$2(4) - 5y = 15,$$
$$8 - 5y = 15,$$
$$-5y = 15 - 8 = 7,$$
$$y = \frac{7}{-5} = -\frac{7}{5}.$$

Sometimes using this method requires multiplying one or both of the equations by a constant before adding or subtracting. Thus the system

$$\begin{cases} x + 2y = 5, \\ 2x + 3y = 7 \end{cases}$$

becomes

$$\begin{cases} 2x + 4y = 10, \\ 2x + 3y = 7 \end{cases}$$

if the first equation is multiplied by 2. Now subtracting the equations gives

$$y = 3,$$

and from the original first equation

$$x = 5 - 2y = 5 - 2(3) = -1.$$

Of course, the reason for multiplying the first equation by 2 was to get $2x$ appearing in both equations.

See Example 5 for a case in which *both* equations must be multiplied by a constant.

EXAMPLE 4

The Boy Scouts The boy scout troop has \$18 to spend on 15¢ and 20¢ candy bars. They would like to buy exactly 100 bars since there are 100 boys in the troop. Is this possible?

Let x 15¢ bars and y 20¢ bars be bought. Then we want

$$\begin{cases} 15x + 20y = 1800, \\ x + y = 100. \end{cases}$$

Multiplying the second equation by 15 gives

$$\begin{cases} 15x + 20y = 1800, \\ 15x + 15y = 1500. \end{cases}$$

We subtract to find

$$5y = 300,$$

$$y = \frac{300}{5} = 60.$$

The second original equation tells us $x = 40$. They should buy 40 of the 15¢ bars and 60 of the 20¢ bars.

EXAMPLE 5

Kings and Queens A plant makes ceramic mugs in the form of the head of a king or queen. The king takes 3 minutes and the queen 2 minutes to trim after molding. The king takes 4 minutes to paint and the queen 5 minutes. There are 570 minutes of trimming time and 970 minutes of painting time available. How many of each should be made to use all this time?

Let x kings and y queens be made. Then

$$\begin{cases} 3x + 2y = 570 & \text{(trimming)}, \\ 4x + 5y = 970 & \text{(painting)}. \end{cases}$$

In order to use the addition-subtraction method, we multiply the first equation by 4 and the second by 3 in order to get $12x$ in both.

$$\begin{cases} 12x + 8y = 4(570) = 2280, \\ 12x + 15y = 3(970) = 2910. \end{cases}$$

Now we subtract the first equation from the second.

$$\begin{array}{lrcl} & 12x + 15y & = & 2910 \\ (-) & 12x + 8y & = & 2280 \\ \hline & 7y & = & 630. \end{array}$$

Thus $y = 630/7 = 90$. From the original first equation,

$$3x = 570 - 2y = 570 - 2(90) = 570 - 180 = 390,$$

and $$x = \frac{390}{3} = 130.$$

The plant should make 130 kings and 90 queens.

Which Method Is Best?

In general, the addition-subtraction method is easiest whenever either x or y appears with the same coefficient in both equations or when this can be arranged by easy multiplications. If the coefficients are complicated, substitution is probably the better method. When the coefficients are small whole numbers, however, the addition-subtraction method may delay having to deal with fractions until a later step in the problem than substitution.

Exercises 1.6

In problems 1 through 4 solve each system of equations exactly.

1. $\begin{aligned} 2x + 3y &= 13 \\ x - 2y &= -4 \end{aligned}$ **2.** $\begin{aligned} 5x + 2y &= 5 \\ 3x - y &= -8 \end{aligned}$

3. $\begin{aligned} 2x + 3y &= 14 \\ 4x - 2y &= 4 \end{aligned}$ **4.** $\begin{aligned} -3x + 2y &= 3 \\ 2x - 6y &= 5 \end{aligned}$

In problems 5 through 8 solve each system of equations, giving answers to two decimal places, that is, two places to the right of the decimal.

5. $\begin{aligned} 3x + y &= 7 \\ 4x - 3y &= 3 \end{aligned}$ **6.** $\begin{aligned} 2.1x + y &= 3 \\ 3.3x - y &= -1 \end{aligned}$

7. $\begin{aligned} 51x + 13y &= 5 \\ 17x - 12y &= 2 \end{aligned}$ **8.** $\begin{aligned} 21x + 10y &= 43 \\ 19x - 7y &= 25 \end{aligned}$

In problems 9 through 12 solve each system of equations exactly. Graph the two equations and check whether your answer corresponds to the point where their graphs meet.

9. $\begin{aligned} 3x + y &= 13 \\ x + 2y &= 11 \end{aligned}$ **10.** $\begin{aligned} 3x + 2y &= 1 \\ 2x + 3y &= 4 \end{aligned}$

11. $\begin{aligned} x + 2y &= 8 \\ 3x - 2y &= 4 \end{aligned}$ **12.** $\begin{aligned} 4x + 2y &= 2 \\ 2x - 3y &= 9 \end{aligned}$

In each problem below express the given facts as a system of two equations in two unknowns, then solve the system. Graph the two equations and check whether your answer corresponds to the point where their graphs meet.

13. A seed company sells 1-ounce packets containing a mixture of seeds of two types of radish, Champion and Cherry Belle. They want to get 50¢ per ounce for Champion seeds and 35¢ per ounce for Cherry Belle seeds. The packet is to sell for 40¢. Suppose each packet contains x ounces of Champion and y ounces of Cherry Belle.

14. A packer sells pound cans of mixed peanuts and cashews. Peanuts cost him 3¢ per ounce; cashews cost 5¢ per ounce. He wants his cost per can to be 52¢. Suppose he puts x ounces of peanuts and y ounces of cashews in each can.

15. A gasoline company has on hand two types of gasoline. Each gallon of type A contains 45 energy units; each gallon of type B contains 50 energy units. The company wants a mixture containing 47 energy units per gallon. Let each gallon of mixture contain x gallons of A and y gallons of B.

16. An advertiser plans to spend $87,000 on 20 one-minute nighttime TV spots. A one-minute spot costs $5000 before 10 P.M. and $4000 after. The advertiser buys x spots before 10 P.M. and y spots after.

17. A trucker likes to keep his truck loaded to its weight limit of 20,000 pounds and its capacity of 2000 cubic feet. He is making up a load of record changers, each weighing 20 pounds and taking up 1 cubic foot, and dust covers, each weighing 5 pounds and taking up 1 cubic foot.

18. A nonprofit organization has observed that it sells 3 boxes of cookies (50¢ profit each) for each 2 boxes of candy (70¢ profit each) it sells. The organization is ordering the boxes for a sale to raise $580 to buy soccer uniforms.

19. Canned evaporated milk contains 346 calories and 17.6 grams of protein per cup. Skim milk contains 87 calories and 8.6 grams of protein per cup. A nutritionist wants to make up a mixture of the two (not necessarily a cupful) containing the same number of calories but twice as much protein as a cup of whole milk, which has 166 calories and 8.5 grams of protein.

20. A farmer leases land to raise corn to fatten calves. It takes $\frac{1}{4}$ acre of land to feed one calf for a year. He has $20,000 to spend on calves, costing $100 each, and land, at $400 per acre per year.

21. A speculator buys 120 acres of land, paying \$500 an acre for x acres and \$800 an acre for y acres, for a total of \$81,000.

22. Another speculator buys 55 acres of land for x dollars an acre and 75 acres of land for y dollars an acre, for a total of \$100,500. The 75 acres cost half again as much per acre as the 55 acres.

1.7 SYSTEMS OF LINEAR INEQUALITIES

In this section we will consider various real-world problems where one has the opportunity to choose values for two variables, x and y. Conditions impose certain restrictions expressed by inequalities involving x and y. In the last two sections of this chapter we will see how to choose values for x and y that are "best" in the sense that they produce the most profit, least cost, most protein, least fat, most pleasure, least pain, and so on, yet still satisfy these restrictions. For the present, as a first step, we will merely see how to picture explicitly the set of values of x and y from which our "best" choice will be made.

Beets or Lettuce

A farmer has 15 acres and plans to plant beets and lettuce. It costs her \$300 in seed, chemicals, equipment, and outside labor to raise an acre of beets and \$500 for an acre of lettuce. She has \$5000 in capital. Of course she need not plant the whole 15 acres, nor need she invest all her money.

Let us denote by x the number of acres of beets the farmer plants and by y the number of acres of lettuce. Of course, she would like to choose x and y so as to make the most money. Her profit will depend on what she can sell an acre of each crop for at harvest time. Although the expenses of growing lettuce are higher, for example, it might still be the preferable crop if it sells for more per acre than beets. On the other hand, to put all 15 acres into lettuce would cost

$$15(\$500) = \$7500,$$

more than the farmer has.

We will consider problems of maximizing profit such as this in the next section. For the present we will limit ourselves to graphing the set of options open to the farmer. This will let us see at a glance what is *possible*, a first step in determining what is *optimal*.

Recall that we are assuming x acres of beets and y acres of lettuce are planted. Since only 15 acres are available, (number of acres of beets) + (number of acres of lettuce) $\leqslant$ 15,

or

$$x + y \leqslant 15.$$

Since the farmer has \$5000 to spend, and since beets and lettuce cost \$300 and \$500 per acre, respectively, we have

(cost of beets) + (cost of lettuce) $\leqslant$ \$5000,

or

$$300x + 500y \leqslant 5000.$$

This can be simplified to $3x + 5y \leqslant 50$ by dividing through by 100.

As in Section 1.5 we graph the system

(1) $\quad x + y \leqslant 15,$

(2) $\quad 3x + 5y \leqslant 50$

as shown in Figure 1.20.

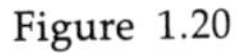

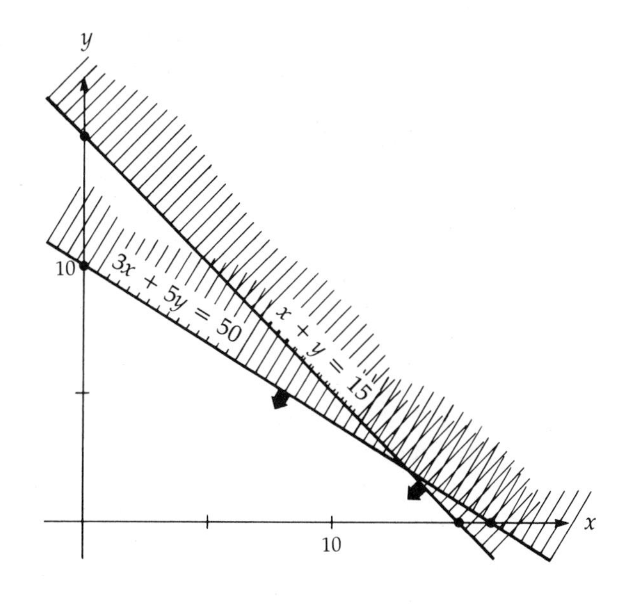

Figure 1.20

Actually, two more linear inequalities are implicit in the situation described above, namely

$$x \geqslant 0 \quad \text{and} \quad y \geqslant 0.$$

This is because the farmer cannot plant negative acreage of either beets or lettuce. Thus we really have the following system:

$$\begin{aligned}
&(1) \qquad & x + y &\leqslant 15,\\
&(2) \qquad & 3x + 5y &\leqslant 50,\\
&(3) \qquad & x &\geqslant 0,\\
&(4) \qquad & y &\geqslant 0.
\end{aligned}$$

This system is graphed in Figure 1.21.

Figure 1.21

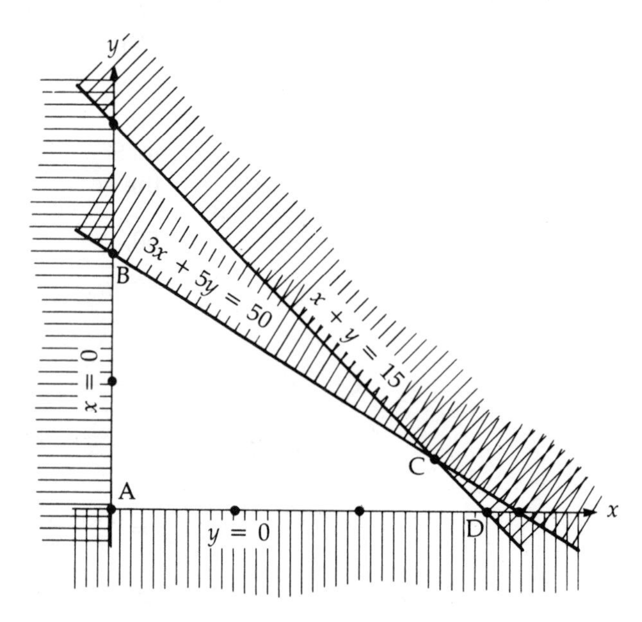

The corners A, B, C, and D of the region graphed in Figure 1.21 are called its *vertices*. Clearly A = (0, 0); and B and D are easily calculated.

For example, B is the solution of the system of equations

$$\begin{cases} 3x + 5y = 50, \\ x = 0, \end{cases}$$

Solving this merely requires setting $x = 0$ in the first equation, yielding

$$5y = 50,$$

and $\quad y = 10.$

We see B = (0, 10).

Likewise, D is determined by the system of equations

$$\begin{cases} x + y = 15, \\ y = 0, \end{cases}$$

which implies D = (15, 0). In fact, the points B and D would most likely already have been found in graphing the lines $3x + 5y = 50$ and $x + y = 15$.

Only C is any trouble to find. It corresponds to the solution of the system

$$\begin{cases} x + y = 15, \\ 3x + 5y = 50. \end{cases}$$

We use one of the methods from Section 1.6. From the first equation $x = 15 - y$; substituting this in the second equation gives

$$\begin{aligned} 3(15 - y) + 5y &= 50, \\ 3(15) - 3y + 5y &= 50, \\ 45 + 2y &= 50, \\ 2y &= 50 - 45 = 5, \\ y &= \frac{5}{2} = 2.5. \end{aligned}$$

Then $x = 15 - y = 15 - 2.5 = 12.5$ and C = (12.5, 2.5).

IMPLICIT CONDITIONS

Very often a quantity involved in a real-life problem is subject to some constraint that is so clearly demanded by common sense that it is left unstated. If we are told, for example, that z people died in automobile crashes in 1973, we need not be told that $z \geqslant 0$.

In this example another unstated condition on z is that it is a whole number. Such a condition is often harder to treat mathematically. It may be best to ignore it in a mathematical analysis until the end of the problem.

Suppose we are told that Mr. Kaltenbach buys all the apples he can for a dollar at 6¢ each. If he buys w apples we have

$$6w \leqslant 100,$$

or

$$w \leqslant \frac{100}{6} = 16\tfrac{2}{3}.$$

We now invoke the common-sense fact that w is a whole number and conclude that w is 16.

EXAMPLE 1

Beef and Beans A nursing-home dietitian plans a lunch of beef chuck and red kidney beans. The beef contains 86 calories and 7 grams of protein per ounce, and the beans contain 230 calories and 15 grams of protein per cup. The dietitian wants each lunch to contain at least 700 calories and at least 25 grams of protein.

Suppose he serves x ounces of beef and y cups of beans. Then

$$\text{(beef calories)} + \text{(bean calories)} \geqslant 700$$

or

$$86x + 230y \geqslant 700.$$

Also

$$\text{(beef protein)} + \text{(bean protein)} \geqslant 25 \text{ grams},$$

or

$$7x + 15y \geqslant 25.$$

Since it makes no sense for x or y to be negative, we have the system

(1) $86x + 230y \geqslant 700,$

(2) $7x + 15y \geqslant 25,$

(3) $x \geqslant 0,$

(4) $y \geqslant 0.$

The unhatched area in Figure 1.22 is the graph of this system. Here the

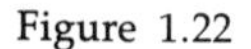
Figure 1.22

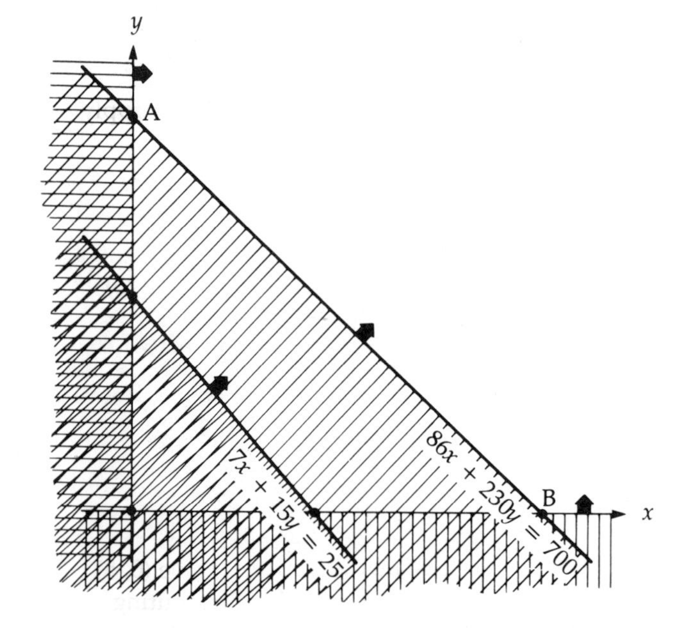

region of possibilities is unbounded. Its vertices are A, which can be calculated to be (8.14, 0) by solving the system

$$\begin{aligned} &(1) \qquad 86x + 230y = 700, \\ &(3) \qquad x = 0, \end{aligned}$$

and B, which can be calculated to be (0, 3.04) by solving the system

$$\begin{aligned} &(1) \qquad 86x + 230y = 700, \\ &(4) \qquad y = 0. \end{aligned}$$

Notice that inequality (2) contributes nothing to the shape of the region. This simply means that any combination of beef and beans containing sufficient calories automatically meets the protein requirements of the problem.

EXAMPLE 2

The Artist An artist needs at least 5 ounces of pink paint, which she is mixing from tubes of red and white. She has only 3 ounces of red she can use, but an unlimited supply of white.

Let us assume she uses x ounces of red paint and y ounces of white paint, and graph the region of possibilities. Since at least 5 ounces are to be mixed, we have

$$x + y \geqslant 5.$$

The limitation on the supply of red paint can be expressed as

$$x \leqslant 3.$$

Adding the obvious observation that neither x nor y can be negative, we have the system (Figure 1.23)

$$\begin{cases} x + y \geqslant 5, \\ x \leqslant 3, \\ x \geqslant 0, \\ y \geqslant 0. \end{cases}$$

Vertex A is found to be (0, 5) by substituting $x = 0$ into $x + y = 5$. Vertex B is found by substituting $x = 3$ into $x + y = 5$. Then we have

Figure 1.23

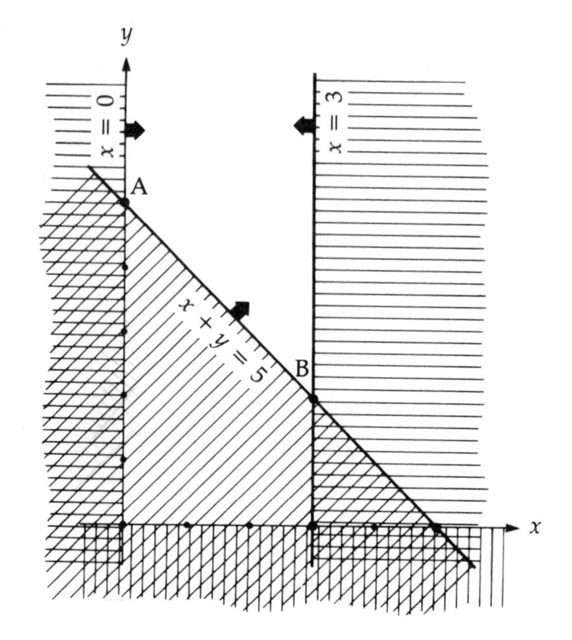

$$3 + y = 5,$$

$$y = 5 - 3 = 2.$$

Thus B = (3, 2).

EXAMPLE 3

The Fruit Basket The produce manager of a grocery is making up fruit baskets to sell as gifts. They are to sell for no more than \$5, and contain only apples and oranges. She wants to get 25¢ per orange, 20¢ per apple, and 50¢ for the basket. No more than 19 pieces of fruit will fit in the basket. Suppose she uses x oranges and y apples.

The price limitation tells us (converting to cents)

$$25x + 20y + 50 \leqslant 500,$$

or

$$25x + 20y \leqslant 450.$$

We divide by 5 to get

$$5x + 4y \leqslant 90.$$

The limitation to 19 fruit says

$$x + y \leqslant 19.$$

Figure 1.24

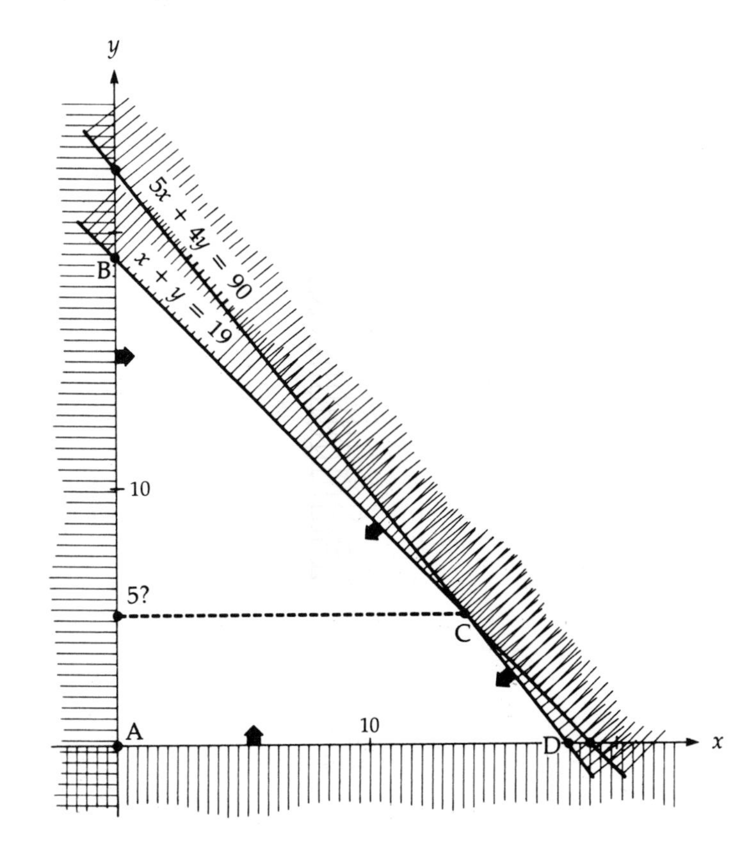

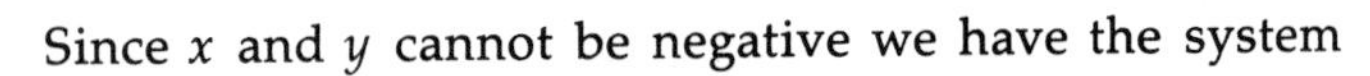

Since x and y cannot be negative we have the system

$$\begin{cases} 5x + 4y \leqslant 90, \\ x + y \leqslant 19, \\ x \geqslant 0, \\ y \geqslant 0. \end{cases}$$

as shown in Figure 1.24.

Clearly A = (0, 0), while the vertices B and D were found in the course of graphing the lines $5x + 4y = 90$ and $x + y = 19$ to be (0, 19) and (18, 0).

The vertex C can be found by solving the system

$$\begin{cases} x + y = 19, \\ 5x + 4y = 90, \end{cases}$$

but after making a careful graph it might be worth trying to determine the coordinates of C from it. Eyeballing the picture suggests that the y-coordinate of C might be 5. Since $x + y = 19$, this makes $x = 14$. We plug these values into

$$\begin{aligned} 5x + 4y &= 5(14) + 4(5) \\ &= 70 + 20 = 90. \end{aligned}$$

Our guess was correct: C = (14, 5).

EXAMPLE 4

Cutting Grass A high school student needs \$58 to buy a part for his car. He cuts lawns for \$5 and cleans gutters for \$7. It takes him 3 hours to mow a lawn and 5 hours to clean a set of gutters, and he only has 39 hours he can work.

Suppose he cuts x lawns and cleans y sets of gutters. To make the \$58 he needs,

$$5x + 7y \geqslant 58.$$

Because only 39 hours are available, we have

$$3x + 5y \leqslant 39.$$

Since x and y must be nonnegative, we have the system (Figure 1.25)

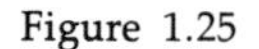
Figure 1.25

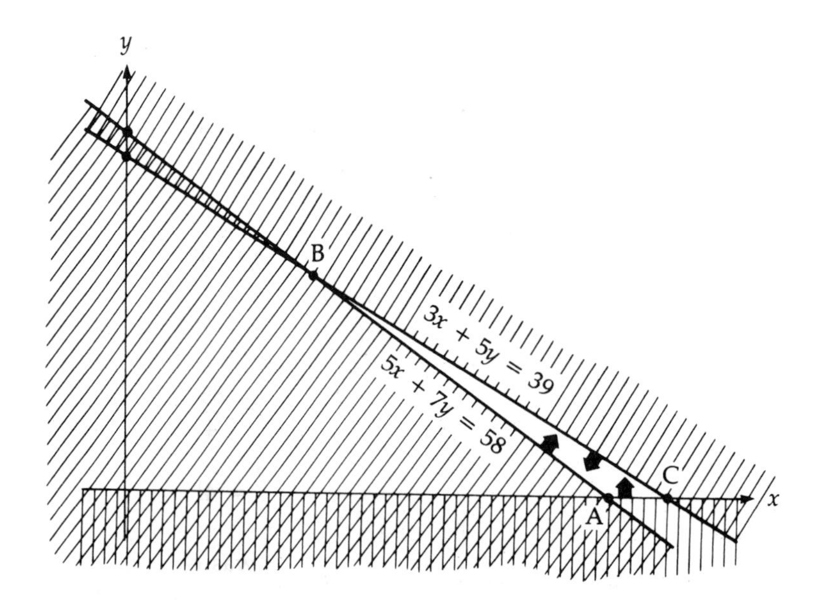

$$\begin{cases} 5x + 7y \geq 58, \\ 3x + 5y \leq 39, \\ \quad\quad x \geq 0, \\ \quad\quad y \geq 0. \end{cases}$$

In the course of graphing we found A = $(11\frac{3}{5}, 0)$ and C = (13, 0). From the figure a reasonable guess for B is (4, 5). However, for $x = 4$ and $y = 5$,

$$5x + 7y = 5(4) + 7(5) = 20 + 35 = 55 \neq 58.$$

Our guess is incorrect, and we fall back on solving the system

$$\begin{cases} 5x + 7y = 58, \\ 3x + 5y = 39. \end{cases}$$

Multiplying the first equation by 3 and the second by 5 gives

$$\begin{cases} 15x + 21y = 3(58) = 174, \\ 15x + 25y = 5(39) = 195. \end{cases}$$

We subtract the first new equation from the second.

$$\begin{array}{rr} & 15x + 25y = 195 \\ (-) & \underline{15x + 21y = 174} \\ & 4y = \ \ 21. \end{array}$$

We see $y = \dfrac{21}{4} = 5\frac{1}{4}$.

The first of the original equations now gives

$$5x = 58 - 7y = 58 - 7\left(\frac{21}{4}\right)$$

$$= \frac{232}{4} - \frac{147}{4} = \frac{85}{4}.$$

Thus $$x = \frac{1}{5} \cdot \frac{85}{4} = \frac{17}{4} = 4\tfrac{1}{4}.$$

The guess B = (4, 5) was close, but actually B = $(4\frac{1}{4}, 5\frac{1}{4})$.

Exercises 1.7

Graph each of the systems of inequalities in problems 1 through 8 and calculate the coordinates of the vertices of the region of possibilities.

1. $x + y \leqslant 5$, $x \leqslant 4$, $x \geqslant 0$, $y \geqslant 0$

2. $x + 2y \geqslant 4$, $y \geqslant 1$, $x \geqslant 0$, $y \geqslant 0$

3. $x + 2y \leqslant 10$, $x \geqslant 0$, $x \leqslant 7$

4. $2x + 3y \geqslant 12$, $y \leqslant 5$, $x \geqslant 0$, $y \geqslant 0$

5. $x + 2y \leqslant 6$, $2x + y \leqslant 6$, $x \geqslant 0$, $y \geqslant 0$

6. $x + 3y \leqslant 6$, $3x + 2y \leqslant 6$, $x \geqslant 0$, $y \geqslant 0$

7. $2x + 3y \geqslant 12$, $x + y \geqslant 5$, $x \geqslant 0$, $y \geqslant 0$

8. $3x + 3y \geqslant 15$, $4x + 2y \geqslant 12$, $x \geqslant 0$, $y \geqslant 0$

Translate the statements of each problem into a system of linear inequalities. Include any inequalities required by common sense, even if not corresponding to any explicit statement. Graph the system and indicate the vertices of the region of possibilities on your graph. Calculate the coordinates of these vertices.

9. A boy has 18¢ to spend on 2¢ bubble gum balls and 3¢ jawbreakers. Of course, he doesn't have to spend it all. He must buy at least 3 bubble gum balls for his 3 sisters. He buys x bubble gum balls and y jawbreakers.

10. A baker decides to make some doughnuts and rolls. Each doughnut takes 4 tablespoons of flour and each roll takes 5 tablespoons of flour. He has only 40 tablespoons of flour and only enough cinnamon for 8 doughnuts (there is no cinnamon in the rolls). There is plenty of all the other ingredients. He makes x doughnuts and y rolls.

11. To belong to the Parkside Grade School Book Club a student must read at least 10 books a month. At least 3 of these must be fiction and at least 3 nonfiction. Lisa, who is in the club, read x fiction and y nonfiction books last month.

12. Ed is trying to quit smoking. He will not smoke more than a total of 12 cigarettes and cigars in any one day, including not more than 10 cigarettes or 8 cigars. Tuesday he smoked x cigarettes and y cigars.

13. A certain machine puts toothpaste into tubes. It can fill 50,000 of the 1.5-ounce tubes or 30,000 of the 3-ounce tubes per hour. The machine can be used at most 80 hours per week, and 6,500,000 ounces of toothpaste is available per week. Suppose one week the machine makes 1.5-ounce tubes for x hours and 3-ounce tubes for y hours.

14. A part-time toymaker makes toy trains and toy trucks. A train takes him 5 minutes to put together and 7 minutes to paint. A truck takes him 6 minutes to put together and 10 minutes to paint. The toymaker devotes at most 5 hours per week to putting the toys together and at most 10 hours per week to painting. One week he makes x trains and y trucks.

15. A truck has a capacity of 10,000 pounds and 1000 cubic feet. A load is being made up of television sets of two types: 30-pound portables taking up 7 cubic feet, and 50-pound consoles taking up 15 cubic feet. Suppose x portables and y consoles are carried.

16. Some students are trying to make at least $100 by collecting scrap paper and glass. They can only make one trip in the truck they have borrowed, which will carry 10,000 pounds. They can sell the paper for $15 per ton and the glass for $25 per ton. They collect x tons of paper and y tons of glass.

17. A piano tuner can find as many jobs tuning uprights and spinets as she wants. She charges $25 for an upright, which takes her 3 hours, and $15 for a spinet, which takes her 2 hours. She wants to take in at least $200 per week, working no more than 40 hours.

18. A man has up to $10,000 to invest. The bank pays 5% interest, while the savings and loan pays $5\frac{1}{2}$%. He wants at least $400 interest per year.

19. A chain supermarket manager finds he always sells at least four times as many cans of a well-known brand of soup as of the chain's brand. He has room to display and sell up to 2000 cans per month.

20. An auto dealer makes $500 for each full-sized car she sells and $300 for each intermediate. She must sell at least 100 cars per year to retain the franchise and take in at least $45,000 per year to pay her expenses.

1.8 LINEAR PROGRAMMING

In this section we will consider problems where some sort of optimal choice of values for x and y is to be made, the variables x and y being restricted to a region defined by linear inequalities. It turns out that if there is a "best" choice to be made, it will correspond to one of the vertices of this region.

Mixed Nuts

A canner puts up 1-pound cans of mixed nuts in two versions. The standard can contains 8 ounces of peanuts and 8 ounces of other nuts. The deluxe can contains 4 ounces of peanuts and 12 ounces of other nuts.

The cannery machinery can process only 1000 cans per day, and suppliers can provide up to 450 pounds of peanuts and 600 pounds of other nuts per day. The profit to the canner is 20¢ per standard can and 25¢ per deluxe can.

Of course, the canner would like to maximize his profit. It is not obvious how to do this. One might think that since the profit is greater on the deluxe can, then producing as many deluxe cans as possible would give the most profit. Since each deluxe can contains $\frac{3}{4}$ pound of other nuts, to produce 1000 deluxe cans, the maximum the machinery can handle, would require $(\frac{3}{4})(1000) = 750$ pounds of other nuts.

Only 600 pounds of other nuts are available, however. Let us suppose y deluxe cans are produced, using up the entire 600 pounds, then

$$\tfrac{3}{4}y = 600,$$

and $$y = \tfrac{4}{3}(600) = 800.$$

We see that there are enough other nuts for 800 deluxe cans. These 800 cans would require $(\frac{1}{4})(800) = 200$ pounds of peanuts, which are available. Of course, no standard cans could be made, since all the other nuts would be needed for the deluxe. The profit on the 800 deluxe cans would be

$$800(\$.25) = \$200.$$

This is not the maximum profit possible. We need to analyze the problem in a more orderly way. Let us suppose that x standard and y deluxe cans are produced per day. We will start by deciding what possible values x and y can assume.

Since the plant can produce at most 1000 cans per day, we have

$$\begin{pmatrix}\text{number of}\\ \text{standard cans}\end{pmatrix} + \begin{pmatrix}\text{number of}\\ \text{deluxe cans}\end{pmatrix} \leq 1000,$$

or

$$x + y \leq 1000.$$

Because each standard can contains $\frac{1}{2}$ pound of peanuts while each deluxe can contains $\frac{1}{4}$ pound of peanuts, and since only 450 pounds of peanuts are available, we have

$$\begin{pmatrix}\text{pounds peanuts}\\ \text{in standard can}\end{pmatrix}\begin{pmatrix}\text{number of}\\ \text{standard cans}\end{pmatrix} + \begin{pmatrix}\text{pounds peanuts}\\ \text{in deluxe can}\end{pmatrix}\begin{pmatrix}\text{number of}\\ \text{deluxe cans}\end{pmatrix} \leq 450,$$

or

$$\tfrac{1}{2}x + \tfrac{1}{4}y \leq 450.$$

We multiply by 4 to avoid fractions:

$$2x + y \leq 1800.$$

In the same way the availability of at most 600 pounds of other nuts implies

$$\begin{pmatrix}\text{pounds other nuts}\\ \text{in standard can}\end{pmatrix}\begin{pmatrix}\text{number of}\\ \text{standard cans}\end{pmatrix} + \begin{pmatrix}\text{pounds other nuts}\\ \text{in deluxe can}\end{pmatrix}\begin{pmatrix}\text{number of}\\ \text{deluxe cans}\end{pmatrix} \leq 600,$$

or

$$\tfrac{1}{2}x + \tfrac{3}{4}y \leq 600.$$

Again we avoid fractions by multiplying by 4:

$$2x + 3y \leq 2400.$$

Adding the common-sense requirements that x and y be non-negative, we have the system

(1) $\quad x + y \leq 1000,$

(2) $\quad 2x + y \leq 1800,$

(3) $\quad 2x + 3y \leq 2400,$

(4) $\quad x \geq 0,$

(5) $\quad y \geq 0.$

We graph the region of possibilities in Figure 1.26.

Any point inside or on the boundary of the unhatched region represents a possible choice of x and y for the canner. The question is which of these points yields the greatest profit. In fact,

Figure 1.26

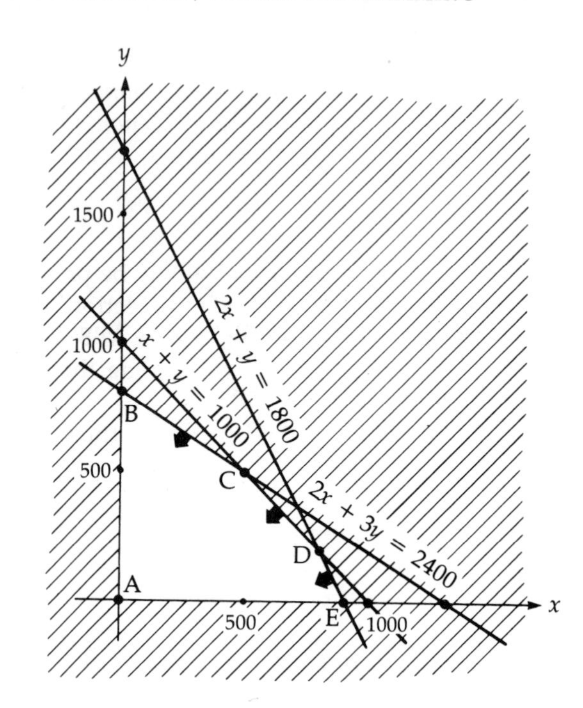

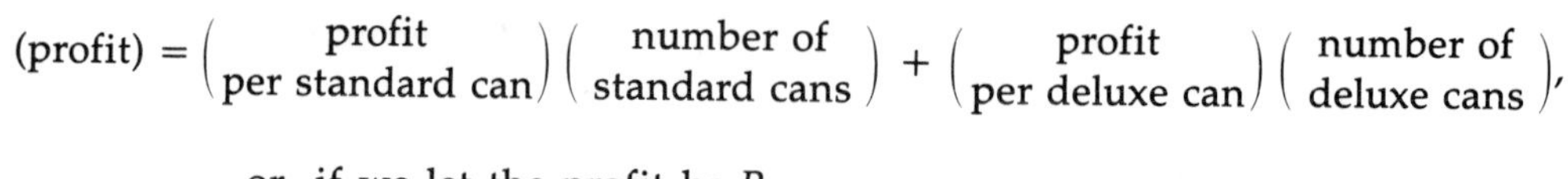

$$(\text{profit}) = \begin{pmatrix}\text{profit}\\ \text{per standard can}\end{pmatrix}\begin{pmatrix}\text{number of}\\ \text{standard cans}\end{pmatrix} + \begin{pmatrix}\text{profit}\\ \text{per deluxe can}\end{pmatrix}\begin{pmatrix}\text{number of}\\ \text{deluxe cans}\end{pmatrix},$$

or, if we let the profit be P,

$$P = .20x + .25y.$$

For example, the point (400, 400) can be seen from our graph to be within the region of possibilities. This corresponds to making 400 standard and 400 deluxe cans. This combination would result in a profit

$$P = .20(400) + .25(400) = \$180.$$

In this problem the profit is a *linear function* of x and y. This merely means that the profit is given by an expression of the form

$$ax + by + c,$$

where a, b, and c are constants. (Here $a = .20$, $b = .25$, and $c = 0$.)

The canner's problem is one of *linear programming,* and can be solved by using a theorem (that is, a statement that can be proved) of linear programming, namely,

THEOREM *If the region of possibilities is determined by linear inequalities, and if a quantity to be maximized or minimized is given by a linear function of x and y, then the maximum or minimum, if it exists, occurs at a vertex of the region of possibilities.*

This theorem will not be proved in this book, although it will be made plausible in the next section. In the problem at hand the region of possibilities has the vertices A, B, C, D, and E shown. Each of these can be found explicitly by solving the corresponding system of two linear equations.

Clearly A = (0, 0), and the vertex B is the intersection of the lines with equations

$$(3) \qquad 2x + 3y = 2400,$$

$$(4) \qquad x = 0.$$

Substituting $x = 0$ into (3), we see

$$3y = 2400$$

and $$y = 800.$$

Thus B = (0, 800).

The vertex C is the intersection of the lines having equations

$$(1) \qquad x + y = 1000,$$

$$(3) \qquad 2x + 3y = 2400.$$

From (1) we see $x = 1000 - y$. We substitute this into (3) to get

$$2(1000 - y) + 3y = 2400,$$

$$2000 - 2y + 3y = 2400,$$

$$y = 2400 - 2000 = 400.$$

Then $x = 1000 - y = 1000 - 400 = 600$. We see C = (600, 400).

Likewise, D is the solution of the system

$$(1) \qquad x + y = 1000,$$

$$(2) \qquad 2x + y = 1800.$$

Subtraction works well here.

$$\begin{array}{rrcl} & 2x + y & = & 1800 \\ (-) & x + y & = & 1000 \\ \hline & x & = & 800. \end{array}$$

From (1), $y = 1000 - x = 1000 - 800 = 200$. We see D = (800, 200).

Finally, we get E by substituting $y = 0$ into $2x + y = 1800$ to get $2x = 1800$ and $x = 900$. Thus E = (900, 0).

Let us now calculate the profit for each of the vertices, using the formula

$$P = .20x + .25\,y.$$

VERTEX	x	y	$P = .20x + .25y$
A	0	0	$.20(0) + .25(0) = \$0$
B	0	800	$.20(0) + .25(800) = \$200$
(*) C	600	400	$.20(600) + .25(400) = \$220$
D	800	200	$.20(800) + .25(200) = \$210$
E	900	0	$.20(900) + .25(0) = \$180$

Since our theorem says that the maximum profit must occur at a vertex, we compare the profits $0, $200, $220, $210, and $180. The (*) marks the largest profit, $220, which occurs at C = (600, 400). This corresponds to making 600 standard and 400 deluxe cans each day.

EXAMPLE 1

The Jail Lunch A jailer plans to serve a lunch of hamburger and bread. He is required by law to serve at least 700 calories and 3 milligrams of iron, but would like to minimize his cost. He can buy hamburger for 9¢ per ounce (105 calories and .8 milligram iron) and bread for 1.5¢ per slice (63 calories and .1 milligram iron). Suppose he serves x ounces of hamburger and y slices of bread. Then, because of the law, we must have

$$(\text{calories in hamburger}) + (\text{calories in bread}) \geq 700$$

or

$$105x + 63y \geq 700,$$

and

$$(\text{iron in hamburger}) + (\text{iron in bread}) \geq 3 \text{ milligrams},$$

or

$$.8x + .1y \geq 3.$$

Multiplying by 10 to avoid the decimals gives

$$8x + y \geqslant 30.$$

Since neither x nor y can be negative, we have the system

(1) $$105x + 63y \geqslant 700,$$

(2) $$8x + y \geqslant 30,$$

(3) $$x \geqslant 0,$$

(4) $$y \geqslant 0.$$

The corresponding region is graphed in Figure 1.27.

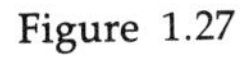

Figure 1.27

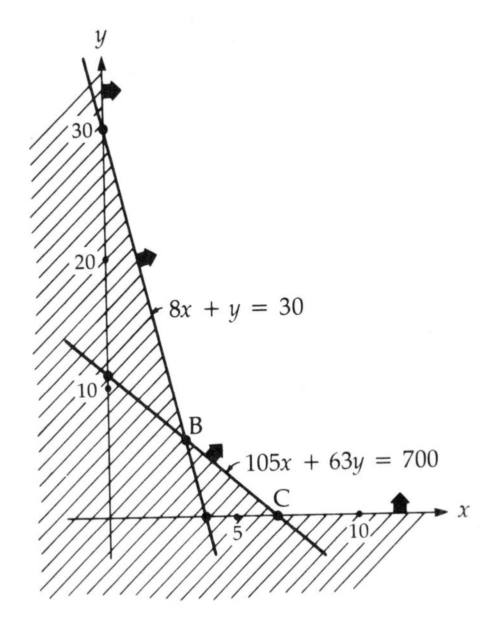

The vertices are calculated to be A = (0, 30), B = (3.0, 6.0), and C = (6.7, 0)

Now the jailer's cost per lunch is

$$(\text{cost of hamburger}) + (\text{cost of bread}) = 9x + 1.5y.$$

We compute this cost at each vertex.

VERTEX	x	y	Cost $= 9x + 1.5y$
A	0	30	$9(0) + 1.5(30) = 45¢$
(*) B	3.0	6.0	$9(3.0) + 1.5(6.0) = 36¢$
C	6.7	0	$9(6.7) + 1.5(0) = 60.3¢$

We see that the cheapest legal lunch consists of 3 ounces of hamburger and 6 slices of bread.

EXAMPLE 2

The Knives A manufacturer makes two kinds of knives, the Camper model and the Trapper model. According to their contract with a large retailer, they must supply at least 5000 knives. The Trapper has a special bone handle of which only 4000 can be obtained. Both knives are made on a special machine, and only 15,000 hours of machine time are available. The Camper takes 2 hours to make on the machine; the Trapper, 3 hours. The profit is 75¢ per Camper and $1 per Trapper. How many of each should be made to maximize the profit?

Let x Campers and y Trappers be made. Because of the contract

$$x + y \geqslant 5000.$$

The limit on bone handles implies

$$y \leqslant 4000.$$

Because of the available machine time we have

$$2x + 3y \leqslant 15{,}000.$$

Adding the conditions that x and y cannot be negative gives the system

(1) $\quad x + y \geqslant 5000,$

(2) $\quad y \leqslant 4000,$

(3) $\quad 2x + 3y \leqslant 15{,}000,$

(4) $\quad x \geqslant 0,$

(5) $\quad y \geqslant 0.$

It is graphed in Figure 1.28.

The vertices are easily computed. We get A by substituting $y = 0$ into $x + y = 5000$, getting A = (5000, 0). Likewise we find D = (7500, 0) by plugging $y = 0$ into $2x + 3y = 15{,}000$.

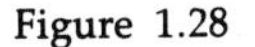

Figure 1.28

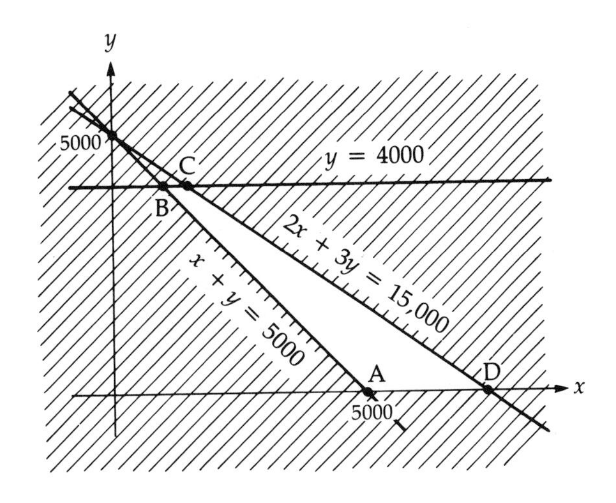

To calculate B, we substitute $y = 4000$ into $x + y = 5000$. We find $x = 1000$, so $B = (1000, 4000)$.

Finally, we get C by putting $y = 4000$ into $2x + 3y = 15{,}000$, finding

$$\begin{aligned} 2x &= 15{,}000 - 3y \\ &= 15{,}000 - 3(4000) \\ &= 15{,}000 - 12{,}000 = 3000; \end{aligned}$$

so $x = 1500$, and $C = (1500, 4000)$.

Now the profit to be maximized is $P = .75x + y$.

VERTEX	x	y	$P = .75x + y$
A	5000	0	$.75(5000) + 0 =$ \$3750
B	1000	4000	$.75(1000) + 4000 =$ \$4750
C	1500	4000	$.75(1500) + 4000 =$ \$5125
(*) D	7500	0	$.75(7500) + 0 =$ \$5625

We find the maximum profit of \$5625 is made by making 7500 Campers and no Trappers.

Exercises 1.8

In problems 1 through 4 find the maximum and minimum of the function given on the region defined by the system of inequalities and give the coordinates of the vertices where these occur.

1. Function $2x + 3y$; region defined by $x + y \geq 1$, $3x + 4y \leq 12$, $x \geq 0$, $y \geq 0$.

2. Function $3x + 4y$; region defined by $2x + 3y \leq 12$, $x \leq 5$, $x \geq 0$, $y \geq 0$.

3. Function $5x - 6y$; region defined by $x + 2y \geq 4$, $x \geq 1$, $x + y \leq 6$.

4. Function $-x - y$; region defined by $x + 3y \leq 12$, $y \geq -1$, and $x - y \geq 1$.

5. A brush salesperson generally sells out his whole case every day. He can carry at most 40 pounds. Men's brushes weigh 1 pound each, yield \$1 profit each to him, and his case holds 50 of them. Women's brushes weigh $\frac{1}{2}$ pound, yield 80¢ profit, and also fit 50 to the case. How many of each should he take to maximize his profit?

6. A trucker is making up a load of 30-pound crates of apples and 40-pound crates of pears. His profit is \$2 per crate of apples and \$2.50 per crate of pears. His truck holds up to 5000 pounds and only 80 crates of pears are available. How many crates of each fruit should he take to maximize his profit?

7. A toothbrush is made on two machines, one to stamp out the handle and the other to insert the bristles. Two kinds of brushes are made. Type A takes 10 seconds to stamp out and 20 seconds for the bristles. Type B takes 8 seconds to stamp out and 25 seconds for the bristles. On a certain day the stamping machine is available for 8 hours and the bristle machine for 10 hours. How can the maximum number of brushes be made?

8. A farmer has contracted to grow at least 2000 pounds of beets and 3000 pounds of beet greens. An acre of type A beet yields 1000 pounds of beets and 1000 pounds of greens. An acre of type B beet yields 500 pounds of beets and 1500 pounds of greens. How can the farmer minimize the beet acreage?

9. A bakery is to make loaves of two types of bread. The Crustygood loaf uses 3 cups of wheat flour and 2 cups of rye flour, and brings a profit of 10¢. The Ovenfresh loaf uses 2 cups of wheat flour and 2.5 cups of rye flour, and brings a profit of 14¢. Only 12 cups of wheat flour and 10 cups of rye flour are available. How many of each loaf should be made to maximize the profit?

10. A gardener is mixing up her own chemical fertilizer from two commercial brands. Each ounce of Brand A contains 5 units of nitrogen, 4 units of phosphorus, and 2 units of potassium. Each ounce of Brand B contains 4 units of nitrogen, 2 units of phos-

phorus, and 3 units of potassium. The mixture should contain at least 80 units of phosphorus, at least 60 units of potassium, and as little nitrogen as possible. How many ounces of each brand should be used to do this?

11. A children's shoe store is trying to sell two types of shoes, Fastrunners and Zips. They have 80 pairs of Fastrunners and 50 pairs of Zips on hand. The profit is $1 on a pair of Fastrunners and $1.20 on a pair of Zips. The store has advertised that 3 free balloons will be given away with each pair of Fastrunners and 5 balloons with each pair of Zips, and only 300 balloons are on hand. How many pairs of each should the store sell to maximize its profit?

12. A cafeteria dietitian is experimenting with a mixture of orange and grapefruit juice. A serving of the mixture should have at least one ounce of each juice in it, and no more than 5 ounces of orange juice or 4 ounces of grapefruit juice. There are 8 calories and 9 units of vitamin C in an ounce of orange juice and 9 calories and 10 units of vitamin C in an ounce of grapefruit juice. The mixture is to have no more than 72 calories and as much vitamin C as possible. How many ounces of each juice should it contain?

13. A health food store is making cookies from a mixture of oatmeal and wheat germ. They find that unless at least half oatmeal is used in the mixture the cookies won't hold together. Each batch contains at most 2 cups of the mixture. How can they maximize the protein in a batch of cookies? A cup of oatmeal has 11 grams of protein and a cup of wheat germ has 17 grams.

14. An engineer needs a mixture of iron and lead and aluminum weighing at least 8.5 grams per cubic centimeter. The mixture must be at least one-third iron for strength. If a cubic centimeter of iron weighs 7.9 grams and costs .1¢, a cubic centimeter of lead weighs 11.3 grams and costs .2¢, and a cubic centimeter of aluminum weighs 2.7 grams and costs .1¢, how can she minimize her cost? Let a cubic centimeter of the mixture contain x cubic centimeters of iron, y cubic centimeters of lead, and the rest aluminum.

1.9 MORE LINEAR PROGRAMMING

In this section we show a graphical method of solving linear programming problems that allows us to avoid computing the coordinates of all the vertices of the region of possibilities. We will also see *why* it is reasonable to expect the "best" values of x and y to correspond to a vertex.

The Magazine Racks

A grocery store manager has racks to sell copies of two magazines by his checkout stands. He has at most \$500 to buy the magazines each month, and expects to sell out each month. Each copy of *Woman's World* costs him 25¢ and gives 10¢ profit. Each copy of *Home and Garden Digest* costs him 10¢ and gives 5¢ profit. His *Woman's World* racks hold at most 1500 copies, and his *Home and Garden Digest* racks hold at most 3000 copies. The racks are provided by the publishers, who allow only their own magazines in their racks.

Suppose he buys x copies of *Woman's World* and y copies of *Home and Garden Digest*. Since he has at most \$500 to spend,

$$(\text{cost of } \textit{Woman's World}) + (\text{cost of } \textit{H\&GD}) \leqslant \$500,$$

or (converting to cents)

$$25x + 10y \leqslant 50{,}000.$$

We divide by 5 to get

$$(1) \qquad 5x + 2y \leqslant 10{,}000.$$

By the rack limitations

$$(2) \qquad x \leqslant 1500,$$

$$(3) \qquad y \leqslant 3000.$$

Since x and y cannot be negative we have the system

$$\begin{array}{ll} (1) & 5x + 2y \leqslant 10{,}000, \\ (2) & x \leqslant 1500, \\ (3) & y \leqslant 3000, \\ (4) & x \geqslant 0, \\ (5) & y \geqslant 0. \end{array}$$

This system is graphed in Figure 1.29.

We see that the region of possibilities has five vertices: A, B, C, D, and E. The profit is given by

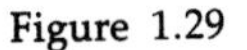
Figure 1.29

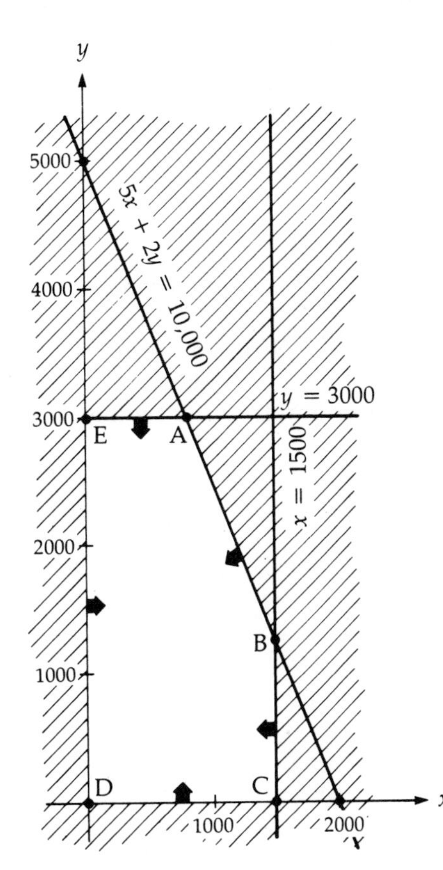

$$P = (\text{profit from } \textit{Woman's World}) + (\text{profit from } \textit{H\&GD})$$

or $$P = 10x + 5y.$$

Our next step would be to compute the coordinates x and y at each of the five vertices and then evaluate the profit at each one. There is a graphical method of choosing the vertex yielding the maximum profit, however, that does not require all this work.

Let us consider the set of pairs (x, y) yielding a given *fixed profit*. What number we take this fixed profit to be makes no difference to our method. Any convenient value may be chosen.

For example, if $x = 1000$ and $y = 0$, we find a profit of

$$10x + 5y = 10(1000) + 5(0) = 10,000.$$

(The point (1000, 0) was chosen as one easily graphed.) *Any* pair (x, y) satisfying

$$10x + 5y = 10,000$$

will yield a profit of 10,000 cents. Since this is a linear equation, its graph is a straight line. Only one more point (besides (1000, 0)) is needed to determine its graph. For example, if $x = 0$, then

$$5y = 10{,}000,$$

and

$$y = 2000.$$

We add the graph of the equation

$$10x + 5y = 10{,}000$$

to our graph of the region of possibilities (Figure 1.30).

Figure 1.30

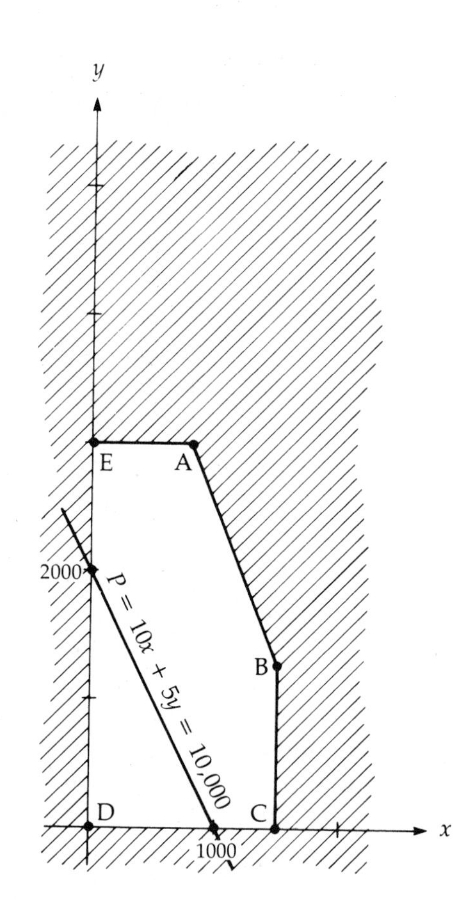

What if a different fixed profit had been chosen, say 20,000 instead of 10,000? We easily find the two points (2000, 0) and (0, 4000) on the line

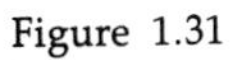

Figure 1.31

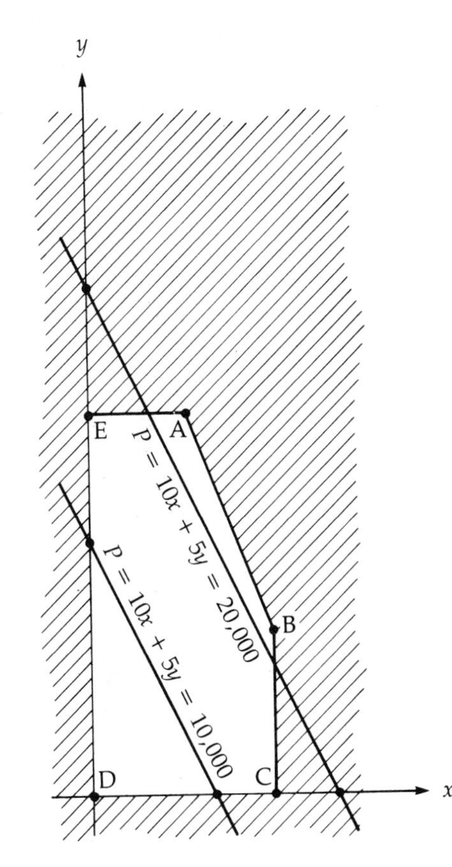

$$10x + 5y = 20{,}000,$$

and add its graph to our diagram (see Figure 1.31).

The graph of

$$10x + 5y = 20{,}000$$

appears to be a line parallel to the graph of

$$10x + 5y = 10{,}000.$$

It is. In fact, for any constant value P the graph of

$$10x + 5y = P$$

is a line parallel the graph of

$$10x + 5y = 10{,}000.$$

Figure 1.32

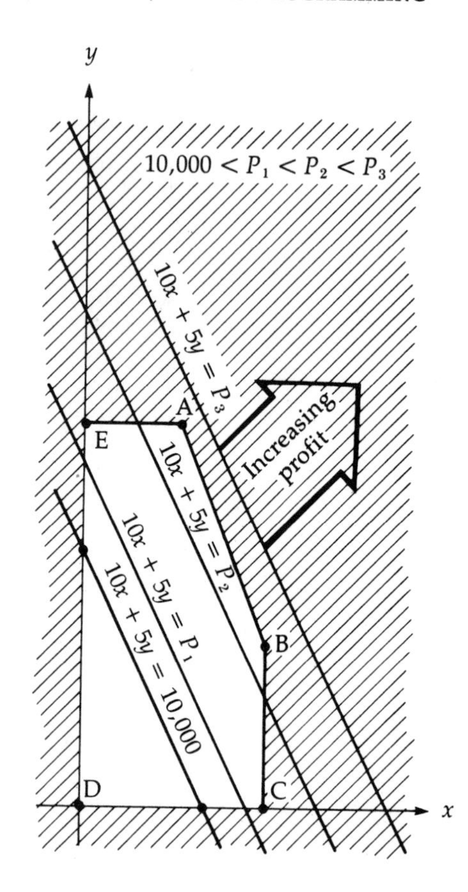

Increasing the value of P merely shifts the line upward and to the right (Figure 1.32).

Now the question as to which point in the region of possibilities yields the greatest profit is easily answered. Imagine the line

$$10x + 5y = 10{,}000$$

shifted parallel to itself upward and to the right. We must move the line as far as possible in this direction so that it still contains at least one point of the region of possibilities. From Figure 1.32 we see this occurs when the line goes through the vertex A. Thus the maximal profit occurs at A.

Since A is at the intersection of the lines

(1) $$5x + 2y = 10{,}000,$$

(3) $$y = 3000,$$

we solve this system of equations. We have

$$\begin{aligned} 5x + 2(3000) &= 10{,}000, \\ 5x &= 10{,}000 - 6000 = 4000, \\ x &= \frac{4000}{5} = 800. \end{aligned}$$

The vertex is (800, 3000). The manager should buy 800 *Woman's World*'s and 3000 *House and Garden Digest*'s. Then his profit will be

$$10(800) + 5(3000) = 23{,}000¢ = \$230.$$

Let us check the result by the method of the previous section. We will skip showing the calculations, just noting that we easily compute B = (1500, 1250), C = (1500, 0), D = (0, 0), and E = (0, 3000). Then $10x + 5y$ gives the profit corresponding to each of these

VERTEX	Profit = $10x + 5y$
B = (1500, 1250)	10(1500) + 5(1250) = 21,250¢ = \$212.50
C = (1500, 0)	10(1500) + 5(0) = 15,000¢ = \$150.00
D = (0, 0)	10(0) + 5(0) = 0¢ = \$0.00
E = (0, 3000)	10(0) + 5(3000) = 15,000¢ = \$150.00

None of the profits listed matches the \$230 we found at vertex A.

One advantage of this method is obvious. Not only did we find the maximal profit without checking the profits at points B, C, D, and E; it was not even necessary to compute the coordinates of these points.

The example also makes plausible the theorem of Section 1.8 that if a linear function has a maximum or minimum on a region defined by linear inequalities, then this occurs at a vertex. It amounts to saying that if you hit a region with straight sides with a straight line, the first place you will touch is a corner.

EXAMPLE

Concrete Blocks A trucker delivers 30-pound and 50-pound concrete blocks to a construction company. His contract calls for the delivery of at least 10,000 blocks weighing at least 200 tons each week. It costs the trucker 8¢ to deliver a 30-pound block and 10¢ to deliver a 50-pound block. How can he minimize his cost?

Suppose the trucker delivers x 30-pound and y 50-pound blocks per week. By his contract

$$\begin{pmatrix}\text{number of}\\ \text{30-pound blocks}\end{pmatrix} + \begin{pmatrix}\text{number of}\\ \text{50-pound blocks}\end{pmatrix} \geqslant 10{,}000,$$

or

$$x + y \geqslant 10{,}000. \tag{1}$$

Also

$$\begin{pmatrix}\text{weight of}\\ \text{30-pound blocks}\end{pmatrix} + \begin{pmatrix}\text{weight of}\\ \text{50-pound blocks}\end{pmatrix} \geqslant 200 \text{ tons},$$

or (since a ton is 2000 pounds)

$$30x + 50y \geqslant 200(2000) = 400{,}000,$$

or

$$3x + 5y \geqslant 40{,}000. \tag{2}$$

We have the system

$$\begin{aligned} &(1) & x + y &\geqslant 10{,}000,\\ &(2) & 3x + 5y &\geqslant 40{,}000,\\ &(3) & x &\geqslant 0,\\ &(4) & y &\geqslant 0. \end{aligned}$$

The trucker's cost is given by

$$(\text{cost of 30-pound blocks}) + (\text{cost of 50-pound blocks}) = 8x + 10y.$$

If $x = 10{,}000$ and $y = 0$, for example, then the cost is

$$8(10{,}000) + 10(0) = 80{,}000¢.$$

The same cost results from all pairs (x, y) such that

$$8x + 10y = 80{,}000.$$

We find another point on this line by setting $x = 0$, so

$$\begin{aligned} 10y &= 80{,}000,\\ y &= 8000. \end{aligned}$$

Thus (10,000, 0) and (0, 8000) are on the line

$$8x + 10y = 80{,}000.$$

Now we graph this line along with the region of possibilities (Figure 1.33).

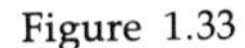
Figure 1.33

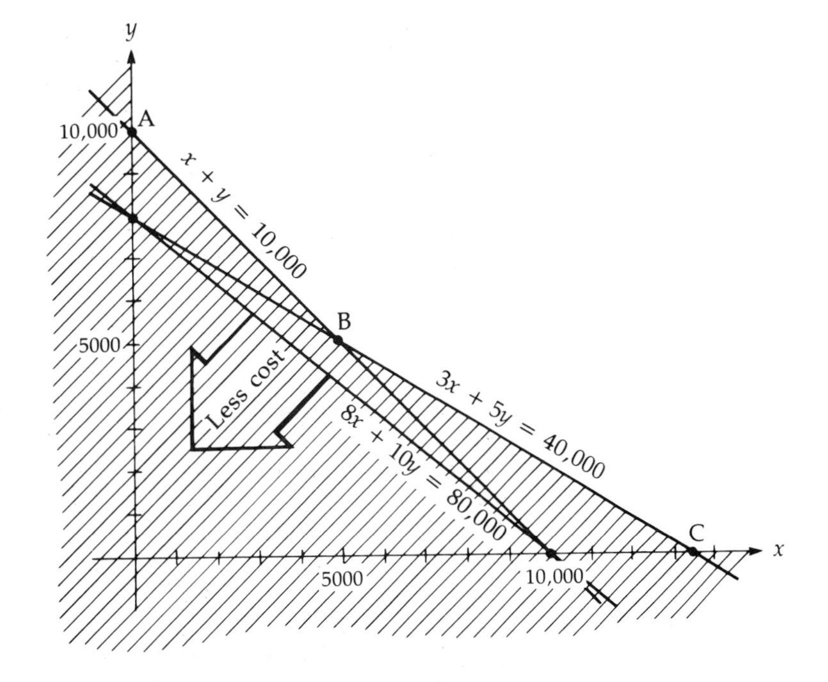

Since decreasing x or y lowers the trucker's cost, we seek the line parallel to $8x + 10y = 80{,}000$ farthest to the lower left but still containing a point of the region of possibilities. Although it is a close decision, we see from the graph that such a line would go through the vertex B. We calculate the coordinates of B by solving the system

$$\begin{aligned} &(1) \qquad x + y = 10{,}000, \\ &(2) \qquad 3x + 5y = 40{,}000. \end{aligned}$$

We multiply (1) by 3 and subtract it from (2).

$$\begin{array}{lr} & 3x + 5y = 40{,}000 \\ (-) & \underline{3x + 3y = 30{,}000} \\ & 2y = 10{,}000. \end{array}$$

Thus $y = 5000$ and $x = 10{,}000 - 5000 = 5000$. The trucker should carry 5000 blocks of each type. His cost will then be

$$8(5000) + 10(5000) = 90{,}000¢ = \$900.$$

The General Method

We summarize the method of this section.

1. Write the conditions of the problem as a system of linear inequalities in x and y and graph this system.
2. Choose a convenient fixed value of the quantity to be maximized or minimized and graph the straight line of all pairs (x, y) yielding this fixed value.
3. Imagine moving the line graphed in 2 parallel to itself to a position which is as far as possible in the desired direction and yet still touching a point of the region of possibilities. This point is the solution vertex.
4. Calculate the coordinates of the vertex determined in 3 by solving the corresponding system of two linear equations.

LIMITATIONS OF GRAPHING
The method summarized here depends upon the construction of a graph, in which some imprecision is unavoidable. It may be impossible in step 3 to tell which of two or more points is the solution. In this case the values of the function to be maximized or minimized should be calculated at each vertex that appears to be a possible solution.

It is even possible that each of two vertices produces the same maximal or minimal value. In this case any point on the line segment joining these vertices (including, of course, the vertices themselves) is a solution.

Exercises 1.9

1. A company running a contest has promised to give away at least 6000 prizes with a total value of at least \$100,000. The prizes are \$20 toasters and \$15 electric drills. These cost the company \$12 and \$8 each, respectively. How many of each should the company give away to minimize its cost?

2. A lighting engineer planning to illuminate a store decides at least 30 light fixtures totaling at least 1500 watts are needed. Available are

40-watt and 55-watt fixtures, costing \$20 and \$27 each, respectively. How many of each should be used to minimize the cost?

3. A bookstore owner is ordering a shipment of bibles (2 pounds each) and dictionaries (3 pounds each). He wants them sent the cheapest way, namely, by U.S. Mail in one package. This means the order can weigh no more than 70 pounds. The box the shipper uses can hold at most 30 books, mixed any way. How can the bookseller's profit be maximized if he makes \$1.40 per bible and \$1.80 per dictionary?

4. A factory is set up to make two kinds of calendars. A children's calendar takes 5 minutes to print, 3 minutes to cut, and 8 minutes to assemble. A scenic calendar takes 8 minutes to print, 2 minutes to cut, and 4 minutes to assemble. The factory's profit is 20¢ and 25¢ for each type of calendar, respectively. How many of each should be made to maximize the profit, if 40,000 minutes of printing time, 12,000 minutes of cutting time, and 30,000 minutes of assembling time are available?

5. A tree farmer has \$14,300 to buy blue spruce and Norway spruce seedlings and 1.2 million square feet on which to plant them. A blue spruce seedling requires 10 square feet of space, costs 13¢, and will provide her with a profit of \$1.35. A Norway spruce seedling requires 12 square feet of space, costs 11¢, and will yield a profit of \$1.41. How many of each should she buy and plant to maximize her profit?

6. An auto dealer sells American and imported cars. His profit averages \$400 on an American car and \$500 on an import. He must sell at least 1000 American cars per year to keep his franchise and he is allowed no more than 500 of the imports by the factory. He simply cannot handle the paperwork of selling more than 1300 cars per year. How many of each should he sell to maximize his profit?

7. A store sells 9′-by-12′ and 9′-by-9′ rugs, making a profit of \$40 and \$35 on each kind, respectively. The factory allows it up to 2000 rugs totaling at most 20,000 square yards annually. How many of each should it sell to maximize its profit?

8. A chocolate chip cookie contains 52 calories and 8 grams of protein, and makes a child smile 7 times when eating it. A health cookie contains 20 calories and 30 grams of protein, and makes a child smile 2 times when eating it. A teacher plans to serve some cookies to his first grade class. They should total at least 1800 grams of protein, and no more than 2600 calories. How many of each kind should he serve to maximize smiles?

9. The winner of a television quiz program is allowed to choose as a prize as many identical items from a large department store as he wants, as long as the sum of the price per item (in dollars) and the number of items does not exceed 100. Suppose x items worth y dollars each are chosen. Graph the region of possibilities and find the total value of the prize at each vertex. What is the value of the prize for $x = y = 50$? How can this be?

THE COMPUTER

CHAPTER TWO

In this chapter we begin with a brief history of computing and computers and then provide a short overview of the nature of a modern digital computer. Before we consider a programming language, it is necessary to learn the rudiments of developing and reading a flow chart. We then provide independent introductions to two computer languages. The first, BASIC, is an easily learned language designed to be used on interactive terminals (to communicate directly with the computer). The second, FORTRAN IV, was designed to use punched cards as the medium of communication with the computer. We will actually introduce a variation of FORTRAN IV called WATFIV.

If you are not familiar with some of the terms used in the last paragraph take heart, we will explain them all in due course. But first, how did it all begin?

2.1 A BRIEF HISTORY AND INTRODUCTION

Computing was the second step of mathematical evolution, the first being counting. The way in which societies expressed their numbers had a telling effect on the way the merchants, astronomers, astrologists, and mathematicians computed.

HIEROGLYPHIC SYMBOLS	
1	|
10	∩
100	𓍢
1000	𓆼
10000	𓂭

In ancient Egypt, where the numeration system (that is, the method of naming numbers) was hieroglyphic, computation was very difficult, particularly multiplication. To express a number, say 347, an Egyptian would just write down the right number of each sort of symbol. For this example it would be

𓍢𓍢𓍢 ∩∩∩∩ |||||||

Addition would be easy, but multiplication could only be done by a very tedious process.

The Greeks, much the superior mathematicians, had an alphabetic numeration system that continued to be used during the Dark Ages by scientists. A separate letter was used to designate each of the numbers 1 to 9, 10, 20, . . . , 90, 100, 200, . . . , 900. A number like 169 would be expressed as $\rho\xi\theta$.

If you thought it was hard to learn *your* multiplication tables, think what a Greek student had to learn—a total of $27 \cdot 27 = 729$ different products.

Roman numerals, which are still used on certain buildings and in some motion picture credits, were the common numeration system for over 1000 years. They are slightly more sophisticated than the Egyptian hieroglyphics because position is used to denote subtraction.

GREEK NUMERALS								
1	2	3	4	5	6	7	8	9
α	β	γ	δ	ϵ	ϛ	ζ	η	θ
10	20	30	40	50	60	70	80	90
ι	κ	λ	μ	ν	ξ	ο	π	ϟ
100	200	300	400	500	600	700	800	900
ρ	σ	τ	υ	φ	χ	ψ	ω	ϡ

ROMAN NUMERALS			
1	I	100	C
5	V	500	D
10	X	1000	M
50	L		

Recall that IX stands for 9 and XI is 11. So the year of publication of this book is written

MCMLXXIX

Not until our current place-value system, which had its origin in ancient Mesopotamia, came into common usage was there any hope for the development of calculating machines other than the Chinese abacus. In ancient times two merchants would have their slaves count and compute, probably using pebbles, until they agreed. The word "calculus" means "pebble" in Latin, so the origin of the words "calculation" and "calculator" lies here.

Notched sticks were often used for keeping accounts. In fact, the custom of recording financial transactions as notches cut in "tally" sticks persisted in the English Exchequer until 1826. However, it was during the fifteenth and sixteenth centuries that our present Hindu-Arabic place-value system, which uses only the symbols, 0, 1, 2, 3, 4, 5, 6, 7, 8, 9, came into common usage. During this time a computation contest was held between an "abacist" (one who used the abacus) and an "algorist" (one who used our notation and a computational algorithm such as we use today). Although the abacus won, the use of counting boards and counters continued to diminish.

The next century saw two startling inventions. The first, in 1617, was by John Napier, a Scottish baron, who invented the theory of logarithms and applied them to a device referred to as Napier's bones or Napier's rods. Whereas the abacus required the user to do the carrying from one column of markers to another, Napier's rods took care of the carrying process automatically during a multiplication problem. This device was the forerunner of the slide rule, which has only in the last decade been replaced by the pocket calculator.

The first mechanical calculator, a cogwheel adding machine, was the invention of a brilliant young mathematician who may have let his religious fanaticism turn him into the greatest might-have-been in

Figure 2.1

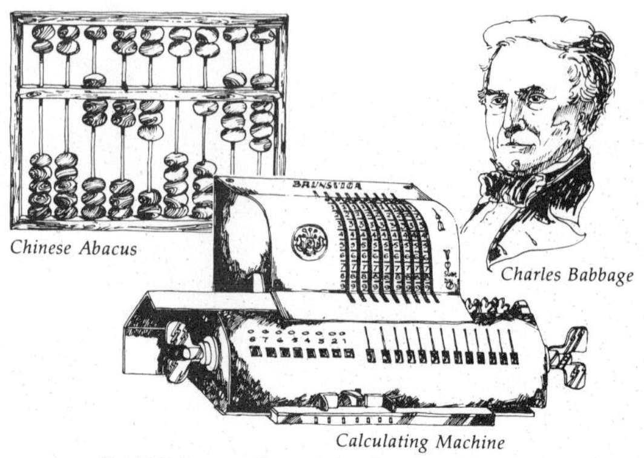

British Crown Copyright. Science Museum, London

history. At 18, Blaise Pascal* (1623–1662) invented a calculating machine that could add and subtract to assist his father in auditing government accounts. This was two years after his first published mathematical research resulting in a theorem which the nineteenth century mathematician Sylvester called a "cat's cradle." (Have any of you read Kurt Vonnegut's book by that name?) Pascal subsequently gave his machine to Queen Christine of Sweden, and it has been lost.

Only about 30 years later a machine that not only did addition and subtraction, but also multiplication, division, and the extraction of roots, was invented by perhaps the world's greatest universal genius, Gottfried Wilhelm Leibniz (1646–1716). Besides being an independent discoverer with Newton of the calculus, Leibniz made fundamental contributions to law, literature, logic, history, religion, statecraft, metaphysics, and philosophy.

It was not until 1820 that Thomas de Colmar invented an adding machine that was fast and efficient enough to be a commercial success and that set the basic design for commercial calculators for 50 years. At that time Swedish and American patents were granted to different inventors which led to the Brunsviga-type machines, and to the Comptometers and Burroughs machines.

In spite of the increased speed of calculation and the commercial success of these machines, their speed was often exaggerated. In 1946 the American army staged a competition which involved only ordinary arithmetic operations between a private using a modern desk calculator and one of their Japanese clerks using an abacus. The clerk won every time.

However, the modern electronic calculators, first introduced in 1961, are clearly on their way, as a result of the recent technology of miniaturization, toward affecting education and commerce in ways not

*See E. T. Bell's book *Men of Mathematics* for a very readable account of the life of this child prodigy (Simon and Schuster, New York, 1938).

seen since the acceptance of our numeration system in the fifteenth and sixteenth centuries. Many of these calculators, even some of the pocket-size ones, are genuine digital computers that can be programmed very nearly like large machines.

How have the large machines evolved? When were they first conceived, when invented, and how have they changed? The answers seem to us quite surprising.

Over 100 years before the first workable large-scale computing machine was built, an Englishman named Charles Babbage (1791–1871) envisioned an "analytical engine" that would do any calculations for which instructions about the mathematical operations could be given (Figure 2.2). Moreover, his basic design and the logic of organization of

Figure 2.2 Up to his death in 1871, Charles Babbage had spent many years in the design of this machine for automatically solving mathematical equations. The basic principles of the engine are similar to those underlying the first electronic computers.

the components are entirely workable today. He was a classic example of a man born before his time, since the machine shops of his day could not produce the cogwheels and levers he needed in quantity. He not only invented the railroad cowcatcher and the first tachometer, but he spent years in improving lathes and gear-cutting tools, and no doubt even conceived of the fundamental idea of mass production, universal interchangeable parts, which was 50 years in coming.

We will not describe in detail the principles that Babbage enunciated which have been used recently in creating enormous computing machines. One indication of his insight is of independent interest, however. He had to find a mechanical method of controlling the operations of his machine. Fortunately a weaver, Joseph Jacquard, had in 1801 begun to use punched cards to control weaving patterns on his looms. Babbage realized this was just what he needed.

> Charles Babbage had a totally eccentric fetish for accuracy. For example, after reading the noted lines
>
> Every moment dies a man
> Every moment one is born,
>
> he wrote the poet, Lord Tennyson, the following: "It must be manifest that if this were true, the population of the world would be at a standstill." Babbage proposed the more precise:
>
> Every moment dies a man
> Every moment $1\frac{1}{16}$ is born.

It was not until 20 years after Babbage's death, however, that Herman Hollerith devised ways of recording data on punched cards and invented machines that could read the cards and compute with the information. The Census Bureau had all but despaired of having the census figures for 1890 before the 1910 census was to be taken (making them all but obsolete) when Hollerith invented an electromagnetic machine activated by punched cards. He said the idea occurred to him when he saw a railroad conductor use a ticket punch.

As a result of Hollerith's invention, the actual census of 1890 took a third of the time to tabulate that the 1880 census had taken, in spite of a 25% increase in the population. In 1896 the Tabulating Machine Company was organized by Hollerith. After a later consolidation, it became a corporation probably more familiar to you, International Business Machines Corporation (IBM).

The first digital computers were designed and built in 1939 and 1940. At first electric relays were used for computation and storage, and in 1944, after 5 years, the largest electromechanical calculator ever built, containing 3300 relays and weighing 5 tons, was finished. It could multiply two 23-digit numbers in 6 seconds. In 1946 vacuum tubes were used instead of relays, and the speed of computation was increased many thousands of times. Moreover, the concept of a computer system began to emerge.

The transistor (invented in 1947 at Bell Laboratories) together with research and development in the use of magnetics (magnetic disc and drum storage, magnetic cores, and magnetic tapes), provided another generation of computers that began emerging in the late 1950s. Whereas the vacuum tube computer could multiply two 10-digit numbers in 1/40 of a second in 1946 and in 1/2000 of a second in 1953, by 1959 the semiconductor computer took only 1/100,000 of a second.

As if this increase in speed and concurrent decrease in size and weight of the machines was not remarkable enough, the last two decades have been incredible. All aspects of the computer have been transformed. The core memory is now composed of tiny silicon chips mounted on $\frac{1}{2}$ inch modules with each pair of modules containing more than 4000 bits of data. Moreover, tiny "miracle chips" less than $\frac{1}{4}$ inch square and quite flat (small enough to go through the eye of a large needle), which have a calculating capacity equal to that of a room-size computer of only 25 years ago, are in common use.* The speed is almost unimaginable—up to millions of calculations a second, and new experimental devices can raise that to billions. Ultraviolet light is used to draw the circuit lines. With this process the 1250 pages of the Old and New Testaments of the Bible can be printed on a wafer $1\frac{1}{2}$ inches square. In the future, electron beams will probably be used to draw even finer circuit lines.

Early magnetic discs packed 1000 bits of data on a square inch, whereas today the number is several million. Moreover, the new research on "magnetic bubbles" will likely increase the density even further. Laser and electrophotographic printing have recently been combined by IBM to produce a printer that prints 13,000 lines per minute (over 6 times as fast as their previous fastest impact printer). See Table 2.1 (page 88).

All these factors and more are combining to give us smaller and faster computers with larger storage and greater flexibility of use. The small components are being used in such everyday items as washing machines, cameras, and microwave ovens. Some home computers, not just calculators, but full-scale computers, cost less than $1000—well under the cost of a good piano.

From medicine to manufacturing to warehouse and retail inventories to space exploration to home and school, the computer is becoming an ever more pervasive factor in both our work and our play. There is no doubt that what was to have been the nuclear age has really become the computer age.

Before providing an overview of the components of a present day

*See the February 25, 1978 issue of *Time* for a look into the present and future uses of computers.

TABLE 2.1. The chart shows how data processing costs and time have declined during the past two decades. It represents a mix of about 1700 computer operations, including payroll, discount computation, file maintenance, table lookup, and report preparation. Figures show costs of the period, not adjusted for inflation.

	1955	1960	1965	TODAY
Cost	$14.54	$2.48	$.47	$.20
Processing time	375 seconds	47 seconds	37 seconds	5 seconds
Technology	Vacuum tubes Magnetic cores Magnetic tapes	Transistors Channels Faster cores Faster tapes	Solid Logic Technology Large, fast disk files New channels Larger, faster core memory Faster tapes	Monolithic memory Monolithic logic Virtual storage Larger, faster disk files New channels Advanced tapes
Programming	Stored program	Overlapped input/output Batch processing	Operating system Faster batch processing	Virtual storage Advanced operating systems Multiprogramming Batch/on-line processing

computer, we quote from an extraordinarily prophetic statement which was made by a committee from the British Association that examined Babbage's analytical engine after his death over 100 years ago.

> Apart from the question of its saving labour in operations now possible, we think that the existence of an instrument of this kind would place within reach much which, if not actually impossible, has been too close to the limits of human endurance to be practically available.

The Nature of a Computer

Before learning the basic features of a computer, you should consider the four aspects of today's computers that are responsible for their intrusion into almost every corner of our life.

1. **Speed** By performing more than 100,000 operations per second, and in some cases millions, computers are able to obtain and manipulate data at a previously unbelievable speed.
2. **Accuracy** Gone are the days when errors in results could frequently be traced to the computer. There *may* be an error every trillion or so operations, but most errors are a result of GIGO (Garbage In, Garbage Out).

3. **Ease of Communication** Although English is still not, nor will it likely be soon, a means of communicating with most computers, the simplicity of many computer languages makes access to computers much easier than before. Also, the increasing accessibility of computers through time-sharing terminals (typewriterlike machines through which one has direct access to a computer) and easy-access batch processing (the use of punched Hollerith cards as input to a computer) improve the chance that nontechnicians like us will make use of computers.
4. **Relevance** The increasing technology of our society and the rapidly expanding population of the world makes the speed and storage capacity of present-day computers almost a necessity. Certainly nineteenth-century England, where Babbage first conceived of a digital computer, had neither a sufficiently complex society nor a large enough population to really make it relevant. Many recent scientific advances could not have been made without computers, including space travel.

Basic Tasks and Components

The components of a computer reflect the interplay needed to perform its three basic tasks. These tasks are

1. Accept input
2. Work with information
3. Prepare output

All of these tasks are performed in several ways, in many cases simultaneously, and usually use more than one of the five basic components of a computer. These components (already known to Babbage) are (see Figure 2.3)

1. Input devices
2. Memory or storage
3. Arithmetic unit
4. Control unit
5. Output devices

Usually viewed as hardware, we will describe them more fully in the next subsection. First we will say a little more about the general way the computer works.

1. **Input** The computer is capable of receiving input through many devices, for example, punched cards, punched tape, magnetic tape, typewriterlike console, electronic impulse

How the computer works

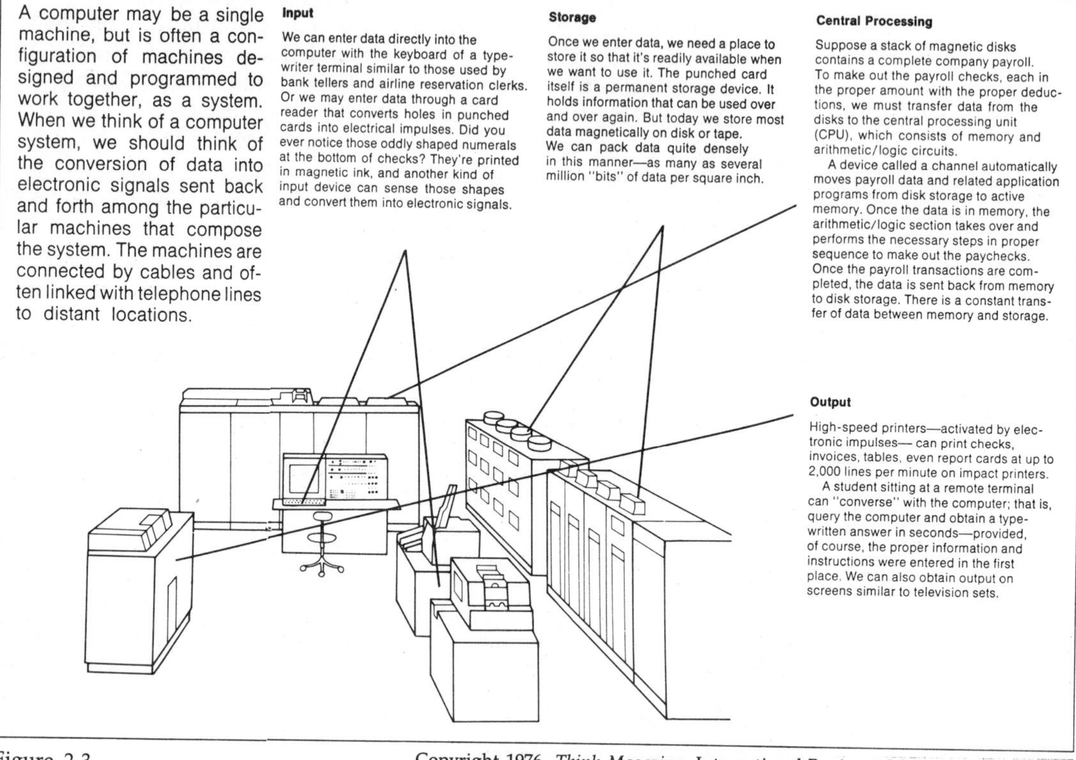

A computer may be a single machine, but is often a configuration of machines designed and programmed to work together, as a system. When we think of a computer system, we should think of the conversion of data into electronic signals sent back and forth among the particular machines that compose the system. The machines are connected by cables and often linked with telephone lines to distant locations.

Input

We can enter data directly into the computer with the keyboard of a typewriter terminal similar to those used by bank tellers and airline reservation clerks. Or we may enter data through a card reader that converts holes in punched cards into electrical impulses. Did you ever notice those oddly shaped numerals at the bottom of checks? They're printed in magnetic ink, and another kind of input device can sense those shapes and convert them into electronic signals.

Storage

Once we enter data, we need a place to store it so that it's readily available when we want to use it. The punched card itself is a permanent storage device. It holds information that can be used over and over again. But today we store most data magnetically on disk or tape. We can pack data quite densely in this manner—as many as several million "bits" of data per square inch.

Central Processing

Suppose a stack of magnetic disks contains a complete company payroll. To make out the payroll checks, each in the proper amount with the proper deductions, we must transfer data from the disks to the central processing unit (CPU), which consists of memory and arithmetic/logic circuits.

A device called a channel automatically moves payroll data and related application programs from disk storage to active memory. Once the data is in memory, the arithmetic/logic section takes over and performs the necessary steps in proper sequence to make out the paychecks. Once the payroll transactions are completed, the data is sent back from memory to disk storage. There is a constant transfer of data between memory and storage.

Output

High-speed printers—activated by electronic impulses— can print checks, invoices, tables, even report cards at up to 2,000 lines per minute on impact printers.

A student sitting at a remote terminal can "converse" with the computer; that is, query the computer and obtain a typewritten answer in seconds—provided, of course, the proper information and instructions were entered in the first place. We can also obtain output on screens similar to television sets.

Figure 2.3

Copyright 1976, *Think Magazine*, International Business Machines Corporation

(from sister machines), and even sound waves (Figure 2.4). Both instructions and data have to be input into the machine and many different languages can be used. Some of the most common are FORTRAN, WATFIV, BASIC, COBOL, APL, PL/1, RPG, ALGOL, and ASSEMBLER. In addition, each computer has its own Job Control Language (JCL) and internally used machine language. We will learn a little about BASIC and a simplified version of FORTRAN called WATFIV.

2. **Working** The computer works with information in two basic ways, by moving information around and by performing arithmetic. Of course, the manner in which information is

Figure 2.4

British Crown Copyright. Science Museum, London

Punched Card

Punched Tape

Core Storage Plane

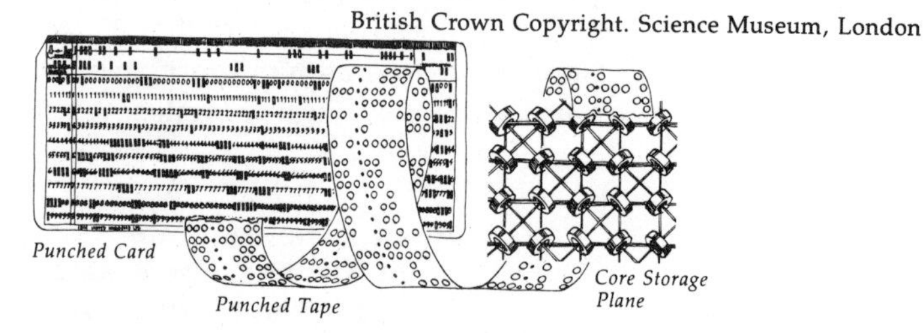

transformed or moved may depend upon the outcome of certain computations.

3. **Output** The common forms of output are the same as the inputs mentioned. In addition, typed copy, graphs, television pictures, and even music can be obtained as output.

See Figures 2.5–2.9 and 2.11.

Hardware and Software

The heart (and brain) of any computer is the central processing unit (CPU). This contains the panel with all the little lights that flash on and

Figure 2.5 A high-speed IBM printer with a punched card reader in the background.

E. Robert Stefl, Illinois State University

Figure 2.6 A device for placing data and programs directly onto magnetic tape.

E. Robert Stefl, Illinois State University

off, which, although really unnecessary, is a means of reassuring us that the computer is actually working. The CPU contains the control unit, the arithmetic unit, and the internal memory. Nearby will be machines called disc drives and tape drives, which are used to run discs (which are similar to phonograph records) and magnetic tape reels. Information that is seldom used or is too extensive for the internal memory is stored on these devices. For example, the IBM 3336 discs contain room for 100 million characters and the standard magnetic tape reel holds over 20 million characters. Because of the speed of the CPU and the relative slowness of the input-output devices, the stream of work through the computer system is generally from input device to disc to CPU to disc to output device. In fact, the input and output devices are attached to a part of the CPU called a compiler, which translates from the input language into machine language before the instructions go to the disc, and then translates back after the job is done. This is so slow that it is done at the same time the CPU is working on someone else's job.

Other standard input-output hardware are card readers, card

punchers, tape readers and punchers, printers, plotters, and terminals, both cathode ray tube and teletype-like consoles. Other devices not attached to the CPU, but a necessary part of any computer center, include key punch machines and card sorters.

Software is the term used to describe programs of various sorts. Operating systems, compilers, and library programs are examples of software. An operating system is the program that furnishes the CPU with instructions on which programs to run next, what data or subroutines from storage it will need, and so on. The increasing sophistication of software packages may be the greatest contribution of the 1970s to computing.

Now that you know a little about the basics of computer history and design, you should start to become familiar on a personal basis with your computer. This requires you to communicate with him. (We use this pronoun advisedly since we think of a computer as a big, unimaginative, computationally fast, but slow-witted friend who, sometimes to our chagrin, does *exactly what we tell him to do.*) Since our communication is usually for the purpose of asking for the solution of a problem, it behooves us to analyze a little some of the problems that our friend is very adept at solving. For this reason we look next at a process called flow-charting.

Figure 2.7 Main console of an IBM system 370 and the terminal used by the machine operator to communicate directly with the central processing unit.

E. Robert Stefl, Illinois State University

 2.2 FLOW CHARTS

Before starting to actually program, it is wise to consider what we want to program. That is, we need to be able to describe in the computer's language what it is we want the computer to do. As an intermediate step, it is almost always best to describe the algorithm we want the computer to perform by means of a flow chart. In many cases this just expresses how you would do the problem if you had to do it without the computer.

E. Robert Stefl, Illinois State University

Figure 2.8 Two tape-drive machines used to transfer information between the computer (usually disc storage) and magnetic tape. The one on the right has a magnetic tape reel in place.

Figure 2.9 A bank of disc drives.

E. Robert Stefl, Illinois State University

Simple Flow Charts

Suppose we want to multiply two numbers together. What procedure, or algorithm, do we go through? We might proceed like this:

1. Get pen and paper.
2. Read the first number, write it down, and think of it as A.
3. Read the second number, write it down, and think of it as B.
4. Multiply A times B and call the result PROD.
5. Write the answer PROD.
6. Stop; the job is done.

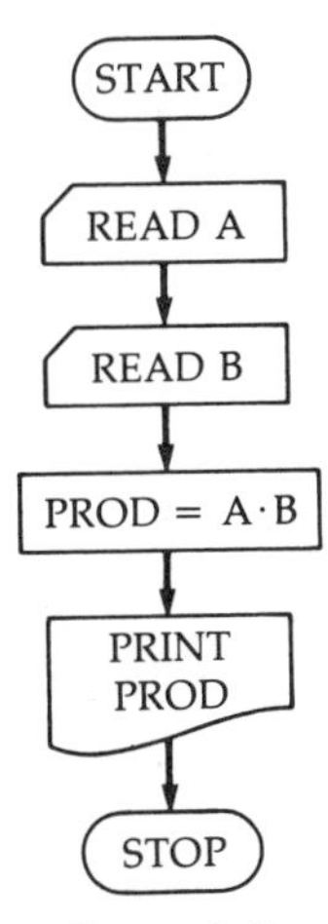

Figure 2.10

This algorithm is described in a more easily understood and a more visually pleasing way by the flow chart in Figure 2.10.

Notice that the symbols enclosing the words are not all the same. We are using the American standard flow-chart symbols, which we list in Figure 2.12 along with their names and intended interpretations.

We shall assume that each flow chart will contain a START

Figure 2.11 A high-speed printer.

E. Robert Stefl, Illinois State University

symbol and a STOP symbol. We shall also assume that all information will be put into the machine by using punched cards and that the output will be on paper. When we discuss BASIC we will see that there are other ways of inputting information besides using punched cards. However, for the purpose of flow-charting it helps to assume that the data are in a specific place such as on a card. Whenever a variable name is used, we do what the instruction says regardless of whether or not the variable already has a value. For example,

is interpreted to mean that a card containing a value is to be read and that this value is assigned to the variable name B regardless of whether or not B had been previously assigned a value. (That is, if B had been previously assigned a value, we forget that value and take B to have the new value read.)

Figure 2.12

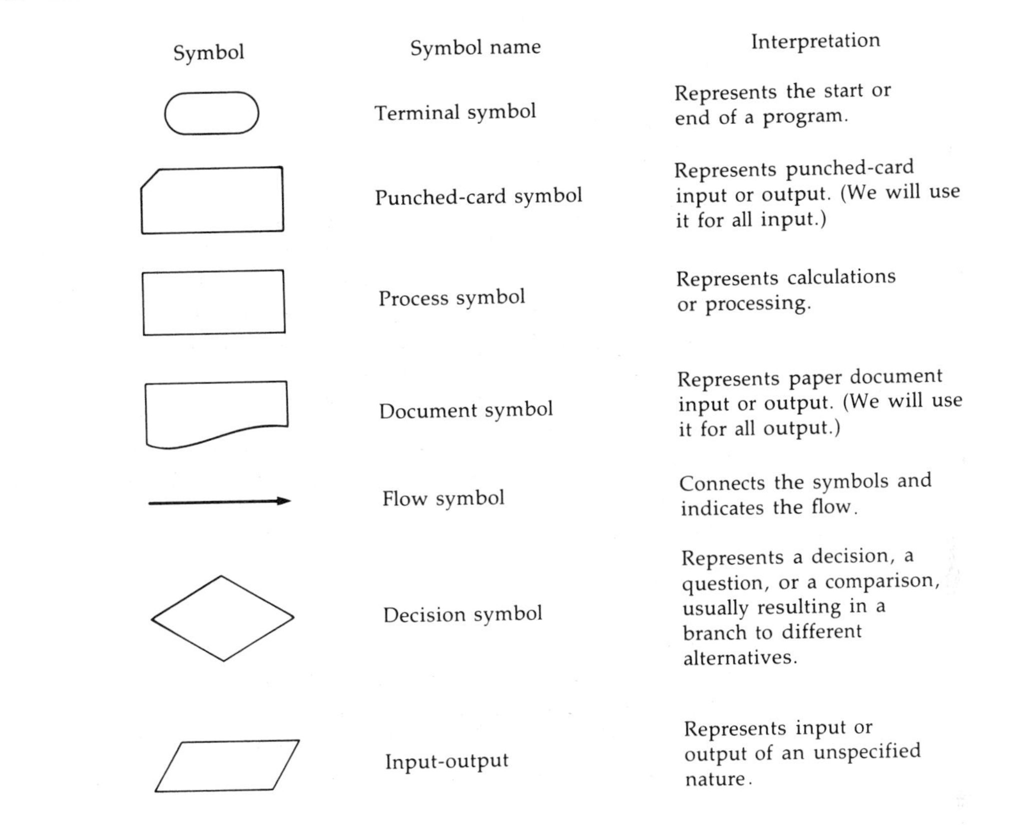

Symbol	Symbol name	Interpretation
	Terminal symbol	Represents the start or end of a program.
	Punched-card symbol	Represents punched-card input or output. (We will use it for all input.)
	Process symbol	Represents calculations or processing.
	Document symbol	Represents paper document input or output. (We will use it for all output.)
	Flow symbol	Connects the symbols and indicates the flow.
	Decision symbol	Represents a decision, a question, or a comparison, usually resulting in a branch to different alternatives.
	Input-output	Represents input or output of an unspecified nature.

The statement PROD = A·B is an assignment statement and means assign to the variable PROD the product of the present values of the variables A and B. (You are probably used to seeing x and y used as variables but not words or abbreviations of words like PROD. In flowcharting we often use a descriptive word to name our variable even if the language in which we eventually write the program does not allow us to use the same word.) If PROD had been previously assigned a value, forget it, and replace that old value by the product of the present values of A and B. A statement like

$$D = D + 2$$

is perfectly legitimate, and simply means that 2 is to be added to the present value of the variable D and this new value assigned to the variable D. Naturally, any statement of the form

$$\text{NAME} = \text{EXPRESSION}$$

can be used when NAME is a variable and EXPRESSION is an arithme-

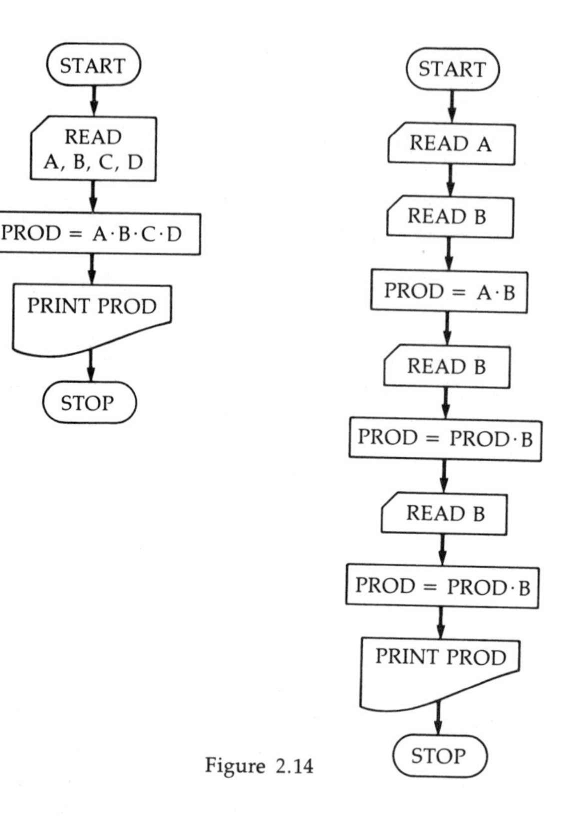

Figure 2.13

Figure 2.14

tic expression or an expression involving certain functions (such as log, sin, cos). This will become more meaningful as we get to other examples.

Consider next the problem of multiplying four numbers together. Of course, we could just read all four numbers, compute the product, print the output, and stop, as suggested in the flow chart in Figure 2.13. But we would not really do a problem this way, especially if the numbers were like 47.53, 591.6, 63.7, and 347.2. We would probably read the first two numbers, multiply them together, multiply that result by the third number, and finally multiply that result by the fourth number. That is, the actual process is more nearly represented by the flow chart in Figure 2.14. Notice that we did not use variable names C and D since we could use B again as a result of the conventions we noted earlier in this section.

Doesn't this look tedious? If we had 10 numbers to multiply together, the flow chart would be too long for the page. There must be an easier way—and of course there is. This is where the *decision symbol* is used, and where we first meet *looping*. Before describing looping, which we do in the next subsection, we will do a few examples which illustrate branching and how to ask the computer to make decisions. We then give you a chance to write some simple flow charts.

EXAMPLE 1

Some years ago the fine for speeding in a certain town was $2 per mile over the speed limit of 30 mph, plus $5 for court costs. In addition, if the driver was going more than 45 mph he or she was to be cited for careless driving and fined an additional $20. Write a flow chart to show how the fine should be computed.

Solution Of course, there is no unique flow chart. We give two possible flow charts in Figure 2.15.

Figure 2.15

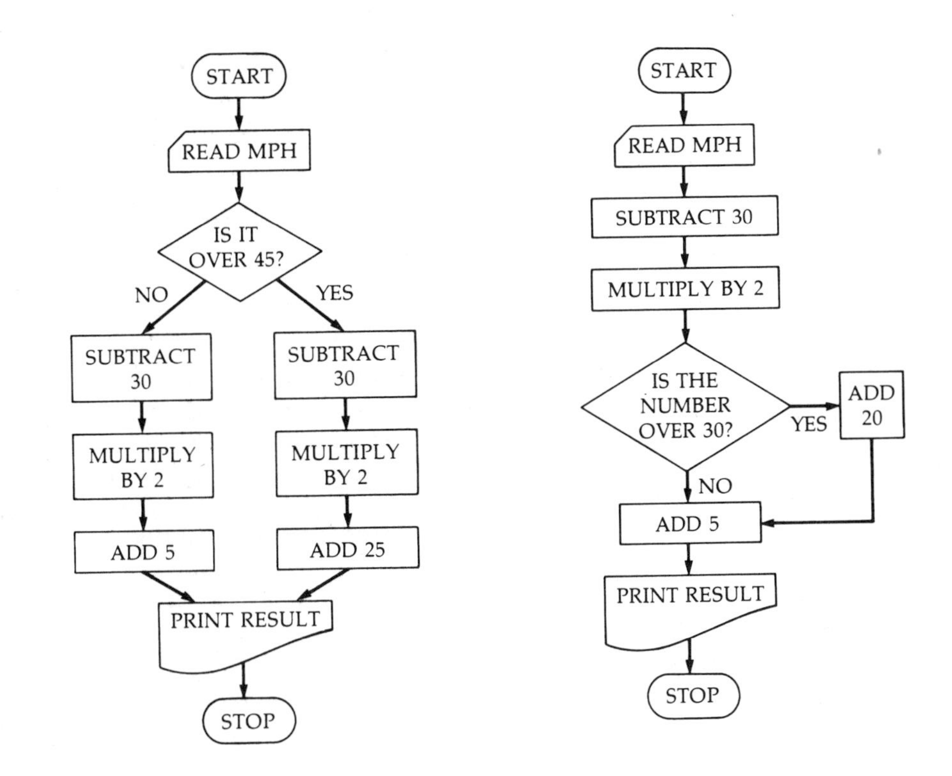

In both flow charts we subtract 30 in order to find the excess over the speed limits, multiply by 2 to find the mileage fine, add 5 for court costs, and ask a question, the answer to which is either yes or no. As a result of the answer to this question, we either add 20 or we do not.

The first flow chart looks more complicated, or at least has more flow symbols, than the second. In the second diagram we have changed the question of whether the speed is over 45 mph to a question about twice its excess over 30.

The flow chart in Figure 2.16 appears to be more efficient than either previous one. We have consolidated the computation into one symbol and asked about the size of the fine just for speeding. The flow chart in Figure 2.16 requires us to be able to compute the maximum fine

Figure 2.16

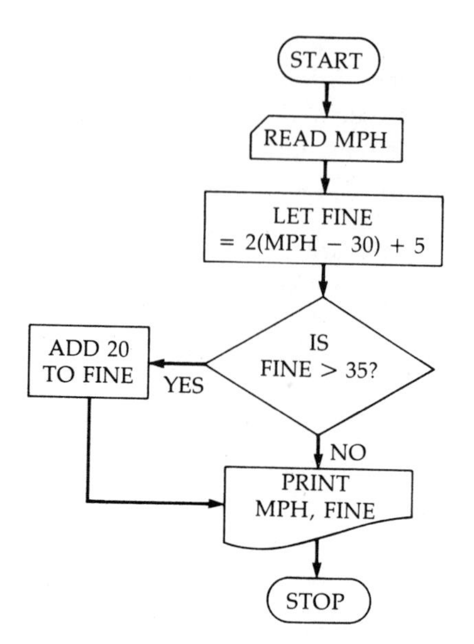

for a speeder who is not to be cited for careless driving. We show in Figure 2.17 one other example of a flow diagram for this example which does not require the knowledge of this maximum fine. Either Figure 2.17 or Figure 2.16 is preferred to the previous flow charts because of the efficiency of the program that would be likely to be written from the diagram. The diagrams in Figure 2.15 are not wrong, they merely imply that the problem is longer than it should be.

Figure 2.17

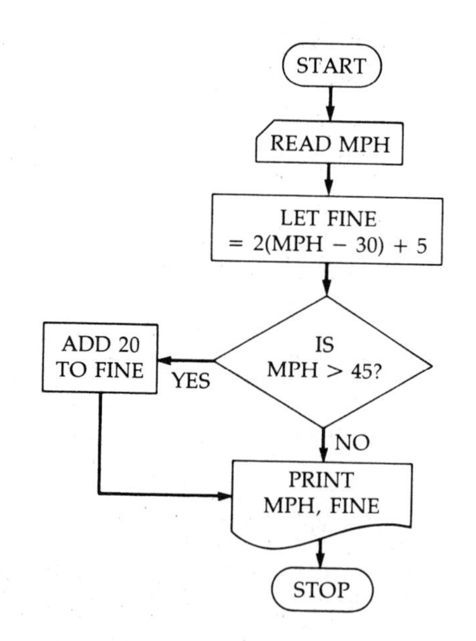

Our next example is a little more complicated because we have to make two decisions rather than one.

EXAMPLE 2

We want to compute the withholding tax for an employee's paycheck. We must first determine the amount of the paycheck. If the paycheck (for a week) is for less than \$100, there is no withholding. If it is for an amount between \$100 and \$300, only 15% is withheld for taxes. But if the check is for more than \$300, the withholding will be 22%.

In the flow chart in Figure 2.18 we first ask if the amount was less

Figure 2.18

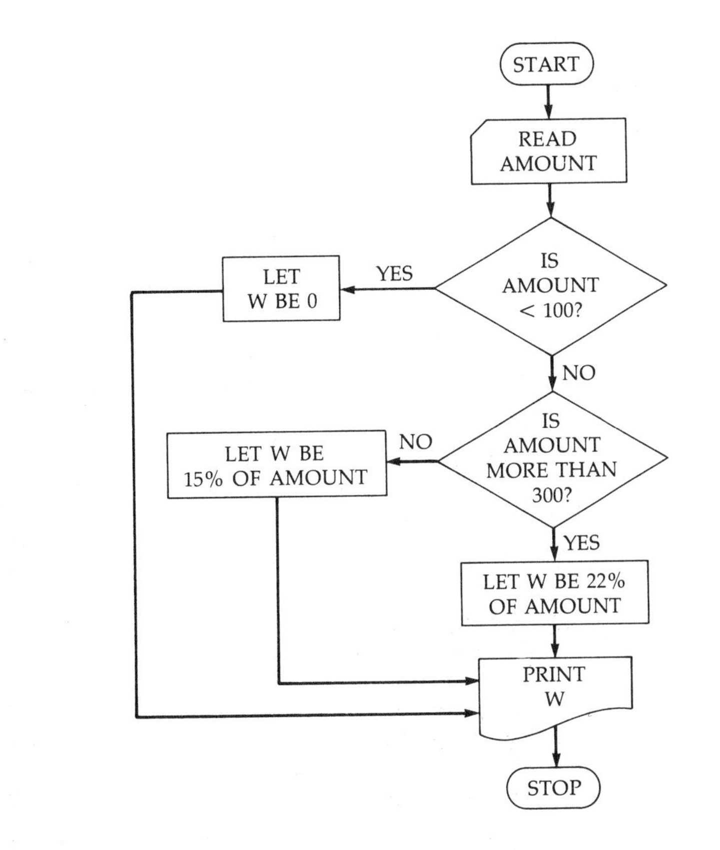

than 100. If it is, we let the withholding W be zero; otherwise we go on to ask whether the amount is over 300. If it is not over 300, and from the previous discussion we know it is not less than 100, then we want to deduct withholding at the 15% rate. Of course, if the amount is over 300 we want to compute W at the 22% rate. In any case, we print out the withholding.

EXAMPLE 2

Alternative Two changes could be made to the solution of Example 2 given in Figure 2.18. The first has to do with what is called *initializing*. When we initialize, we set a variable equal to a convenient value to begin with even though the variable may take on another value later in the program. In Figure 2.19, we have initialized *W* at zero to begin with. This will be a particularly useful device, as we will see in the next section.

Figure 2.19

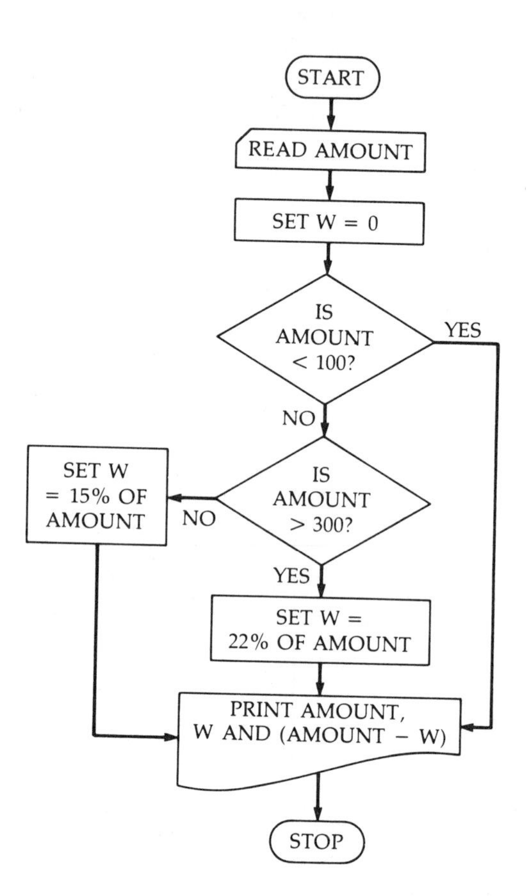

The second change has to do with what we print out at the end. In Figure 2.18 we only printed out the amount of withholding tax *W*. Surely we would want to know the gross amount of the check (the amount we started with) and the amount after the withholding was deducted. We have shown these additional figures being printed out in the flow chart in Figure 2.19.

A word of warning here. Some languages allow a computation to take place in the PRINT statement and some do not. If you are flow-charting for a language that does not allow a computation in the PRINT

Figure 2.20

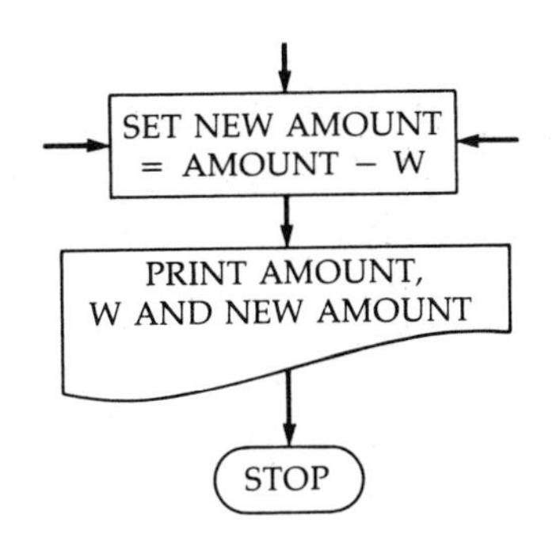

statement you should replace the last two symbols in the flow diagram in Figure 2.19 with the three-symbols diagram in Figure 2.20.

EXAMPLE 3

Write a program that reads a checking account balance and the amount of the check just written. Let the output be the amount of the check and the new balance. If the check is larger than the current balance in the checking account, let the output be zero and the old checking account balance.

Solution We will use AB to stand for the balance in the checking account and C to stand for the amount of the check. See Figure 2.21.

Figure 2.21

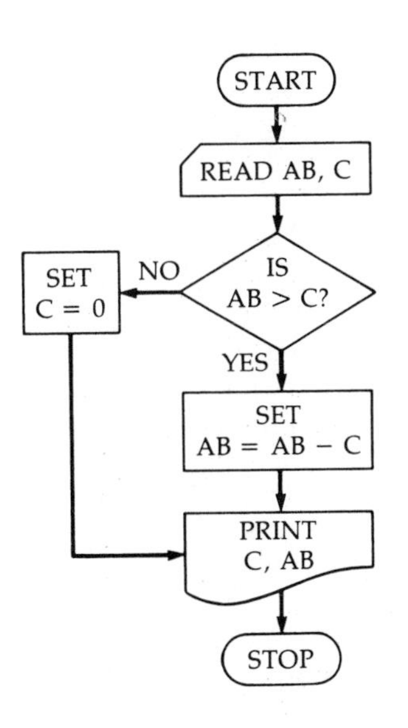

A final word about decision symbols. Any query that can be answered yes or no can be put into a decision symbol. We will wait until we get to actually programming in a language before we worry about how to arithmetize the question in such a way that the computer can answer it and take appropriate action. (There is one exception to this when we introduce counters in the next section.)

Exercises 2.2

In problems 1–10, write a flow chart that will show how to compute the required output from the given input.

1. **(a)** Output is the volume of a rectangular box with given height, width, and length.
   (b) Given three numbers, output the largest.
2. The volume of a cylinder is given by the formula

   $$\text{Volume} = 3.1416 \times \text{radius}^2 \times \text{height}$$

   Given the radius and height, find as output the volume.
3. Given the number of gallons and the price per gallon of gasoline, output the number of gallons and the total cost.
4. Given four numbers, find how to compute their sum as output.
5. Given the total cost and the number of ounces in a jar of coffee, output the cost per ounce.
6. Given the cost of an eight-pack of cola and a six-pack of lemon soda, output the cost per bottle of each and which is cheaper per bottle.
7. Take the balance due on a customer's account and the amount of this customer's check. If the customer still owes money after deducting the amount of the check from the balance due, add to the balance now due on the account a service charge of the larger of $.50 and 1% of the balance due. Output this new balance due and the service charge.
8. Read a checking account, savings account, and the amount of a check. Transfer from savings to checking if the check is larger than the checking account. Output the amount paid on the check and the new balances in the checking and savings account. (The amount paid on the check will be zero if the check exceeds the amount in both accounts put together.)
9. A certain town with a 25-mph speed limit charges $3 for each mile per hour over the speed limit up to 10 mph and $5 for each mile per hour over that. Moreover, there is a $10 court cost fee added, and

any one driving over 50 mph is given an additional fine of $50 for negligent driving. Output the speed and the fine for a given speed input.

10. Complicate the withholding problem given in Example 2 by introducing the effect of dependents. Suppose each dependent allows for a reduction of $10 in the amount of salary considered for withholding. Now, with given input of the number of dependents and the gross amount of the check, show how to get the output of the gross amount, the number of dependents, the amount of withholding, and the gross amount minus the withholding.

11. Write a flow chart that describes the steps involved in solving a linear programming problem.

12. Write a flow chart that describes how to enter a car, start it, and drive away from the curb.

13. Choose some activity that you do every week and write a flow chart describing how to do it.

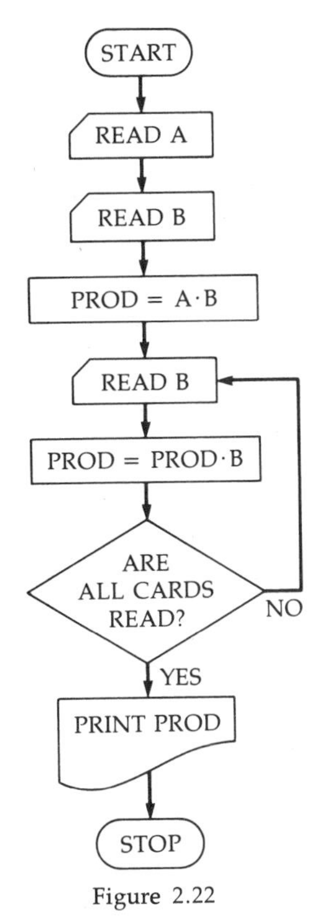

Figure 2.22

2.3 FLOW CHARTS WITH LOOPING

Let us return to the problem of multiplying four numbers together, which we flow-charted in Figure 2.14. As you can see by looking at Figure 2.22, by using one decision symbol and reusing previous flow-chart symbols we can make the total flow chart with one symbol less. This is not much of a saving here, but notice that our flow diagram can be used even if there are 100 factors. Moreover, as we will see when we start to program, the length of our program will be comparably shorter. Reusing symbols in this way is called *looping*.

By initializing a variable to a convenient value we can effect a simplification in the flow chart and the resulting program. By using initializing we can write a flow diagram more typical of what a professional programmer would write; this is given in Figure 2.23.

The problem with looping is that the computer must be told whether all the cards have been read. This can be done in a number of ways, only one of which is illustrated here; other ways will come up in subsequent examples and in the Appendix. A *counter* is an integer variable that we use to keep track of how many times we have gone through the loop. Moreover, we can arrange things so that the same algorithm can be used regardless of the number of factors. We simply tell the computer how many to expect and it acts accordingly. The flow chart illustrating this is given in Figure 2.24. After N is read, its value is reduced by 1 each time a card is read. All the cards have been read when

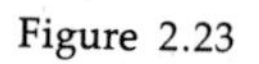

Figure 2.23

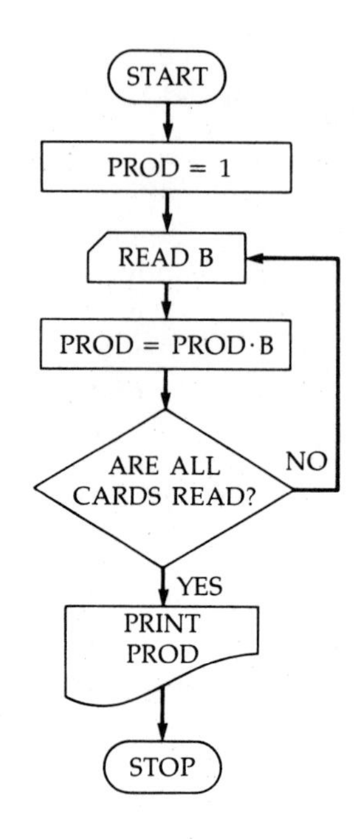

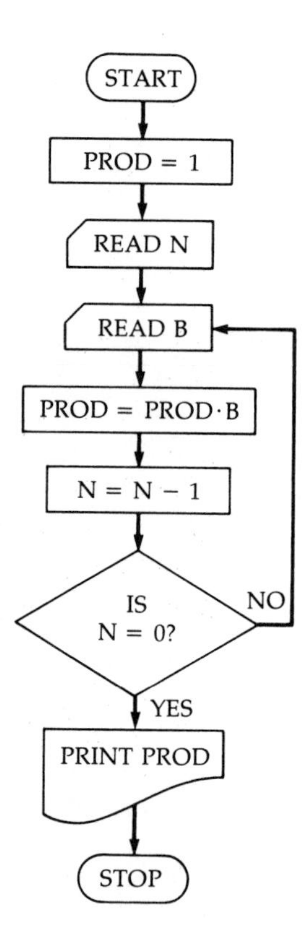

Figure 2.24

the value of zero is reached, so the product is printed and the program terminated.

EXAMPLE 1

Construct an algorithm that will read five numbers and output the largest of them. See Figure 2.25. To show what happens with a specific set of five numbers, let us take the values 3.6, 3, 3.9, 2.4, 3.95.

The computer would go like this:

START

K = 1,

X = 3.6

Y = 3.6

K = 2

X = 3

IS 3.6 < 3? NO

IS 2 ⩾ 5? NO

K = 3

X = 3.9

IS 3.6 < 3.9? YES

Y = 3.9

IS 3 ⩾ 5? NO

K = 4

X = 2.4

IS 3.9 < 2.4? NO

IS 4 ⩾ 5? NO

K = 5

X = 3.95

IS 3.9 < 3.95? YES

Y = 3.95

IS 5 ⩾ 5? YES

PRINT 3.95

STOP

Figure 2.25

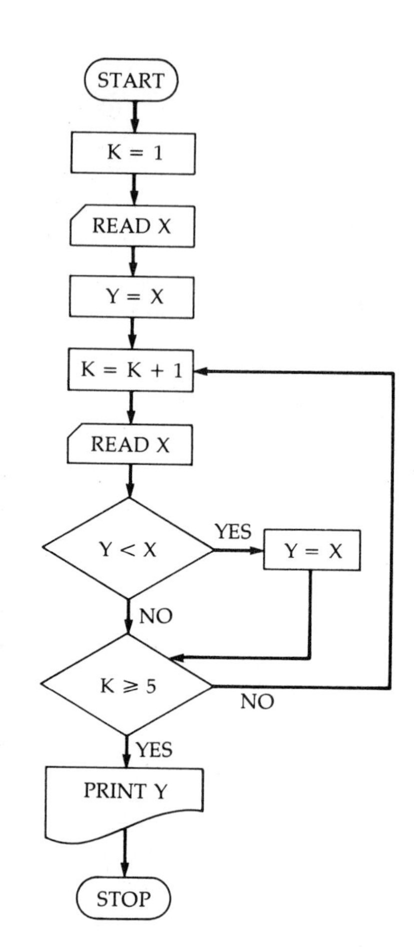

Begin.

Set the counter equal to 1.

Read the first number.

Y is going to be the largest of the numbers read so far. At this time it is clearly the same as this X.

Advance the counter by 1.

Read the X corresponding to this counter number.

Compare Y and X; if Y larger go on; if X larger, give Y this new value.

Have we read all the values? If not we loop by going back. Otherwise we have our answer.

Print the answer.

Stop.

EXAMPLE 2

The Slick Wick Company has 10 employees and we wish to compute their wages. The employees are paid at different hourly rates and the number of hours they work is often different. We have assigned numbers to the employees. We wish to record their number, the number of hours worked, the hourly wage, and the gross wages. See Figure 2.26.

Figure 2.26

Begin.

Set the counter equal to 1.

Read the hours and rate for the employee with number *N*.

Compute the employee's gross earnings.

Print the desired information.

If *N* is not 10 or larger add one to the counter and compute the wages for the next employee.

Once we have all 10 employees figured, stop.

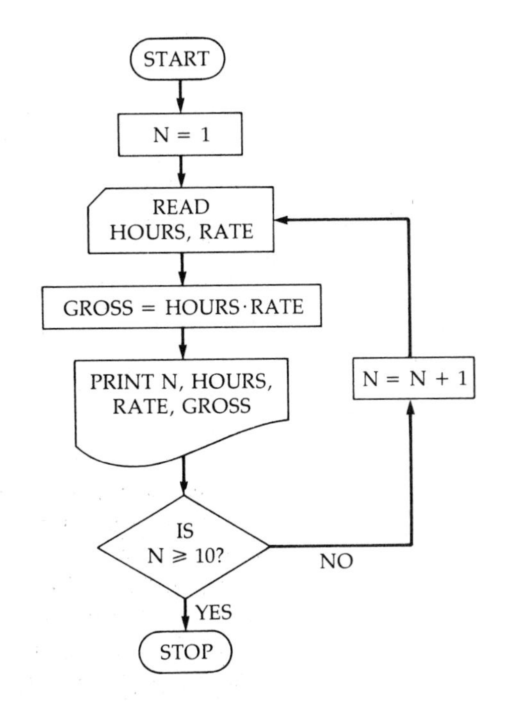

Exercises 2.3

1. Describe what each of the following flow charts does. Determine what the output will be if the given input is used.

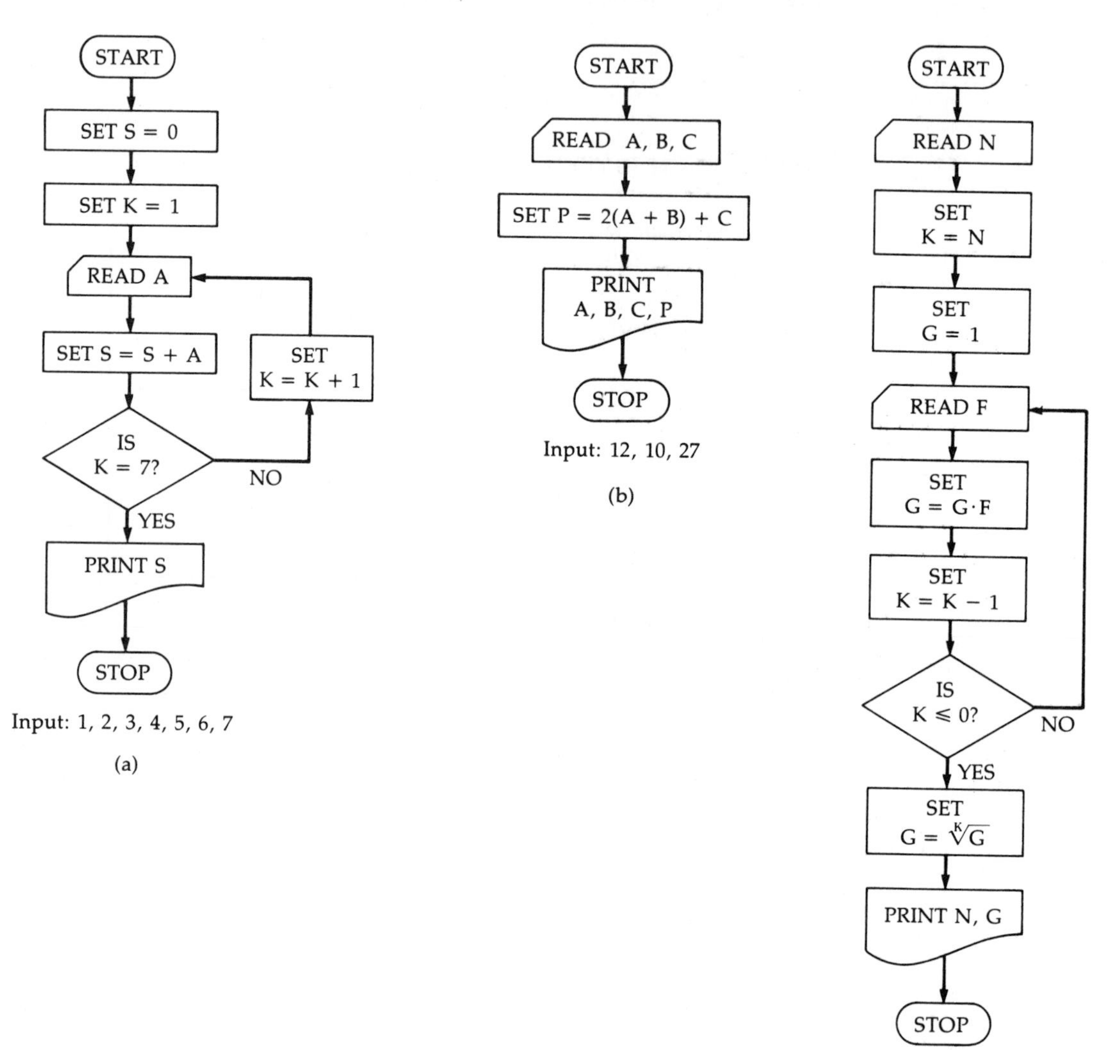

Write flow charts for the following exercises.

2. Construct an algorithm that will read 10 numbers and output the smallest of them.

3. Find the average of 10 numbers, that is, their sum divided by 10.
4. You own five rectangular lots. Find the areas of each.
5. Find the average of M numbers. Your algorithm should read M first and then the M numbers. (The average is the sum divided by M.)
6. Referring to problem 4, you pay taxes of $1.50 per square foot on the five lots (assume the length and width of the lots are given in feet). Construct an algorithm that outputs the number of the lot, its length, width, and area, and the amount of taxes on the lot.
7. Do problem 6 for K lots instead of five (K will need to be read in at the beginning).
8. The employer in Example 2 decides to deduct withholding tax. He wishes to compute DEDUCT as 17% of gross wages, subtract this amount from gross wages, and show the result as net wages, outputting these two figures in addition to the other four.
9. Find the number of positive values, the number of zero values, and the number of negative values in a list of N numbers. First input N and then the values.
10. A bank wishes to charge each customer that had a balance of under $100 on any day during the month a service charge of $1.00. For a study the managers are making, the bank wishes to have the minimum as well as the service charge printed out. The input should be the number of days in the month and the daily balances for each day in the month.

2.4 AN INTRODUCTION TO BASIC

If you have available for use terminals connected directly to a computer, you may wish to learn the computer language called BASIC. BASIC is short for *B*eginners *A*ll-purpose *S*ymbolic *I*nstruction *C*ode, and is an easy-to-learn language developed at Dartmouth College for use at remote time-sharing terminals. These terminals are connected directly, usually by means of a telephone line, to a large central computer, which may have from 50 to 200 such terminals connected to it. (See Figure 2.27.)

The terminals are often called interactive since it appears to the user that he has the computer to himself. Before you can use this language you will need to learn how to gain access to the computer via the terminal. This is the *log-on* and *log-off* process, which is peculiar to each institution or computer center.

Figure 2.27 A remote terminal.

E. Robert Stefl, Illinois State University

Before logging-on, however, we need to describe a little about the language and how it is used, and a few basic commands. Languages have several components such as words, sentences, and punctuation marks. We must learn what words are acceptable to our machine, and then how to combine them into sentences or, as we shall call them, instructions.

Arithmetic and Algebraic Expressions and Operations

Every language has a grammar. Moreover, words and sentences are formed in certain ways. In any computer language one of the first concerns is how to write numbers and algebraic expressions. To write an algebraic expression you need to be able to write variables, to have symbols for the four arithmetic operations, and to take powers.

First of all, what kind of numerical words, usually called numbers, can we use in our conversations with the computer? The length of the numbers allowable depends somewhat on the computer available, but we will be specific. There are three types of numeric constants:

1. Integer
2. Decimal
3. Exponential

Although each type has its own specific rules, which govern its use, these rules apply to all:

1. A comma cannot be used to delimit thousands or millions. (Thus 1,296,300 must be written 1296300.)
2. When a positive or negative sign is not explicitly shown, the number is assumed to be positive.
3. Although any number of digits can appear in a numeric constant, a maximum of 14-digit accuracy is used in any computation.
4. Any nonzero constant must be in the range 3.13152×10^{-294} to 1.26501×10^{322}. (These bounds may vary depending on the computer.)

Now we describe briefly each type of numeric constant.

1. **Integer** An integer is any whole number, positive or negative or zero, written *without* a decimal point.

ACCEPTABLE	NOT ACCEPTABLE
25001	25,001
−49	−49.00
+125736897	83.
0	5673+

2. **Decimal** A decimal constant is any whole number, fraction, or mixed number, positive or negative or zero, written *with* a decimal point. Leading zeros to the left of the decimal point and trailing zeros to the right of the decimal point are ignored.

ACCEPTABLE	NOT ACCEPTABLE
−5.03	839,675
1.915326145	$5.35\frac{1}{2}$
.000001	49,376.891
46.3	636.36.7
+3025.098	89.76+
0.	

Before looking at the last type of numeric constant, we need to recall how to write numbers in what some people call *scientific notation*.

Basically, it is a shorthand notation for representing numbers with many significant zeros. For example,

Instead of	543000000000,	we can write	5.43×10^{11}.
Instead of	.0000000035560,	we can write	3.556×10^{-9}.
Instead of	.013694,	we can write	136.94×10^{-4}.

These are perfectly respectable numbers but, except for the last one, are long to write and difficult to judge in size when not written in a form using a power of 10. In BASIC these three numbers could be written like this:

Instead of	5.43×10^{11},	we can write 5.43E11,	or 54.3E+10
Instead of	3.556×10^{-9},	we can write 3.556E−9,	or 0.3556E−8
Instead of	136.94×10^{-4},	we can write 136.94E−4,	or .00013694E2

Thus you see that the E stands for "times 10 to the power."

3. **Exponential** An exponential constant must satisfy the following three rules.
   (a) A number, either an integer or decimal constant, must precede the E.
   (b) The exponent (that is the number following the E) must consist of one, two, or three digits and may be preceded by a positive or negative sign. If no sign is present, a positive sign is assumed. (However, see general rule 4 mentioned before.)
   (c) Decimal points are not permitted in the exponent.

ACCEPTABLE	NOT ACCEPTABLE
−2.317E130	E+120
8E+25	3.76E1432
4.418765E−23	3.965E13.5

We note in passing that it is also possible to have "string constants," but we defer discussion of these to the Appendix.

Just as constants are fixed values, variables represent values that are not fixed. We again consider only numeric variables, relegating a brief discussion of "string variables" to the Appendix.

Simple Numeric Variable A simple numeric variable represents a numeric value that may change during a program execution. These variables are named by a single alphabetic character (that is, one of the

26 letters of our alphabet) or an alphabetic character followed by a numeric character (one of the digits 0, 1, 2, 3, 4, 5, 6, 7, 8, 9).

WHY CAPS?
Notice that we only use capital letters when we write words that correspond to computer input and output. This is because *there are no lowercase letters available on our terminals.*

ACCEPTABLE	NOT ACCEPTABLE
A	A12
T	PROD
S8	PV
Q	STOP
L1	2L
R3	S30
Y	G *

Although we do not do much with them in this chapter, for completeness and for possible use in later chapters we define subscript variables.

Subscript Variables A subscripted variable locates the value of a particular element in an array, and is written as a simple variable followed by a maximum of three subscripts enclosed in parentheses. The subscripts may be a numeric constant, a simple or subscripted numeric variable, or an arithmetic expression (see below).

ACCEPTABLE	NOT ACCEPTABLE
A(1)	A12(3)
B(N, 7)	B(N, 3, 5, 17)
C3(5)	A.2(6)
A(B2(3))	3B(17, 8)
X(2, K + 3, L + M)	
J(L − K, A(5), 4)	

A final word about subscripts before proceeding to expressions. Subscripts must have values between 1 and 10, except by special arrangement as described in the Appendix. If a subscript has a nonintegral value, the fractional part is ignored and only the integer part is used.

Now that we have names for variables and constants, we must learn how to combine them into expressions by the usual arithmetic operations of addition, multiplication, subtraction, division, and exponentiation. The following table shows which symbol is used by BASIC to denote the various arithmetic operations.

OPERATION	ALGEBRAIC SYMBOL	BASIC SYMBOL
Addition	+	+
Subtraction	−	−
Multiplication	· or ×	*
Division	÷ or /	/
Exponentiation	None	↑ or **

For example,

ALGEBRAIC	BASIC	RESULT
3 + 7	3 + 7	10
8.3 − 5	8.3 − 5	3.3
3 · 7 or 3 × 7	3 * 7	21
2.1 ÷ 3 or 2.1 / 3	2.1 / 3	.7
6^3	6 **3 or 6 ↑ 3	216
$4^{1.5}$	4 **1.5 or 4 ↑ 1.5	8

Notice that two algebraic operations cannot appear side by side (for example, 5++7 is not allowed). Parentheses must be used if a minus sign is used to denote a negative value, and operations cannot be implied.

ACCEPTABLE	NOT ACCEPTABLE
17 * (−3)	17 * −3
(17 + 3) * (18 − 25)	(17 + 3)(18 − 25)

Just as in ordinary algebra, we have an order of precedence that allows us to decrease the number of parentheses we need. For example, multiplication is done before addition, so 6 * 3 + 14 means 18 + 14 or 32 and not 6 times 17. If we want to denote the latter, we would have to write 6 * (3 + 14). However, parentheses can always be used. Although not necessary, we could write (6 * 3) + (2 * 7) rather than 6 * 3 + 2 * 7.

Formally, we say that the operations are performed *from left to right* according to the following order of precedence.

1. Exponentiation
2. Multiplication and division in the order written
3. Addition and subtraction in the order written

If parentheses occur, the operations called for inside the parentheses are done first. Some examples should clarify what we mean.

EXAMPLES

```
1. 3 * 5 + 2 * 15 / 6 - 3**2
   3 * 5 + 2 * 15 / 6 - 9
    15  +  30 / 6 - 9
        15 + 5 - 9
          20 - 9
            11

2. 39.6 / 1.2 / .03          39.6 / (1.2 / .03)
     33.0 / .03                  39.6 / 40.
     1100.0                         .99

3. 4  * 16 ↑ 1.5 / 3.
    4 * 64.0 / 3.
     256.0 / 3.
      85.333333
```

Algebraic expressions follow the same rules, as the following illustrates.

ALGEBRAIC EXPRESSION	BASIC EXPRESSION
$3x^2 + xy - (x + y)^2$	3 * X ↑ 2 + X * Y − (X + Y) ↑ 2
$z - \dfrac{x^2y^3}{r}$	Z − X**2 * Y**3 / R
$1.9a^{2.6-(x_3)/y}$	1.9 * A ↑ (2.6 − X3 / Y)

What algebraic expression corresponds to the expression

A**3 / C**2 * B**3 + B * A**(−2) / C**3 ?

The translation would proceed like this:

$$a^3 \div c^2 \cdot b^3 + b \cdot a^{-2}/c^3,$$

$$\frac{a^3}{c^2}b^3 + \frac{ba^{-2}}{c^3}.$$

Notice that 2**3**2 could be interpreted two ways, depending on how the exponents are associated. In one case,

$$(2**3)**2 = (2^3)^2 = 8^2 = 64$$

and in the other case,

$$2**(3**2) = 2^{(3^2)} = 2^9 = 512.$$

Quite a difference. Different machines compute 2**3**2 differently, so to avoid any possible confusion, we will always place parentheses in the proper places. Naturally, even when the order of precedence should be clear, you may insert parentheses to be sure the computer does what you want. Work through these additional examples.

BASIC EXPRESSION	ALGEBRAIC EXPRESSION
3 * X ↑ 2 + 2 * X + 19.6	$3x^2 + 2x + 19.6$
19.75 * X ↑ 10 − 1.36 * X ↑ 3 * Y ↑ 5 / (4.3 * Z ↑ 6.5)	$19.75x^{10} - 1.36x^3y^5 \div 4.3z^{6.5}$
5 * X ↑ 2 − 2 * (X ↑ 3 + 3 * X * Y) + X ↑ 5 / Y ↑ 4	$5x^2 - 2(x^3 + 3xy) + x^5/y^4$

The other kinds of expressions that occur in BASIC are *comparative expressions* or *logical comparisons*. Equality and inequalities are used for these. We also present them now in tabular form for future reference (Table 2.2).

Table 2.2

BASIC SYMBOL	MEANING	ALGEBRAIC SYMBOL
=	Equal	$=$
<	Less than	$<$
< = or = <	Less than or equal to	$\leqslant$ or $\leqq$
>	Greater than	$>$
> = or = >	Greater than or equal to	$\geqslant$ or $\geqq$
< > or > <	Not equal to	$\neq$

We will use these symbols to make statements about which we can ask whether or not they are true. That is, we will use them to make decisions about branching in our program corresponding to the branching that we did in flow-charting.

For example, the comparison

$$x^2 + 3 < x^3 - 4$$

may or may not be true depending on the value of x that we have. If $x =$ 2,

$$x^2 + 3 < x^3 - 4 \qquad \text{is false,}$$

since

$$2^2 + 3 < 2^3 - 4$$

says $$7 < 4,$$

which is not true. However, if $x = 3$,

$$x^2 + 3 < x^3 - 4 \qquad \textit{is}\ \text{true,}$$

since

$$3^2 + 3 < 3^3 - 4$$

says $$12 < 23,$$

which is true.

Usually, although not always, we will be making simple comparisons like

$$K \geqslant 5, \quad \text{or} \quad K = 0,$$

or the like, similar to the questions we asked when we were writing flow charts.

Exercises 2.4

1. State whether the following numbers are integer, decimal, exponential, or none of these.

(a) 561	**(b)** 43,987,651	**(c)** 436.75E29
(d) 51.00	**(e)** −4398.01	**(f)** 596403
(g) 596078395	**(h)** 4378.56594	**(i)** 1.379000E−76
(j) 5,678.43	**(k)** −436.	**(l)** −4937

2. State whether the following variables are simple or subscript numeric or neither.

(a) A3	**(b)** C(1, D)	**(c)** H.5
(d) AB	**(e)** K(1, 5L, 6)	**(f)** 5D
(g) X9	**(h)** T10	**(i)** Z:
(j) Y	**(k)** X(1, 3, 5, 7)	**(l)** L(2, K(3, 6), D)

3. Compute the following numbers and state whether the result is an integer or a decimal constant.

   (a) 4 * 2 ↑ 3 (b) 9 * 5 **2 / 15
   (c) 3.2 * 11.3 / 4.0 * 5.7 (d) 2.1 **2 * 3 + 8.4 / 2.1
   (e) 10. * 3 **2 / 6 (f) (3 * 2) ↑ 2 / (5 * 2)

4. Compute the following numbers.

   (a) 2 * (3 + 4) + 3 ↑ 2 − 8 * (5 − 2)
   (b) 2 * 2 + 3 **2 * 4 − 3 * 2 **4 / 2
   (c) 3 * 5 **2 + (2 **3) **2 / 4 * 3 / 2
   (d) 4 / 3 * 12 * 5 / 2 − 3 * 5 / 2 + 3 * 2 ↑ 3 * 2

5. Express the following algebraic expressions in traditional form.

   (a) 3 * A ↑ 2 + 15 * A + 43
   (b) 5 * X **2 * Y **3 − 14.6 * X * Y **2 + 5.93 * Y
   (c) A **3 / C * B **4 + B * 2 * A **(2 + C) / C * (−2)
   (d) B **3 / A * 2 **3 * B **(−2)
   (e) 3 * A ↑ 2 + 4 * A − 17
   (f) X ↑ 2 * Y + 3 * X * Y ↑ 2 − 17 * X ↑ 3 * Y ↑ 2

6. Write the following algebraic expressions in the BASIC language.

   (a) $15x^3 + 3x + 7$
   (b) $3x^2 - 17x + 5$
   (c) $13x^2y^2 - 9.5x^3y + 43xy^3$
   (d) $(x^2 + 2y)(x - 3y)(5 + 4y^2)$
   (e) $x_1y_1 + x_2y_2 + x_3y_3$
   (f) $3x^ay - 17x^3y^b + (43x - y^a)(4x^b + 2y)$
   (g) $a^2b^3 \div 2c^2 - 3c^{-2} \div b^5$
   (h) $\dfrac{3x^2y}{z} - \dfrac{17xy}{3z^2} + y^3z^{15}$

In problems 7 and 8 express the given comparison in the BASIC language. (See the box "Inequalities in Plain English" in Section 1.1.)

7. (a) x does not exceed $3y$.
   (b) k is not less than 5.
   (c) The square of x exceeds the cube of y by 16.
   (d) x is more than y.

8. (a) Twice y is not more than 15.
   (b) $2t$ exceeds zero.
   (c) The fourth power of t is not less than 3 more than the cube of r.
   (d) 3 times x does not equal 17 more than twice y.

In problems 9 and 10 express the given quantity as an expression in the BASIC language.

9. **(a)** The length of the hypotenuse of a right triangle with legs of lengths x and y.
(b) The volume of a cylinder in terms of its height h and radius r.

10. **(a)** The surface area of a rectangular box of given height, length, and width.
(b) The volume of a cube that holds twice as much as a given cube whose edges have length a.

11. A grocer sells apples for 29¢ per pound and pears for 33¢ per pound. If he pays x¢ per pound for apples and y¢ per pound for pears, express in BASIC the profit on m pounds of apples and n pounds of pears.

12. A used-car dealer figures to make 10% profit on all cars she sells for $5000 and 15% on all cars she sells for $7000. If she sells r $5000 cars and s $7000 cars, express her profit in BASIC.

13. A large manufacturer buys sheet steel and iron ingots. The sheets are c feet square and weigh 6.7 pounds per square foot. The ingots are cubes of length b feet on each side and weigh 78.9 pounds per cubic foot. Express in BASIC the total weight of 130 sheets of steel and 15t iron ingots.

14. A large grocer buys x pounds of oranges, y pounds of apples, and z pounds of bananas at a cost of a¢, b¢, and c¢ per pound. If he sells apples for 35¢ per pound, oranges for 20¢ per pound, and bananas for 23¢ per pound, express the grocer's profit as an expression in BASIC.

2.5 SUMMARY OF SOME BASIC COMMANDS

Now that we have learned how to write numbers, variables, and expressions, how do we put them together to write a program? In this section we describe the minimal number of commands that you need to start to write a program.

EXAMPLE 1

Recall the first flow chart we wrote down in Section 2.2. We show it again in Figure 2.28. A BASIC program that would correspond to this flow chart would be

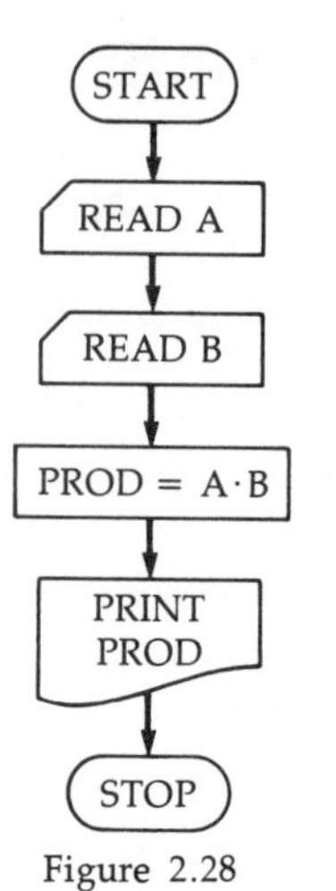

Figure 2.28

```
10 READ A
20 READ B
30 LET P = A * B
40 PRINT P
50 DATA 3.6
60 DATA 7.2
70 END
```

First notice that each *line is numbered.* This number also serves as the *address* for the line. We use multiples of 10 so that if we forget an instruction we can insert it easily. You will see how this works in the next example. The line number can have no more than five digits, that is, it must be between 1 and 99999.

READ DATA

The READ instruction tells the computer to search down the program to the first DATA statement not completely used (that is, a statement with DATA as its initial word which has not had all the numbers that follow it previously assigned to a variable), and to begin assigning to the variables in the READ statement the unused numbers in the DATA statement, reading from the left. For example, if we had a program with

```
   .
   .
   .
 30 READ X, Y, Z
   .
   .
   .
 70 READ A, B1, T, X
 72 READ B2, Z, C, W9
   .
   .
   .
120 DATA 6, 3, 5.1
122 DATA 2, 4, 6, 8, 11.3, 6.4
123 DATA 9, 176.8, 19
   .
   .
   .
```

when the computer reached the statement with line number 30 it would set

$$X = 6, \quad Y = 3, \quad Z = 5.1.$$

Later, when it got to statement 70, it would pass over line 120 and read the data from line 122 to set

$$A = 2, \quad B1 = 4, \quad T = 6, \quad \text{and} \quad X = 8.$$

(Notice the value of X is changed.) When line 72 is executed, the computer remembers it has read the 8 in line 122, so it sets

$$B2 = 11.3 \quad \text{and} \quad Z = 6.4.$$

Since there are still variables to be assigned values, it seeks the next DATA statement and sets

$$C = 9 \quad \text{and} \quad W9 = 176.8.$$

There is no comma after READ or DATA, just a space, but the variables and data values are separated by a comma. There is no punctuation at the end of the line.

LET

This tells the computer we are going to give a variable a value. After LET should come a space, then a variable name, then the equal symbol, and finally an expression or number. For example,

```
50  LET C = 6 * 3 / 5 ↑ 2
```

would assign C the value $6 \cdot 3 \div 5^2$ or $\frac{18}{25}$. That is, it would set C = .72. If A = B = 3 and C = D = 2, then

```
37  LET T2 = (A * C) + B ↑ D
```

would assign T2 the value $3 \cdot 2 + 3^2 = 6 + 9 = 15$.

Thus = in this context stands for an assignment. The statement

```
LET X = EXPRESSION
```

means that we assign X the value obtained when the expression is computed. Such a statement as

```
60  LET X = X + 2
```

makes sense under this definition. For, suppose X had the value 7 just before this command. Then the expression would be evaluated as 7 + 2 or 9, and this value would be assigned to X. Thus *after* execution of this line 60, X would have the value 9.

We also note that in most BASIC systems the use of LET is optional. This means that

```
60  LET X = X + 1
```

could also be written as

```
60  X = X + 1
```

You should check your computer to see what it allows.

PRINT

This instruction is very useful and, because of its versatility, its use presents a little more complication. In Example 1, since

$$A = 3.6, \qquad B = 7.2,$$

we have

$$P = (3.6)(7.2) = 25.92,$$

so the value printed by

```
40  PRINT P
```

would be 25.92.

In BASIC, a computation can be performed in the PRINT instruction. Thus

```
50 PRINT X, Y, X**Y
```

will print the values of X, Y, and X^Y. If $X = 3$ and $Y = 4$, then as a result of this instruction the computer will print

```
3             4             81
```

We should say a word about the form and accuracy of numbers printed by the PRINT statement. There are four cases to consider.

(a) Exact integers of less than 10 digits are printed as usual.
(b) Nonintegers between -1 and 1 whose six most significant digits immediately follow the decimal point are printed as ordinary decimals.
(c) Nonintegers whose integer portion is less than 7 digits are written as ordinary decimals (rounded off if necessary).
(d) All other numbers are written in scientific notation with six digits before E and one before the decimal point.

In (b), (c), and (d) only 6 significant digits are retained.

EXAMPLES

Number:	314159283	314159.28	−.31415928	.003141592	3141592838
Printout:	314159283	314159.	−.314159	3.14159E−3	3.14159E+9

The PRINT statement can be used to print both the value of variables and literal strings. A *literal string* is just a sentence or phrase from the symbols available on our terminal. The literal string is enclosed in double quotes and is printed out just the way it appears. For example,

```
40 PRINT "THE SKY IS BLUE"
```

will yield

```
THE SKY IS BLUE
```

If we were to replace the statement 40 that we have, namely,

```
40  PRINT P
```

by

```
40  PRINT "THE VALUE OF THE PRODUCT IS", P
```

the computer would print

```
THE VALUE OF THE PRODUCT IS          25.92
```

If a semicolon is used instead of a comma, then the natural spacing which is built into the language is overridden and less space is left between symbols. To see what we mean, the printout from

```
40  PRINT A, B, P
```

would be

```
 3.6              7.2              25.92
```

whereas

```
40  PRINT A; B; P
```

would give a printout of

```
 3.6  7.2  25.92
```

The best way to see what happens is to experiment. Part of the fun of BASIC is that you can interact with the computer and try different things.

END

This must be the statement with the highest-numbered line. A colleague always gives it line number 10000, since he never goes higher than that, but we recommend that you number it in sequence as we have done. It simply must be last.

Before introducing more commands, let us take a look at a few more examples.

EXAMPLE 2 A program for Figure 2.13 is as follows.

```
10  READ A, B, C, D
20  LET P = A * B * C * D
30  PRINT P
40  DATA 47.53, 591.6, 63.7, 347.2
50  END
```

EXAMPLE 3 Write a program that reads two numbers, forms their sum and product, raises the product to an exponent equal to the sum, and prints out the three numbers computed. Take the numbers to be 5 and 4.

Solution First we make a flow chart (Figure 2.29) and then write the program

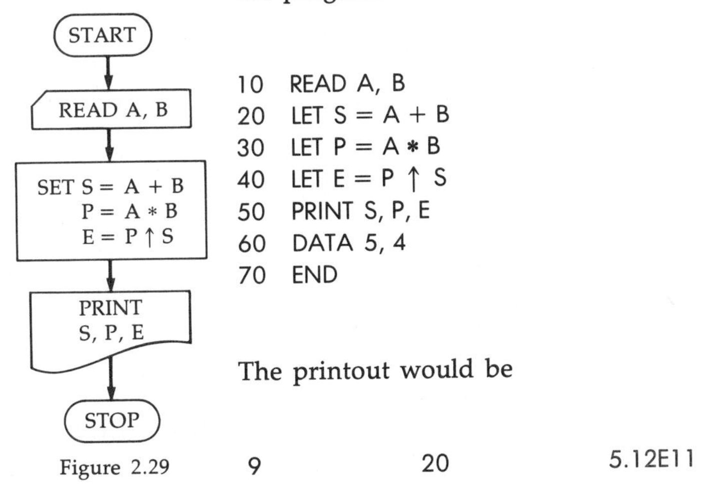

Figure 2.29

```
10  READ A, B
20  LET S = A + B
30  LET P = A * B
40  LET E = P ↑ S
50  PRINT S, P, E
60  DATA 5, 4
70  END
```

The printout would be

```
9          20          5.12E11
```

Exercises 2.5

1. Write a program that computes the average (arithmetic mean) of two numbers. Print out the two numbers and the average.
2. Write a program that computes the geometric mean of two positive numbers. (This is the square root of their product.) Output three numbers.
3. Write a program that finds the length of the hypotenuse of a right triangle given the lengths of the two sides. Output the three lengths. Try your program by hand for a right triangle with sides of length 5 and 12.
4. If a circle of radius r has area $3.1416r^2$ and circumference twice the area divided by r, write a program that outputs r, the circumference, and the area.
5. Write a program that computes the area of a face and the volume of a cube whose edges are of length 6.3 feet. Print out the length of the edge, the area of a face, and the volume.
6. The formulas for the surface area A and the volume V of a sphere of radius r are

   $$A = 4\pi r^2$$

   and

   $$V = \tfrac{4}{3}\pi r^3.$$

   Write a program that computes the surface area and volume of a sphere of radius 18.37 feet.
7. Write a program for problem 5 of Section 2.2. (Use a cost of $3.89 per pound.)
8. Write a program for problem 6 of Section 2.2. (Suppose cola is 89¢ and lemon soda is 69¢ per pack.)

9–14. Write a program that computes and prints the quantities described in problems 9–14 in Section 2.4.

2.6 RUNNING A BASIC PROGRAM

Before you can log-on a terminal you will need to be told and shown how to gain access to the computer. It often requires dialing a telephone number, and generally requires you to have a user number and password. Once you have logged on, the computer will type

```
RECOVER / SYSTEM:
```

You will type

```
BASIC
```

and push the carriage return. The computer will respond

```
OLD, NEW, OR LIB FILE:
```

and you will type

```
NEW
```

Again the computer will respond, this time with

```
FILE NAME:
```

Now you must make up a file name. It must begin with a letter and should not be more than seven symbols long. After the first symbol, you may use letters or numbers. For example, PRODUCT, FAT, GLOB, F12, T30S are all possible file names, but 12X, TRIANGLE, and $YAK are not.

The computer now types

```
READY
```

and away you go writing your program. Do not forget to push the carriage return when done typing a line. When you have typed it in, type RUN (or RNH) and the program will run with a heading (or with no heading).

Here is what happened when we tried to do problem 3 in Section 2.5:

```
RECOVER / SYSTEM: BASIC
NEW, OLD, OR LIB FILE: NEW
FILE NAME: HYPOT
READY

10  REM  THIS PROGRAM COMPUTES THE HYPOTENUSE
20  REM  OF A RIGHT TRIANGLE GIVEN THE LEGS
30  READ, A, B
40  LET H = A ↑ 2 + B*
40  LET H = (A ↑ 2 + B ↑ 2) ↑ .5
50  PRINT A, B, H
60  END
30  READ A, B
55  DATA 5, 12
RUN

    79 / 06 / 21.       16.54.23

PROGRAM      HYPOT

    5             12             13
```

Shall we talk about this a little? First of all, we called our program HYPOT for hypotenuse. Then we typed in our program line by line. We will say a few words about each line.

10 A new command we did not talk about. This is just a remark statement. It is used by programmers to tell them what they are doing so that when they return to this program later, they can read what the program does instead of having to figure it out. The computer passes right over these statements without doing anything.

20 Nothing new, just a continuation because the previous line was too short to fit everything into it.

30 We typed the READ command, but made a mistake because we put a comma after it. We didn't notice it here.

"Let's make this perfectly clear!
I am the programmer. You are the programmee!"

Joyce Richardson, MICC DIGIT

40 Made a typing error, so we pushed the carriage return and started again.

40 This new line with number 40 replaces the old one. Notice that we are using the fact that raising a number to the $\frac{1}{2}$ power (or .5) is the same as taking the square root.

50 We got the PRINT statement right.

60 The last statement.

30 In reading what we had done, we caught our error and corrected it by retyping the right command statement.

55 Noticed we forgot to put the DATA statement in. Any number between 50 and 60 could be used (or any unused number larger than 0 actually).

Our program is complete. To run the program, type RUN (and do not forget to push the carriage return or nothing happens).

The computer typed the heading, which in this case consists of the date and time (to nearest second) and the program name. Then the program was run with the result shown.

If you make mistakes, as we did, it is usually a good idea to have the computer list the program so you can give it a quick look to see if it is correct. The instruction and what happens in this case is

```
LIST

    79 / 06 / 21          16.56.47

PROGRAM     HYPOT

10  REM  THIS PROGRAM COMPUTES THE HYPOTENUSE
20  REM  OF A RIGHT TRIANGLE GIVEN THE LEGS
30  READ A, B
40  LET H = (A ↑ 2 + B ↑ 2) ↑ .5
50  PRINT A, B, H
55  DATA 5, 12
60  END

READY
```

If you want the program run or listed without the heading (it is a little faster), type RNH and LNH for run–no heading and list–no heading, respectively. Notice that different values for A and B can be used just by typing a new line 55, and then typing RUN or RNH.

When you are finished with the computer, be sure to log-off by typing BYE and pushing the carriage return.

Exercises 2.6

1–14. Run the programs from problems 1–14 in Section 2.5. Make up your own data.

2.7 TRANSFERS AND LOOPING

We now discuss the commands used to make decisions and how looping is done. There are two kinds of transfer, a *conditional transfer* (if something is true, then go someplace) and an *unconditional transfer* (no matter what, go where told). The latter is easier, so we explain it first.

GOTO (or *GO TO*)

This simply transfers control to another statement. It is always of the form

GOTO a number.

So, for example,

```
50 GOTO 120
```

would have the effect of sending the computer to statement number 120 whenever it got to statement 50. That is, whenever instruction 50 is executed, the next instruction to be executed is always the instruction with line number 120.

IF THEN

This is always of the form

IF relational expression THEN instruction address.

For example,

```
IF R = 0 THEN 60
```

or

```
IF X ↑ Y> = Y ↑ X THEN 35
```

When the relational expression in the IF part of the statement is false, the computer ignores the THEN part of the statement and proceeds to execute the next line.

INPUT

Instead of the READ statement, which requires the data to be part of the program, we can use a statement called INPUT. Like the READ statement, the INPUT statement assigns a value to each variable listed in it. However, the INPUT statement asks for values of the variable (or variables) to be typed in while the program is running. When an INPUT statement is executed, the computer types a question mark and waits for the values of the variable (or variables) to be typed in (separated by commas).

We will give an example and then expand it to show how a program can grow in sophistication as we add more instructions to the program.

EXAMPLE

Write a program that will find the volume V of a sphere of radius R. We figure we can estimate easily the number of beans in the spherical bowl if we know its volume.

Solution Let's call the program SFERVOL for *sphere's volume*. Assuming we have already typed BASIC, our conversation with the computer looks like this:

```
NEW, SFERVOL

READY

10  INPUT R
20  LET V = (4 / 3) * 3.1416 * R ↑ 3
30  PRINT R, V
40  END
RNH

?  1.3
  1.3                    9.20279

RUN COMPLETE
```

Now the computer is ready to go again; we can start writing a new program or run this one again. If we want to run this program for several different values of the radius, it would save time if we could get the computer to run through again rather than ending. We can do this by inserting an unconditional transfer. We also add a remark statement so that we don't forget what we are doing.

```
03  REM   THIS PROGRAM FINDS THE VOLUME V OF A SPHERE OF RADIUS R
10  INPUT R
20  LET V = (4 / 3) * 3.1416 * R ↑ 3
30  PRINT R, V
36  GOTO 10
40  END
RNH

?1.3
  1.3                    9.20279
?3
  3                      113.098
?1
  1                      4.1888
```

You should notice two things. First, the numbers 1.3, 3, 1 are so close together you can hardly tell what is happening. Second, this is going to go on forever. How can we get the program to stop? Every system has a means of terminating the program (besides pulling the plug). You should find out what yours is. (Very often it is to type STOP or push the S key while the computer is typing.)

We handle both of these problems and also make the input and output a little fancier by introducing additional print statements. PRINT with no variable after it will give a blank line; this fact helps the spacing between lines. We can use the fact that we know the volume of a sphere of zero radius to allow us to use the setting of the radius equal to zero as a command to stop. This uses the conditional transfer. The other changes are for appearance and help the user interpret what the input and output are. A good program would then be

```
03  REM  THIS PROGRAM FINDS THE VOLUME V OF A SPHERE OF RADIUS R
06  PRINT "WHAT IS THE RADIUS OF THE SPHERE";
10  INPUT R
15  IF R = 0 THEN 40
20  LET V = (4 / 3) * 3.1416 * R ↑ 3
22  PRINT
29  PRINT "THE SPHERE OF RADIUS"; R; "HAS VOLUME"; V
30  PRINT
36  GOTO 06
40  END
RNH

WHAT IS THE RADIUS OF THE SPHERE? 1.3

THE SPHERE OF RADIUS  1.3  HAS VOLUME 9.20279

WHAT IS THE RADIUS OF THE SPHERE? 2

THE SPHERE OF RADIUS  2  HAS VOLUME 33.5104

WHAT IS THE RADIUS OF THE SPHERE? 0

RUN COMPLETE
```

If you did not like the numbers you have as addresses, for example if they are too close together to get other instructions between them, you can type RESEQ (which is short for resequence) and the computer will spread your statement numbers out, usually increasing by steps of 10 beginning at 100.

Before doing this and seeing the result, we will rewrite our program one more time. It shows a program with rather more done to it than usual.

```
03  REM  THIS PROGRAM FINDS THE VOLUME V OF A SPHERE OF RADIUS R
04  REM  TYPE 0 TO TERMINATE THE PROGRAM
06  PRINT "WHAT IS THE RADIUS OF THE SPHERE";
07  PRINT "(TYPE 0 IF YOU WISH THE PROGRAM TO TERMINATE)";
10  INPUT R
15  IF R = 0 THEN 40
16  IF R > 0 THEN 20
17  PRINT "REALLY MY GOOD FELLOW, DON'T YOU KNOW THE RADIUS"
18  PRINT "MUST BE POSITIVE OR ZERO? TRY AGAIN."
19  GOTO 06
```

```
20  LET V = (4 / 3) * 3.1416 * R ↑ 3
22  PRINT
29  PRINT "THE SPHERE OF RADIUS"; R; "HAS VOLUME"; V
30  PRINT
36  GOTO 06
40  END
```

Now when we type RESEQ and then LIST, we get

```
00100  REM  THIS PROGRAM FINDS THE VOLUME V OF A SPHERE OF RADIUS R
00110  REM  TYPE 0 TO TERMINATE THE PROGRAM
00120  PRINT "WHAT IS THE RADIUS OF THE SPHERE";
00130  PRINT "(TYPE 0 IF YOU WISH THE PROGRAM TO TERMINATE)";
00140  INPUT R
00150  IF R = 0 THEN 00250
00160  IF R > 0 THEN 00200
00170  PRINT "REALLY MY GOOD FELLOW, DON'T YOU KNOW THE RADIUS"
00180  PRINT "MUST BE POSITIVE OR ZERO? TRY AGAIN."
00190  GOTO 00120
00200  LET V = (4 / 3) * 3.1416 * R ↑ 3
00210  PRINT
00220  PRINT "THE SPHERE OF RADIUS"; R; "HAS VOLUME"; V
00230  PRINT
00240  GOTO 00120
00250  END
```

Finally, after all this work, we would hate to lose the program. In order to have it available next time we want to find the volume of spheres, we type

```
SAVE, SFERVOL
```

and the program will be saved in our library. When we want it back, we type

```
OLD, SFERVOL

READY
```

types the computer, and away we go again.

By the way, if you cannot remember the exact name of a program you can type

```
CATLIST
```

and the computer will list the names of all the programs that are stored in your user library.

Exercises 2.7

1–10. Write and run programs for problems 1–14 in Section 2.3.

Write and run a program for each of the following exercises. If data are needed and your instructor does not give you special data, provide some numbers yourself.

11. Do problem 7 in Section 2.2.

12. Do problem 8 in Section 2.2.

13. Do problem 9 in Section 2.2.

14. Do problem 10 in Section 2.2.

2.8 A NEW LOOP COMMAND AND A LAST EXAMPLE

In this last section on BASIC we begin with a simple example and illustrate the RESEQ command again. We introduce the FOR NEXT statements for simplified looping and show how it applies to a more general version of our initial example.

Suppose we have a contract to paint the exterior surface of some

large boxlike storage bins. We need to know what the surface area is so that we can determine how much paint we will need. To begin with suppose there are only four of these "boxes."

A flow chart for finding the surface area for unknown dimensions of the bins would be as shown in Figure 2.30.

Figure 2.30

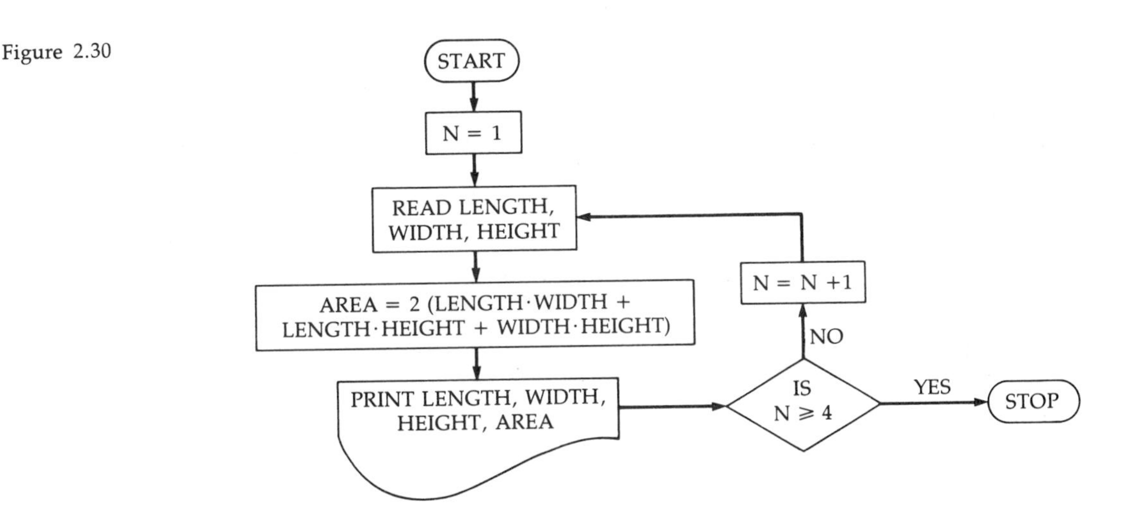

The left-hand BASIC program that follows was typed. After the RESEQ instruction explained in Section 2.7 was performed, it looked like the program on the right. Notice that the addresses in the instructions were changed to correspond to the change in the line numbers.

```
PROGRAM   AREA

10   LET N = 1
20   INPUT L, W, H
30   LET A = 2(L
30   LET A = 2 * (L * W + L * H + W * H)
40   IF N > = 4 THEN 100
34   PRINT L, W, H, A
50   LET N = N + 1
60   GOTO 20
100  END
READY
```

```
PROGRAM   AREA

00100   LET N = 1
00110   INPUT L, W, H
00120   LET A = 2 * (L * W + L * H + W * H)
00130   PRINT L, W, H, A
00140   IF N > = 4 THEN 00170
00150   LET N = N + 1
00160   GOTO 00110
00170   END
READY.
```

Now, when we type RUN and type in values, the printout would look like

```
PROGRAM   AREA

? 4,5,6
 4              5              6              148
? 98.5,45,6
 98.5           45             6              10587
? 13.68,67.87,64.57
 13.68          67.87          64.57          12388.3
? 5,4,1.3456
 5              4              1.3456         64.2208

RUN COMPLETE.
```

The looping command we introduce now is very easy to use and very powerful. We show this in our last example.

FOR NEXT

These instructions delimit the size of the loop. The FOR instruction alerts the computer that a loop is starting, and the NEXT instruction informs the computer that the end of the loop has been reached. Another run through will be performed if they have not yet all been done. The instructions take the forms

```
FOR snv = I TO J
 .
 .
 .
NEXT snv
```

Here snv stands for any simple numeric variable, which must be the same in both instructions. I and J stand for any numbers or any arithmetic expressions. The variable is increased by 1 each time the NEXT instruction is reached until the value of J is reached. That is, when snv = J, the computer executes the loop for the last time and continues on to the instruction after NEXT instead of going back to the FOR instruction.

For example, suppose the following is part of a larger program:

```
   .
   .
   .
40 LET K = 3
50 LET S = K * K
60 PRINT S
62 IF K = 11 THEN 80
65 LET K = K + 1
70 GOTO 50
80 GOTO 120
   .
   .
   .
```

This subprogram clearly computes and prints the squares of the integers from 3 to 11 inclusive. The program could be shortened and simplified by replacing these seven instructions by the following five instructions:

```
.
.
.
40 FOR K = 3 TO 11
50 LET S = K * K
60 PRINT S
70 NEXT K
80 GOTO 120
.
.
.
```

The computer, upon reaching line 40, would set

$K = 3$,	$S = 3 \cdot 3 = 9$,	print 9,	increase K and restart with
$K = 4$,	$S = 4 \cdot 4 = 16$,	print 16,	increase K and restart with
$K = 5$,	$S = 5 \cdot 5 = 25$,	print 25,	increase K and restart with
.	.	.	.
.	.	.	.
.	.	.	.
$K = 11$,	$S = 11 \cdot 11 = 121$,	print 121,	

but now K is at its maximum value so the program would execute the next instruction, which is a transfer to line 120, and so on.

Now we give a program for computing the surface area of an arbitrary number of boxes. In this program we use subscripted variables, which are explained in the Appendix.

```
PROGRAM        AREA1

100  PRINT "THIS IS A PROGRAM TO COMPUTE THE SURFACE"
110  PRINT "AREA OF BOXES, I.E., PARALLELEPIPEDS."
120  PRINT
130  PRINT "HOW MANY BOXES DO YOU HAVE?"
140  INPUT N
150  PRINT
160  FOR I = 1 TO N
170  PRINT "TYPE THE LENGTH, WIDTH, HEIGHT OF BOX NUMBER"; I
180  INPUT L(I), W(I), H(I)
190  NEXT I
200  PRINT
210  PRINT
220  PRINT "BOX NO.", "LENGTH", "WIDTH", "HEIGHT", "AREA"
230  FOR I = 1 TO N
240  LET A = 2 * (L(I) * W(I) + L(I) * H(I) + W(I) * H(I))
250  PRINT
260  PRINT I, L(I), W(I), H(I), A
270  NEXT I
280  PRINT
290  END
```

Using the same input as in the program AREA, the output would look like this when the program is run:

```
PROGRAM   AREA1

THIS IS A PROGRAM TO COMPUTE THE SURFACE
AREA OF BOXES, I.E., PARALLELEPIPEDS.

HOW MANY BOXES DO YOU HAVE?
?  4
```

```
TYPE THE LENGTH, WIDTH, HEIGHT OF BOX NUMBER  1
?  4, 5, 6
TYPE THE LENGTH, WIDTH, HEIGHT OF BOX NUMBER  2
?  98.5, 45, 6
TYPE THE LENGTH, WIDTH, HEIGHT OF BOX NUMBER  3
?  13.68, 67.87, 64.57
TYPE THE LENGTH, WIDTH, HEIGHT OF BOX NUMBER  4
?  5, 4, 1.3456

BOX NO.     LENGTH      WIDTH       HEIGHT      AREA

1           4           5           6           148

2           98.5        45          6           10587

3           13.68       67.87       64.57       12388.3

4           5           4           1.3456      64.2208
```

Exercises 2.8

1. If your social security number is *abc-de-fghi,* let m be $+1$ or -1 according to whether c is even or odd, but let n be $+m$ or $-m$ as f is even or odd, respectively. Let

$$y = m(gh.i)^{n(a.b)} \quad \text{and} \quad x = (abc.de)^f y^3,$$

where each dot is a decimal point and multiplication is denoted by putting one factor in parenthesis. Output your social security number, $m,n,y,$ and x.

2. Write a program for problem 9, Section 2.3. Make up your own numbers, but be sure N (the number of numbers in the list) is at least 15 and that there are at least 4 representatives in the list of each type of number.

3. Write a program that computes the sum of the cubes of the first N numbers, that is, computes $1^3 + 2^3 + \cdots + N^3$. (The input will be N.)

4. Write a program that computes the sum of the first N integers and then squares the result, that is, computes $(1 + 2 + \cdots + N)^2$.

5. Write a program that inputs N and computes both the numbers computed in problems 3 and 4 above. Output N, $(1^3 + \cdots + N^3)$, and $(1 + \cdots + N)^2$. Run the program for various values of N. What mathematical statement do you believe to be true as a result of your computations?

6. Let $a_1 = 1$ and a_n be the sum of the first n integers (so $a_3 = 6$, $a_5 = 15$, and so on). Compare $a_1 + a_2 + \cdots + a_n$ and $n(n + 1)(n + 2)$. Can you deduce an equation?

7. Find the quotients when $7^{2n} + 16n + 63$ is divided by 64 and 128 for $n = 0$ to 70. What conjecture would you make?

8. Compare $1^2 + 2^2 + \cdots + n^2$ and $(n + 1)(2n + 1) / 6$. What equation can you deduce? Is your equation true for $n = 25$? 47?

9. Redo some of your previous programs by using a loop with the FOR and NEXT commands rather than the logical IF.

10. A famous theorem relates the size of the arithmetic mean and the geometric mean of a sequence of numbers. If there are n numbers, the arithmetic mean is their sum divided by n and the geometric mean is the nth root of their product. Write a program that will input n and the n numbers, compute the two means, print them out, and state which is the larger. Run your program for different values of n and the numbers. Can you guess the theorem?

11. Professor Ness decides he will bicycle through Europe next summer and wants a table of conversion from kilometers to miles. He finds that 1 mile is 1.6093 kilometers, and decides he wants a conversion for each mile from 1 to 10. Write a program for him that will output the desired table with headings.

12. A student decides to accompany the professor of problem 11, but wants the table to also include the conversion into miles of all distances in kilometers that are multiples of 5 up to 100. Rewrite the program from problem 11 to include these values at the bottom of the table.

13. If roofing costs $22.97 per square (enough for 100 square feet of roof), write a program that determines the cost of material for roofing m houses (m is not more than 10) when the input is m and the areas in square feet of the roofs of the m houses. Output a table with the number of the house, its area, and the cost for that house. Also output the total cost and the average cost for each house.

14. The Sky-High Roofing Company needs a program to help it estimate the cost of roofing various-size houses. Roofing is $23.78 a square. (A "square" is enough roofing to shingle 100 square feet of roof.) The company wants a program that will input the number of houses under consideration (never more than 20); for each house the number of rectangular roof areas, the number of ridges, and the number of valleys; for each house the dimensions of the rectangular roof areas and the length of the ridges and valleys.

Valleys cost $1.98 per linear foot and ridges cost $1.69 per linear foot to complete.

As output we want the number of the house, the total roof area, the number of feet of valleys, the number of feet of ridges, and the total cost for that house in tabular form. Finally, we also want the total cost for roofing, for the valleys and for the ridges, a grand total, and an average per house. There needs to be enough description of the input and the output so the secretaries can input data and the owner can read the output.

2.9 AN INTRODUCTION TO FORTRAN (WATFIV)

How do we communicate with a computer? Naturally we must use a language of some kind. It would be nicest to simply be able to converse with the computer as we do with each other. Unfortunately, although work is progressing that someday will make this a reality, for now we are stuck with the need to learn the computer's language. Actually, for our purposes, he is rather simple minded, so we will be able to get by with a rather simple language called *WATFIV*. It is a version of FORTRAN IV that has a simplified input and output scheme (FORTRAN comes from FORmula TRANslation). Moreover, we are not even going to learn the full language, but merely enough to find out a little about the joys and frustrations of dealing with a computer via punched cards. FORTRAN was developed for programming scientific problems and is excellent for programs that are computational in nature. It is not as good for some other purposes, such as sorting and listing, however.

What does WATFIV stand for and was there ever a 'WATT-ONE', 'WATT-THREE' or 'WATT-FOUR'? Actually, the first language of this type was WATFOR, which was a simplified version of FORTRAN. It was developed at the University of Waterloo in Canada, and stands for "Waterloo FORTRAN." When FORTRAN IV came along and WATFOR was modified, it seemed only natural to call it WATFIV!

Languages have several components such as words, sentences, and punctuation marks. We must learn what words are acceptable to our machine and then how to combine them into sentences or, as we shall call them, *instructions*.

Numbers, Variables, and Arithmetic Operations

What kind of numerical words, usually called numbers, can we use in our conversations with the computer? The length of the numbers allowable depends somewhat on the computer available, but we will be specific.

There are two types of numbers that we use to express data in FORTRAN: (1) *integers* or *fixed-point numbers* and (2) *real numbers* or *floating-point numbers.*

An integer may have from 1 to 9 digits and may have a plus or minus sign. There must be *no decimal point* and commas are not used.

ACCEPTABLE	NOT ACCEPTABLE
+193	19.0
10	19578.
−329745098	31+
0	83594780935
53701	936,854,321

The rules for floating-point numbers are slightly more complicated. A floating-point number

1. Must have a decimal point
2. Must have from 1 to 7 digits and may have a plus or minus sign
3. May be followed by the letter E, which in turn must be followed by a one- or two-digit signed or unsigned integer

ACCEPTABLE	NOT ACCEPTABLE
3.560	356
+.5673985	50.35−
−49	4394.56789
13.59E+17	13.36E153
.1596E26	3,361.5

"What does E mean?" you may ask. "Do any of you remember writing numbers in what is usually called 'scientific notation' "?, we will ask. Since most of you haven't heard of it, or have forgotten, we will explain. Basically, it is a shorthand notation for representing numbers with many zeros either before or after the number. For example;

Instead of	543000000000.,	we can write	5.43×10^{11}.
Instead of	.000000035560,	we can write	3.556×10^{-8}.
Instead of	.013694,	we can write	136.94×10^{-4}.

These are perfectly respectable real numbers, but, except for the last example, do not have seven or fewer digits. In FORTRAN

Instead of	543000000000.,	we can write	5.43E11	or	543.0E+9.
Instead of	.000000035560,	we can write	3.556E−8	or	0.35560E−7.
Instead of	.01369,	we can write	136.9E−4	or	.0001369E2.

> The rule for moving the decimal point is really quite simple. If you move the decimal point to the left you increase the exponent by the number of places moved, and if you move the decimal point to the right you decrease the exponent by the number of places moved. For example,
>
> 6579030000. = 657.903E7, (7 places left)
>
> 8963.97E12 = 8.96397E15, (3 places more)
>
> .00036517 = 36.517E−5, (5 places right)
>
> .0398E−9 = 3.98E−11, (2 places more).

Thus you see that the E stands for "times ten to the power." There is an additional restriction that we must place on the one- or two-digit integer that follows the E. It must not be larger than 75 nor smaller than −78. (These bounds will vary depending on the computer you have available.)

Numbers such as we have been describing are called *constants*. We also need to have names for unknown or variable quantities. As for constants, there are two types of such names: integer and floating-point.

1. A name is composed of from 1 to 6 of the 36 possible characters 0, 1, . . . , 9, A, B, C, . . . , Y, Z.
2. The first character in a name must be an alphabetic character, that is, one of A, B, C, . . . , Y, Z.
3. Names associated with integer numbers must start with I, J, K, L, M, or N.
4. Names associated with floating-point numbers must *not* begin with I, J, K, L, M, or N (and hence must begin with A, B, C, . . . G, H, O, P, . . . , Y, Z).

INTEGER NAMES	
ACCEPTABLE	NOT ACCEPTABLE
MN13X	4MIN
NONE	R123
K	MIN?3
J379	NONSENSE
LAX	A
NSUM	SUM

FLOATING-POINT NAMES	
ACCEPTABLE	NOT ACCEPTABLE
ABLE	8XY
WAGE	NSUM
SUM	VARIATION
HOURS	X4.3
RISE	X!Y
OH54	

If you should make a mistake and use the wrong kind of variable name, the computer may reject your program or it may proceed and compute,

WHY CAPS?
Notice that we only use capital letters when we write words which will be read by the computer. This is because these are the only letters a keypunch machine will print.

but change the data to a form consistent with your naming. For example, if you have KAT for the name of a number that the computer finds to be 13.79, the computer will read it as 13 and ignore the decimal part. Similarly, if you call 179 by the name REX, the computer will actually read it as 179.0 by adding the decimal point itself.

Now that we have names for variables and constants, we must learn how to combine them by the usual arithmetic operations of addition, multiplication, subtraction, division, and exponentiation into expressions. We have *real expressions* and *integer expressions* corresponding to the two types of numbers we allow. The symbols used to denote operations in these expressions are the same and are as follows:

OPERATION	ALGEBRAIC SYMBOL	FORTRAN SYMBOL
Addition	+	+
Subtraction	−	−
Multiplication	· or ×	*
Division	÷ or /	/
Exponentiation	none	**

Examples:

ALGEBRAIC	FORTRAN	RESULT
	Integers	
3 + 7	3 + 7	10
3 − 7	3 − 7	−4
3 · 7	3 * 7	21
6 ÷ 3	6 / 3	2
6^3	6**3	216
	Real numbers	
3.2 + 7.1	3.2 + 7.1	10.3
7.1 − 3.2	7.1 − 3.2	3.9
5.1 · 3.2	5.1 * 3.2	16.32
5.1 ÷ 3.2	5.1 / 3.2	1.59375
$5.1^{3.2}$	5.1**3.2	1.837488E2

Just as in ordinary algebra, we have an order of precedence that allows us to decrease the number of parentheses we need. For example, multiplication is done before addition, so 6 * 3 + 14 means 18 + 14, or 32, and not 6 times 17. If we want to denote the latter, we would have to write 6 * (3 + 14). However, parentheses can always be used. Although not necessary, we could write (6 * 3) + (2 * 7) rather than 6 * 3 + 2 * 7.

Formally, we say that the operations are performed *from left to right* according to the following order of precedence.

1. Exponentiation
2. Multiplication and division in the order written
3. Addition and subtraction in the order written

If parentheses occur, the operations called for inside the parentheses are done first. Some examples should clarify what we mean.

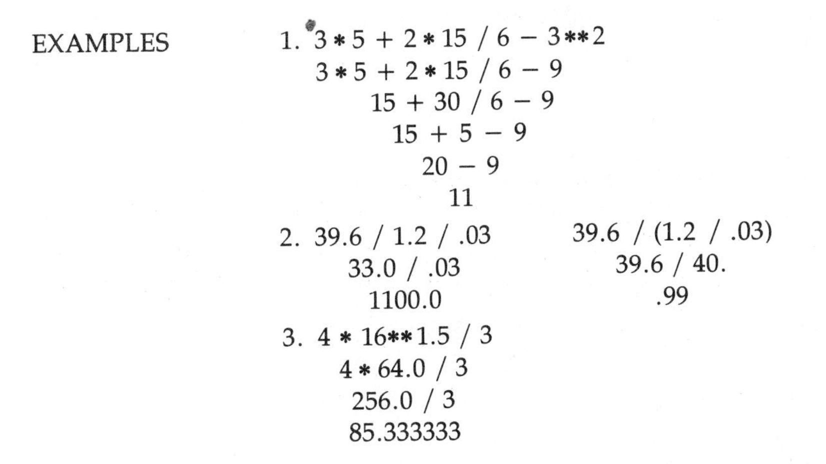

EXAMPLES

1. 3 * 5 + 2 * 15 / 6 − 3**2
   3 * 5 + 2 * 15 / 6 − 9
   15 + 30 / 6 − 9
   15 + 5 − 9
   20 − 9
   11

2. 39.6 / 1.2 / .03
   33.0 / .03
   1100.0

   39.6 / (1.2 / .03)
   39.6 / 40.
   .99

3. 4 * 16**1.5 / 3
   4 * 64.0 / 3
   256.0 / 3
   85.333333

Algebraic expressions follow the same rules:

ALGEBRAIC EXPRESSION	FORTRAN EXPRESSION
$3x^2 + xy - (x + y)^2$	3 * X**2 + X * Y − (X + Y)**2
$z - \frac{x^2y^3}{r}$	Z − X**2 * Y**3 / R

What algebraic expression corresponds to the expression

A**3 / C**2 * B**3 + B * A**(−2) / C**3?

The translation would proceed like this:

$$(a^3 \div c^2) \cdot b^3 + b \cdot a^{-2} / c^3,$$

$$\frac{a^3}{c^2}b^3 + \frac{ba^{-2}}{c^3}.$$

Notice that 2**3**2 could be interpreted two ways, depending on how the exponents are associated. In one case,

$$(2**3)**2 = (2^3)^2 = 8^2 = 64,$$

and in the other case,

$$2**(3**2) = 2^{(3^2)} = 2^9 = 512.$$

Quite a difference. Different machines compute 2**3**2 differently, so to avoid any possible confusion, we will always place parentheses in the proper places. Naturally, even when the order of precedence should be clear, you may insert parentheses to be sure the computer does what you want.

Exercises 2.9

1. State whether the following numbers are integer, floating point, or neither.

   (a) 561 **(b)** 43,987,651
   (c) 436.75E29 **(d)** 51.00
   (e) −4398.01 **(f)** 596403
   (g) 596078395 **(h)** 4378.56594
   (i) 1.379000E−76 **(j)** 5,678.43
   (k) −436. **(l)** −4937

2. State whether the following variables are integer, floating point, or neither.

   (a) KNEAL **(b)** N12
   (c) HORSE **(d)** S1M6
   (e) ISM **(f)** Y19.5
   (g) CLARENCE **(h)** 43 RUE
   (i) ZCAR! **(j)** X
   (k) PL / 1 **(l)** PENCILS

3. Compute the following numbers and state whether the result is an integer or a floating-point number.

   (a) 4 * 2**3 **(b)** 9 * 5**2 / 15
   (c) 3.2 * 11.3 / 4.0 * 5.7 **(d)** 2.1**2 * 3 + 8.4 / 2.1
   (e) 10. * 3**2 / 6 **(f)** (3 * 2)**2 / (5 * 2)

4. Compute the following numbers.

(a) 2 * (3 + 4) + 3**2 − 8 * (5 − 2)
(b) 2 * 2 + 3**2 * 4 − 3 * 2**4 / 2
(c) 3 * 5**2 + (2**3)**2 / 4 * 3 / 2
(d) 4 / 3 * 12 * 5 / 2 − 3 * 5 / 2 + 3 * 2**3 * 2

5. Express the following algebraic expressions in traditional form.

(a) 3 * A**2 + 15 * A + 43
(b) 5 * X**2 * Y**3 − 14.6 * X * Y**2 + 5.93 * Y
(c) A**3 / C * B**4 + B * 2 * A** (2 + C) / C * (−2)
(d) B**3 / A * 2**3 * B** (−2)
(e) 3 * A**2 + 4 * A − 17
(f) X**2 * Y + 3 * X * Y**2 − 17 * X**3 * Y**2

6. Write the following algebraic expressions in the FORTRAN language.

(a) $15x^3 + 3x + 7$
(b) $3x^2 - 17x + 5$
(c) $13x^2y^2 - 9.5x^3y + 43xy^3$
(d) $(x^2 + 2y)(x - 3y)(5 + 4y^2)$
(e) $x_1y_1 + x_2y_2 + x_3y_3$
(f) $3x^ay - 17x^3y^b + (43x - y^a)(4x^b + 2y)$
(g) $a^2b^3 \div 2c^2 - 3c^{-2} \div b^5$
(h) $\dfrac{3x^2y}{z} - \dfrac{17xy}{3z^2} + y^3z^{15}$

In problems 7 and 8 express the given quantity as an expression in the FORTRAN language.

7. (a) The length of the hypotenuse of a right triangle with legs of lengths x and y
(b) The volume of a cylinder in terms of its height h and radius r

8. (a) The surface area of a rectangular box of a given height, length, and width
(b) The volume of a cube that holds twice as much as a given cube whose edges have length a

9. A grocer sells apples for 29¢ per pound and pears for 33¢ per pound. If he pays x¢ per pound for apples and y¢ per pound for pears, express in FORTRAN the profit on m pounds of apples and n pounds of pears.

10. A used-car dealer wants to make 10% profit on all cars he sells for $5000 and 15% on all cars he sells for $7000. If he sells r $5000 cars and s $7000 cars, express his profit in FORTRAN.

11. A large manufacturer buys sheet steel and iron ingots. The sheets are c feet square and weigh 6.7 pounds per square foot. The ingots are cubes of length b feet on each side and weigh 78.9 pounds per cubic foot. Express in FORTRAN the total weight of 130 sheets of steel and $15t$ iron ingots.

12. A large grocer buys x pounds of oranges, y pounds of apples, and z pounds of bananas at a cost of a¢, b¢, and c¢ per pound. If she sells apples for 35¢ per pound, oranges for 20¢ per pound, and bananas for 23¢ per pound, express the grocer's profit as an expression in FORTRAN.

2.10 SUMMARY OF SOME FORTRAN COMMANDS

We are now ready to begin to program. To do this, we need to be able to translate from a flow chart into a user language.

Before you can actually punch cards and have your program run by the computer, the instructor will need to give you instructions. The program "deck" will generally have three kinds of cards: Job Control Language (JCL) cards, which are peculiar to each computer installation; the source deck or actual FORTRAN program; and some data cards (although data cards are not necessary). The instructor will describe how to punch the JCL cards and where they are to be placed in the deck of cards. In Figure 2.31 we illustrate a deck with two JCL cards ($JOB and $ENTRY) and one data card, as well as a five-statement program. The *columns* are numbered on the punch card by the small numerals running from 1 to 80.

The actual program, or *source deck*, is key-punched with these rules in mind: The statements are always started in column 7. Columns 1–5 are statement-number columns (more about this later), and column 6 should always be blank. (Actually, column 6 is always blank only because our statements are not going to be very long. It is used when more than one card is needed to represent a single instruction and is called a "continuation" column. We will not be using it.)

When you use a key-punch machine to type your cards, the machine both punches the cards and types at the top of the card the letter or symbol that has been punched. Later we will describe how to use the key-punch machine to type your source deck.

Figure 2.31

```
         26.31     84.75
$ENTRY
         END
         STOP
         PRINT, A, B, P
         P = A*B
         READ, A, B
$JOB

00000000000000000000000000000000000000000000000000000000000000000000000000000000
1 2 3 4 5 6 7 8 9 10 11 12 13 14 15 16 17 18 19 20 21 22 23 24 25 26 27 28 29 30 31 32 33 34 35 36 37 38 39 40 41 42 43 44 45 46 47 48 49 50 51 52 53 54 55 56 57 58 59 60 61 62 63 64 65 66 67 68 69 70 71 72 73 74 75 76 77 78 79 80
11111111111111111111111111111111111111111111111111111111111111111111111111111111
22222222222222222222222222222222222222222222222222222222222222222222222222222222
33333333333333333333333333333333333333333333333333333333333333333333333333333333
44444444444444444444444444444444444444444444444444444444444444444444444444444444
55555555555555555555555555555555555555555555555555555555555555555555555555555555
66666666666666666666666666666666666666666666666666666666666666666666666666666666
77777777777777777777777777777777777777777777777777777777777777777777777777777777
88888888888888888888888888888888888888888888888888888888888888888888888888888888
99999999999999999999999999999999999999999999999999999999999999999999999999999999
1 2 3 4 5 6 7 8 9 10 11 12 13 14 15 16 17 18 19 20 21 22 23 24 25 26 27 28 29 30 31 32 33 34 35 36 37 38 39 40 41 42 43 44 45 46 47 48 49 50 51 52 53 54 55 56 57 58 59 60 61 62 63 64 65 66 67 68 69 70 71 72 73 74 75 76 77 78 79 80
```

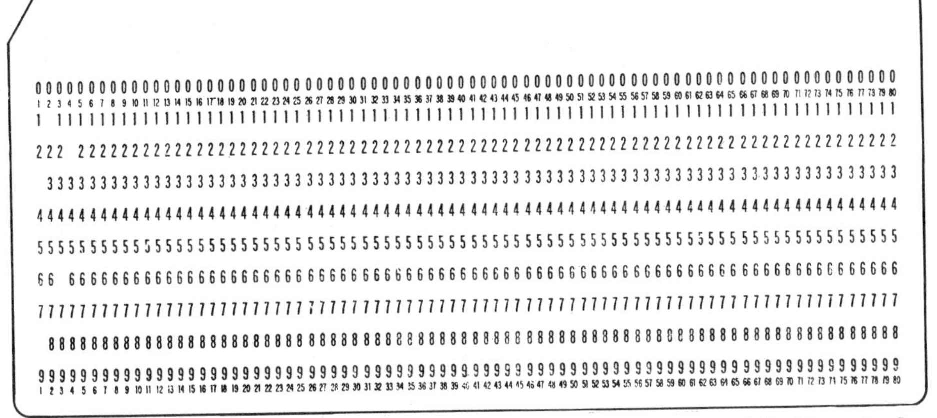

Courtesy Office Electronics, Inc.

READ and PRINT Statements

The basic input statement is of the form

READ, names

where names is a sequence of one or more variable names separated by commas. The comma after READ is a necessary part of the statement, but no comma appears after the last name in the sequence. (If your computer does not understand WATFIV, and you must use FORTRAN, consult the Appendix for the basics of how to write FORMAT statements.)

The numbers to be read by the computer and assigned to these variables are punched (with + or − if desired) on the input cards and are separated by blank columns or commas. A card may contain one or more numbers (integer or floating point). Data are read and numbers are assigned to the variables in the same order as the variables appear on the READ instruction and the numbers appear on the data card. If more variables occur in the READ card than there are numbers on the first card, the computer goes to the next data card. Also, whenever a new READ instruction is encountered the computer reads from a data card not previously read.

Two examples should serve as illustrations.

EXAMPLE 1

Suppose the first READ statement is

```
READ, X, IDA, Q
```

and the first data card has on it

```
39.7, 113, 65.93
```

When the program is executed and this statement is reached, the computer will assign X the value 39.7, IDA the value 113, and Q the value 65.93 in subsequent work.

EXAMPLE 2

Suppose that the two READ statements

```
READ, X, Y, Z
READ, OX, NUM, V
```

are either consecutive, or have other non-READ statements between them, and that the first data cards are as follows:

First card:	23.94, 19.17
Second card:	20.1, 197
Third card:	14.37, 95, 149.5

Then the variables X, Y, Z are assigned the values 23.94, 19.17, and 20.1, respectively, while OX, NUM, V are assigned the values 14.37, 95, 149.5, respectively. Notice that 197 is never used and that NUM is the only integer variable.

The basic output statement is of the form

PRINT, names

where names is a sequence of one or more variable names, and the same comments about commas made for READ apply here. As a result of this statement, the values currently associated with the variable names in the sequence are printed in the order in which the names occur in the statement. In addition, the PRINT statement can be used to output any string of acceptable symbols (including letters, numerals, and certain other special symbols). Notice that expressions cannot be placed in the PRINT statement.

In the following examples, the PRINT statement is given and what the computer will type follows it immediately.

EXAMPLE 3

```
PRINT, X, JANA, PEG
```

```
3.75     14     97.6
```

(assuming the current values of the variables X, JANA, and PEG are 3.75, 14, and 97.6, respectively.)

EXAMPLE 4

```
PRINT, 'THE BOSS IS'
```

```
THE BOSS IS
```

EXAMPLE 5

```
PRINT, 'THE RADIUS IS', R, 'THE VOLUME IS', V
```

```
THE RADIUS IS 3.9     THE VOLUME IS 49.7
```

(assuming that the current values for R and V are 3.9 and 49.7, respectively.)

EXAMPLE 6

```
PRINT, X, Y, X**Y
```

This is not an acceptable PRINT statement because in FORTRAN a computation cannot be performed in the PRINT process.* You would need to compute X^Y in a separate statement, assign it to a variable name and then print that variable name. For example,

```
Z = X**Y
PRINT, X, Y, Z
```

would print

```
6.25    .5    2.5
```

if X and Y had the values 6.25 and .5, respectively.

Assignment Statements

Assignment statements have the general form

$$x = \text{expression}$$

where x is a variable name (either integer or floating point) and expression stands for any valid arithmetic expression written as we did in Section 2.9. Thus the equal sign represents an assignment just as it did in our flow charts. Some examples are

```
PROD = X * Y
K = K + 1
B = C**2 * A**3 / D + 16 * A**2 / C**2
VOLUME = 4 / 3 * 3.1416 * RADIUS**3
```

STOP and END Statements

Every program must contain these statements. The STOP statement is translated into machine language and stops the execution of the pro-

*On some computers a computation can be performed in the PRINT statement.

gram in the central processing unit. The END statement informs the compiler that the final FORTRAN statement of the program has been reached. Thus there may be several STOP statements in a program with several branches, but each program has exactly one END statement—the last.

A First Program

We can now write programs that correspond to the first two flow charts in Section 2.2. Instructions for using a key-punch machine are given in Table 2.3. A picture of the machine is given in Figure 2.32.

TABLE 2.3 INSTRUCTIONS FOR KEY-PUNCH MACHINE

Notice these components of the key-punch machine.

1. On top there is a bin for blank cards with a pressure plate behind.
2. On the upper front is (i) a space (trough) for holding punched cards that have been released, (ii) a window with a red pointer, which points to the number of the column peing punched, (iii) a space where the card goes immediately after being punched (and where you place a card if you wish to copy it), and (iv) a space where the card being punched is placed and registered.
3. On the bench in front is a work space to the left and the keyboard to the right.
4. Under this bench is the on-off switch.

To begin to punch cards you should

1. Be sure there are blank cards in the bin on top.
2. Be sure all toggle switches are up (except the "clear" switch, which will not stay up).
3. Turn on the machine (the switch below keyboard bench).
4. Push "Feed" (a card should drop down).
5. Push "Feed" again. This will drop another card down and "register" the first card. (If you only want to do one card, push "REG" key to register the first card.)
6. Now you are ready to type (that is, punch out) the card. *Remember:* JCL cards start in column 1 and program instructions in column 7. Each program line is typed onto a separate card.
7. When you are through typing on one card and wish to type on the next card, push "REL" key (that is, release). The card you just punched will move left, the second card will become the first, and a new second card will drop down.
8. When you are all finished, push the "clear" toggle switch up and all the cards will be removed up to the holding trough.

Notice that what you have typed can be read across the top of the card.

Figure 2.32

E. Robert Stefl, Illinois State University

```
READ, A
READ, B
PROD = A * B
PRINT, PROD
STOP
END
```

```
READ, A, B, C, D
PROD = A * B * C * D
PRINT, PROD
STOP
END
```

Recall that each of these lines represents what is punched onto a single card, and the first letter always occurs in column 7. The actual program cards are called the *source deck*. Here the first program has six cards in its source deck and the second program has only five cards in its source deck.

Exercises 2.10

1. Write a program that computes the average (arithmetic mean) of two numbers. Print out the two numbers and the average.

2. Write a program that computes the geometric mean of two positive numbers. (This is the square root of their product.) Output three numbers.

3. Write a program that finds the length of the hypotenuse of a right triangle given the lengths of the two sides. Output the three lengths. Try your program by hand for a right triangle with sides of length 5 and 12.

4. If a circle of radius r has area $3.1416r^2$ and circumference twice the area divided by r, write a program that outputs r, the circumference and the area.

5. Write a program that computes the area of a face and the volume of a cube whose edges are of length 6.3 feet. Print out the length of the edge, the area of a face, and the volume.

6. The formulas for the surface area A and the volume V of a sphere of radius r are

$$A = 4\pi r^2$$

and

$$V = \tfrac{4}{3}\pi r^3.$$

Write a program that computes the surface area and volume of a sphere of radius 18.37 feet.

7. Write a program for problem 5 of Section 2.2. (Use a cost of \$3.89 per pound.)

8. Write a program for problem 6 of Section 2.2. (Suppose cola is 89¢ and lemon soda is 69¢ per pack.)

For problems 9–14 write a program that computes and prints the quantities described in problems 7–12 in Section 2.9.

2.11 ADDRESSING, TRANSFERS, AND LOOPING

In this section we describe how to program decisions, why some instructions have numbers associated with them (addresses), and how to save time and shorten programs by looping.

Addressing and the GO TO Instruction

As mentioned earlier, columns 1 to 5 of the card are used for statement numbers. Thus up to 99999 statements can be numbered. Not all statements need to be numbered, since the computer reads the cards in the order in which you submit them. Thus, for efficiency reasons, statements should only be numbered when necessary. The numbers can also occur in any order. The examples we give subsequently should make things clear.

The unconditional branch statement in FORTRAN is of the form

GO TO k

where k is the number of an executable statement in the program.

For example, the following program would have the effect of printing the current value of VOL over and over.

```
15  PRINT, VOL
    GO TO 15
```

The 15 that occurs as the number of the PRINT statement can have the 1 in any of columns 1 through 4, with the 5 in the next column. (Remember column 6 is blank.)

When we combine the GO TO statement with a logical IF (see below), we obtain a *conditional transfer*. As we will see next, this is what we need in order to program the problems we flow-charted in Section 2.3.

Relational Operators, Logical Expressions, and the Logical IF

As you know, the standard symbols for comparing the sizes of numbers are $=$, $<$, $\leq$, $>$, and $\geq$. When these symbols are placed between arithmetic expressions, statements (or "logical expressions") are formed about which it makes sense to ask the question "Is it true or false?" when specific values are substituted for the variables. For example, $2x^2y > x^2 + 3y$ may or may not be true depending upon the values of x and y. If $x = 4$ and $y = 2$, the statement is true, but if $x = y = 1$, it is false.

These relational operators, as they are sometimes called, are available in FORTRAN according to Table 2.4. Notice that the periods are an essential part of the operator.

TABLE 2.4 TABLE OF RELATIONAL OPERATORS

FORTRAN OPERATOR	MEANING	ALGEBRAIC SYMBOL
.EQ.	Equal to	$=$
.LT.	Less than	$<$
.LE.	Less than or equal to	$\leqslant$
.GT.	Greater than	$>$
.GE.	Greater than or equal to	$\geqslant$
.NE.	Not equal to	$\neq$

Some acceptable logical expressions follow.

FORTRAN	ALGEBRAIC
X**2 − 3 * X + 5 .GT. Z * Y − 13	$x^2 - 3x + 5 > zy - 13$
ED + DAVE .NE. ANN	ED + DAVE $\neq$ ANN
HOURS .LE. 40	Hours $\leqslant$ 40

The logical IF statement has the general form

IF (expression) statement

where expression is any logical expression and statement is any executable statement that we have had before. The effect of this statement is to execute the statement if the expression is true and to go on to the next instruction, that is, do nothing, otherwise. Notice the parentheses are an essential part of the instruction.

For example,

```
IF (X + Y .EQ. 0) Z = 1 + X
```

has the effect of changing the current value of Z to 1 + X provided X + Y is indeed zero, but leaves Z as it was in case X + Y is not zero. In either case, the computer next goes to the instruction on the card following.

For us, a more typical example is given by

```
IF (HOURS .GT. 40) GO TO 7
```

which transfers to the executable statement with number 7 provided HOURS is greater than 40, but goes on to the next instruction if HOURS is less than or equal to 40.

Some Examples

EXAMPLE 1

Suppose we wish to decide whether or not x^y or y^x is larger for two given values of x and y.

We first make a flow chart (Figure 2.33). Our program would be as follows, where we comment at the right. Notice that we have printed in such a way that the first number is the base, the second is the exponent, and the third is the first raised to the second.

READ, X, Y	Reads the numbers x and y
A = X**Y	Computes x^y
B = Y**X	Computes y^x
IF (A .GT. B) GO TO 3	Compares and branches to 3 if $x^y > y^x$
PRINT, Y, X, B	x^y must not be larger
STOP	Notice there are two
3 PRINT, X, Y, A	STOP statements, one for
STOP	each branch
END	The end of the program

Figure 2.33

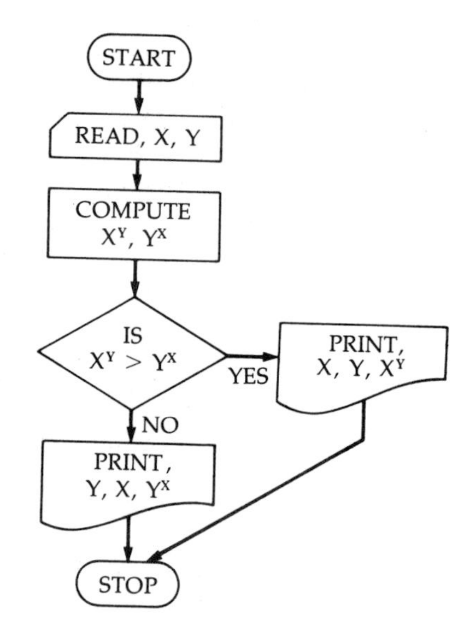

An interesting question often asked some calculus students is "Which is larger, e^{π} or π^{e}?" Here $\pi \simeq 3.14159$ is the ratio of the circumference to the diameter of a circle, and $e \simeq 2.71828$ is the base of the natural logarithms, or the amount of money you would have in a year if you put $1 into a bank at 100% interest per year compounded instantaneously. If you try it by hand, the numbers are fairly close.

Two additional comments should be made. The first is that we could have written the fourth statement as IF (X**Y .GT. Y**X) GO TO 3, which would seem to make the second and third statements unnecessary. This is not true, however, because the statement

```
PRINT, Y, X, Y**X
```

is not acceptable. A computation is *not* allowed in the PRINT statement, as we mentioned earlier. The program could be written as follows, though it is no shorter.

```
   READ, X, Y
   IF (X**Y .GT. Y**X) GO TO 3
   B = Y**X
   PRINT, Y, X, B
   GO TO 1              (The STOP statement could be left here.)
3  B = X**Y             (Could also still use A here.)
   PRINT, X, Y, B
1  STOP
   END
```

After running this program, how do we know that the two numbers we have been comparing are really unequal? If not, how might the program be changed so that the output will tell about the case in which they are the same?

EXAMPLE 2 We show next a program for Example 2 in Section 2.3.

```
    N = 1
12  READ, HOURS, RATE
    GROSS = HOURS * RATE
    PRINT, N, HOURS, RATE, GROSS
    IF (N .GE. 10) GO TO 2
    N = N + 1
    GO TO 12
 2  STOP
    END
```

Exercises 2.11

1. Write a program for Example 1 in Section 2.3

2–10. Write programs for problems 2–10 in Section 2.3.

In problems 11 and 12 express the given comparison in the FORTRAN language. (See the box "Inequalities in Plain English" in Section 1.1.)

11. **(a)** x does not exceed $3y$.
(b) k is not less than 5.
(c) The square of x exceeds the cube of y by 16.
(d) x is more than y.

12. **(a)** Twice y is not more than 15.
(b) $2t$ exceeds zero.
(c) The fourth power of t is not less than 3 more than the cube of r.
(d) 3 times x does not equal 17 more than twice y.

In problems 13 to 16 determine what each of the programs does.

13.

```
      READ, X, Y, Z
      SUM = X + Y + Z
      SUM = SUM - 20 + Y
      PRINT, Y, SUM
      STOP
      END
```

14.

```
      J = 1
    5 READ, X, Y
      Z = X**2 + Y**2
      Z = Z*Z
      PRINT, X, Y, Z
      IF (J .GE. 8) GO TO 1
      J = J + 1
      GO TO 5
    1 STOP
      END
```

15.

```
      NSUM = 0
      K = 1
    1 NSUM = NSUM + K**3
      IF (K .GT. 15) GO TO 140
      K = K + 1
      GO TO 1
  140 PRINT, K, NSUM
      STOP
      END
```

16.

```
      J = 0
      K = 1
    1 J = J + K
      IF (K .GT. 10) GO TO 2
      K = K + 1
      GO TO 1
    2 J = J**2
      PRINT, K, J
      STOP
      END
```

In problems 17 and 18 find what is wrong with the program.

17.

```
      READ X, Y
   10 Z2 = X**2 * Y
      IF (X N.E. 3) GO TO 13
      X = X + 1, Y = Y + 1
      GO TO 10
   13 PRINT, Z1, Z2
      STOP
      END
```

18.

```
      N = 1
      Y = 2
   10 Z = Y**N
      IF (N * Y LT. 5) GO TO 3
      N = N + 7
      GO TO 10
      STOP
      END
```

19. Key punch and run this program. (Notice no data cards are needed.)

```
      ISM  =  0
      N = 1
      K = 1
    3 ISM = ISM + K**2
      PRINT, ISM
      IF (N .GE. 15) GO TO 2
      N = N + 1
      K = K + 2
      GO TO 3
    2 PRINT, ISM
      STOP
      END
```

(a) What would happen if you replaced ISM by SUM every place it occurs? (Do it and see. You will need to punch only four additional cards.)

(b) What would happen if you replaced the eighth instruction by one which has K = K + 1 on it?

(c) Why is the last number printed twice?

20. Answer the questions asked in the paragraph preceding Example 2.

2.12 A LAST EXAMPLE AND DO-LOOPS

Suppose you are the programmer for the Slick Wick Company, which makes kerosene lantern wicks. Your employer wants to be able to input an employee's hourly rate of pay, the number of hours that the employee worked in the week, and the number of dependents the employee has and to output this information together with the employee's gross pay, withholding tax, deductions for charity (if any), and net pay.

Naturally time and one-half is paid for overtime (hours in excess of 40). The withholding tax is figured in a rather complicated way. The gross earnings for the week are multiplied by 52 and 775 times the number of dependents is subtracted. This figure is then compared to 10,500. If it is not less than 10,500, then the tax is found by taking 20% of the gross earnings. If it is less than 10,500 then the rate of withholding is only 17%. For the sake of simplicity, we will say that each employee contributes 1% of his or her pay after taxes to some local charity.

We describe a flow chart for the program assuming there are five employees (Figure 2.34). H stands for hours worked, R for rate of pay per hour, D is the number of dependents, YGROS is the amount to be compared to 10,500, TAX is the amount of taxes to be paid, DED is the deductions and PAY is the net pay (that is, the amount of the check). The program might look something like this:

```
  K = 1
6 READ, H, R, D
  IF (H .GT. 40) GO TO 1
  G = H * R
  GO TO 2
1 G = 40 * R + 1.5 * (H - 40) * R
2 YGROS = 52 * G - 775 * D
  IF (YGROS .GE. 10500) GO TO 3
  TAX = .17 * G
  GO TO 4
3 TAX = .2 * G
4 DED = .01 * (G - TAX)
  PAY = G - TAX - DED
  PRINT, H, R, D, G, TAX, DED, PAY
  IF (K .GE. 5) GO TO 5
  K = K + 1
  GO TO 6
5 STOP
  END
```

We mention in passing two additional commands. (For more information you should consult the Appendix.) The first allows us to make comments which the computer prints when it prints our program, but which involve no other execution. It is simply passed over during the computation. By putting a C (for Comment) in column 1, the

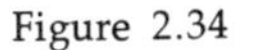
Figure 2.34

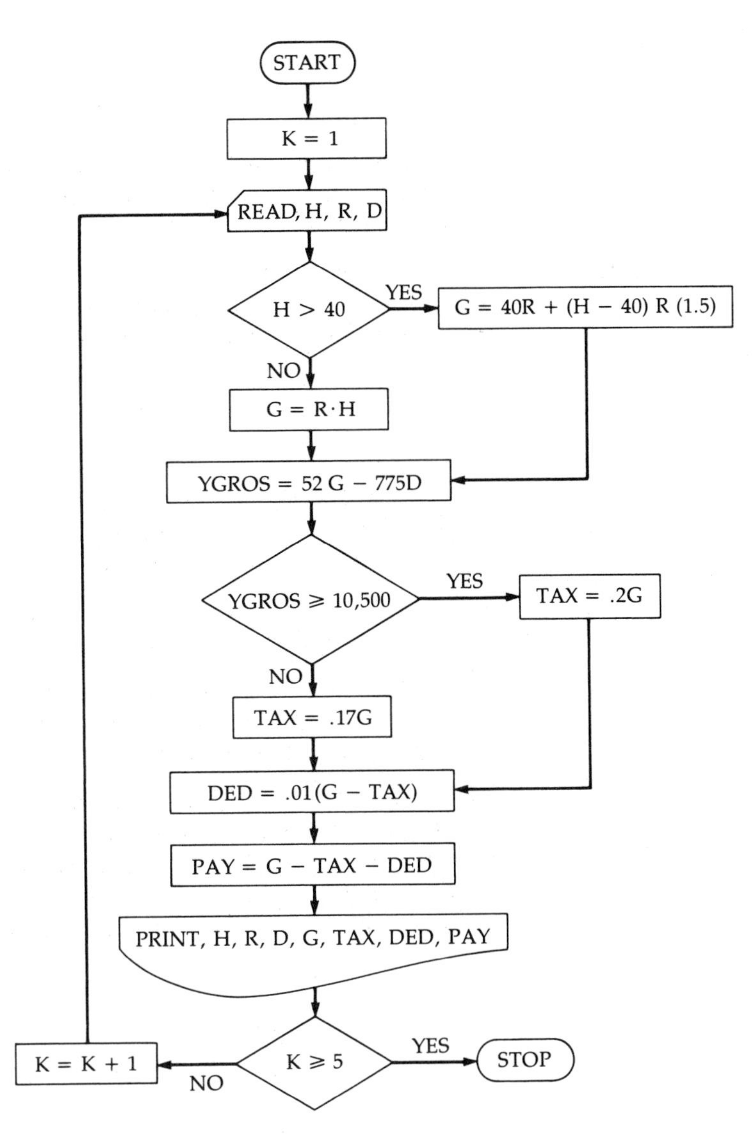

computer is alerted that there is no executable instruction. We can indicate as we go along what the program is doing and we show this in the program that follows.

Finally, we mention a device for looping that eliminates the need for some of the transfer instructions and more clearly identifies the loop. The command is described by

$$\text{DO } n \; i = m_1, m_2$$

where

n	is the number of the statement that ends the DO-loop,
i	is the index variable (integer),
m_1	is the initial value assigned to i,
m_2	is the largest value assigned to i,

and each time through the loop i is increased by 1. For simplicity, we will always use the

CONTINUE

command as the statement with number n. This command tells the computer to either go back through the loop again after increasing i by 1, or if $i = m_2$, to go on to the next instruction.

For example,

```
      .
      .
      .
      DO 70 K = 3, 11
      L = K * K
      PRINT, L
70    CONTINUE
      GO TO 120
      .
      .
      .
```

is a part of a program that computes the squares of the integers from 3 to 11 and prints them out sequentially. The first four instructions are done nine times (once for each value of K from 3 to 11) and the CONTINUE command sends the computer on to the next instruction, which is an unconditional transfer to another part of the program. We now show another formulation of the program for the Slick Wick Company.

```
C       THIS PROGRAM FINDS THE GROSS PAY, THE TAX, THE AMOUNT
C       OF DEDUCTIONS FOR CHARITY AND THE NET PAY WHEN
C       THE WEEKLY HOURS, HOURLY RATE, AND NUMBER OF DEDUCTIONS
C       ARE PROVIDED FOR FIVE EMPLOYEES.
C
C       FIRST WE PRINT HEADINGS FOR THE COLUMNS OF OUTPUT
        PRINT, 'HOURS', 'RATE', 'DEPEND', 'GROSS', 'TAX', 'CHARITY', 'NET'
        PRINT, '       ', '      ', ' ENTS ', '  PAY ', '     ', 'DEDUCT ', 'PAY'
C       ONLY ONE DO-LOOP IN THIS PROGRAM
C
        DO 5 K = 1, 5
        READ, H, R, D
        IF (H .GT. 40) GO TO 1
        G = H * R
        GO TO 2
      1 G = 40 * R + 1.5 * (H - 40) * R
C
C       NOW WE COMPUTE THE ANNUAL GROSS FOR COMPARISON
C
      2 YGROS = 52 * G - 775 * D
        IF (YGROS .GE. 10500) GO TO 3
C       NOW WE COMPUTE WITHHOLDING TAX
        TAX = .17 * G
        GO TO 4
      3 TAX = 2 * G
C       NEXT DEDUCTION FOR CHARITY IS FOUND
      4 DED = .01 * (G - TAX)
C       FINALLY THE NET PAY IS FOUND AND EVERYTHING PRINTED
        PAY = G - TAX - DED
        PRINT, H, R, D, G, TAX, DED, PAY
      5 CONTINUE
        STOP
        END
```

Notice that the portion of this program that corresponds to the program given before is two instructions shorter. Of course, data would have to be provided before we could run either program. In some of the problems that follow, several modifications are suggested. In actual practice, some of the information would be stored permanently (such as the rate of pay and number of deductions), but we simplify by assuming this is read in each week.

Exercises 2.12

Problems 1–4 refer to the example of the Slick Wick Company given in this section. You may modify either program or write your own.

1. Run the program given above with the data provided below. Assume your nine-digit social security number is *abc-de-fghi*.

WORKER	HOURS	RATE	NO. OF DEPENDENTS
1	39	4.75	f
2	45	$a.bc$	g
3	$3d$	5.10	h
4	42	5.50	i
5	50	5.63	2

2. Suppose that your employer wishes to be able to compute the pay for a variable number of employees, not just 5. That is, the number of employees varies so much from week to week that it is important that the program be able to handle any number of employees. What changes must be made? Write an appropriate program.

3. Suppose that withholding is 16% if YGROS $\leqslant$ 9500, 18% if 9500 < YGROS $\leqslant$ 11500 and 20% otherwise. Rewrite the program to provide for this method of computing TAX.

4. Rewrite the program assuming that if the weekly gross earnings are over $250, then only $\frac{3}{4}$% is taken for deductions to charity.

5. If your social security number is *abc-de-fghi*, let m be $+1$ or -1 according to whether c is even or odd, but let n be $+m$ or $-m$ as f is even or odd, respectively.

   Let

$$y = m(gh.i)^{n(a.b)} \quad \text{and} \quad x = (abc.de)^f y^3,$$

   where each dot is a decimal point and multiplication is denoted by putting one factor in parenthesis. Output your social security number, m, n, y and x.

6. Write a program for problem 9, Section 2.3. Make up your own numbers, but be sure N (the number of numbers in the list) is at least 15 and that there are at least 4 representatives in the list of each type of number.

7. Write a program that computes the sum of the cubes of the first N numbers, that is, computes $1^3 + 2^3 + \cdots + N^3$. (The input will be N.)

8. Write a program that computes the sum of the first N integers, and then squares the result, that is, computes $(1 + 2 + \cdots + N)^2$.

9. Write a program that inputs N and computes both the numbers computed in problems 7 and 8 above. Output N, $(1^3 + \cdots + N^3)$, and $(1 + \cdots + N)^2$. Run the program for various values of N. What mathematical statement do you believe to be true as a result of your computations?

10. Let $a_1 = 1$ and a_n be the sum of the first n integers (so $a_3 = 6, a_5 = 15$, and so on). Compare $a_1 + a_2 + \cdots + a_n$ and $n(n + 1)(n + 2)$. Can you deduce an equation?

11. Find the quotients when $7^{2n} + 16n + 63$ is divided by 64 and 128 for $n = 0$ to 70. What conjecture would you make?

12. Compare $1^2 + 2^2 + \cdots + n^2$ and $(n + 1)(2n + 1) / 6$. What equation can you deduce? Is your equation true for $n = 25$? 47?

13. Redo some of your previous programs by using a DO-loop rather than the logical IF.

14. A famous theorem relates the size of the arithmetic mean and the geometric mean of a sequence of numbers. If there are n numbers, the arithmetic mean is their sum divided by n, while the geometric mean is the nth root of their product. Write a program that will input n and the n numbers, compute the two means, print them out, and state which is the larger. Run your program for different values of n and the numbers. Can you guess the theorem?

15. Professor Ness decides he will bicycle through Europe next summer and wants a table of conversion from kilometers to miles. He finds that 1 mile is 1.6093 kilometers, and decides he wants a conversion for each mile from 1 to 10. Write a program for him that will output the desired table with headings.

16. A student decides to accompany the professor of problem 15, but wants the table to also include the conversion into miles of all distances in kilometers that are multiples of 5 up to 100. Rewrite the program from problem 15 to include these values at the bottom of the table.

17. If roofing costs $22.97 per square (enough for 100 square feet of roof), write a program that determines the cost of material for roofing m houses (m is not more than 10) when the input is m and

the areas in square feet of the roofs of the m houses. Output a table with the number of the house, its area, and the cost for that house. Also output the total cost and the average cost for each house.

18. The Sky-High Roofing Company needs a program to help it estimate the cost of roofing various-size houses. Roofing is $23.78 a square. (A "square" is enough roofing to shingle 100 square feet of roof.) The company wants a program that will input the number of houses under consideration (never more than 20); for each house the number of rectangular roof areas, the number of ridges and the number of valleys; for each house the dimensions of the rectangular roof areas and the length of the ridges and valleys. Valleys cost $1.98 per linear foot and ridges cost $1.69 per linear foot to complete.

As output we want the number of the house, the total roof area, the number of feet of valleys, the number of feet of ridges, and the total cost for that house in tabular form. Finally, we also want the total cost for roofing, for the valleys and for the ridges, a grand total, and an average per house. There needs to be enough description of the input and the output so the secretaries can input data and the owner can read the output.

PROBABILITY

CHAPTER THREE

3.1 PROBABILITY

The word "probability" is often used in ordinary speech. A weather forecaster, for example, might say that there was a slight *probability* of snow tomorrow. In this section we will assign a precise mathematical meaning to the word.

Computing Probabilities

Figure 3.1 is the wheel of a carnival game known by various names, including "the big six." Around the perimeter of the stand containing

Figure 3.1

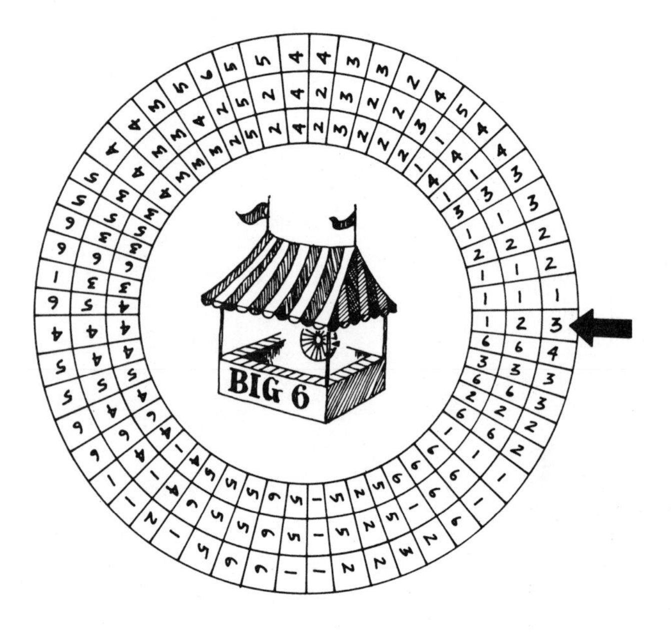

the big six wheel are countertops marked with the numbers from 1 to 6, something like Figure 3.2.

A player may bet on any number from 1 to 6 by placing his money on that number. Suppose he puts a dollar on the number 5. After everyone who wants to has bet, the wheel is spun. The outcome of the player's bet depends on the three numbers on the wheel that the arrow points to when it stops.

If none of these three numbers is a 5, then the player loses his dollar. For example, Figure 3.1 shows the wheel stopped with the arrow pointing to 1-2-3. A player betting on 5 would lose his bet after such a spin.

Figure 3.2

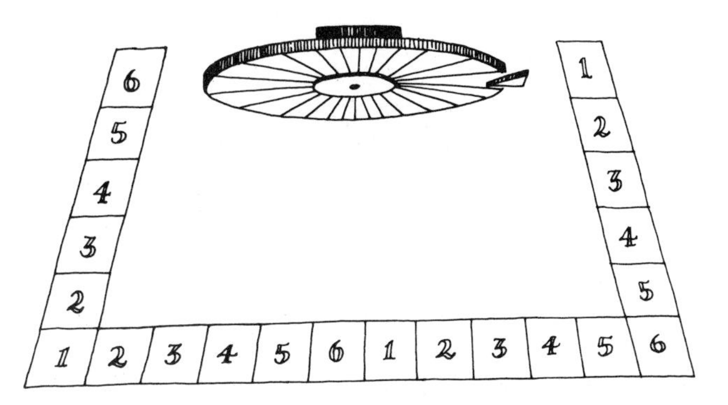

If exactly one of the three numbers is a 5, the player wins whatever he bet, in this case a dollar (and also keeps the dollar he put down).

If exactly two of the numbers are 5's, the player wins double his bet, in this case $2.

If all three numbers are 5's, the player wins triple his bet, $3 in our example.

We summarize the possible payoffs to the player in a table, representing a loss as a negative payoff.

RESULT	PAYOFF
No 5's	−$1
One 5	$1
Two 5's	$2
Three 5's	$3

A player betting on a different number would of course win or lose according to how many of that number the arrow pointed to. The game seems fair enough, especially when the operator of the wheel explains that if, for example, 1-2-3 is the result of a spin, then a player betting a dollar would win a dollar if he had chosen one of these three numbers and lose a dollar if he had bet on 4, 5, or 6.

In order to make a closer analysis of the game, let us count the number of sections of the wheel leading to each possible result. We find that there are 54 triples that might stop by the arrow, of which

Seven contain one 5

Four contain two 5's

Four contain three 5's

Thus $54 - 7 - 4 - 4 = 39$ triples contain no 5's at all. If a number other than 5 is counted, exactly the same distribution of triples will be found.

It seems reasonable to assume that any of the 54 sections of the big six wheel is as likely as any other to end up by the arrow. Since four of the 54 sections contain three 5's, we would expect that over the long run three 5's would be the result of a spin about 4/54 of the time.

The number 4/54 is called the *probability* of the result 5-5-5. Likewise 39 of the 54 triples contain no 5's at all. A player betting on 5 would expect to lose 39 times out of 54, and his probability of losing is 39/54.

For the want of a better word let us call a spinning of the big six wheel an *experiment*, the stopping of a particular section of the wheel next to the arrow an *outcome*, and the occurrence of any one of a particular set of outcomes of interest to a player (for example, three 5's) an *event*. We have illustrated the following general principle:

If an experiment has n equally likely outcomes, and if in the case of exactly m of these a certain event is said to have taken place, then the probability of that event is m/n.

If the probability of an event is 0 it can never happen, while if it is 1 it is a sure thing. A probability of 1/2 is attached to an event, such as the flip of a coin coming up heads, that is equally likely to occur or not occur.

Since the big six player has the probability 39/54 of losing, and since $39/54 > 1/2$, the game may not seem as fair to the player as at first glance. The fairness of the big six wheel needs more examination, however, since a person betting $1 can only lose $1, while he may win as much as $3. The analysis of the big six will be completed in Section 3.2 with the introduction of the notion of *expectation*.

EXAMPLE 1

Coin Flipping A quarter and a penny are flipped. What is the probability that either both will be heads or both tails?

We picture the four equally likely outcomes in Figure 3.3.

The event that the coins agree occurs with exactly two of the four outcomes. Thus $n = 4$, $m = 2$, and the probability sought is $2/4 = 1/2$.

Note that the following analysis of the problem is incorrect. There are three possible outcomes: two heads, two tails, or one head and one tail. Since the desired event corresponds to two of these three outcomes, its probability is 2/3.

The fallacy in this argument is that the three outcomes listed are

Figure 3.3

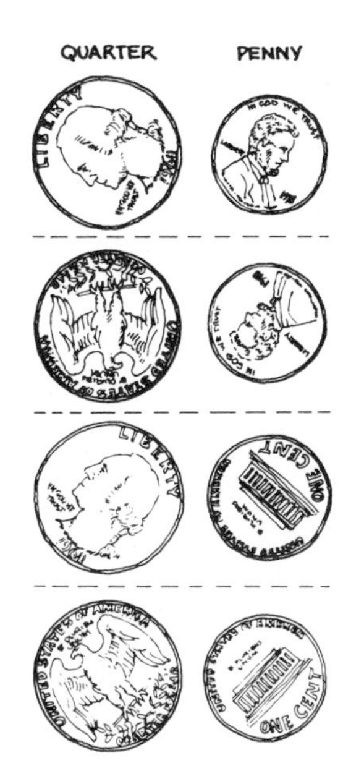

not equally likely. In order to have equally likely outcomes, heads for the quarter and tails for the penny must be counted separately from tails for the quarter and heads for the penny. Of course, the denominations of the coins are irrelevant to the problem, and the probability of a match would still be 1/2 even if both were pennies. Yet the fallacy is even easier to fall into in this case.

EXAMPLE 2

Cutting Cards An ordinary deck of cards is cut. What is the probability of drawing an ace?

Since there are 52 cards in the deck and each is equally likely to be drawn, and since 4 of the cards are aces, we have $n = 52$ and $m = 4$. The probability is $4/52 = 1/13$.

EXAMPLE 3

Drawing to a Flush A woman playing draw poker has been dealt the 4, 5, 10, and queen of hearts and the 7 of spades. She plans to discard the 7 of spades and draw one more card. What is the probability that she draws another heart, thus making a flush (five cards in the same suit)?

The woman cannot draw any of the five cards she was dealt. As far as she is concerned she is equally likely to draw any of the remaining $52 - 5 = 47$ cards. Thus $n = 47$. Since she has four hearts in her hand there are $13 - 4 = 9$ hearts among the cards she might draw. Thus $m = 9$, and her probability of getting a flush is 9/47.

A DECK OF CARDS
A standard deck consists of 52 cards divided into four suits: spades, hearts, diamonds, and clubs. There are 13 cards in each suit: ace (= 1), deuce (= 2), 3, 4, 5, 6, 7, 8, 9, 10, jack, queen, and king. Thus there are four 6's, one of each suit. Spades and clubs are black; hearts and diamonds are red.

Exercises 3.1

1. What is the probability that a big six player betting a dollar wins exactly $2?
2. What is the probability that a big six player wins more than she bet?
3. A big six player puts a dollar on 4 and a dollar on 5. What is the probability that she loses both dollars?
4. What is the probability that the player in problem 3 comes out exactly $3 ahead?
5. What is the probability that the player in problem 3 wins more than $3?
6. A big six player decides to beat the game by betting $1 on each of the six numbers. What is his probability of winning? of breaking even?
7. A penny, a dime, and a quarter are flipped. What is the probability that exactly two heads come up?
8. Three dimes are flipped. What is the probability the three coins will match?
9. A penny is flipped four times. What is the probability it comes up heads all four times?
10. A roulette wheel contains the numbers from 1 to 36, plus 0 and 00. In roulette the 0 and 00 are considered to be neither even nor odd. A player bets that the ball will stop on an even number. What is his probability of winning?
11. A card is dealt from an ordinary deck. What is the probability it is an ace or a king?
12. What is the probability that the card in problem 11 is either an ace or a heart? (Our use of the word "or" is meant to include the possibility that the card is both an ace and a heart: the ace of hearts.)
13. A draw-poker player has been dealt the 6 of spades, 7 of clubs, 8 of clubs, 9 of hearts, and ace of diamonds. He will discard the ace of diamonds and draw one more card. What is the probability it is a 5 or a 10?
14. A man has six black socks, five blue socks, and four brown socks in a drawer. One dark morning he pulls out a black sock. If he cannot see what he is doing, what is the probability the second sock he pulls out will also be black?

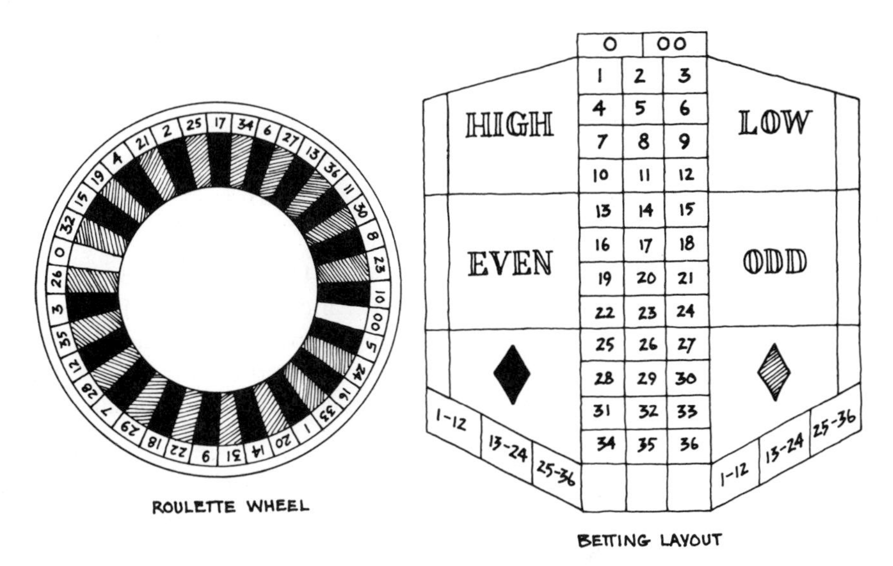

3.2 EXPECTATION

Expectation Playing the Big Six

We now complete our analysis of the big six game. Recall that the big six wheel is made up of 54 divisions, and that if N is any number from 1 to 6, then seven of the divisions contain N exactly once, four divisions contain two N's, four divisions contain three N's, and the remaining 39 divisions contain no N's at all.

Since each of the 54 divisions is equally likely to be the result of a spin of the wheel, one way to analyze the game is as follows. Let us suppose a player bets a dollar on the number 5 (as good as any other number) 54 times, and that each of the 54 divisions of the wheel comes up once. Table 3.1 summarizes the results.

TABLE 3.1

OUTCOME	PAYOFF	NUMBER OF TIMES	TOTAL PAYOFF
No 5's	−1	39	$39(-1) = -39$
One 5	1	7	$7(1) = 7$
Two 5's	2	4	$4(2) = 8$
Three 5's	3	4	$4(3) = 12$

The net result of the 54 plays is $-39 + 7 + 8 + 12 = -\$12$. All in all the player loses \$12. His *average* payoff per play over the 54 spins of the wheel is $-\$12/54 = -22¢$. This is the amount he can expect to win *on the average* each time he plays a dollar. Notice that this average payoff per play is

$$\frac{1}{54}\left\{39(-1) + 7(1) + 4(2) + 4(3)\right\} = \frac{39}{54}(-1) + \frac{7}{54}(1) + \frac{4}{54}(2) + \frac{4}{54}(3).$$

The numbers 39/54, 7/54, 4/54, and 4/54 are exactly the probabilities of the payoffs $-\$1$, \$1, \$2, and \$3, respectively; and the average payoff per play is just the sum of the numbers formed by multiplying each possible payoff by its respective probability.

We define the *expectation* of an experiment with possible payoffs $x_1, x_2, x_3, \ldots, x_n$ to be

$$p_1x_1 + p_2x_2 + \ldots + p_nx_n,$$

where p_1 is the probability of the payoff x_1, and so on.

EXAMPLE 1

The Raffle Ticket Consider, for example, a person who finds a raffle ticket. She learns that 100 tickets were distributed, and that there is a grand prize of \$50, a second prize of \$20, and five third prizes of \$2. Here the possible payoffs are \$50, \$20, \$2, and \$0 (if she doesn't win). Thus $n = 4$ and $x_1 = 50$, $x_2 = 20$, $x_3 = 2$, and $x_4 = 0$. Since there is only one first and one second prize, $p_1 = p_2 = 1/100$. There are five third prizes, so the probability of winning one of them is $5/100 = p_3$. Finally, the remaining 93 tickets will win nothing, so $p_4 = 93/100$.

We see that the expectation of the ticket holder is

$$\begin{aligned} p_1x_1 + p_2x_2 + p_3x_3 + p_4x_4 &= \frac{1}{100} \cdot 50 + \frac{1}{100} \cdot 20 + \frac{5}{100} \cdot 2 + \frac{93}{100} \cdot 0 \\ &= \frac{80}{100} = \$.80, \quad \text{or} \quad 80¢. \end{aligned}$$

If the tickets sold for \$1.50 each, then the payoffs for a person contemplating *buying* one would be (allowing for the cost of the ticket)

$$\begin{aligned} x_1 &= 50 - 1.50 = 48.50, \\ x_2 &= 20 - 1.50 = 18.50, \\ x_3 &= 2 - 1.50 = .50, \\ x_4 &= 0 - 1.50 = -1.50. \end{aligned}$$

The expectation is then

$$\frac{1}{100}(48.5) + \frac{1}{100}(18.5) + \frac{5}{100}(.5) + \frac{93}{100}(-1.5) = \frac{-70}{100} = -70¢.$$

Notice that this is just \$.80 − \$1.50.

EXAMPLE 2

Matching Coins Sam and Sally play a game in which each flips a quarter. If the coins match, Sam takes both coins; Sally takes both coins if they do not match. Let us compute the expectation for Sam. See Figure 3.4.

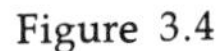

Figure 3.4

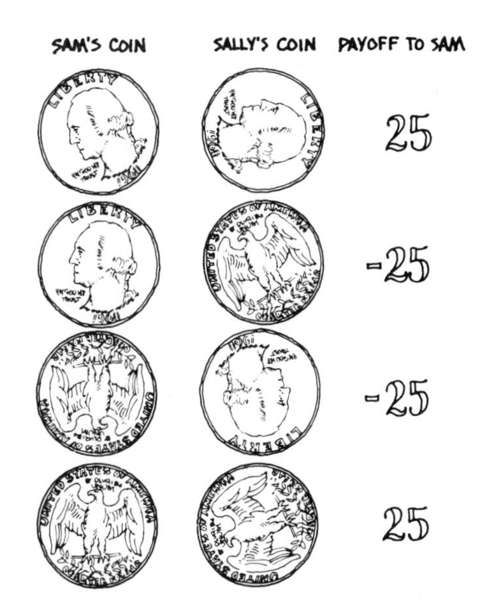

ALL'S FAIR
A game with expectation 0 is said to be *fair*. Neither side has an advantage.

Since each of the four outcomes is equally likely, Sam's expectation is $\frac{1}{4}(25) + \frac{1}{4}(-25) + \frac{1}{4}(-25) + \frac{1}{4}(25) = 0$.

EXAMPLE 3

Soft Ice Cream The concept of expectation need not be restricted to games. A hotdog-stand owner is considering renting a soft ice cream machine for the summer for \$3000. Not counting the leasing cost, he expects that having the machine will increase his profits by \$4000 if it is an average summer, by \$6000 if it is a hot summer, and by only \$500 if it is a cool summer. The weather bureau estimates the probability of an average summer to be .6, of a hot summer to be .15, and of a cool summer to be .25.

WEATHER	PROBABILITY	PAYOFF
Average	$p_1 = .6$	$x_1 = 4000 - 3000 = 1000$
Hot	$p_2 = .15$	$x_2 = 6000 - 3000 = 3000$
Cool	$p_3 = .25$	$x_3 = 500 - 3000 = -2500$

If the machine is leased, the hotdog-stand owner's expectation is

$$p_1x_1 + p_2x_2 + p_3x_3 = .6(1000) + .15(3000) + .25(-2500) = \$425.$$

Since this is positive, leasing the machine appears to be desirable.

EXAMPLE 4

The Tomato Farmer Sometimes a reasonable course of action can be chosen by comparing the expectations of different options. It costs a tomato farmer $500 to plant his field. If he plants on May 1, there is a probability of .4 that a late frost will kill his seedlings and necessitate another sowing. If he waits until May 15, however, the probability of frost is only .05. Other growing costs amount to $2000 in either case. He expects to take in $6000 in sales of early tomatoes if he sows May 1 and the tomatoes survive, but only $5000 for tomatoes sown later. (Assume that any second sowing will be after all danger of killing frost.) If a frost kills tomatoes sown May 15, there will be no time to plant another crop.

First we make a table for a May 1 sowing.

PROBABILITIES SUM TO 1
The probabilities of all the possible outcomes of an experiment must always add up to 1, the probability of a sure thing.
Thus if the probability of frost after May 1 is .4, then the probability of no frost after May 1 is 1 − .4 = .6. Likewise the probability of no frost after May 15 is 1 − .05 = .95.

WEATHER	PROBABILITY	PAYOFF
Late frost	.4	$-500 - 500 - 2000 + 5000 = 2000$
No late frost	.6	$-500 - 2000 + 6000 = 3500$

In this case the farmer's expectation is .4(2000) + .6(3500) = $2900.

For a May 15 sowing the situation is as follows:

WEATHER	PROBABILITY	PAYOFF
Late frost	.05	-500
No late frost	.95	$-500 - 2000 + 5000 = 2500$

Here the expectation is $.05(-500) + .95(2500) = \$2350$. It appears that a May 1 sowing is best.

Exercises 3.2

1. A big six player puts \$1 on 4 and \$1 on 5. What is his expectation?
2. A big six player puts \$1 on each of the six numbers. What is his expectation?
3. Juan flips a quarter of his while Andrea flips two quarters belonging to her. If all three coins match, Juan gets them all, otherwise Andrea gets them all. What is Juan's expectation?
4. The game in problem 3 is played as described except that Juan uses a dime instead of a quarter. What is Juan's expectation now?
5. The game of problem 3 is played with Juan providing a dime and Andrea providing three quarters. If all four coins match, Juan gets them all; otherwise Andrea does. What is Juan's expectation?
6. A player plays \$1 on the roulette wheel described in problem 10 of Section 3.1 on even. If an even number comes up, she wins \$1; otherwise she loses \$1. What is her expectation?
7. The player in the last problem bets \$1 on the number 11. If the ball stops on 11, she wins \$35; otherwise she loses her dollar. What is her expectation?
8. A raffle ticket costs \$2, and 5000 tickets are sold. The prize is a car worth \$6000. What is the expectation of a person buying a ticket?
9. A raffle ticket costs 50¢, and 400 tickets are sold. A grand prize of \$100 and three second prizes of \$20 are to be given. What is the expectation of a person buying a ticket?
10. A company is trying to decide whether or not to introduce a transistor radio built into a beach ball. Costs of introducing the new product are estimated at \$100,000, but if it catches on, the net profit will be \$500,000. The probability that it catches on is estimated to be 1/4. What is the expectation associated with introducing the product?
11. Table 3.2 is a table of odds for a supermarket giveaway. Saying that the "odds against an event are N to 1" means that the probability of that event is $1/(N + 1)$. For example, the probability of winning \$1 in a single store visit is 1/173. What is the expectation of a person making 26 store visits?

TABLE 3.2

PRIZE VALUE	NO. OF PRIZES	ODDS FOR ONE STORE VISIT	ODDS FOR 13 STORE VISITS	ODDS FOR 26 STORE VISITS
$1000.00	50	254,100 to 1	19,546 to 1	9,773 to 1
100.00	450	28,233 to 1	2,172 to 1	1,086 to 1
20.00	1,000	12,705 to 1	977 to 1	489 to 1
5.00	2,500	5,082 to 1	391 to 1	195 to 1
2.00	7,500	1,694 to 1	130 to 1	65 to 1
1.00	74,000	172 to 1	13 to 1	$6\frac{1}{2}$ to 1
Total no. of prizes	85,500	149 to 1	12 to 1	6 to 1

12. Table 3.3 is the odds chart for another supermarket contest. See problem 11 for an explanation of how saying "the odds against winning $2 in one visit are 302 to 1" means the probability of winning $2 is 1/303. Compute the expectation of 13 visits.

TABLE 3.3

PRIZE	NUMBER OF WINNERS	ODDS 1 VISIT	ODDS 13 VISITS	ODDS 26 VISITS
$2000	12	529,167 to 1	40,705 to 1	20,353 to 1
1000	24	264,583 to 1	20,353 to 1	10,176 to 1
200	77	82,468 to 1	6,344 to 1	3,172 to 1
100	154	41,234 to 1	3,172 to 1	1,586 to 1
50	233	27,253 to 1	2,096 to 1	1,048 to 1
25	466	13,627 to 1	1,048 to 1	524 to 1
10	579	10,967 to 1	844 to 1	422 to 1
5	1,159	5,479 to 1	422 to 1	211 to 1
2	20,997	302 to 1	23 to 1	12 to 1
Total number of prizes	23,701	268 to 1	21 to 1	10 to 1

13. Tables 3.4 and 3.5 are the odds charts for two supermarket games being promoted at the same time in the same city as the game in problem 12. Here saying the odds of winning $1 with one ticket are "1 in 76" means the probability is 1/76. Compute the expectation of one ticket or one visit for each game. Which game gives the customer the highest expectation, A, B, or the game in problem 12?

TABLE 3.4 GAME A

GAME	NUMBER OF PRIZES	ODDS 1 TICKET	ODDS 13 TICKETS	ODDS 26 TICKETS
$1,000	54	1 in 150,092	1 in 11,546	1 in 5,772
100	217	1 in 37,350	1 in 2,873	1 in 1,436
20	389	1 in 20,835	1 in 1,602	1 in 801
10	757	1 in 10,706	1 in 823	1 in 412
5	3,093	1 in 2,620	1 in 201	1 in 101
1	100,598	1 in 76	1 in 6	1 in 3
TOTALS	105,108	1 in 77	1 in 6	1 in 3

TABLE 3.5 GAME B

	WEEKLY WINNING ODDS				
RACE	SPONSOR PAYS WINNING GAME CARDS	1 STORE VISIT PER WEEK	2 STORE VISITS PER WEEK	TOTAL WINNING CARDS FOR 1 WEEK	TOTAL WINNING CARDS FOR 13 WEEKS
1st RACE	WIN $2.	one in 168	one in 84	3,000	39,000
2nd RACE	WIN $5.	one in 5,038	one in 2,519	100	1,300
3rd RACE	WIN $10.	one in 10,077	one in 5,039	50	650
4th RACE	WIN $100.	one in 100,769	one in 50,385	5	65
5th RACE	WIN $1,000.	one in 251,923	one in 125,962	2	26
DAILY DOUBLE	WIN $2,000.	one in 503,846	one in 251,923	1	13

3.3 THE MULTIPLICATION PRINCIPLE

Counting Outcomes of an Event

Another carnival and casino game very similar to the big six game is sometimes called chuck-a-luck. In this game three ordinary dice are enclosed in an hourglass-shaped cage. A player bets on a number from 1 to 6. The cage is inverted and the player is paid or loses his bet according to how many of the three dice come to rest with that number facing up. The payoffs are exactly as in the big six, a player losing his bet, or having his bet matched, doubled, or tripled according as the dice show his number not at all, once, twice, or three times.

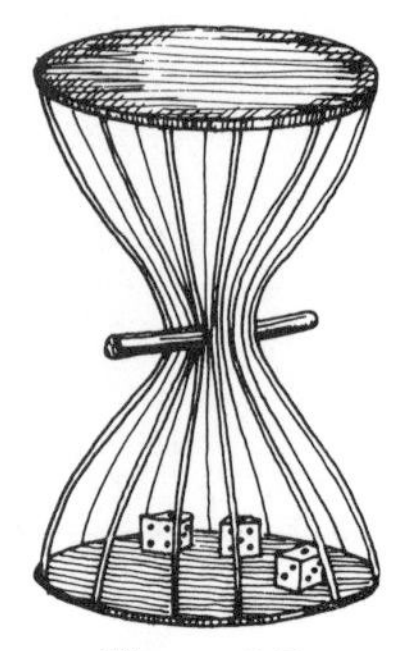
Figure 3.5

A gambler has been losing steadily at the big six wheel and decides to try his luck at chuck-a-luck. Let us analyze this game and see if a player's expectation for it is any better than the −22¢ per dollar bet at the big six wheel.

The situation is similar to that in the coin-flipping problems we have been considering. We need to count the number of ways the three dice can come up, leading to various payoffs. So as not to fall into the fallacy mentioned in Example 1 of Section 3.1, let us imagine that the three dice can be distinguished in some way. We could suppose that they were different colors, say red, white, and blue. We list the possibilities.

RED DIE	WHITE DIE	BLUE DIE
1	1	1
1	1	2
1	1	3
1	1	4
1	1	5
1	1	6
1	2	1
1	2	2
.	.	.
.	.	.
.	.	.

This could take all day! There must be an easier way. First let us count how many ways the three dice can come up in all, putting aside the question of how much is won or lost.

The red die can show any of the numbers from 1 to 6. The white die can also come up in six different ways. Thus the number of possible ways the red and white dice might come up is $6 \cdot 6 = 36$. For each of these 36 ways the blue die can also show six sides. Thus the total number of ways the three dice can come up is $36 \cdot 6 = 216$. This illustrates the *multiplication principle.*

The Multiplication Principle. If n separate experiments are performed, and if the first experiment can come out in k_1 different ways, the second experiment in k_2 different ways, and so on, then the n experiments can come out in a total of $k_1 \cdot k_2 \cdot \cdot \cdot k_n$ different ways.

We have seen that there are $6 \cdot 6 \cdot 6 = 216$ ways for the three dice to come up. Let us suppose our gambler bets $1 on number 5. First let us count how many ways he can lose his dollar because none of the dice shows a 5.

The red die can show a 1, 2, 3, 4, or 6—five possibilities. Likewise there are five ways for the white die and for the blue die to come up. By the multiplication principle the number of combinations of the three dice showing no 5's at all is

$$5 \cdot 5 \cdot 5 = 125.$$

The gambler's probability of losing is 125/216.

Now let us compute the probability of getting exactly one 5. This could be on either the red, white, or blue die. Using blanks to represent a number other than 5, we could have the events

RED	WHITE	BLUE
5	—	—
—	5	—
—	—	5

Since each of the blanks can be any of five numbers, the number of ways to get 5— —, for example, is $5 \cdot 5 = 25$. There are also 25 ways to get —5— and 25 ways to get — —5. There are $25 + 25 + 25 = 75$ ways in all to get exactly one 5, and the probability of this is 75/216.

We can get exactly two 5's as 55—, 5—5, or —55. In each of these cases the blank can be any of five numbers. Thus there are $5 + 5 + 5 = 15$ ways to get exactly two 5's, and the probability of this is 15/216.

Finally, there is only one way to get three 5's, and so the probability of this is 1/216.

NUMBER OF 5's	PROBABILITY	PAYOFF
0	125/216	−1
1	75/216	1
2	15/216	2
3	1/216	3

We see that the gambler's expectation is

$$\frac{125}{216}(-1) + \frac{75}{216}(1) + \frac{15}{216}(2) + \frac{1}{216}(3) = -8¢.$$

Although the gambler can expect to lose his money at a rate of approximately 8¢ per dollar bet, this is at least a slower way to go broke than the 22¢ loss per dollar bet of the big six wheel.

EXAMPLE 1

Design Your New Car An automobile company offers a particular model in any one of 21 body paint colors, eight interior color schemes, and three roof treatments. The company wishes to emphasize in its advertising the number of styling options available to a buyer. How many are there?

We can apply the multiplication principle. There are 21 choices of body paints, eight choices of interiors, and three choices of roofs, for

$$21 \cdot 8 \cdot 3 = 504$$

possible combinations.

EXAMPLE 2

The Shirt Store A shirt store carries shirts in five different colors, three different collar styles, and two different tapers. They find that a customer will have a definite preference in each of these three categories, but that if the shirt they display in their small window agrees with his preference in at least one category, then the customer will enter the store to look around. Assuming that the shirt they put in the window is chosen at random from all the possibilities, what is the probability it will lure a given customer into the store?

First we note that there are

$$5 \cdot 3 \cdot 2 = 30$$

possible combinations of color, collar, and fit. The easiest way to count how many of these agree with the customer's preference in at least one category is to count how many *do not,* that is, how many differ from the customer's choice in all three respects. Such a shirt can have any of four colors, either of two collars, and only one fit. There are $4 \cdot 2 \cdot 1 = 8$ possibilities for such a shirt. Thus there are $30 - 8 = 22$ possible shirts agreeing in at least one respect with the customer's preferences, and the probability he will enter the store is 22/30.

EXAMPLE 3

The Baseball Lineup A baseball manager has decided which nine players he wishes to play in the team's next game, but has not decided in what order they will bat. How many choices has he?

The multiplication principle may be used to solve this problem, but there is a wrinkle we have not seen before. The manager can choose any of nine players to bat first. Suppose he makes this choice. Then *only eight players remain who could bat second.* Likewise, after the first and second batters have been chosen, the choice for the third spot is limited to seven players. Continuing with this argument, we see that the manager has

$$9 \cdot 8 \cdot 7 \cdot 6 \cdot 5 \cdot 4 \cdot 3 \cdot 2 \cdot 1$$

possible lineups. This number may be calculated to be 362,880.

> *FACTORIALS*
> The product of the first n consecutive whole numbers arises in many counting problems and has been given a notation, namely $n!$.
>
> $$n! = 1 \cdot 2 \cdot 3 \cdot \cdot \cdot n$$
>
> In particular $1! = 1$, $2! = 1 \cdot 2 = 2$, $3! = 1 \cdot 2 \cdot 3 = 6$, and $9! = 362{,}880$. We read $n!$ as "n factorial."

EXAMPLE 4

The Absent-Minded Professor A professor wrote five letters of recommendation for a student, then sealed them in envelopes before she had addressed the envelopes. If she addresses the envelopes at random, what is the probability all will go to the correct addresses?

We need to count the number of ways the letters can have been put in the envelopes. Only one of these ways is the right way.

Any of five letters could have gone in the first envelope, then any of four in the second, and so on. The total number of ways is

$$5 \cdot 4 \cdot 3 \cdot 2 \cdot 1 = 5! = 120.$$

The probability of getting them all right is 1/120.

Exercises 3.3

1. If two dice are rolled, how many ways can they come up? What is the probability of no 3's coming up?
2. A simplified version of chuck-a-luck is played with two dice instead of three. A player betting a dollar on, say, the number 3, loses his dollar if no 3's come up, or wins $1 or $2 according as one or two 3's come up. What is the expectation of such a player?

3. Three dice are rolled. If a blank represents a number equal to neither 3 nor 6, what is the probability that the dice come up in each of the following ways, in the order given? (a) ---, (b) 3--, (c) 36-, (d) 33-, (e) 366, (f) 666.

4. Three dice are rolled. What is the probability they come up with: (a) no 3's or 6's, (b) exactly one 3 or 6, (c) exactly one 3 and exactly one 6, (d) exactly two 3's and no 6, (e) exactly two 6's and no 3, (f) two 3's and a 6 or two 6's and a 3, (g) three 3's or three 6's? (See problem 3.)

5. A chuck-a-luck player puts 50¢ on 3 and 50¢ on 6. What is his expectation? (See problem 4.)

6. A chuck-a-luck player puts a dollar on each of the six numbers. What is her expectation?

7. A shirt company makes a certain shirt in neck sizes from 14 to $17\frac{1}{2}$, and sleeve lengths from 32 to 40, in both cases with $\frac{1}{2}$-inch differences between successive sizes. How many different sizes does the company make?

8. An agricultural researcher wants to test the merits of seven different types of hybrid corn. She wants to test each type in northern and southern climates, with and without irrigation, and using herbicide A, herbicide B, both A and B, and no herbicide. How many test plots are needed?

9. Eric and Renee play a game in which Eric puts up a penny and Renee puts up seven quarters. All the coins are flipped. If all come up heads, Eric gets them all; otherwise Renee gets them all. What is Eric's expectation?

10. If eight coins are flipped, what is the probability that exactly seven of them match?

11. In a certain state each auto license number consists of two letters followed by four digits. How many different licenses are possible?

12. A police laboratory has been able to identify a suspect's sex, blood type (out of eight possibilities), and hair color (out of five possibilities). Assuming equal distribution in all these categories, what is the probability a person chosen at random would agree with the suspect in all three respects?

13. How many ways can seven jars of spices be arranged on a shallow shelf?

14. The members of a four-person mathematics department at a small college are asked by the administration to rank themselves annu-

ally in order of merit for the purpose of determining raises. So as not to have a fight, they agree among themselves to turn in a different ranking each year. How many years can they do this without repeating?

15. A man is called by a rock radio station and told he can win \$100 if he can list in the order of their sales the current top five records, the titles of which are given to him. Never having heard of any of them, he orders them at random. What is his expectation?

16. A disc jockey must submit a list of three records, in order, to be played during a 15-minute show. If the three must all be different, and if they must be chosen from the current top 40 records, how many different lists are possible?

3.4 A MULTIPLICATION RULE FOR PROBABILITIES

The Probability of Independent Events

Uncle Herman likes to tease his nephews and nieces with a simple game. He hides a piece of candy, either a jelly bean, gum drop, or chocolate kiss, in one of his hands. Only a child who can guess both what kind of candy Uncle Herman has and which hand it is in gets to eat the candy. Since there are three types of candy and two hands the candy could be in, there are

$$3 \cdot 2 = 6$$

possibilities to guess from. Thus a child's probability of success is 1/6.

Notice that the probability of guessing the candy is 1/3, that the probability of guessing the hand is 1/2, and that (1/3) (1/2) = 1/6. This example illustrates a general principle.

The probability that all of a set of independent events happen is the product of the probabilities of the individual events.

The word "independent" means that the events in question have nothing to do with each other, that whether one event takes place or not has no effect on the probability that another does. Thus in our example we are assuming that the type of candy Uncle Herman has does not affect which hand he puts it in.

EXAMPLE 1

The Bleacher Bet During a hot afternoon at the baseball park, a fan in the bleachers makes the rash statement that his favorite player will get a hit both of his next two times at bat. Another spectator offers to bet him \$10 to \$1 that he won't. The player is a .300 hitter. What is the rash fan's expectation?

Since the player is a .300 hitter, the probability of his getting a hit in any given time at bat is .3. By the multiplication principle for probabilities, his probability of hitting his next two times up is

$$p_1 = (.3)(.3) = .09.$$

The probability he fails is $p_2 = 1 - .09 = .91$. Thus the rash fan's expectation is $.09(10) + .91(-1) = .90 - .91 = -.01$. Over the long run he will lose money making such bets, but at the slow pace of only 1¢ each, on the average.

The Complement of an Event. Let E denote some event. Then by the *complement* of E we mean the event that E does not occur. The complement of E is denoted by E'.

Since the sum of the probabilities of E and E' must be 1, always

$$(\text{probability of } E') = 1 - (\text{probability of } E)$$

and

$$(\text{probability of } E) = 1 - (\text{probability of } E').$$

Very often the probability of E' is easier to compute than that of E.

EXAMPLE 2

The Picky Eaters One of Daisy Duck's three nephews has appeared unexpectedly for supper. The only vegetable Huey likes is corn and the only dessert is apple pie. Dewey likes only peas and cheesecake, while Louie will eat only snap beans and gelatin. Daisy unfortunately has never been able to tell her nephews apart, but wonders what the probability is that the corn and cheesecake she has fixed will please her guest in both respects.

The probability Daisy has made a vegetable to please her nephew is, of course, 1/3, as is the probability he will welcome the cheesecake dessert. The probability she has both vegetable and dessert to her visitor's liking is *not* (1/3)(1/3) = 1/9, however, since these are not independent events. As a matter of fact, if her visitor likes the corn she has fixed then he must be Huey, and Huey likes apple pie for dessert. Thus with corn and cheesecake, Daisy's probability of pleasing her guest both ways is 0.

EXAMPLE 3

The TV Game Show On a certain daytime television game show the contestant with the most points at the end of the program gets to spin the wheel of fortune, which contains 10 numbers. Only if number 10 comes up does the contestant get to pick one of three curtains, behind which are a Cadillac Eldorado, a can of Easy-Off Oven Cleaner, and a bag of Hall's Mentholyptus cough drops. Of course, the prize is the item behind the curtain chosen. If there are five contestants and they are awarded points on the basis of pure luck, what is a given contestant's probability of winning the car?

Since the probability of earning the most points is 1/5, of the wheel's stopping at 10 is 1/10, and of picking the right curtain is 1/3, the probability of winning the car is

$$\frac{1}{5}\frac{1}{10}\frac{1}{3} = \frac{1}{150}.$$

EXAMPLE 4

The Space Capsule Because of the extreme conditions during takeoff and in space, the bulb illuminating the cabin of a space capsule is estimated to be capable of staying in working condition during the whole mission with probability only .95. Thus two spare bulbs are taken along. What is the probability that at least one bulb survives the whole trip in working order?

This question can be answered only if we assume that the events of each of the three bulb's going bad are independent of each other. Under this assumption, we can most easily compute the desired probability in a backhanded way, namely, by first computing that the probability that any one bulb will fail is $1 - .95 = .05$; so the probability that all three fail is

$$(.05)(.05)(.05) = .000125.$$

Thus the probability that not all three fail is

$$1 - .000125 = .000875.$$

Summary

In this section we have found that the probability that one event *and* another event happen is the product of their respective probabilities, provided the events are independent. In the next section we will treat the case when the word "and" is replaced by "or."

Exercises 3.4

1. On the average it rains just two days in July in El Gatos, California. What is the probability that the Fourth of July falls on a Sunday and it rains there that day?
2. The first three batters in the Chicago Cubs lineup are a .250 hitter, a .260 hitter, and a .300 hitter. What is the probability that all three get hits leading off the first inning, assuming these are independent events?
3. In the situation described in problem 2, what is the probability that none of the three batters gets a hit?
4. A roulette player betting on even can run $1 into $16 by winning four times in a row, letting his winnings ride each time. What is his probability of doing this? (See problem 10, Section 3.1.)
5. A roulette player playing even decides to bet a dollar and let her winnings ride until she either loses or has $16. What is her expectation? (See problem 4.)
6. A student takes a five-question multiple-choice test in which each question has five possible answers. What is the probability he gets all the answers correct by just guessing?
7. What is the probability the student in problem 6 gets all the answers wrong?
8. In an ordinary deck of cards, only the four kings and the heart and spade jacks have mustaches. A card is drawn. What is the probability (a) it is a king? (b) it shows a mustache? (c) it is a king and shows a mustache?
9. A card is drawn from an ordinary deck. What is the probability (a) it is a king? (b) it is black? (c) it is a king and black? (d) it is a king or black?
10. A quarter slot machine has three wheels. The probability that a cherry comes up on any one wheel is 1/20. If getting three cherries pays $10, what is the expectation of a person putting a quarter in the machine?

11. A person's chance of being exposed to a certain flu virus one winter is .6, and, if exposed, the person will catch the flu with probability .5 if not vaccinated in the fall. One person in 10 has been vaccinated and is immune. What is the probability a random person will get the flu?

12. No letter carrier likes to deliver to the Smiths' house since there is probability 1/3 that the Smiths' dog will bite him, a probability 1/2 that he will stumble on their broken-down steps, and a probability 1/4 that the Smith boy will squirt him with his water pistol. What is a letter carrier's probability of delivering the mail to the Smiths without an unpleasant incident?

13. It costs 25¢ to play a carnival dart game. A player gets to throw three darts and must break a balloon with each throw to win. The prize is $5. An observer notes that one throw in five breaks a balloon. What is the expectation of a player?

14. What is the probability that a .300 hitter in baseball gets four hits in four times at bat?

15. What is the probability that a .300 hitter in baseball gets no hits in four times at bat?

16. A nurseryman plants three seeds in each pot, even though he only wants one plant. He hopes at least one seed will germinate, and pinches off any extras. If the probability that a given seed germinates is .6, what is the probability he will get a plant?

17. The nurseryman in problem 16 is trying to grow a very difficult plant, seeds of which germinate only 20% of the time. How many seeds should he put in each pot to have at least half a chance of getting a plant?

18. A greenhouse grower finds that 30% of her chrysanthemum cuttings do not root. Of those that do, only 80% develop into plants that can be offered for sale. Of those plants she offers for sale, 70% actually sell. What is the probability a given cutting will turn into a plant that is sold? How many cuttings should she start with to sell 1000 plants?

19. The Central City electric plant has a probability of .05 of failing during any given night, but even if there is a failure the plant can switch into power from a large cooperative system with probability .8. The plant supplies power to St. Agnes Hospital, which has an emergency generator. This generator could supply the hospital in the case of an outage with probability .9. What is the probability that the hospital is without power on any given night?

20. Mrs. Mondello finds that if she brings home a new pair of tennis

shoes for her daughter Lisa the probability is .5 that they won't fit, .6 that Lisa won't like the color, and .8 that they won't be endorsed by the correct television star. What is the probability she brings home shoes that Lisa likes?

21. Mr. King decides that if he runs a red light the probability is only .3 that he will cause an accident and only .2 that a police officer will see him do it. What is the probability he gets away with it?

3.5 AN ADDITION RULE FOR PROBABILITIES

The Probability of One Event OR *Another*

One reason the big six game may appeal to a player is that, although a player betting \$1 can lose only \$1, he may win \$2 or \$3. Let us compute the probability that the latter happens. We saw in Section 3.1 that the big six wheel is made up of 54 triples, and that, given any particular number from 1 to 6, four of these triples contain the number twice and four of these triples contain the number three times. Thus the probability of winning more than one's bet is $(4 + 4)/54 = 8/54$.

Notice that this answer is

$$4/54 + 4/54 = (\text{probability of 2 matches}) + (\text{probability of 3 matches}).$$

We have illustrated a general principle, namely,

The probability that at least one of two or more mutually exclusive events happens is the sum of the individual probabilities of the events.

"Mutually exclusive" means that the events in question have the property that, if one of them happens, then none of the others can. A person betting \$1 on the big six can win \$2 or win \$3—*but cannot win both \$2 and \$3 on the same spin*. Thus winning \$2 and \$3 are mutually exclusive events, and the addition principle could be applied legitimately to the above problem. Later we will illustrate further the idea of being mutually exclusive.

EXAMPLE 1

Roulette Roulette is a casino game played with a horizontal wheel divided into 38 compartments marked with the numbers from 1 through 36 and also 0 and 00. (The 0 and 00 compartments assure the house its advantage. Some wheels have also a 000 compartment, but we

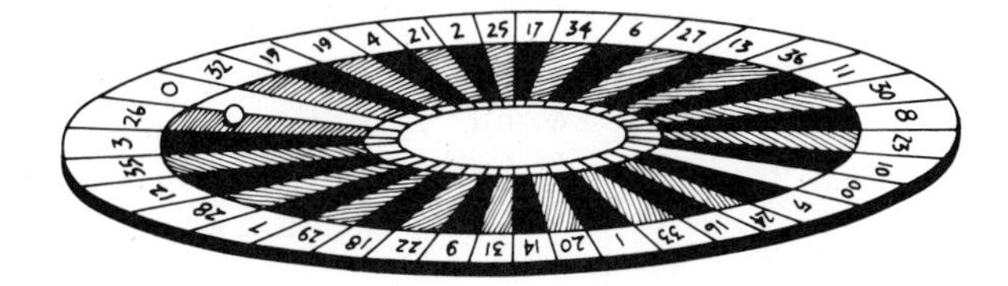

will consider a wheel without 000.) The wheel is spun; at the same time a small ball is rolled along the wheel in the opposite direction. Eventually the ball comes to rest in one of the compartments, which one determining the outcomes of the various bets. (See p. 179.)

If a player bets on a particular number, he is paid 35 times his bet if this number comes up, otherwise he loses his bet. Clearly his probability of winning is 1/38.

A player may also bet on odd or even, but the 0 and 00 are considered neither odd nor even, and all such bets lose if 0 or 00 come up. A person betting on even has his bet matched if an even number comes up, otherwise he loses his bet. Since 18 of the numbers from 1 to 36 are odd and 18 even, the probability of winning with such a bet is 18/38.

Other bets are possible at roulette, but we will not consider them.

A gambler who was born on June 25, 1936 decides in scientific fashion to put a dollar on each of the numbers 6, 25, and 36. What is his probability of winning?

We want the probability that

(6 comes up) or (25 comes up) or (36 comes up).

Since only one number can come up these are mutually exclusive events. Thus his probability of winning is 1/38 + 1/38 + 1/38 = 3/38.

Now suppose a player bets \$1 on the number 13 and \$1 on even. What is his probability of winning at least one of his bets?

We want the probability that

(13 comes up) or (an even number comes up).

Since 13 is not even these are mutually exclusive events. The probability is 1/38 + 18/38 = 19/38 that at least one of them occurs.

Finally, let us consider a player putting a dollar on both 13 and odd. What is his probability of winning at least one bet?

Although we are after the probability that

(13 comes up) or (odd comes up),

the answer is *not* 1/38 + 18/38 = 19/38, which is the sum of the probabilities of these two events. The addition principle does not apply because the events are not mutually exclusive. In fact, if the player wins his bet on 13, he will also win on odd.

To answer the question posed we can go back to a direct count of the ways the player can win at least one of his bets. This will happen exactly when one of the 18 odd numbers comes up; the probability of this is 18/38.

A Refined Addition Principle

We will give a rule applying to cases like that just considered, where the probability of one or the other of two events *that are not mutually exclusive* is desired. We will only consider the case of *two* events since the situation becomes much more complicated if there are more.

Refined Addition Principle. The probability that at least one of two events happens is the sum of their individual probabilities minus the probability that they both occur.

Since if two events are mutually exclusive the probability they both occur is 0, this reduces to the principle stated on page 196 in such a case.

Let us consider the person betting on both the number 13 and odd at roulette. His probability of winning both bets is exactly the probability the number 13 comes up: 1/38. Thus our refined principle says the probability he wins at least one bet is

$$\begin{aligned}(\text{probability of } 13) + (\text{probability of odd}) - (\text{probability of } 13) &= 1/38 + 18/38 - 1/38 \\ &= 18/38.\end{aligned}$$

EXAMPLE 2

Some Card Probabilities We will illustrate our two addition principles with some card questions. A card is drawn from an ordinary deck. What is the probability it is either black or a red queen?

Half of the cards are black, so the probability of drawing a black card is 1/2. Two of the 52 cards are red queens, and the probability of getting one of them is 2/52. Since no card is both black and a red queen, the answer to our question is 1/2 + 2/52 = 7/13.

Now we ask for the probability that a drawn card is either black or a 7.

Here the refined principle is needed. The probability the card is black is 1/2. The probability it is one of the four 7's is 4/52. The probability it is both black and a 7 is 2/52, since there are two black 7's. Our answer is $1/2 + 4/52 - 2/52 = 7/13$.

Compound Events

In order to save space and time we will abbreviate the probability of the event E by $P(E)$. Thus our calculation of the probability of drawing a black card or a seven could have been written

$$P(\text{black or } 7) = P(\text{black}) + P(7) - P(\text{black and } 7)$$
$$= \frac{1}{2} + \frac{4}{52} - \frac{2}{52} = \frac{7}{13}.$$

An event defined as a combination of other events is called a *compound event*. Drawing a queen from a deck of cards is a *simple event*, but drawing either a queen or a black card is a compound event, as is drawing both a queen and a black card. Different formulas relate the probability of the compound event to the probability of its component events depending on whether the words "and" or "or" are used in its definition.

Events using "and" were treated in Section 3.4. The rule developed there is the multiplication rule. In our new notation this becomes

$P(E \text{ and } F) = P(E)P(F)$, if E and F are independent events.

For example,

$$P(\text{queen and black}) = P(\text{queen})\,P(\text{black}) = \frac{4}{52} \cdot \frac{26}{52} = \frac{1}{26}.$$

Events using "or" were treated in this section. The addition rule applies. In our new notation this becomes

$P(E \text{ or } F) = P(E) + P(F)$, if E and F are mutually exclusive events.

In the case where E and F are not mutually exclusive, a more complicated rule was developed.

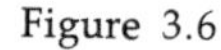

$$P(E \text{ or } F) = P(E) + P(F) - P(E \text{ and } F).$$

For example,

$$P(\text{queen or black}) = P(\text{queen}) + P(\text{black}) - P(\text{queen and black})$$
$$= \frac{4}{52} + \frac{26}{52} - \frac{1}{26} = \frac{7}{13}.$$

Clearly great attention must be paid to the small words "and" and "or" in order to compute the probabilities of compound events properly.

Given the problem of finding the probability of a compound event, it is useless just to pick a formula at random and then try to apply it. Only the correct formula will lead to the right answer. The "tree" shown in Figure 3.6 may be useful in choosing the proper formula. (It is called a tree because it branches.)

Figure 3.6

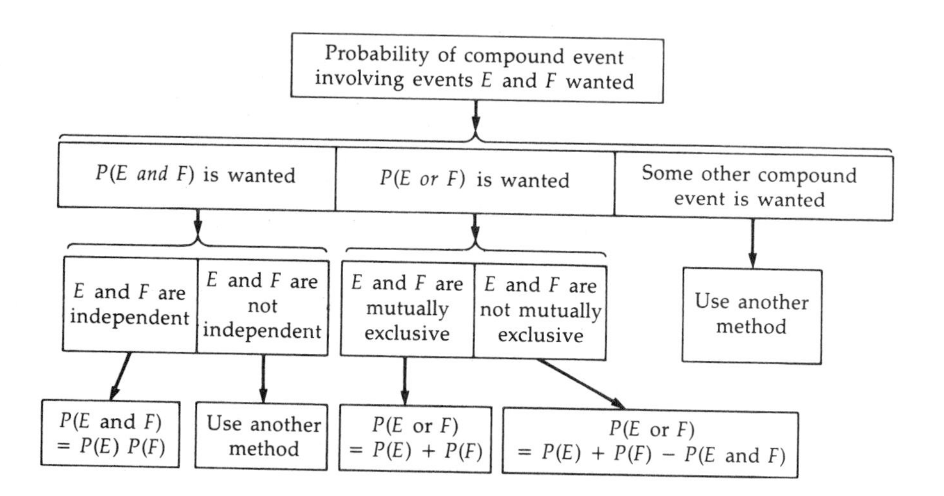

Examples Using the Various Formulas

In a certain country half the people are men and half women. One person in 20 is left-handed, and one in 100 is pregnant. We will assume that left-handedness is independent of sex and pregnancy. (That is, although it is conceivable that a man has either more or less chance of being left-handed than a woman, or that left-handed women are either more or less likely to be pregnant than women in general, we assume neither of these is the case.)

A person is chosen at random. First we ask what the probability is that the person is male and left-handed.

Since maleness and left-handedness are independent, the correct formula is

$$P(E \text{ and } F) = P(E)P(F).$$

Thus

$$P(\text{male and left-handed}) = P(\text{male})P(\text{left-handed}) = \frac{1}{2} \cdot \frac{1}{20} = \frac{1}{40}.$$

Next we ask what the probability is the person is male and pregnant. Clearly the answer is 0 and not

$$P(\text{male})P(\text{pregnant}) = \frac{1}{2} \cdot \frac{1}{100} = \frac{1}{200}.$$

The product formula does not apply since maleness and pregnancy are not independent.

Likewise, if we ask what the probability is that the person is female and pregnant, the answer is *not*

$$P(\text{female})P(\text{pregnant}) = \frac{1}{2} \cdot \frac{1}{100} = \frac{1}{200}.$$

Actually, if the person is pregnant she must be a woman, so

$$P(\text{female and pregnant}) = P(\text{pregnant}) = \frac{1}{100}.$$

Now we ask for the probability the person is male *or* pregnant. These are mutually exclusive events, so the formula

$$P(E \text{ or } F) = P(E) + P(F)$$

is valid. We have

$$P(\text{male or pregnant}) = P(\text{male}) + P(\text{pregnant}) = \frac{1}{2} + \frac{1}{100} = \frac{51}{100}.$$

Finally we ask for the probability that the person is female *or* left-handed. Since femaleness and left-handedness are not mutually exclusive, we must use the formula

$$P(E \text{ or } F) = P(E) + P(F) - P(E \text{ and } F).$$

We have

$$P(\text{female or left-handed}) = P(\text{female}) + P(\text{left-handed}) - P(\text{female and left-handed}).$$

Since femaleness and left-handedness are independent, we have

$$P(\text{female and left-handed}) = P(\text{female})P(\text{left-handed}).$$

Thus

$$P(\text{female or left-handed}) = \frac{1}{2} + \frac{1}{20} - \frac{1}{2} \cdot \frac{1}{20} = \frac{21}{40}.$$

Exercises 3.5

1. A person is chosen at random from the country just described. What is the probability the person is female and left-handed?
2. What is the probability the person of problem 1 is male or left-handed?
3. What is the probability the person of problem 1 is left-handed and pregnant?
4. What is the probability the person in problem 1 is left-handed and not pregnant?
5. What is the probability the person in problem 1 is pregnant but not left-handed?
6. What is the probability the person in problem 1 is not pregnant and a woman? [*Hint:* First compute the probability of the complement of this event.]
7. What is the probability the person in problem 1 is not pregnant and a man?
8. What is the probability the person in problem 1 is not pregnant or a man?
9. To win a certain lottery a person must guess a number correctly (out of 20 possibilities) and also a color correctly (out of 5 possibilities). What is the probability of at least one right guess?

10. Assume 4 out of any 10 people have blue eyes and 1 in 11 has a mustache. What is the probability a person chosen at random will have blue eyes or a mustache, assuming these are independent events?

11. A roulette player bets on 15, 17, 31, and even. What is her probability of winning at least one bet?

12. What probability of winning at least one bet does a player have betting on 1, 3, 5, 10, 18, and even at roulette?

13. We remind the reader that only the four kings and the jacks of hearts and spades have mustaches in an ordinary deck. What is the probability a drawn card (a) is a queen or shows a mustache? (b) is a jack or shows a mustache? (c) is a king or shows a mustache? (d) is red or shows a mustache?

14. In a certain town exactly half the adults are female and 1/3 of the adults belong to a church. Assuming these two events are independent, what is the probability an adult chosen at random is female or belongs to a church?

15. If 1/2 of the persons accused of drunk driving in a certain court are assessed a fine (and perhaps sent to jail), if 1/10 of those accused are sent to jail (and perhaps assessed a fine), and if 1/15 of those accused are both fined and jailed, what is the probability a person accused of drunk driving is either fined or jailed?

16. Of the adults in a certain town, 1/2 are men, 1/3 have drivers' licenses, and 1/8 are women drivers. What is the probability an adult chosen at random (a) is either a woman or has a driver's license? (b) is a man with no license? (c) is a man or has no license?

17. Of the adults in a certain town 1/10 own Chevrolets, 1/12 own Fords and 1/15 own Plymouths. If no one has more than one car, what is the probability an adult chosen at random owns either a Chevrolet, Ford, or Plymouth?

18. The small print in the advertising for a giveaway contest states that an entrant's probability of winning the grand prize is 1/10,000, of winning one of the 10 second prizes is 1/1000, and of winning one of the 50 third prizes is 1/200. If these are all the prizes and no one can win more than one of them, what is an entrant's probability of winning at all?

19. After 10,000 raffle tickets are sold, 10 are drawn to determine who wins 10 bicycles. Then these are replaced and 5 more are drawn to see who wins 5 motorcycles. It is possible to win both a bicycle and a motorcycle. What is a ticket holder's probability of winning something?

20. A man is given two chances of drawing a red card from a deck of cards. His first draw is replaced and the cards shuffled before his second draw. What is his probability of drawing at least one red card?

21. A woman draws twice from a deck of cards, with her first draw replaced and the cards shuffled before her second draw. What is the probability she draws at least one heart?

22. One winter it was estimated a person had probability .2 of catching the Hong Kong flu, .15 of catching the Asian flu, and .3 of catching the Russian flu. Assume catching one of these gives immunity to the others. What is the probability of catching one?

3.6 PERMUTATIONS AND COMBINATIONS

We have seen that in many cases the problem of computing the probability of an event depends upon counting the number of ways in which it can occur. In this section we will learn more sophisticated ways of counting.

EXAMPLE 1

Why a Flush Beats a Straight Let us consider a problem from poker, a game that is unusual in that it has no fixed set of rules. Not only are hundreds of different forms of poker, such as Five-Card Stud, Seven-Card Stud, Draw, and High-Low, played, but in many groups the dealer may call for any form he or she wants to, even one that has never been played before, at the start of any deal.

In spite of the differences there are certain near-invariants, the presence of which enables one to distinguish poker from pinochle, canasta, or tetherball. Poker is played with a deck of ordinary playing cards (sometimes with a joker or jokers added), and the winner of a particular hand is that player among all those who have matched all bets made who has the "best" five-card poker hand. Which of a group of five-card hands is "best" is determined (except in Low or High-Low games) by a conventional arrangement of types of hands, running from no-pair (worst) to a straight flush, or, if wild cards are involved, five-of-a-kind (best). This arrangement is determined by the principle that *of two types of hands, the less likely one is ranked higher.*

Two types of hands fairly high in the poker hierarchy are a *straight* (five cards in order, for example, 7 of hearts, 8 of hearts, 9 of spades, 10 of clubs, and jack of diamonds) and a *flush* (five cards in the same suit). Since a flush is considered better than a straight, we may presume that a person is more likely to be dealt a straight than a flush. We will confirm this by counting the number of possible hands of each of these two types.

The multiplication principle can be used to count the number of straights. Let us start by counting the number of straights with the highest card of a particular denomination, for example, all straights of the form 7-8-9-10-J. Since the 7 may be in any of the four suits, as may the 8, 9, 10, and jack, there are

$$4 \cdot 4 \cdot 4 \cdot 4 \cdot 4 = 4^5 = 1024$$

straights with a jack as high card. (It is true that some of these straights are actually straight-flushes, but since we will also include the straight-flushes when we count the flushes, our comparison will still be valid.)

Now, in poker the cards are considered to run in the order

$$2 - 3 - 4 - 5 - 6 - 7 - 8 - 9 - 10 - J - Q - K - A,$$

with the ace above the king. Thus any of the *nine* cards from 6 through ace may be the high card in a straight. (Many players also accept A-2-3-4-5 as a straight, but for the purposes of this count we will not.) We see there are a total of

$$9 \cdot 1024 = 9216$$

possible straights.

Counting the flushes will involve us in a new problem. Let us start by counting the flushes in a particular suit, say clubs. There are 13 clubs, and any five of them constitutes a club flush. It appears that the multiplication principle again applies. Imagine the five cards of a club flush laid out before us. The first card can be any of the 13 clubs. Having chosen it, there are 12 possibilities left for the second card, and so on. The total number of club flushes appears to be

$$13 \cdot 12 \cdot 11 \cdot 10 \cdot 9.$$

The error in the above count is that we have counted the same hand more than once. For example, the hand counted as the 5-7-9-10-K of clubs is counted again as the 7-5-9-10-K of clubs, and, in fact,

counted once for each rearrangement of these five cards. To correct our count let us determine how many such rearrangements there are.

We need to count how many ways we can arrange any particular group of five cards. This is really a problem we have considered before. Any of the five cards may be put in the first position. This leaves four possibilities for the second position, three for the third position, and so on. The total number of arrangements is

$$5 \cdot 4 \cdot 3 \cdot 2 \cdot 1 = 5! = 120.$$

Now, because our previous count of club flushes, which gave the result $13 \cdot 12 \cdot 11 \cdot 10 \cdot 9$, counted each such flush 120 times, the actual number of club flushes is

$$\frac{13 \cdot 12 \cdot 11 \cdot 10 \cdot 9}{120} = 1287.$$

Of course, the same count holds in any of the four suits. We see that there are in all

$$4 \cdot 1287 = 5148$$

flushes. Since this is fewer than the 9216 straights, it is reasonable that in poker a flush beats a straight.

Permutations

Our first count of the club flushes was a correct one if we wished to consider two hands as different when their cards are arranged in a different order. Such an arrangement is called a *permutation*. In a permutation the order of the objects is important. Suppose we have n objects and wish to count the number of arrangements of any r of them. We wish to count the number of lists of r objects we can make, using any of the original n objects in each list. There are n objects that we could list first. After choosing the first item there are $n - 1$ choices left for second place. Continuing in this way until we have filled all r places on the list, we see we have

$$\underbrace{n(n-1)(n-2)\cdots}_{r \text{ factors}}$$

choices.

To determine the last (or rth) factor in this product, notice that the first is $n = (n+1) - 1$, the second is $n - 1 = (n+1) - 2$, and so on. Thus the rth factor is $(n+1) - r = n - r + 1$, and the number of permutations is

$$n(n-1)(n-2)\cdots(n-r+1).$$

The number of permutations of n objects, taken r at a time, is

$$n(n-1)(n-2)\cdots(n-r+1).$$

For example, the number of possible club flushes, counted separately according to the order of the cards, is

$$\underbrace{13 \cdot 12 \cdot 11 \cdot 10 \cdot 9}_{5 \text{ factors}} = 154{,}440.$$

EXAMPLE 2

The Graduation Speaker Mr. Snavely is Superintendent of Schools in Tarrytown and speaks at each high school graduation there. He likes to include three jokes in each of his speeches, one at the beginning, one in the middle, and one at the end. Unfortunately he only knows 10 jokes, but he tries to mix them up and at least present them in a different order each time he speaks. How many speeches can he give before he has to tell the same three jokes in the same order again?

Here we want the number of permutations of $n = 10$ objects (the jokes) taken $r = 3$ at a time. This is

$$\underbrace{10 \cdot 9 \cdot 8}_{3 \text{ factors}} = 720.$$

EXAMPLE 3

The Rats A team of medical researchers is interested in whether, when a person takes more than one medicine, the *order* in which the medicines are taken makes any difference. They decide to experiment with rats, giving them two different medicines, chosen from a group of nine, in all possible orders, and then studying their reactions. How many rats will be needed to perform the experiment?

Here we have $n = 9$ medicines, from which we should choose $r = 2$ in all possible ways. The number of permutations of nine things taken two at a time is

$$\underbrace{9 \cdot 8}_{\text{2 factors}} = 72.$$

EXAMPLE 4

The Class Officers The senior class at Parkdale High School has 20 members. A class president, vice president, secretary, and treasurer need to be chosen. In how many ways is this possible?

Here we must count the number of permutations of $n = 20$ things taken $r = 4$ at a time. This is

$$\underbrace{20 \cdot 19 \cdot 18 \cdot 17}_{\text{4 factors}} = 116{,}280.$$

EXAMPLE 5

The Decorator An interior decorator must choose patterns for the walls, rug, and furniture for a room. He has seven patterns to choose from and doesn't want to repeat a pattern. He decides to make a sketch of each possible group of choices. How many sketches will he have to make?

Here we are after the number of permutations of $n = 7$ things taken $r = 3$ at a time. This is

$$\underbrace{7 \cdot 6 \cdot 5}_{\text{3 factors}} = 210.$$

Combinations

In some cases we wish to count sets of objects where the order of the objects within the set is unimportant to us. Suppose we have n objects and wish to count the number of ways we can choose r of these, with two sets of r objects considered the same if they differ only in the order in which the objects are listed.

To fix our ideas, let us start with an example with $n = 5$ and $r = 3$. From a basketball team of five players, three players are to be chosen to attend a sports banquet. How many ways can this be done?

Say the players are Al, Bo, Clu, Don, and Ed. We have already seen that if order matters there are

$$5 \cdot 4 \cdot 3$$

ways of choosing the three players. But order *doesn't* matter. As far as the players are concerned, choosing Al, Bo, and Clu is exactly the same as choosing Clu, Bo, and Al.

Any given set of three players can be arranged in $6 = 3!$ orders, taking any of the three first, then choosing one of the remaining two to be second.

Since the count $5 \cdot 4 \cdot 3$ counts each group of three players $3!$ times, the correct number of groups of three that can be chosen, ignoring order, is

$$\frac{5 \cdot 4 \cdot 3}{3!} = 10.$$

In general, r objects can be chosen from among n things, if order matters, in $n(n-1) \cdots (n-r+1)$ ways, as we have already seen.

Now any particular group of r objects will be counted in this number

$$\underbrace{r(r-1)(r-2) \cdots 2 \cdot 1}_{r \text{ factors}} = r!$$

times, since this is the number of ways of choosing r things, from among r things.

We see that the number

$$\frac{n(n-1) \cdots (n-r+1)}{r!}$$

counts the number of unordered subsets, or *combinations*, of r of the n objects.

This number can be expressed more compactly in the following way:

$$\frac{n(n-1) \cdots (n-r+1)}{r!} = \frac{\overbrace{n(n-1) \cdots (n-r+1)\,(n-r)(n-r-1) \cdots 2 \cdot 1}^{n!}}{r! \quad \underbrace{(n-r)(n-r-1) \cdots 2 \cdot 1}_{(n-r)!}}$$

$$= \frac{n!}{r!(n-r)!}.$$

The number of combinations of n objects, taken r at a time, is

$$\frac{n!}{(n-r)!r!}$$

For example, the number of club flushes (with the order of the five cards unimportant) is

$$\frac{13!}{(13-5)!5!} = 1287.$$

A CURIOUS NOTATION

The most common symbolism for the quantity $\frac{n!}{(n-r)!r!}$ is $\binom{n}{r}$. This reminds us of a fraction but is not, the horizontal bar being missing. For example,

$$\binom{5}{3} = \frac{5!}{(5-3)!3!} = 10.$$

A Gentle Reminder. We remind you that the distinction between a permutation and a combination is that the order of the objects in a permutation is considered important, while in a combination the order is ignored.

EXAMPLE 6

The Baseball Team A baseball manager has 20 players on his roster. We ask two questions: (a) How many batting orders are possible? (b) How many starting teams are possible?

Since a baseball team contains 9 players (ignoring the designated hitter rule) we are choosing $r = 9$ objects from a set of $n = 20$. In question (a) the order of the players is important, and we want the number of permutations of 20 players, taken 9 at a time. This is $20 \cdot 19 \cdot 18 \cdot 17 \cdot 16 \cdot 15 \cdot 14 \cdot 13 \cdot 12$.

In question (b) no particular order is to be specified, and we want the number of combinations of 20 players, taken 9 at a time. This is

$$\binom{20}{9} = \frac{20!}{(20-9)!9!}.$$

EXAMPLE 7

The Field Trip From the Parkdale High School senior class of 20 students mentioned in Example 4, a group of four students is to be chosen to make a field trip to the Ajax paper mill. How many ways can this be done?

In this case it makes no difference in what order the four students are chosen, and so we want the number of combinations of $n = 20$ objects (students) taken $r = 4$ at a time. This is

$$\binom{n}{r} = \binom{20}{4} = \frac{20!}{(20-4)!4!} = \frac{20!}{16!\,4!} = \frac{20 \cdot 19 \cdot 18 \cdot 17 \cdot 16!}{16!\,4!} = \frac{20 \cdot 19 \cdot 18 \cdot 17}{4 \cdot 3 \cdot 2 \cdot 1} = 4845.$$

EXAMPLE 8

Rats Again The researchers in Example 3 conclude that the order in which rats take two medicines is unimportant. They decide to investigate the way three separate medicines reinforce or interfere with each other when taken together. Each rat is to be given three medicines and its reactions noted. How many ways can this be done, given that there are nine different medicines?

Now order is no longer important, so we want the number of combinations of $n = 9$ things taken $r = 3$ at a time. This is

$$\binom{n}{r} = \binom{9}{3} = \frac{9!}{(9-3)!3!} = \frac{9!}{6!\,3!} = \frac{9 \cdot 8 \cdot 7 \cdot 6!}{6!\,3!} = \frac{9 \cdot 8 \cdot 7}{3 \cdot 2 \cdot 1} = 84.$$

A Calculation Note

Although if n is moderately large $n!$ can be huge ($20! >$ 2,000,000,000,000,000,000), yet with judicious cancellation the numbers involved in permutation and combination counts can often be handled even by those without access to calculators. For example,

$$\binom{20}{9} = \frac{20!}{(20-9)!9!} = \frac{20!}{11!\,9!}$$

$$= \frac{20 \cdot 19 \cdot 18 \cdot 17 \cdot 16 \cdot 15 \cdot 14 \cdot 13 \cdot 12 \cdot 11!}{11!\,9!}$$

$$= \frac{20 \cdot 19 \cdot 18 \cdot 17 \cdot 16 \cdot 15 \cdot 14 \cdot 13 \cdot 12}{9 \cdot 8 \cdot 7 \cdot 6 \cdot 5 \cdot 4 \cdot 3 \cdot 2}.$$

A little cancellation (for example, $18 = 9 \cdot 2$, $15 = 5 \cdot 3$) reduces this to

$$5 \cdot 19 \cdot 17 \cdot 2 \cdot 2 \cdot 13 \cdot 2 = 167{,}960.$$

Exercises 3.6

1. Compute $n!$ for $n = 4$ and $n = 6$.
2. Compute $n!$ for $n = 5$ and $n = 7$.
3. Compute $n(n - 1) \cdots (n - r + 1)$ for $n = 7$ and $r = 3$.
4. Compute $n(n - 1) \cdots (n - r + 1)$ for $n = 8$ and $r = 2$.
5. Compute $n!/(n - r)!$ for $n = 7$ and $r = 3$.
6. Compute $n!/(n - r)!$ for $n = 8$ and $r = 2$.
7. Compute $n!/(n - r)!r!$ for $n = 7$ and $r = 4$.
8. Compute $n!/(n - r)!r!$ for $n = 9$ and $r = 4$.
9. Compute $\binom{10}{7}$.
10. Compute $\binom{9}{3}$.
11. An artist with 10 paintings in his studio wishes to choose 3 for an exhibition. How many ways can he do this?
12. The garden club has 10 members and needs to elect a president, secretary, and treasurer. How many ways could they do this?
13. A *straight-flush* is a poker hand that is simultaneously a straight and a flush. How many are there?
14. If we count the A-2-3-4-5 as a straight, how many straights are there? How many straight-flushes?
15. The *World Almanac* says there are 5108 flushes and 10,200 straights. How can these figures be reconciled with the 5148 flushes and 9216 straights we counted in Example 1?
16. How many poker hands (of five cards) are there?
17. In the poker game of Five-Card Stud, a player is dealt a card face down, then four more cards face up. Since there is a round of betting after each up-card, the order in which the five cards are dealt matters. How many ways can a player be dealt a Five-Card Stud hand?

18. In Seven-Card Stud a player gets two cards face down, then four cards face up, then a final down-card. The order of the first two cards doesn't matter, but the order of the other cards does. How many ways can a player be dealt a Seven-Card Stud hand?

19. In one version of the casino game of Keno, 10 numbers are chosen at random from among the 40 numbers from 1 to 40. How many ways can this be done?

20. A Keno player (see the last problem) chooses 5 numbers from 1 to 40. If all 5 of his numbers are among the 10 chosen at random then he wins a prize. Given that he has chosen his five numbers, how many ways can he win? What is his probability of winning?

21. The mathematics department at a small college is told by the administration that it must provide 5 faculty members in caps and gowns to take part in graduation. If the department contains 11 members, how many ways can this be done?

22. A mail-order nursery offers a selection of six different miniature roses for $11.95. They grow 15 different varieties. How many selections are possible?

23. A stockbroker attending a three-day convention decides to wear a different tie each day of the convention. He owns 12 ties. How many ways can he do this if the order matters?

24. A fashion designer is to present five different dresses at a group show. She has nine new dresses to choose from. If the order makes a difference, how many ways can this be done?

25. A new nation needs to design its flag. It is decided to have three horizontal stripes of different colors. Eight colors are under consideration. How many possible flags are there?

26. A craftsman is constructing a totem pole containing five different animals. He can carve nine different animals. How many different poles are possible?

27. A credit card company stuffs various offers for merchandise in with its bills. They currently have eight different offers available and plan to stuff three in each envelope. How many possibilities are there?

28. A fast-food chain always has two of its food items at special reduced prices at any given time. They sell 22 different items. How many possibilities are there?

29. There are to be five cars of dignitaries in the Davisville Labor Day parade, and the parade organizer must put them in the proper order so as not to offend anyone. How many orders are possible?

30. Miss Gold is presenting a recital of her dancing class, featuring seven different dance numbers. She must decide what order to present them in. How many choices does she have?

31. The Lion's Club has promised to provide a starting judge, timer, and supervisor for the high school track meet from among its 30 members. How many ways can they do this?

32. In a matching test a student had to match a number with each of the letters A, B, C, D, E, F, and G. The numbers had to be chosen from among 1, 2, 3, 4, 5, 6, 7, 8, and 9, and no number could be used twice. How many possibilities were there?

33. When three people play the game Clue, each is dealt six cards from a deck of 21 cards. How many hands are possible?

34. In the game of Eyes-Ears-Nose-and-Throat each player is dealt six of the 12 different cards used in the game. How many hands are possible?

3.7 MORE PROBABILITY PROBLEMS

We have learned various techniques for computing probabilities, some depending on counting. In this section we will consider additional problems, the solutions of which will often require more than one of these techniques. Groups of exercises will be inserted where appropriate within this section, rather than saved for the end.

EXAMPLE 1

Craps In the game of craps two dice are rolled. The outcome of the game depends on the *total* of the numbers showing on the dice. We will not analyze the whole game of craps, but only certain probabilities connected with this total.

Since each die can come up six ways, there are $6 \cdot 6 = 36$ possible rolls. To compute the probability of rolling a 4 we notice that this total can occur just three ways: as 1-3, 2-2, and 3-1. Thus the probability of rolling a 4 is $3/36 = 1/12$. Likewise a 9 can occur as 3-6, 4-5, 5-4, and 6-3; so the probability of rolling a 9 is $4/36 = 1/9$.

Exercises 3.7

1. Find the probabilities of rolling a total of 2, 3, 5, 6, 7, 8, 10, 11, and 12 with two dice.

2. A craps player wins if he rolls a 7 or 11 on his first roll. What is the probability of this?
3. A craps player loses if he rolls a 2, 3, or 12 on his first roll. What is the probability of this?
4. Three dice are rolled. What is the probability that they total 4?
5. What is the probability that three dice total 7?
6. If three dice are rolled what number is most likely to be the total?
7. Two dice are rolled. What is the probability the total is even?

EXAMPLE 2

The Football Parlay In some cities it is possible to bet (illegally) on professional football games. Suppose the Steelers are playing the Bears and the Steelers are favored by 12 points. Then a person betting on the Steelers would be considered to have won that bet only if the Steelers beat the Bears by more than 12 points, while a person betting on the Bears would win if the Bears won the game or lost by fewer than 12 points. If the Steelers win by exactly 12 points, all bettors on the game lose.

In a "parlay" a bettor picks teams in a number of games and must be correct in each case to win anything. Let us suppose a bettor picks five games, and that he is no more or less skillful at predicting the results than the person who decides the point spread. Let us also suppose that in 10% of all games the outcome exactly matches the point spread and all bettors lose.

On a particular game, then, the probability is $1 - .1 = .9$ that a bettor *can* win by picking the right team. If he does this half the time his probability of winning on a particular game is $\frac{1}{2}(.9) = .45$.

A bettor trying to win a five-game parlay must be right five times in a row, and the probability of this is $(.45)(.45)(.45)(.45)(.45) = .018$.

If the point spreads are given with half-points, for example if the Steelers are favored by $12\frac{1}{2}$ points, then a bettor has more of a chance, since ties are impossible. If a bettor has probability 1/2 of picking any particular game under such a system, then his probability of winning a five-game parlay is $(1/2)^5 = 1/32 = .03125$.

Exercises 3.7

8. A mediocre parlay bettor has probability only .3 of picking any particular game, even though half-point spreads are used. What is his probability of winning a five-team parlay?

9. A player betting \$1 on a 10-game parlay wins \$600 if all 10 picks are correct. If the player has probability .5 of predicting each game, what is his expectation?

10. If a player has probability .5 of picking any particular game, what is the probability she correctly picks exactly four games in a five-game parlay?

11. A very well-informed bettor picks football games with probability .6 of being right on any particular game. What is her probability of picking five games correctly?

12. A parlay using half-point spreads pays \$50 on a six-team parlay, \$100 on a seven-team parlay, and \$200 on an eight-team parlay, in each case for a \$1 bet. Which bet gives the highest expectation to a player with probability .5 of picking any given game correctly?

13. Which of the parlays in problem 12 gives the highest expectation to a player having probability .4 of picking any given game?

14. Which of the parlays in problem 12 gives the highest expectation to a player having probability .6 of picking any given game?

EXAMPLE 3

Draw Poker A poker player has been dealt the 5, 7, and 10 of hearts, and the jack and king of clubs. He is considering discarding the two clubs in hopes of drawing two more hearts, thus making a flush. What is his probability of success?

There are $52 - 5 = 47$ cards he might draw, of which $13 - 3 = 10$ are hearts. Thus the probability the first card he draws is a heart is 10/47. If he does draw one heart, then there are only nine hearts left among the remaining 46 cards. Thus the probability of drawing a second heart is 9/46. We see the probability of drawing two hearts is

$$(10/47)(9/46) = .042.$$

Now consider a player holding three aces, a 7, and a jack. She throws away the 7 and jack and draws two new cards. What is the probability she draws the fourth ace?

The easiest way to solve this problem is to first calculate her probability of *not* drawing another ace. There are $52 - 5 = 47$ cards she might draw, only one of which is an ace. Thus the probability that her first draw is not an ace is 46/47. Having drawn one non-ace, the probability she draws another is 45/46 (since only 46 cards are left, 45 of them non-aces). Thus she has probability $(46/47)(45/46) = 45/47$ of not drawing an ace, and probability $1 - 45/47 = 2/47$ of drawing the fourth ace.

Exercises 3.7

15. A poker player with two spades and one card in each of the other suits decides to keep the two spades and discard the other three cards, drawing three new cards. What is his probability of making a flush?

16. A player holding the 5 of diamonds, 6 of hearts, 7 of spades, jack of clubs, and ace of clubs decides to discard the two clubs and draw two more cards. What is his probability of making a straight?

17. A player has two kings and three other cards (not kings). If he keeps the two kings and draws three more cards, what is his probability of drawing at least one more king?

18. A draw poker player has the 3 of hearts, 4 of spades, 5 of spades, 6 of clubs, and 9 of spades. He decides that either a flush or straight will win the hand. Which gives him a better chance, drawing two cards for a spade flush, or drawing one card for a straight?

19. A jar contains five red balls, six green balls, and eleven white balls. Two balls are drawn from it. What is the probability that both are white?

20. In a raffle 100 tickets are sold. There are three prizes. What is the probability that a person buying two tickets wins something?

EXAMPLE 4

A Two-Two Split in Bridge In the game of bridge each of the four players gets 13 cards. At one point in the play however, one of the players, the *declarer,* gets to see his partner's cards, but not those of the other two players.

If the declarer sees a total of nine hearts in his and his partner's hands, he knows his opponents must have the other four hearts. It may be important to the declarer whether these four hearts are distributed with exactly two in each of the opponent's hands. Let us compute the probability of this.

Call the opponents A and B. Clearly it suffices to compute the probability that A holds exactly two hearts. Since 26 cards are unknown to the declarer, and since A has some 13 of them, A can have any of

$$\binom{26}{13} = \frac{26!}{(26-13)!13!} = \frac{26!}{(13!)^2}$$

hands.

How many of these hands have two hearts? There are four hearts to choose from, and so

$$\binom{4}{2} = \frac{4!}{(2!)^2}$$

possibilities for the two hearts. Likewise the remaining 11 cards must be chosen from the 22 nonhearts outstanding, and this can be done in

$$\binom{22}{11} = \frac{22!}{(11!)^2}$$

ways. By the multiplication principle the number of hands A can have with two hearts is

$$\frac{4!}{(2!)^2} \cdot \frac{22!}{(11!)^2}.$$

We see that the probability A has a hand with two hearts is

$$\left(\frac{4!}{(2!)^2}\right)\left(\frac{22!}{(11!)^2}\right)\Big/\left(\frac{26!}{(13!)^2}\right).$$

By a raft of cancellations this horrible expression can be reduced to

$$\frac{18 \cdot 13}{23 \cdot 25} = .406.$$

THE YARBOROUGH
The second Earl of Yarborough, who died in 1897, had the honor of attaching his name to any bridge hand containing no card higher than a 9, that is, with no 10's, jacks, queens, kings, or aces.

Exercises 3.7

21. How many bridge hands are there? (Order is not important.)

22. What is the probability of being dealt a bridge hand of 13 cards, all in the same suit?

23. How many different bridge hands consist of only black cards?

24. The declarer can see 11 spades in his or his partner's hands. What is the probability each opponent has one spade?

25. The declarer and her partner have only seven clubs. What is the probability each opponent has three clubs?

26. What is the probability of being dealt a Yarborough?

EXAMPLE 5

Counting Other Poker Hands A *full house* in poker consists of three cards of one rank and two of another, for example, three kings and two 7's. How many full houses are there?

First let us count the full houses involving particular ranks, for example of the form K-K-K-7-7. There are four ways to choose the three kings (leave out any of the four suits) and $4!/(4-2)!2! = 6$ ways to choose the two 7's. Thus there are $4 \cdot 6 = 24$ full houses of any two specified ranks. There are $13!/(13-2)! = 13 \cdot 12$ ways to choose the two ranks. (Order matters, since K-K-K-7-7 is different from K-K-7-7-7.) We find $24 \cdot 13 \cdot 12 = 3744$ full houses in all.

Exercises 3.7

27. A *three-of-a-kind* is a poker hand of the form x-x-x-y-z, where the x's represent cards of the same rank but y and z are of ranks different from x and each other. How many three-of-a-kinds are there?

28. *Two-pairs* is the name given to a hand of the form x-x-y-y-z, where x, y, and z represent different ranks. How many two-pair hands are there?

29. How many *one-pair* hands (of the form x-x-y-z-w) are there?

30. How many *four-of-a-kind* (of the form x-x-x-x-y) hands are there?

31. A bridge hand (of 13 cards) is said to have a *void* if it contains no cards at all in some suit. How many bridge hands have a void in hearts?

32. How many bridge hands have at least two voids? (See problem 31.)

STATISTICS

CHAPTER FOUR

4.1 THE MEAN AND MEDIAN

Probability versus Statistics

The previous chapter was concerned with *probability*. Now we turn to some applications of elementary *statistics*. The distinction between these two studies is roughly as follows: Probability is used to predict what is going to happen, but statistics is used to explain what something means *after* it happens.

We used considerations of probability in Sections 3.1 and 3.2, for example, to predict that a person playing the big six wheel will usually lose, and in fact can expect to lose about 22¢ for each dollar bet. Next we give an example of a statistics problem.

Definition of the Mean

An agricultural chemical company wishes to test a compound that they hope will increase corn yields when applied to the soil before seeding. The company plants corn on a number of test plots of the same size, treating some of the plots with the chemical. The plots are handled identically in all other respects, and the number of bushels of corn from each plot is recorded.

The treated plots yield 320, 295, 313, 324, and 260 bushels, and the untreated plots yield 298, 270, 327, 304, 310, and 316 bushels.

What are we to make of this? Some of the treated plots produce more than some of the untreated, but some also produce less. The numbers need to be organized or summarized in some way in order to be meaningful.

Let us compute the average yield for the treated and untreated plots. Since there are five treated plots, we add together their yields and divide by 5 to find an average yield of

$$\frac{320 + 295 + 313 + 324 + 260}{5} = 302.4 \text{ bushels.}$$

Likewise, the average yield of the six untreated plots is

$$\frac{298 + 270 + 327 + 304 + 310 + 316}{6} = 304.2 \text{ bushels.}$$

Now we see that our tests indicate that the soil treatment may even decrease yields slightly.

In statistics the word *mean* is used for what is more commonly called the *average* in ordinary speech. The *mean* of the n numbers $x_1, x_2, \ldots, x_n$ is

$$\frac{x_1 + x_2 + \cdots + x_n}{n}.$$

Graphing Grouped Data

A set of numbers can also be made meaningful by converting it to a picture of some sort. Although there are various ways to do this, we will consider only one.

Suppose as part of a nutritional experiment a group of rats are weighed. Their weights, in ounces, are 3.16, 2.89, 3.22, 2.45, 2.66, 2.33, 2.61, 3.49, 3.62, 3.63, 3.88, 4.47, 2.88, 2.52, 3.16, 3.44, 3.69, 4.11, 3.78, 3.91, and 3.25. In order to make some sense out of these numbers, let us group them according to size. Suppose we count the rats with weights between 2 and 2.5 ounces, between 2.5 and 3 ounces, and so on. Letting x represent the weight of a rat, we get the following table:

RANGE OF x	NUMBER OF RATS
$2 \leq x < 2.5$	2
$2.5 \leq x < 3$	5
$3 \leq x < 3.5$	6
$3.5 \leq x < 4$	6
$4 \leq x < 4.5$	2

We will now make a *bar graph* by drawing a horizontal bar above each interval on the x-axis that we used in our grouping. The bar will be drawn at a height equal to the number of values of x in that interval. See Figure 4.1.

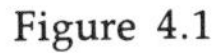
Figure 4.1

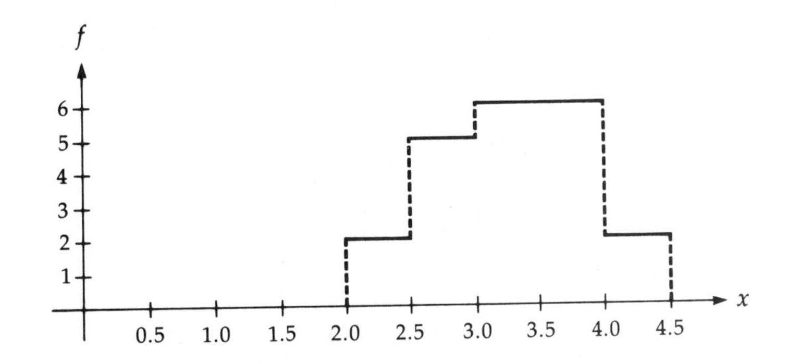

The number of values of x in each interval is called the *frequency* of that interval. Thus the frequency of the interval $2.5 \leqslant x < 3$ is 5.

Our use of intervals of length .5 was entirely arbitrary. Care must be taken, however, not to choose intervals that either are so short as to allow only one or two values of x in each or so long as to allow only a very small number of bars.

A glance at our bar graph tells us that the weights of the rats are concentrated in the neighborhood of 3.5 ounces.

The Mean of Grouped Data

If the number of values of x in a set of measurements is large, the calculation of the mean will entail a tedious addition. There is a shortcut that can be used with grouped data, although it produces only an approximation to the mean as defined earlier in this section.

Consider the weights of the rats mentioned above. We found two rats with weights between 2.0 and 2.5. If we simply assume the weights of these two rats to be at the midpoint of this interval, 2.25, then together they would weigh $2 \cdot 2.25$ ounces. Likewise, the five rats with weights between 2.5 and 3.0 should total about $5 \cdot 2.75$ ounces. Continuing in this fashion, we estimate the total weight of all the rats to be

$$2(2.25) + 5(2.75) + 6(3.25) + 6(3.75) + 2(4.25) = 68.75 \text{ ounces.}$$

Since there are 21 values of x we estimate the mean to be $68.75/21 = 3.27$ ounces.

If n measurements are grouped into equal intervals with midpoints $\bar{x}_1, \bar{x}_2, \ldots,$ and if f_1 is the number of measurements in the interval with midpoint $\bar{x}_1$, if f_2 is the number of measurements in the interval with midpoint $\bar{x}_2$, and so on, then the *mean of the grouped data* is

$$\frac{f_1\bar{x}_1 + f_2\bar{x}_2 + \cdots}{n}.$$

We note that $f_1 + f_2 + \cdots$ should be n, a fact that can be used to check our count. We also note that the formula $\bar{x} = (a + b)/2$ may be used to compute the midpoint $\bar{x}$ of the interval $a \leqslant x < b$. For example, the interval $2 \leqslant x < 2.5$ has the midpoint $(2 + 2.5)/2 = 2.25$.

EXAMPLE 1

The Test A biology test produces grades of 56, 92, 98, 43, 75, 76, 82, 61, 81, 55, 30, 36, 83, 91, 80, 75, 72, 60, 51, 89, 90, 41, 76, 15, 61, 78, and 87. Let us use grouped data to estimate the mean. Intervals of length 10 are very convenient and lead to Table 4.1.

TABLE 4.1

INTERVAL	MIDPOINT $\bar{x}$	f	$\bar{x}f$
$10 \leq x < 20$	15	1	15
$20 \leq x < 30$	25	0	0
$30 \leq x < 40$	35	2	70
$40 \leq x < 50$	45	2	90
$50 \leq x < 60$	55	3	165
$60 \leq x < 70$	65	3	195
$70 \leq x < 80$	75	6	450
$80 \leq x < 90$	85	6	510
$90 \leq x < 100$	95	4	380
Total		$27 = n$	1875

We calculate the grouped mean to be 1875/27 = 69.4. The true mean, found by adding up the 27 scores, is 1834/27 = 67.9.

EXAMPLE 2

More Rats Earlier we grouped the weights of 21 rats as follows:

INTERVAL	FREQUENCY f
$2 \leq x < 2.5$	2
$2.5 \leq x < 3$	5
$3 \leq x < 3.5$	6
$3.5 \leq x < 4$	6
$4 \leq x < 4.5$	2

We now add the midpoints $\bar{x}$ and values of $\bar{x}f$ to this table and sum the $\bar{x}$ column.

INTERVAL	MIDPOINT $\bar{x}$	f	$\bar{x}f$
$2 \leq x < 2.5$	2.25	2	4.50
$2.5 \leq x < 3$	2.75	5	13.75
$3 \leq x < 3.5$	3.25	6	19.50
$3.5 \leq x < 4$	3.75	6	22.50
$4 \leq x < 4.5$	4.25	2	8.50
Total		21	68.75

We find that the grouped mean is 68.75/21 = 3.27.

The Median

The *median* of a set of measurements is found by arranging them according to size and then choosing the middle one. For example, the 27 test scores of Example 1 may be ordered as 15, 30, 36, 41, 43, 51, 55, 56, 60, 61, 61, 72, 75, 75, 76, 76, 78, 80, 81, 82, 83, 87, 89, 90, 91, 92, 98. The middle number in this list is the fourteenth one (it is both preceded and followed by 13 numbers), namely 75. Thus 75 is the median score.

If we have an even number of measurements, there is a slight complication, since there are two numbers nearest the middle. In this case we define the median to be the average of these two numbers. Thus, to find the median of the six numbers 3, 1, 6, 8, 2, 5, we order them as 1, 2, 3, 5, 6, 8 and compute the median to be $(3 + 5)/2 = 4$.

It can be shown that if n numbers are arranged in order, then the middle number is in the position $(n + 1)/2$ if n is odd, while the middle two numbers are in the positions $n/2$ and $(n/2) + 1$ if n is even.

EXAMPLE 3

Pungo Bungo The island of Pungo Bungo has 101 inhabitants: the king, whose annual income is $1,000,000, and 100 subjects each having an annual income of $200. The mean annual income on the island is

$$\frac{1{,}000{,}000 + 100 \cdot 200}{101} = \$10{,}099.$$

A person reading this statistic might form the erroneous impression that the inhabitants of the island were an affluent bunch. Of course, the mean is thrown off by the king's income. This points up the advantage of the median, which is not affected by a few extreme measurements. If the incomes of the 101 island residents were listed in order, the middle income would be $200, and this is the median annual income.

On the other hand, a U.S. government official reading that the median annual income on Pungo Bungo is only $200 might decide American aid is justified, when actually a redistribution of the island's wealth might make more sense. In this case the mean would give a truer picture.

Exercises 4.1

1. The heights of the five starters for the Pottsville Junior College basketball team are 6′4″, 5′11″, 6′7″, 6′5″, and 6′2″. Compute the mean and median heights, in inches.
2. In six consecutive weeks an insurance salesman had sales of

\$50,000, \$80,000, \$25,000, \$0, \$60,000, and \$110,000 in life insurance. Compute his mean and median weekly sales.

3. In four different weeks a supermarket found that the numbers of items that a certain supplier provided that were damaged and unsalable were 15, 21, 16, and 12. Compute the weekly mean and median.

4. Seven groups of 10 seeds were tested for germination by a seed company with 7, 8, 6, 9, 9, 10, and 8 seeds germinating. Compute the mean and median for the groups of 10.

5. In a physics experiment the resistance of a certain wire is measured by each student in a laboratory, with readings of 1.51, 1.53, 1.51, 1.49, 1.52, 15.3, and 1.48 ohms. Compute the mean and median.

6. The average precipitation in inches in each of the 12 months in Albany, New York, is 2.5, 2.2, 2.7, 2.8, 3.5, 3.3, 3.5, 3.1, 3.6, 2.8, 2.7, 2.8. The corresponding figures for Des Moines, Iowa, are 1.3, 1.1, 2.1, 2.5, 4.1, 4.7, 3.1, 3.7, 2.9, 2.1, 1.8, 1.1. Compute the mean and median for each city. Which city gets more rain?

7. The areas in square miles of the 16 counties in Nevada are 4883; 7884; 723; 17,162; 3570; 4182; 9702; 5621; 10,649; 2010; 3765; 18,064; 6001; 262; 6375; and 8904. Group these according to $0 \leqslant x < 2000$, $2000 \leqslant x < 4000$, and so on. Compute the mean of the grouped data. Draw a bar graph.

8. Taking the density of the earth as 1, the other planets have densities $x =$ .68, .94, .60, .71, .24, .12, .28, and .26. Group these data (including earth) as $0 \leqslant x < .2$, $.2 \leqslant x < .4$, and so on. Compute both the mean of the grouped data and the true mean.

9. Scores on a test were 78, 61, 53, 81, 70, 61, 69, 68, 72, 72, 51, 91, 90, 66, 87, 95, 94, 78, and 86. Group the data by $0 \leqslant x < 10$, and so on. Compute the grouped mean and draw a bar graph.

10. The people at a party were surveyed as to how much money each was carrying, with the results (in dollars) 5.53, 26.00, 13.25, .35, 2.00, 45.67, 19.16, 2.01, 5.12, 35.65, 4.15, 9.50, 4.95, 6.75, 11.66, 14.14, and 1.81. Group the data as $0 \leqslant x < 10.00$, and so on. Compute the grouped mean, true mean, and median. Draw a bar graph.

11. The heights of the 16 highest mountains in Alaska are (in feet) 20,320, 18,008, 17,400, 16,523, 16,421, 16,237, 15,885, 15,700, 15,638, 15,300, 15,015, 14,831, 14,730, 14,573, 14,565, and 14,530. Compute the grouped mean, using intervals of 1000 feet.

12. The 1970 populations of the 12 cities in Nevada with more than 6000 people are 15,468; 6,501; 7,621; 16,395; 125,787; 36,216; 24,447; 72,863; 24,187; 10,886; 8,970; and 13,981. Group them in groups of 6000–8000, 8000–10,000, and so on and compute the grouped mean.

4.2 STANDARD DEVIATION

EXAMPLE 1

The Mean Is Not the Whole Story Professor Gleason has given two tests to his mathematics class. The first test covered linear programming and the second covered probability. The scores on the first test were 71, 78, 83, 95, 51, 55, 67, 77, 89, 96, 79, 74, 56, 70, 86, 83, 75, 81, 62, 64, 76, 68, 87, and 65; on the second test they were 75, 86, 53, 68, 89, 95, 51, 44, 81, 98, 50, 55, 85, 41, 83, 95, 62, 93, 52, 86, 66, 93, 48, and 88.

The mean on the first test was 1788/24 = 74.5; it was 1737/24 = 72.4 on the second. Evidently the two tests came out about the same. At least this is how it appeared until the professor grouped the scores x on the two tests and drew bar graphs. See Table 4.2 and Figures 4.2 and 4.3.

TABLE 4.2

RANGE OF x	FREQUENCY (FIRST TEST)	FREQUENCY (SECOND TEST)
$40 \leq x < 50$	0	3
$50 \leq x < 60$	3	5
$60 \leq x < 70$	5	3
$70 \leq x < 80$	8	1
$80 \leq x < 90$	6	7
$90 \leq x < 100$	2	5

The graphs tell quite different stories. The scores in the first test are neatly grouped about the mean, 74.5, while the scores on the second test are clustered in two places. Even though the mean was 74.2, there was only one score in the 70's. It appears that about half the class understood the material very well, while the other half understood it very poorly.

Average Distance from the Mean

The example in the last section shows that the mean of a group of numbers does not tell everything important about their distribution.

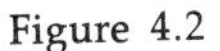

Figure 4.2

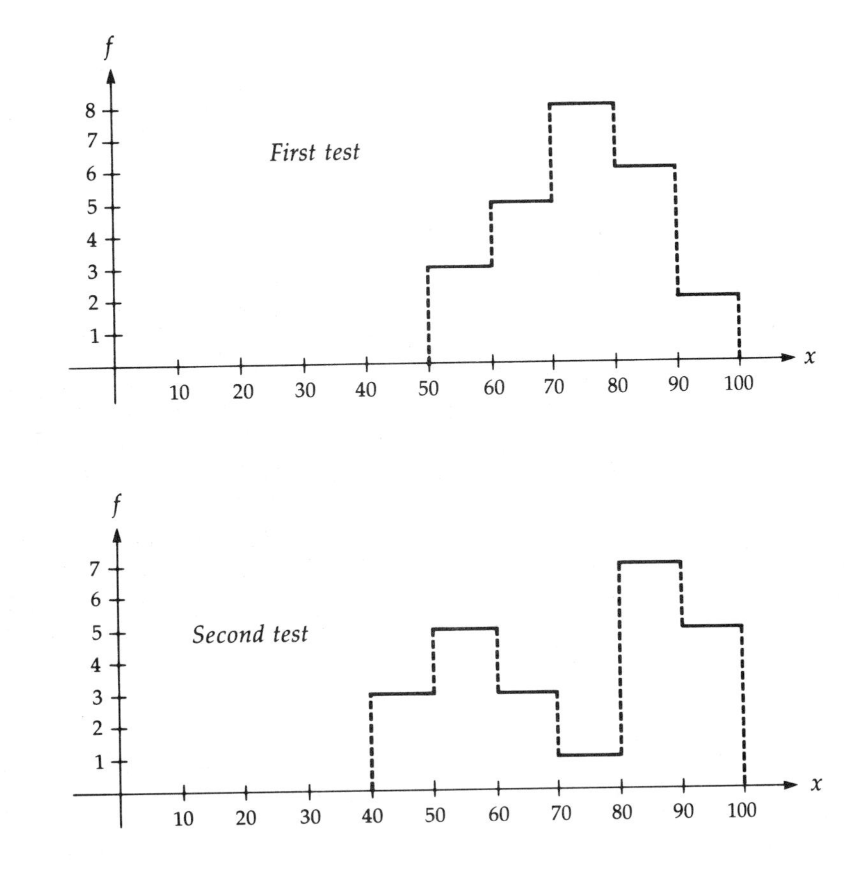

Figure 4.3

Sometimes it is also useful to know whether or not the numbers are clustered closely about their mean.

We will develop the definition of a quantity that will measure how closely a set of numbers is grouped about its mean. The definition is a natural one, but a little complicated, so we will first introduce a new notation in order to simplify our formula.

Let us suppose we have a group of numbers, any of which we might denote by x. These might be measurements of some kind or test scores, say. By Σx we will mean the sum of the various values of x.

If, for example, $x = 4.6, 4.7, 4.0$, and 4.5 are the weights in ounces of four candy bars, then

> *WHY Σ?*
> Σ is the capital Greek letter sigma, and should remind us of the *s* in "sum."

$$\Sigma x = 4.6 + 4.7 + 4.0 + 4.5 = 17.8$$

If we let m denote the mean of the values of x, then in our example

$$m = \frac{\Sigma x}{4} = \frac{17.8}{4} = 4.45,$$

and, in general, if x takes on n values, then

$$m = \frac{\Sigma x}{n}.$$

Now, we are interested in how the values of x are grouped around their mean m. More precisely, we want to know how far, on the average, x tends to be from m.

We might try simply averaging the values of $x - m$, but this runs into a snag, as we see by going back to our example.

x	m	$x - m$
4.6	4.45	.15
4.7	4.45	.25
4.0	4.45	−.45
4.5	4.45	.05

Note that $\Sigma(x - m) = .15 + .25 + (-.45) + .05 = 0$, and so the mean of the numbers $x - m$ is $0/4 = 0$ also. The problem is that $x - m$ only denotes the (positive) distance from x to m when x *exceeds* m; when x is less than m the distance is $m - x$.

The way we will get around this problem is to average the values of $(x - m)^2$ instead. This quantity has the advantage of giving the square of the distance from x to m no matter which side of m the number x is on.

In our example we have

x	$x - m$	$(x - m)^2$
4.6	.15	.0225
4.7	.25	.0625
4.0	−.45	.2025
4.5	.05	.0025
Total		.2900

We find the mean of the numbers $(x - m)^2$ to be

$$\frac{\Sigma (x - m)^2}{n} = \frac{.2900}{4} = .0725.$$

THE SQUARE ROOT TABLE

The square root table (Table I at the end of the book) may be used to compute directly the square roots of numbers between 1 and 1000. For example,

$$\sqrt{76} = 8.718 \text{ and } \sqrt{760} = 27.568$$

(using $n = 76$ and the $\sqrt{n}$ and $\sqrt{10n}$ columns).

We may have to approximate:

$$\sqrt{768} \text{ is about } \sqrt{770} = 27.749.$$

Any other number N can be written as a number j between 1 and 1000 times an even power of 10, say 10^{2k}. Then

$$\sqrt{N} = \sqrt{j} \cdot 10^k.$$

(This is because $\sqrt{N} = \sqrt{j \cdot 10^{2k}} = \sqrt{j}\sqrt{10^{2k}} = \sqrt{j}\sqrt{(10^k)^2} = \sqrt{j} \cdot 10^k$.)

For example,

$$\sqrt{18{,}216} = \sqrt{182 \cdot 10^2} = \sqrt{182} \cdot 10 = 13.416 \cdot 10 = 134$$

and

$$\sqrt{.56} = \sqrt{56 \cdot 10^{-2}} = \sqrt{56} \cdot 10^{-1} = 7.483 \cdot 10^{-1} = .7483.$$

Of course, access to a calculator with a square root key provides an even simpler way to find square roots.

Since the number .0725 is the mean of the *squares* of the distances from x to the mean m, we take the square root $\sqrt{.0725} = .269$ as our measure of the average distance from x to the mean m.

In general, if we have a set of n measurements, denoted by x, and if $m = \Sigma x/n$ is the mean of x, then the *standard deviation* of x, which is denoted by s, is defined to be the square root of the mean of the numbers $(x - m)^2$; that is,

$$s = \sqrt{\frac{\Sigma(x - m)^2}{n}}.$$

For example, the standard deviation of the four numbers 4.6, 4.7, 4.0, and 4.5 is .269.

EXAMPLE 2

Measuring a Brick Five first-graders are given the project of measuring the length of a brick in centimeters, and come up with the measurements $x = 17, 16, 17, 16$, and 15. We compute the mean $m = (\Sigma x)/n = 81/5 = 16.2$. The following table shows the computation of the numbers $(x - m)^2$.

x	$x - m$	$(x - m)^2$
17	.8	.64
16	− .2	.04
17	.8	.64
16	− .2	.04
15	−1.2	1.44
		$2.80 = \Sigma(x - m)^2$

We see the standard deviation $s = \sqrt{2.80/5} = \sqrt{.56} = .7483 = .75$.

A group of five high school seniors was also given the brick to measure. Their measurements were $x = 16.2, 16.3, 16.2, 16.4$, and 16.0. We compute $m = 81.1/5 = 16.22$ and construct a similar table.

x	$x - m$	$(x - m)^2$
16.2	−.02	.0004
16.3	.08	.0064
16.2	−.02	.0004
16.4	.18	.0324
16.0	−.22	.0484
		$.0880 = \Sigma(x - m)^2$

Here we find $s = \sqrt{.0880/5} = \sqrt{.0176} = .13$. The greater accuracy of the high school students is reflected in the smaller standard deviation: .13 versus .75.

Another Formula

The formula $s = \sqrt{\Sigma(x - m)^2/n}$ has the disadvantage that m must be computed separately before it can be used. The following formula can be proved to give the same result:

$$s = \frac{\sqrt{n\Sigma(x^2) - (\Sigma x)^2}}{n}.$$

This formula may appear a little forbidding to someone just learning the Σ notation, so we will explain it in some detail. The difference between $\Sigma(x^2)$ and $(\Sigma x)^2$ comes in the order in which the squaring and summing operations are performed. Of course, operations inside parentheses are always to be done first. Thus $\Sigma(x^2)$ is computed by squaring the various values of x, then summing these squares, while $(\Sigma x)^2$ is computed by first adding the values of x, then squaring the total.

The computation of $\sqrt{n\Sigma(x^2) - (\Sigma x)^2}/n$ could be organized as shown in Table 4.3. Notice in particular that in this formula the division by n comes *after* the square root is taken, whereas in our original formula for the standard deviation the division by n comes first. *It makes a difference.*

TABLE 4.3

OPERATION	RESULT
Square each value of x	x^2
Sum values of x^2	$\Sigma(x^2)$
Multiply $\Sigma(x^2)$ by n	$n\Sigma(x^2)$
Sum values of x	Σx
Square Σx	$(\Sigma x)^2$
Subtract $(\Sigma x)^2$ from $n\Sigma(x^2)$	$n\Sigma(x^2) - (\Sigma x)^2$
Take square root	$\sqrt{n\Sigma(x^2) - (\Sigma x)^2}$
Divide by n	$\frac{\sqrt{n\Sigma(x^2) - (\Sigma x)^2}}{n} = s$

We demonstrate the use of our new formula for s with the brick measurements of the first graders and high school students; see Table 4.4.

TABLE 4.4

FIRST GRADERS		HIGH SCHOOL STUDENTS	
x	x^2	x	x^2
17	289	16.2	262.44
16	256	16.3	265.69
17	289	16.2	262.44
16	256	16.4	268.96
15	225	16.0	256.00
$81 = \Sigma x$	$1315 = \Sigma(x^2)$	$81.1 = \Sigma x$	$1315.53 = \Sigma(x^2)$

First graders:

$$s = \frac{\sqrt{n\Sigma(x^2) - (\Sigma x)^2}}{n}$$

$$= \frac{\sqrt{5(1315) - (81)^2}}{5}$$

$$= \frac{\sqrt{6575 - 6561}}{5}$$

$$= \frac{\sqrt{14}}{5}$$

$$= \frac{3.742}{5}$$

$$= .7484$$

High school students:

$$s = \frac{\sqrt{n\Sigma(x^2) - (\Sigma x)^2}}{n}$$

$$= \frac{\sqrt{5(1315.53) - (81.1)^2}}{5}$$

$$= \frac{\sqrt{6577.65 - 6577.21}}{5}$$

$$= \frac{\sqrt{.44}}{5}$$

$$= \frac{.6633}{5}$$

$$= .1327$$

Exercises 4.2

1. A consumer testing organization played transistor radios until the batteries gave out. Batteries from company A lasted 9, 10, 6, 7, and 10 hours. Compute the mean m and standard deviation s of these numbers.

2. Batteries from company B lasted 8, 8.5, 9, 8, 8.1, and 8.2 hours. (See problem 1.) Compute m and s.

3. The laws of a certain state say that if packages are marked to contain a given weight, then a sample of five packages with mean m and standard deviation s must have the quantity $m - s$ not more than 1% less than the marked weight. Five "one-pound" cans of pumpernickel rye are found to weigh 1.01, .99, .99, 1.01, and 1.01 pounds. Does this sample comply with the law?

4. A postcard survey of new car buyers found they had had their previous cars 2, 3, 4, 3, 3, 5, 3, 2, 4, and 3 years. Compute the mean and standard deviation.

5. The temperature at four midnights in January was 10°, 4°, −5°, and 0°. Compute the mean and standard deviation.

6. Grade-school students measured the length of a one-inch block in centimeters, getting 2.54, 2.54, 2.52, 2.55, 2.55, 2.52, and 2.56. Find the mean and standard deviation.

7. The assessed valuations of the five houses on Maple Street are $32,000, $45,000, $39,000, $41,000, and $33,000. Find the mean and standard deviation.

8. The assessed valuations of the six houses on Rte. 22 just outside the city limits are $75,000, $20,000, $36,000, $49,000, $22,000, and $31,000. Find the mean and standard deviation.

9. The physics laboratory course was broken into two groups, and each group made a measurement with a potentiometer. Group A got readings of 62.3, 60.0, 61.5, and 60.8. Group B got readings of 63.1, 62.9, 62.8, and 63.5. Which group had the lower standard deviation?

10. The students in two laboratory groups got final grades of A, B, C, C, and C (group 1), and B, A, D, B, and C (group 2). Using A = 4, B = 3, and so on, compute the mean and standard deviation for each group.

11. Professor Esposito is teaching two sections of the same course. In the first test he gave, the grades in the first section were 92, 45, 88, 94, 61, 52, 76, 61, 88, and 63. The grades in the second section were 21, 76, 82, 45, 93, 79, 68, 66, 73, and 55. Make a bar graph for each of the two sections, grouping the scores as $0 \leq x < 10$, $10 \leq x < 20$, and so on. Estimate from the graphs which group has the higher standard deviation. Compute the standard deviation for each section.

12. The marriage rates (per 1000 people) in Ohio, Illinois, Michigan, Pennsylvania, and Indiana are 9.5, 10.5, 9.7, 8.2, and 11.4; for California, Oregon, Washington, Nevada, and Arizona they are 7.7, 8.8, 12.0, 180.3, and 12.6. Compute the mean, median, and standard deviation for the two groups of states.

4.3 THE STANDARD NORMAL DISTRIBUTION

Continuous Probability Distributions

In the previous chapter we considered problems of computing the probability of certain events. We might ask, for example, for the probability that if two dice are rolled their total is 6. If we let x be the total of the two dice, then x might be any of the whole numbers from 2 to 12. The set of probabilities that x is equal to 2, 3, . . . , 12 is an example of a *discrete probability distribution*. The word "discrete" indicates that x can only take on some one of a set of separate, individual values.

Now we consider a question that sounds very similar but is really quite different. An adult man is to be chosen at random. What is the probability that he is *exactly* 6 feet tall?

If the word "exactly" is taken literally, if 6.1-foot or even 6.0001-foot men are to be rejected, then the answer seems to be 0. The problem is that there are infinitely many possible heights, as opposed, say, to the 11 possible totals when two dice are rolled. If x represents the height of the man chosen at random, then x may range through a *continuous* (as opposed to discrete) set of values.

Still, we should be able to make some sense out of the question asked. Suppose the heights of a large number of men were measured. These measurements were then grouped and a bar graph drawn. It might look something like Figure 4.4. Here the heights are grouped into

Figure 4.4

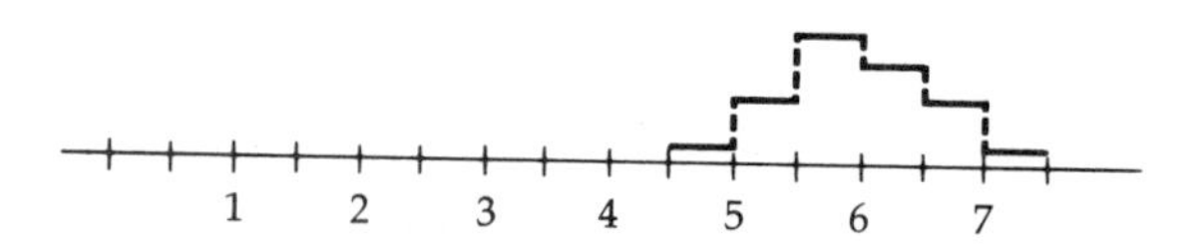

6-inch intervals. If one wanted to know the probability that the height of a man chosen at random was between 6 feet and 6.5 feet, a good estimate would be the proportion of the men measured falling in this interval, that is, the ratio of the height of the bar over this interval divided by the sum of the heights of all the bars.

Of course if the probability of choosing a man between 6 feet and 6.1 feet were wanted, a more detailed grouping would be necessary.

There is a way to construct a graph that would provide answers to all such questions, namely by using a *continuous probability curve*. For the example we are considering such a curve might look like Figure 4.5.

Notice that the graph, rather than being a steplike series of bars, is smooth. The interpretation of the curve is the following. The probability that the height x of a randomly selected man is between a and b is

Figure 4.5

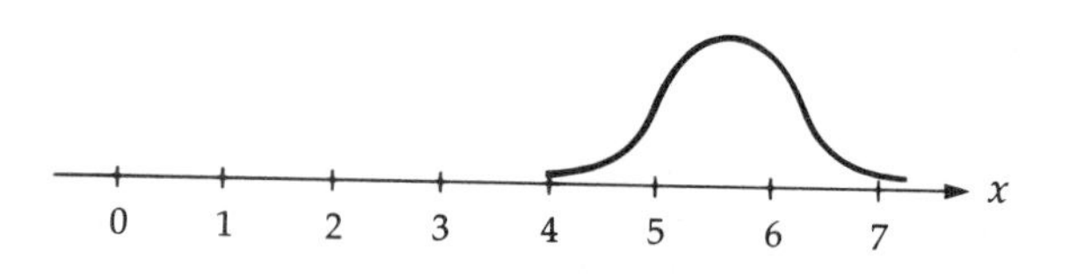

Figure 4.6

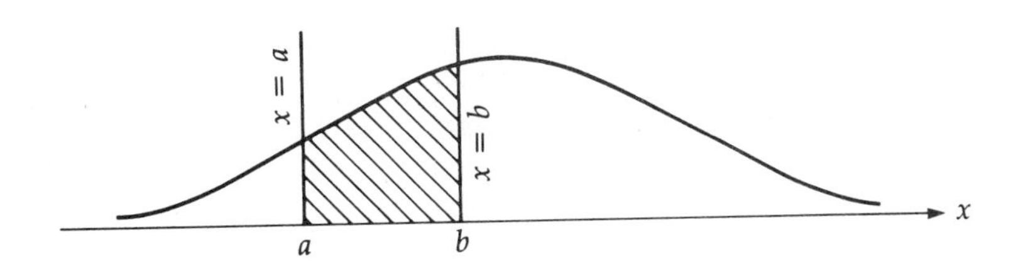

equal to the area under the curve, above the x-axis, and between the lines $x = a$ and $x = b$. See Figure 4.6. For a given distance between a and b, this area will be larger where the curve is higher and smaller where the curve is lower. Since everyone must have *some* height, the total area under the curve must be 1.

> If a measurement x can take on any of the infinite set of values in some interval, then x is said to have a *continuous probability distribution*. Such a distribution is pictured by a *continuous probability curve* having the property that the probability that x lies in any interval equals the area under the curve and above that interval. The total area under a continuous probability curve above the x-axis is 1.

EXAMPLE 1

A Fishy Example To further illustrate the idea of a continuous distribution, let us assume that the graph in Figure 4.7 gives the probability distribution for the lengths in centimeters of bass caught in Angle Lake. The numbers in boxes under the curve give the areas of the corresponding strips under the curve and above the x-axis of width 10.

Figure 4.7

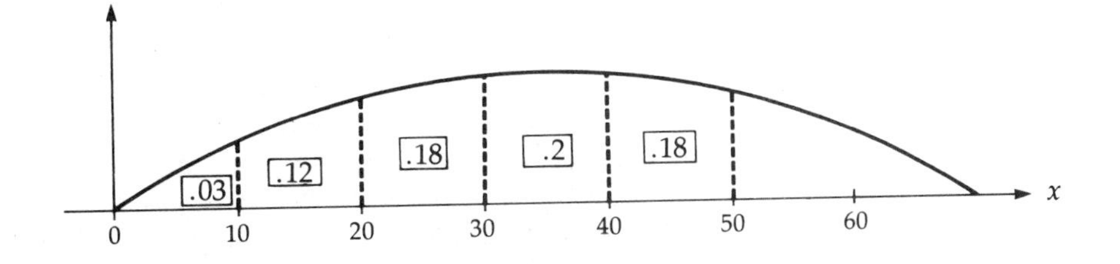

QUESTION If I catch a bass at Angle Lake, what is the probability that its length is between 20 and 30 centimeters?

ANSWER .18.

QUESTION What is the probability that a caught bass will have length between 20 and 40 centimeters?

ANSWER By adding the corresponding areas we get .18 + .2 = .38.

QUESTION What is the probability that a caught bass is more than 50 centimeters long?

ANSWER Although this area is not given explicitly, we know that the total area under the curve is 1. Thus the desired probability is

$$1 - (.03 + .12 + .18 + .2 + .18) = .29.$$

QUESTION What is the probability that a caught bass will be between 20 and 25 centimeters long?

ANSWER Although not enough information is given on the graph to answer this exactly, we might estimate this probability to be slightly less than .09, which is half of .18.

The Normal Curve

We will study only one type of continuous probability curve. The *normal curve* is one that arises in many situations. If x represents, say, the measurements of the length of a stick made by high school students, the weights of four-year olds at a nursery school, the distances from the bull's-eye of a number of thrown darts, the number of hours until each of a sample of light bulbs burns out, or the temperature at noon on January 3 at a particular weather station over a number of years, then the probability that x lies in a given interval can be expressed as the area under the corresponding portion of some normal curve.

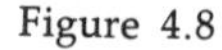
Figure 4.8

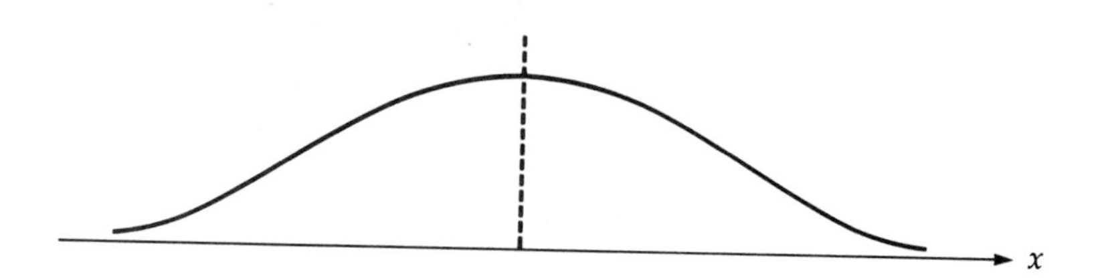

A *normal curve* is a bell-shaped curve symmetric about the vertical line through its highest point. See Figure 4.8. The value of x at the highest point on a normal curve is its *mean* m; this is the same mean that one would expect to calculate from a large number of measurements of x. Different normal curves have different means. If, for example, we ask a group of high school students to measure the length x_1 of a short stick and x_2 of a longer stick, the distribution of x_1 and x_2 would be as shown in Figure 4.9, with $m_1 < m_2$.

Different curves can also have different standard deviations. If x_3 represents the measurements of the short stick by a group of third graders, we might expect the same mean measurement as for x_1 but

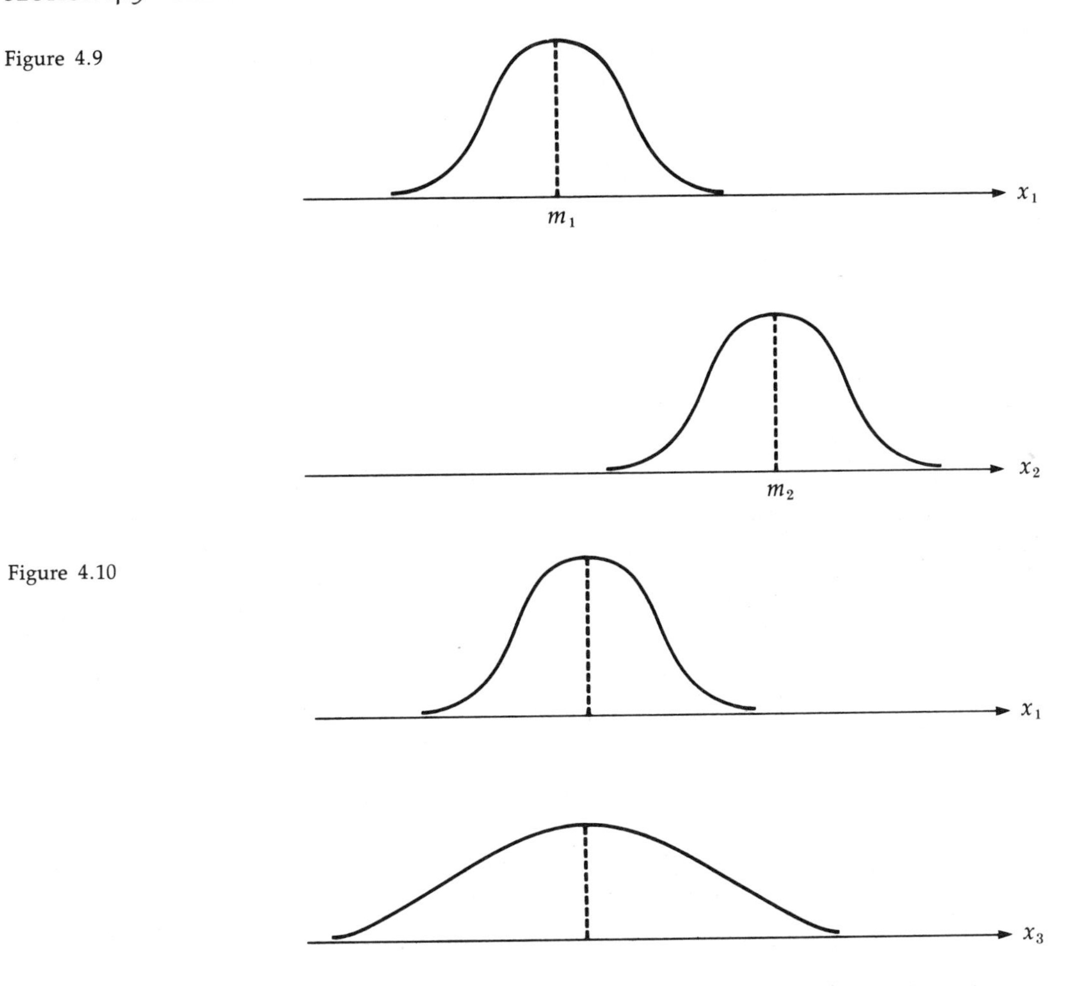

Figure 4.9

Figure 4.10

more of a spread of values. If x_1 and x_3 have standard deviations s_1 and s_3, we would expect $s_1 < s_3$. See Figure 4.10.

There is only one normal curve having any particular mean and standard deviation.

Using Table II

Table II in the back of this book, called Areas Under the Normal Distribution, gives the area under portions of the particular normal curve with mean $m = 0$ and standard deviation $s = 1$. This is called the *standard* normal curve. We will see in Section 4.4 how to find the area under portions of normal curves with different values for m and s.

Suppose we want the area under the particular normal curve with

$m = 0$ and $s = 1$ between $x = 0$ and $x = 1.63$. In Table II we find 1.6 along the left-hand column and .03 along the top (because $1.63 = 1.6 + .03$). The number in the table where the row of 1.6 and the column of .03 intersect, .4484, is the desired area. See Figure 4.11.

Figure 4.11

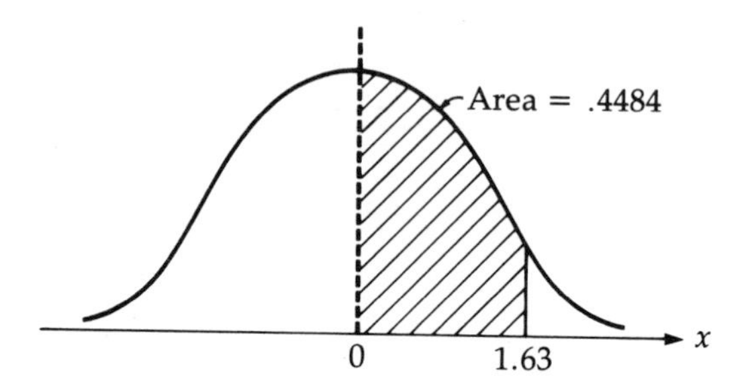

By the interpretation of a continuous probability curve, the probability that x lies between 0 and 1.63 is .4484. We remind the reader of the notation introduced at the end of Section 3.5, that of denoting the probability of the event E by $P(E)$. Thus we could write

$$P(0 < x < 1.63) = .4484.$$

TABLE II
Looking up a number in Table II gives the area under the standard normal curve between $x = 0$ and the number looked up.

In the same way we see

$$P(0 < x < 2.0) = .4772,$$

and

$$P(0 < x < 2.01) = .4778.$$

The table only gives areas between $x = 0$ and some positive value of x, but the areas over other intervals can still be calculated from these. For example, suppose $P(1 < x < 2)$ is wanted. Note that the area under the curve between $x = 1$ and $x = 2$ is equal to the area between $x = 0$ and $x = 2$ minus the area between $x = 0$ and $x = 1$. (See Figure 4.12.) Thus $P(1 < x < 2) = P(0 < x < 2) - P(0 < x < 1) = .4772 - .3413 = .1359$.

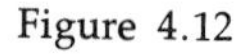
Figure 4.12

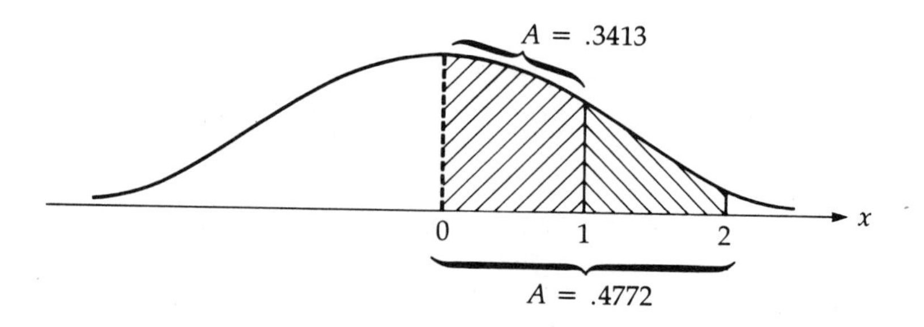

Figure 4.13

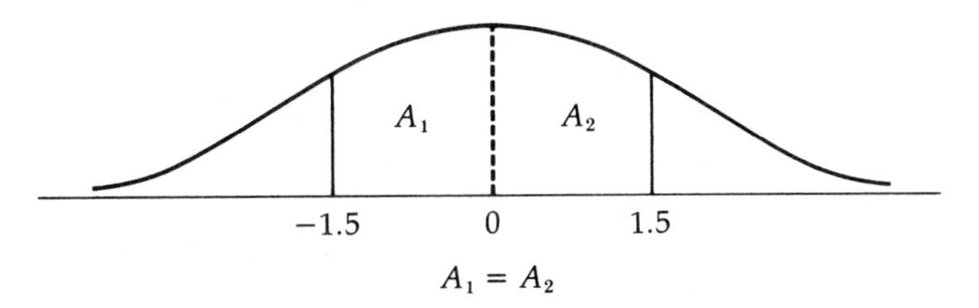

Areas to the left of $x = 0$ can be determined by using the symmetry of the normal curve about its mean, $m = 0$. If $P(-1.5 < x < 0)$ is desired, for example, we simply note that the area under the curve between $x = -1.5$ and $x = 0$ is exactly the same as the area between $x = 0$ and $x = 1.5$. (See Figure 4.13.) Thus $P(-1.5 < x < 0) = P(0 < x < 1.5) = .4332$.

THE AREA UNDER A POINT

Since there is no positive area under a point, it makes no difference whether inequalities defining areas under a continuous probability curve are strict or not. Thus

$$P(0 < x < 1.5) = P(0 \leqslant x < 1.5) = P(0 < x \leqslant 1.5) = P(0 \leqslant x \leqslant 1.5) = .4332.$$

EXAMPLE 2

The Tigers Club The amount x in the treasury of the Tigers Club varies between positive and negative values (right now they owe Mrs. Cline $6 for a broken window). The distribution is normal with mean 0 and standard deviation $1. What is the probability that at any given time the club has less than $2?

This is $P(x < 2) = P(x < 0) + P(0 < x < 2)$. See Figure 4.14. Now, by the symmetry of the standard normal curve about 0, half the area

Figure 4.14

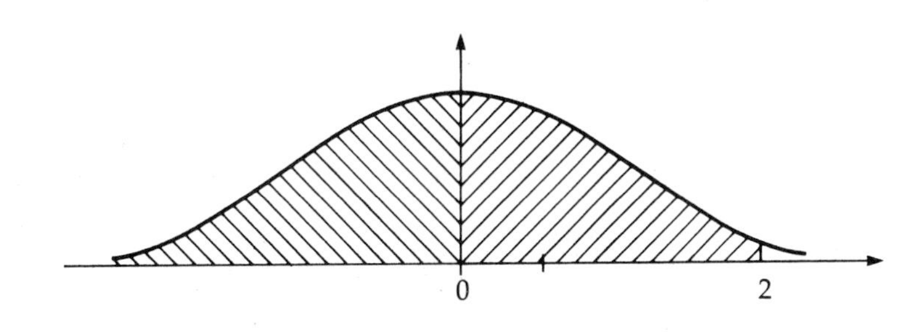

under it lies to the left of $x = 0$. Since the total area under the curve is 1, we conclude $P(x < 0) = 1/2$.

From Table II, $P(0 < x < 2) = .4772$. Thus

$$P(x < 2) = .5 + .4772 = .9772.$$

EXAMPLE 3

The Kindergarten Some of the children entering kindergarten at Wilson Grade School are above the recommended weights for their heights and some are below. In fact, if x is the amount, positive or negative, by which an entering student exceeds his or her recommended weight, then x is normally distributed with mean 0 and standard deviation 1 kilogram.

What is the probability an arbitrarily chosen entering student is no more than 1 pound above or below the recommended weight?

Since 1 pound is .45 kilograms, we want $P(-.45 < x < .45)$. Using the symmetry of the normal curve and Table II, we have

$$P(-.45 < x < .45) = 2P(0 < x < .45) = 2(.1736) = .3472.$$

EXAMPLE 4

The Cold Curve Suppose the December temperatures in a certain cave in Bear Claw, Minnesota, are normally distributed with mean $m = 0$ and standard deviation $s = 1$. What is the probability that a given temperature reading exceeds 2.61°?

We want $P(x > 2.61)$. Since the total area under the normal curve must be 1 (this is true of any continuous probability curve), and since by symmetry half of this area is to the right of $x = 0$, we have $P(x > 2.61) = P(x > 0) - P(0 < x < 2.61) = .5000 - .4955 = .0045$. (See Figure 4.15.)

Figure 4.15

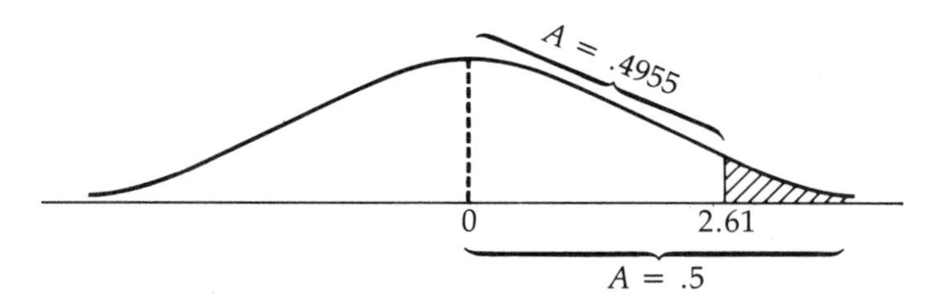

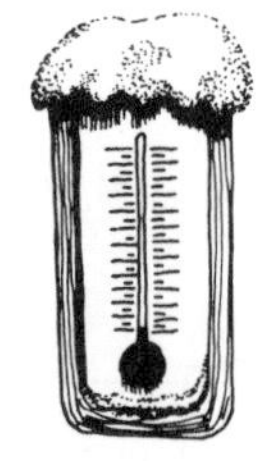

What is the probability the temperature is between $-1.3°$ and 1.77°? We want

$$\begin{aligned} P(-1.3 < x < 1.77) &= P(-1.3 < x < 0) + P(0 < x < 1.77) \\ &= P(0 < x < 1.3) + P(0 < x < 1.77) \\ &= .4032 + .4616 = .8648. \end{aligned}$$

Exercises 4.3

1. Use Table II to find the area under the standard normal curve between 0 and x for the following values of x: 1.1, 1.13, 2.86, .07.

2. Use Table II to find the area under the standard normal curve between 0 and x for the following values of x: 2.3, 3.08, 2.19, .3.

In problems 3 through 14 assume x has the standard normal distribution.

3. Find $P(0 < x < 2.61)$, $P(0 < x < 1.9)$, $P(0 \leq x < 3.05)$, and $P(0 < x \leq 1.92)$.

4. Find $P(0 < x < 1.61)$, $P(0 < x \leq 2.99)$, $P(0 < x < .16)$, and $P(0 \leq x \leq 1.90)$.

5. Find $P(-1.6 < x < 0)$, $P(-2.91 \leq x < 0)$, $P(-3.01 \leq x \leq 0)$, and $P(-.06 < x < 0)$.

6. Find $P(-1.1 < x < 0)$, $P(-2.33 < x < 0)$, $P(-3 \leq x < 0)$, and $P(-.04 < x < 0)$.

7. Find $P(-1.55 < x < 2.4)$, $P(-.22 < x \leq 1.04)$, $P(-2.36 \leq x < 2)$.

8. Find $P(-2.34 < x < .19)$, $P(-.06 \leq x < .67)$, $P(-.25 < x \leq 1.85)$.

9. Find $P(x > 2.46)$, $P(x \geq 1.97)$, $P(x < -.67)$, $P(x \leq -1.1)$.

10. Find $P(x > 2.56)$, $P(x \geq .05)$, $P(x < -.02)$, $P(x < -2.37)$.

11. Find $P(x > -.15)$, $P(x \geq -2.54)$, $P(x < 2.22)$, $P(x \leq 3.01)$.

12. Find $P(x > -2.2)$, $P(x < 3.03)$, $P(x \leq 1.95)$, $P(x \geq -2.44)$.

13. Find $P(1.19 < x < 2.33)$, $P(-2.37 < x \leq 2.46)$, $P(-2.4 \leq x \leq .07)$, $P(-3 < x < -.3)$.

14. Find $P(-1.97 < x \leq 1.3)$, $P(-2.54 < x < 3)$, $P(1.11 \leq x \leq 1.14)$, $P(-2 < x < -.2)$.

15. The following graph shows the probability distribution for the cost x in dollars to repair autos involved in accidents on I-75 in 1979.

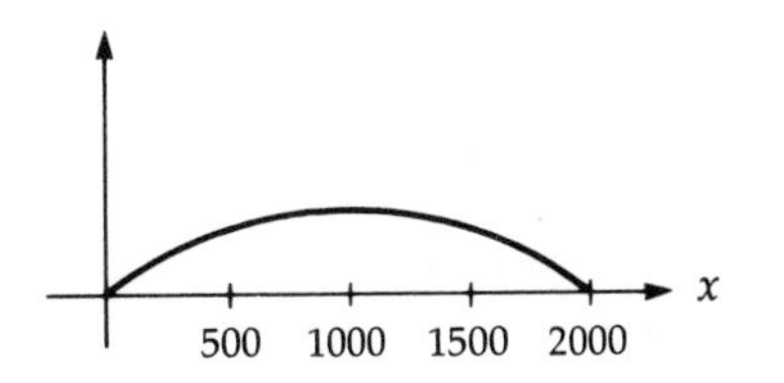

It is given that $P(x < 1500) = .88$, $P(1000 < x < 1500) = .3$, and $P(500 < x < 1000) = .4$. Find (a) the probability an accident costs between \$500 and \$1500 to repair, (b) the probability an accident costs less than \$500 to repair, (c) the probability an accident costs more than \$1500 to repair.

16. Suppose the graph shows a probability distribution for x such that $P(x < 1) = .4$, $P(1 < x < 2) = .3$, and $P(0 < x < 2) = .6$. Find (a) $P(0 < x < 1)$ (b) $P(x > 2)$ (c) $P(x < 0)$.

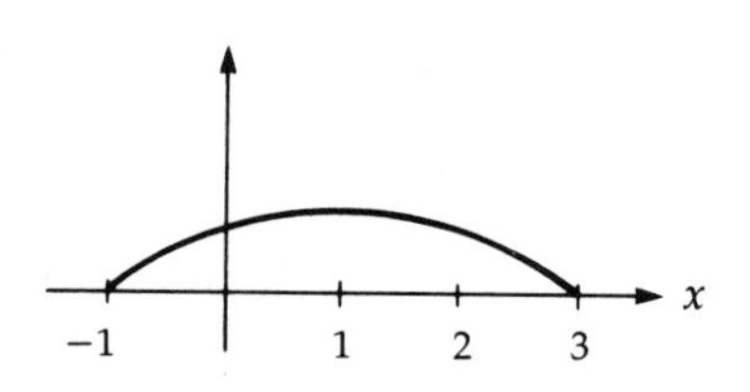

17. A study of mistakes by supermarket clerks at the cash register finds the mean error for an hour of checkout work is \$0 (this means the overcharges balance the undercharges) with standard deviation \$1. Assuming a normal distribution, what is the probability a checker during an hour's time (a) overcharges, but by no more than 50¢? (b) undercharges, but by no more than 50¢? (c) overcharges by some amount? (d) does no worse for the store than undercharge by \$1? (e) overcharges by more than \$1?

18. What is the probability the temperature in the cave described in Example 4 is (a) positive? (b) positive but less than 2°? (c) less than 2°? (d) greater than 3°? (e) between −1° and 0°?

19. If t is normally distributed with mean 0 and standard deviation 1, find (a) the probability that t is between 0 and 1.86 (b) the probability that t is greater than -1.5 (c) the probability that t is less than -1.5.

20. The height of the Ohio River above a certain fixed point is normally distributed with mean 0 and standard deviation 1 meter. What is the probability that this height is (a) more than 2.61 meters? (b) greater than −1 centimeter? (c) between −1.6 and 1.7 meters?

4.4 THE GENERAL NORMAL DISTRIBUTION

In the previous section we learned how to calculate the probability that a variable x having the standard normal distribution would lie in any interval by the use of a table. There is exactly one normal probability curve for each mean m and standard deviation s; for the *standard* normal curve, $m = 0$ and $s = 1$. Although the normal curve gives a reasonable picture of the distribution of many variables arising in the real world, most of these, of course, do not have mean 0 and standard deviation 1. It might appear that a new table, similar to the one we used for the standard normal distribution (Table II), is needed for each pair of values of m and s. Fortunately, however, a mathematical transformation exists by which any normal curve may be related to the standard normal curve.

Other Normal Curves

Suppose a measurement x_1 is distributed with a mean m and standard deviation s other than the 0 and 1 of the curve Table II refers to. We convert from x_1 to the variable x having the standard normal distribution by the formula

$$x = \frac{x_1 - m}{s}.$$

The geometric effect of this is shown in Figure 4.16. In (a) we start with a normal curve with mean $m > 0$ and standard deviation $s < 1$. (A small standard deviation means that the values of x_1 tend to lie close to the mean.) Subtracting m from each value of x_1 gives a new variable $x_1 - m$ with mean zero. Then dividing the numbers $x_1 - m$ by s (which is <1) sends them further from zero and spreads out the curve to a standard normal curve. The interpretation of (b) is similar except that here $m < 0$ and $s > 1$.

If a_1 and b_1 are two values of x_1, then

$$a = \frac{a_1 - m}{s} \quad \text{and} \quad b = \frac{b_1 - m}{s}$$

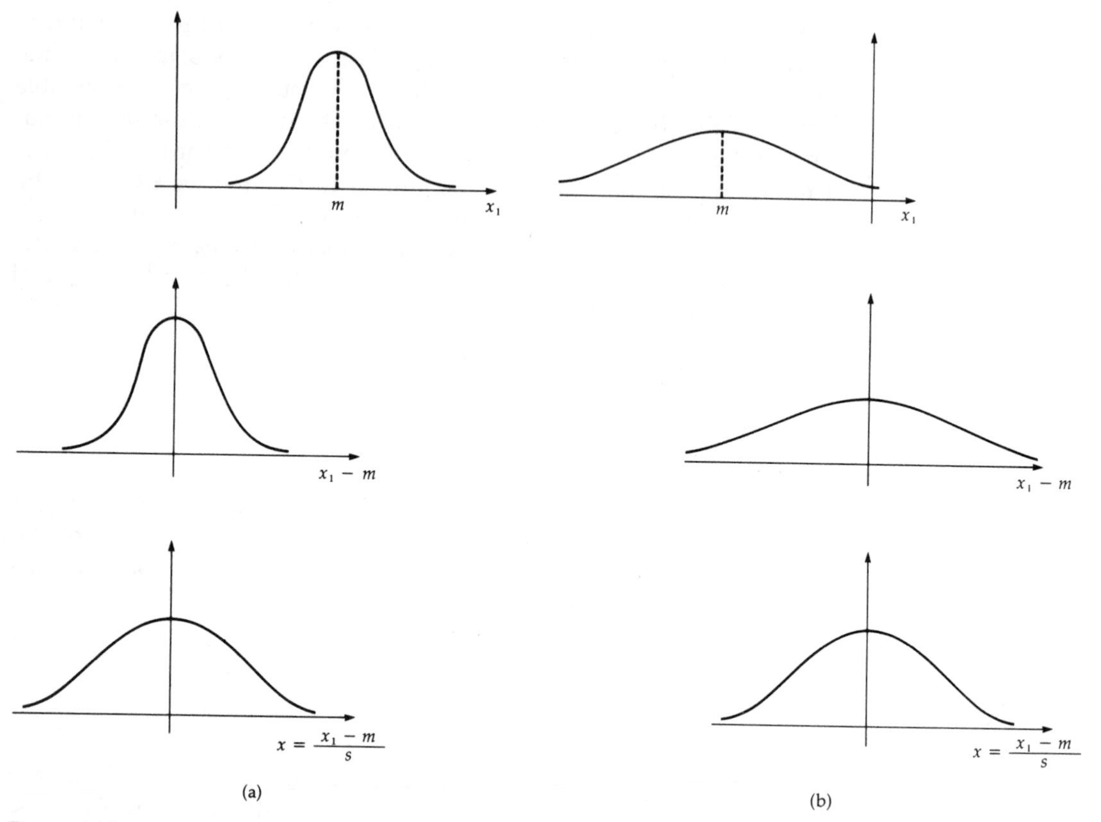

Figure 4.16

are the corresponding points on the standard normal curve; and $P(a_1 < x_1 < b_1) = P(a < x < b)$. The latter quantity can be computed from Table II. Thus the hatched regions in Figure 4.17 have the same area.

Suppose, for example, x_1 is normally distributed with mean $m = 10$ and standard deviation $s = 2$. Suppose we want $P(5 < x < 14)$. When $x_1 = 5$, then $x = (5 - 10)/2 = -2.5$; and when $x_1 = 14$, then $x = (14 - 10)/2 = 2$. Thus

$$
\begin{aligned}
P(5 < x_1 < 14) = P(-2.5 < x < 2) &= P(-2.5 < x < 0) + P(0 < x < 2) \\
&= P(0 < x < 2.5) + P(0 < x < 2) \\
&= .4938 + .4772 = .9710.
\end{aligned}
$$

EXAMPLE 1

A Weight Problem The mean weight of 18-year old American girls is 126 pounds, with a standard deviation of 18 pounds. What is the probability an 18-year old girl weighs less than 100 pounds?

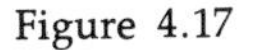

Figure 4.17

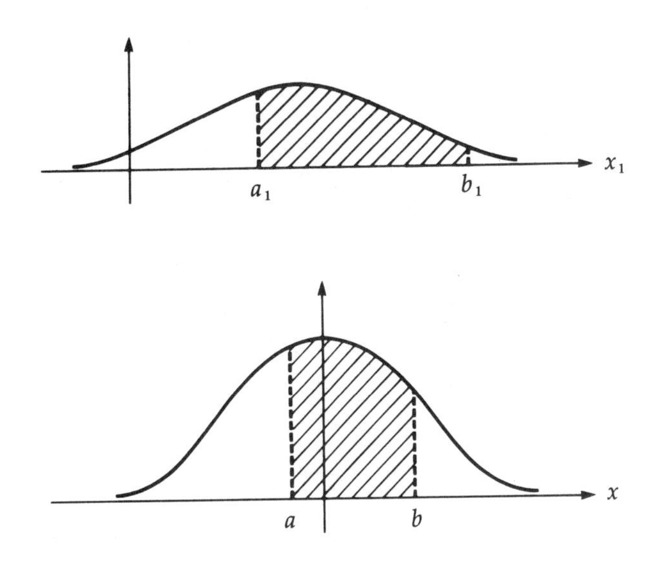

If x_1 is the weight of an 18-year old girl, let us assume x_1 is normally distributed with $m = 126$ and $s = 18$. When $x_1 = 100$, then

$$x = \frac{x_1 - m}{s} = \frac{100 - 126}{18} = -1.44.$$

Then

$$P(x_1 < 100) = P(x < -1.44) = P(x > 1.44)$$
$$= P(x > 0) - P(0 < x < 1.44)$$
$$= .5000 - .4251 = .0749.$$

EXAMPLE 2

The Big Tomato A tomato grower finds that the mean weight of the tomatoes he picks is 8 ounces, with standard deviation 2 ounces. What is the probability that an arbitrarily chosen tomato will weigh more than 11 ounces, assuming a normal distribution?

Let x_1 be the weight of a tomato; we want $P(x_1 > 11)$. Since $m = 8$ and $s = 2$, we have

$$x = \frac{x_1 - 8}{2}.$$

In particular, when $x_1 = 11$, $x = (11 - 8)/2 = 1.5$. Then

$$P(x_1 > 11) = P(x > 1.5) = P(x > 0) - P(0 < x < 1.5)$$
$$= .5000 - .4332 = .0668.$$

EXAMPLE 3

The Raincoats A manufacturer of raincoats designs its "medium" coat for men no more than 3 inches taller or shorter than 5 feet 10 inches. If the mean height of adult men is 5′10″ with standard deviation 2.5″, what proportion of men will be able to wear a medium raincoat?

Let x_1 be the height of an adult male, in inches. Then we want $P(67 < x_1 < 73)$. Since x_1 has mean $m = 70$ and standard deviation $s = 2.5$, we have

$$x = \frac{x_1 - 70}{2.5}.$$

When $x_1 = 67$,

$$x = \frac{67 - 70}{2.5} = -1.2,$$

and when $x_1 = 73$,

$$x = \frac{73 - 70}{2.5} = 1.2.$$

Thus

$$\begin{aligned} P(67 < x_1 < 73) &= P(-1.2 < x < 1.2) \\ &= 2P(0 < x < 1.2) \\ &= 2(.3849) = .7698. \end{aligned}$$

The raincoat should fit about 77% of the adult male population.

Working a Problem Backwards

The manager of a dormitory cafeteria finds from her records that the mean number of people who show up for Sunday morning breakfast is 120, with standard deviation 20. If she wants to have probability .95 of not running out of food, how many breakfasts should she have prepared?

Let x_1 be the number showing up for breakfast. Then we wish to find a value of b_1 such that $P(x_1 < b_1) = .95$.

Since x_1 has mean 120 and standard deviation 20, we have

$$x = \frac{x_1 - 120}{20},$$

Figure 4.18

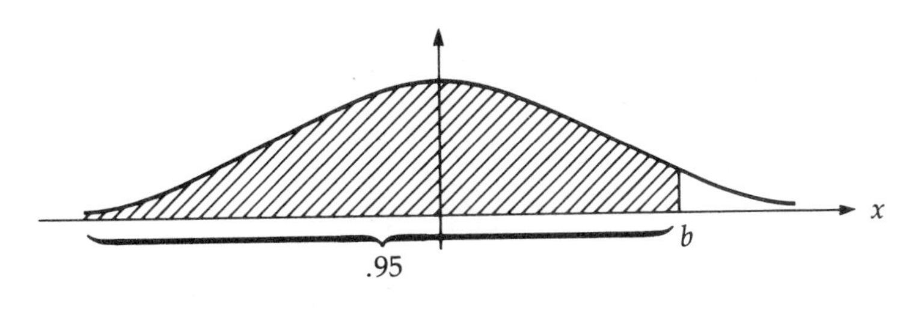

where x has the standard normal distribution. If $b = (b_1 - 120)/20$, then

$$P(x < b) = P(x_1 < b_1) = .95.$$

See Figure 4.18. But then

$$\begin{aligned} P(x < b) &= P(x < 0) + P(0 < x < b) \\ &= .5 + P(0 < x < b) = .95, \end{aligned}$$

so $P(0 < x < b) = .95 - .5 = .45$.

Reading Table II backwards gives $b = 1.64$. (That is, $P(0 < x < 1.64) = .45$.) Then

$$\frac{b_1 - 120}{20} = 1.64,$$

$$b_1 - 120 = 20(1.64) = 32.8,$$

$$b_1 = 120 + 32.8 = 152.8.$$

The cafeteria manager should have 153 breakfasts prepared.

EXAMPLE 4

The Pretzel Boxes A consumer protection agency is to set standards for 1-pound boxes of pretzels. They realize that even if the boxes average 1 pound, some will be over and some under this weight. With modern packing equipment the boxes should average 1 pound with standard deviation .02 pounds. The agency decides to investigate complaints of an underweight box only if the weight is less than a certain fixed weight b_1, this weight being such that the weight of only one box in 25 should be expected to be below b_1. What should b_1 be?

If x_1 is the weight of a box, then x_1 has mean 1 and standard deviation .02. The standard normal variable x satisfies

$$x = \frac{x_1 - 1}{.02}.$$

Figure 4.19

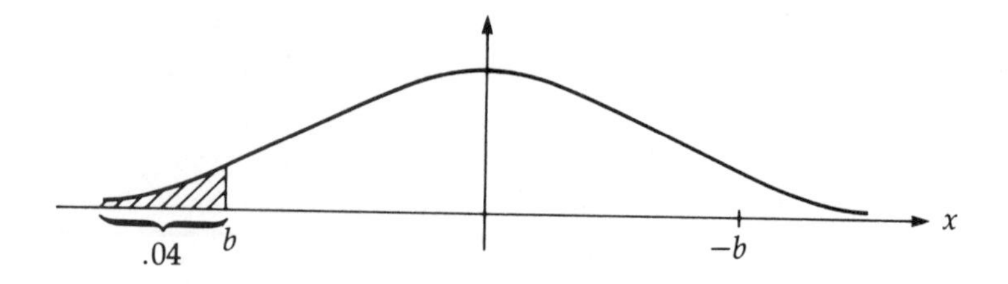

We want b_1 so that $P(x_1 < b_1) = 1/25 = .04$. Now, $P(x < b) = .04$ implies

$$P(0 < x < -b) = P(b < x < 0) = P(x < 0) - P(x < b) = .5 - .04 = .46.$$

See Figure 4.19. From Table II we see $-b = 1.75$, so $b = -1.75$. Now

$$b = \frac{b_1 - 1}{.02},$$

or

$$-1.75 = \frac{b_1 - 1}{.02},$$

$$b_1 - 1 = (.02)(-1.75) = -.035,$$

$$b_1 = 1 - .035 = .965 \text{ pounds.}$$

Exercises 4.4

1. If x_1 has mean 100 and standard deviation 25, what is x when $x_1 = 150$? when $x_1 = 300$? when $x_1 = 90$?
2. If x_1 has mean 5 and standard deviation 2, what is x when $x_1 = 3$? when $x_1 = 12$? when $x_1 = 8$?
3. If for x_1 we have $m = -10$ and $s = 2.5$, what is x when $x_1 = 0$? when $x_1 = -20$? when $x_1 = 5$?
4. If for x_1 we have $m = -3$ and $s = 5$, what is x when $x_1 = 6$? when $x_1 = 0$? when $x_1 = -10$?
5. Suppose x_1 has mean 6 and standard deviation 2. Find $P(6 < x_1 < 10)$, $P(5 < x_1 < 6)$, and $P(5 < x_1 < 10)$.
6. Suppose x_1 has mean 200 and standard deviation 50. Find $P(200 < x_1 < 250)$, $P(100 < x_1 < 200)$, and $P(100 < x_1 < 250)$.
7. Suppose x_1 has standard deviation 4 and mean 21. What is $P(x_1 > 25)$, $P(x_1 < 20)$, and $P(x_1 > 19)$?

8. Suppose x_1 has standard deviation 5 and mean 40. What is $P(x_1 > 50)$, $P(x_1 < 38)$, and $P(x_1 < 44)$?

9. Suppose x_1 has mean 25 and standard deviation 5. Find $P(26 < x_1 < 30)$ and $P(23 < x_1 < 28)$.

10. Suppose x_1 has mean .5 and standard deviation .1. Find $P(.6 < x_1 < .7)$ and $P(.4 < x_1 < .55)$.

11. An analysis of the scores of a certain professional basketball team finds them normally distributed with mean 90 and standard deviation 20. What is the team's probability of scoring more than 100 in a given game?

12. What is the probability that an 18-year old girl weighs less than 150 pounds? (See Example 1.)

13. The life of a certain brand of electric motor is found to have mean 1000 hours with standard deviation 200 hours. What is the probability that a motor lasts less than 600 hours?

14. The fish in a certain lake are found to average 4 pounds with a standard deviation 1. What is the probability that a given fish weighs more than 7 pounds?

15. A survey of a supermarket finds that the amount of money a shopper spends there has mean $15 and standard deviation $6. What is the probability that a given shopper spends more than $20?

16. In the example of the dormitory cafeteria, what is the probability that fewer than 100 people show up for breakfast on a given Sunday morning?

17. If x has the standard normal distribution and $P(0 < x < b) = .4484$, what is b?

18. If x has the standard normal distribution and $P(0 < x < b) = .0675$, what is b?

19. If x has the standard normal distribution and $P(x > b) = .0170$, what is b?

20. If x has the standard normal distribution and $P(x > b) = .0934$, what is b?

21. If x_1 has mean 50 and standard deviation 20 and $P(50 < x_1 < b_1) = .3708$, what is b_1?

22. If x_1 has mean 12 and standard deviation 5 and $P(12 < x_1 < b_1) = .2291$, what is b_1?

23. The manufacturer in Example 3 makes a coat that is too big for 90% of the adult male population. How tall is the shortest man it will fit?

24. The scores on a state-wide high school English test have mean 70 and standard deviation 10. What score should be the dividing line between A and B in order to have 5% A's?

4.5 REPEATED TRIALS

In Section 4.3 the normal distribution was introduced and the claim was made that this distribution seemed to describe many situations arising in the real world. Yet no indication was given as to what the normal distribution really *is* or why it is reasonable to apply it so widely. In this section and the next we will address these problems.

EXAMPLE 1

Gin Rummy Ian and Martha play a game of gin rummy each weeknight before going to bed. They keep a record of who wins each game. It turns out that Ian and Martha each win about half the time, although, of course, over a period of a week or two one of them will often win more than the other.

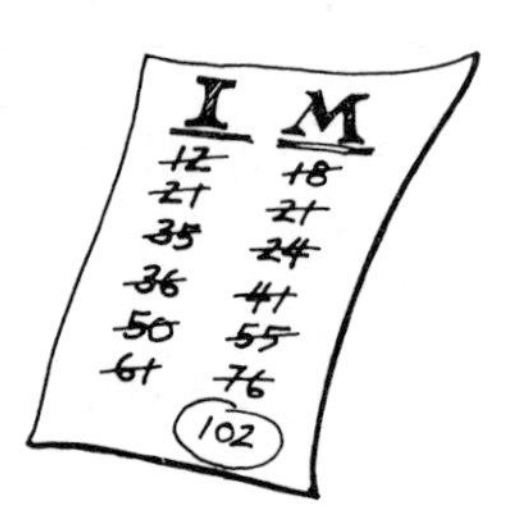

In one week's time they play five games. Usually Ian seems to win two or three of them, but one week he didn't win any. It is possible for him to win any number from zero to five games in a week. Let us compute the probability of each of these occurrences.

First let us compute the probability that Ian wins all five games. His probability of winning any one game is $p = 1/2$. Let us denote by W_1 the event that he wins the first game, by W_2 the event that he wins the second, and so on. We want

$$P(W_1 \text{ and } W_2 \text{ and } W_3 \text{ and } W_4 \text{ and } W_5).$$

Since these are independent events, the multiplication rule for probabilities of Section 3.4 applies. Thus this probability is

$$P(W_1)P(W_2)P(W_3)P(W_4)P(W_5) = \frac{1}{2} \cdot \frac{1}{2} \cdot \frac{1}{2} \cdot \frac{1}{2} \cdot \frac{1}{2} = \frac{1}{32}.$$

If they play every week Ian might expect to win all five games once or twice a year.

Now we compute the probability that Ian wins exactly four

games. This could happen in various ways, namely (letting L_1 mean that he loses the first game, and so on)

$$L_1\ W_2\ W_3\ W_4\ W_5$$
$$W_1\ L_2\ W_3\ W_4\ W_5$$
$$W_1\ W_2\ L_3\ W_4\ W_5$$
$$W_1\ W_2\ W_3\ L_4\ W_5$$
$$W_1\ W_2\ W_3\ W_4\ L_5.$$

Since $P(L_1) = 1/2$ also, the multiplication rule says

$$P(L_1 \text{ and } W_2 \text{ and } W_3 \text{ and } W_4 \text{ and } W_5) = \frac{1}{2}\cdot\frac{1}{2}\cdot\frac{1}{2}\cdot\frac{1}{2}\cdot\frac{1}{2} = \frac{1}{32}.$$

The same calculation shows that the probability of each of the other listed events is also 1/32. These are mutually exclusive events, so that the addition principle tells us

$$P(\text{Ian wins exactly 4 times}) = \frac{1}{32} + \frac{1}{32} + \frac{1}{32} + \frac{1}{32} + \frac{1}{32} = \frac{5}{32}.$$

Now what about Ian's winning exactly three times? This could happen in various ways, such as $W_1\ L_2\ W_3\ W_4\ L_5$ and $L_1\ L_2\ W_3\ W_4\ W_5$. Each of these mutually exclusive events has probability

$$\frac{1}{2}\cdot\frac{1}{2}\cdot\frac{1}{2}\cdot\frac{1}{2}\cdot\frac{1}{2} = \frac{1}{32};$$

the question is how many of them there are.

We are asking how many ways we can choose three objects (the three wins) from a group of five. This is a problem we have already treated in Section 3.6; the answer is

$$\binom{5}{3} = \frac{5!}{3!\ 2!} = 10.$$

Thus

$$P(\text{Ian wins 3 times}) = \binom{5}{3}\frac{1}{32} = \frac{10}{32}.$$

In the same way we compute

$$P(\text{Ian wins 2 times}) = \binom{5}{2}\frac{1}{32} = \frac{10}{32},$$

$$P(\text{Ian wins 1 time}) = \binom{5}{1}\frac{1}{32} = \frac{5}{32},$$

$$P(\text{Ian wins no times}) = \binom{5}{0}\frac{1}{32} = \frac{1}{32}.$$

(In the evaluation of $\binom{5}{0}$ the expression 0! appears. This is defined to be 1.)

The Repeated Trials Formula

Let us generalize the calculation of the previous section. Suppose we are performing an experiment n times. (In the example n was 5.) Suppose the probability of success in any given trial is p. (In the example a "trial" is a game and "success" is Ian's winning, so $p = 1/2$.) Then the probability of failure is $1 - p$. (In the example $1 - p = 1 - 1/2 = 1/2$ also.)

We ask for the probability of exactly r successes out of the n trials. There are various ways these r successes could occur. Perhaps they all come at the start, the pattern being

$$S_1 S_2 \cdots S_r F_{r+1} F_{r+2} \cdots F_n,$$

where S stands for success and F for failure. By the multiplication principle the probability of this event is

$$(*) \quad \underbrace{p \cdot p \cdot p \cdots p}_{r \text{ times}} \cdot \underbrace{(1-p)(1-p) \cdots (1-p)}_{n - r \text{ times}} = p^r(1-p)^{n-r}.$$

In fact, no matter how we arrange the r successes and $n - r$ failures, the probability of that particular event will still be $p^r(1 - p)^{n-r}$, since the multiplication principle will just give us (*) again with the factors rearranged.

Finally, the number of ways we can choose r successes out of the n trials is $\binom{n}{r}$. We see that

---

The probability of exactly r successes from among n trials, if the probability of success on any given trial is p, is

$$\binom{n}{r}p^r(1-p)^{n-r}.$$

EXAMPLE 2

The Melons A melon seed has a probability .8 of germinating. A gardener has room for three melon plants in her garden. The seeds are expensive, but she decides to start five of them inside, in the hope that at least three will germinate. What is the probability of this?

Here $n = 5$ and $p = .8$. The probability that r seeds germinate is

$$\binom{5}{r}.8^r(1-.8)^{5-r} = \binom{5}{r}.8^r\ .2^{5-r}.$$

In particular,

$$P(3 \text{ germinate}) = \binom{5}{3}.8^3\ .2^2 = 10(.8)^3(.2)^2 = .205,$$

$$P(4 \text{ germinate}) = \binom{5}{4}.8^4\ .2^1 = 5(.8)^4(.2)^1 = .410,$$

$$P(5 \text{ germinate}) = \binom{5}{5}.8^5\ .2^0 = .8^5 = .328.$$

ZERO AS AN EXPONENT
Recall that if $a \neq 0$, then $a^0 = 1$.

Since these are the probabilities of mutually exclusive events, the probability that at least three germinate is

$$.205 + .410 + .328 = .943.$$

EXAMPLE 3

The .300 Hitter What is the probability that a .300 hitter in baseball gets exactly 3 hits in 10 times at bat?

Here $p = .3$, $n = 10$, and $r = 3$. The answer is

$$\binom{10}{3}.3^3(1-.3)^{10-3} = \frac{10!}{(10-3)!3!}(.3)^3(.7)^7 = \frac{10!}{7!\ 3!}(.027)(.0824)$$

$$= \frac{10 \cdot 9 \cdot 8 \cdot \not{7!}}{\not{7!} \cdot 1 \cdot 2 \cdot 3}(.00222) = .267.$$

EXAMPLE 4

The Eggs An inspector looks at cartons of one dozen eggs chosen at random. If more than two eggs in a carton have cracks, the whole batch is used for powdered eggs instead of being sold fresh. If 1% of the eggs have cracks, what is the probability that a given carton is rejected?

Rather than compute the probability that 2, 3, 4, 5, 6, 7, 8, 9, 10, 11, or 12 eggs are cracked, it is easier to figure out the probability that 0 or 1 are cracked, so that the carton is *not* rejected.

Taking success as finding an egg to be cracked, we have $p = .01$ and $n = 12$. Then

$$P(r \text{ eggs are cracked}) = \binom{12}{r}(.01)^r(1 - .01)^{12-r} = \binom{12}{r}(.01)^r(.99)^{12-r}.$$

Thus

$$P(\text{none cracked}) = \binom{12}{0}(.01)^0\,(.99)^{12} = \frac{12!}{12!0!}(.99)^{12} = (.99)^{12}.$$

With a calculator we find $(.99)^{12} = .886$.

Likewise

$$P(1 \text{ cracked}) = \binom{12}{1}(.01)^1(.99)^{11} = \frac{12!}{11!\,1!}\,(.01)(.99)^{11}$$
$$= 12(.01)(.99)^{11} = .107.$$

Thus

$$P(\text{none or 1 cracked}) = .886 + .107 = .993,$$

and

$$P(\text{more than 1 cracked}) = 1 - .993 = .007.$$

EXAMPLE 5

The Guesser A student taking a 10-question true-false test has to guess completely at the answers. If he gets seven or more answers right, he will pass the test. What is the probability of this?

Here $n = 10$, $p = 1/2$, and

$$P(r \text{ right answers}) = \binom{10}{r}\left(\frac{1}{2}\right)^r\left(1 - \frac{1}{2}\right)^{10-r} = \binom{10}{r}\left(\frac{1}{2}\right)^r\left(\frac{1}{2}\right)^{10-r} = \binom{10}{r}\left(\frac{1}{2}\right)^{10}.$$

Then

$$P(7 \text{ right}) = \binom{10}{7}\left(\frac{1}{2}\right)^{10} = \frac{10!}{3!7!} \cdot \frac{1}{1024} = .117,$$

$$P(8 \text{ right}) = \binom{10}{8}\left(\frac{1}{2}\right)^{10} = \frac{10!}{2!8!} \cdot \frac{1}{1024} = .044,$$

$$P(9 \text{ right}) = \binom{10}{9}\left(\frac{1}{2}\right)^{10} = \frac{10!}{1!9!} \cdot \frac{1}{1024} = .010.$$

$$P(10 \text{ right}) = \binom{10}{10}\left(\frac{1}{2}\right)^{10} = \frac{10!}{0!10!} \cdot \frac{1}{1024} = .001.$$

Thus

$$P(7 \text{ or more right}) = .117 + .044 + .010 + .001 = .172.$$

He has a little better than one chance in six of passing.

Exercises 4.5

1. The probability is .4 that a thoroughbred colt can be developed into a race horse. What is the probability that exactly three out of a group of six colts can be so developed?
2. A company making flashlight batteries tests them in groups of seven. If more than one is deficient in power, the batch is rejected. If the probability that an individual battery is deficient is .05, what is the probability that a batch is not rejected?
3. A dart thrower hits the bull's-eye once in 10 throws, on the average. What is the probability that she gets exactly two bull's-eyes in six throws?
4. A certain surgical operation is successful 90% of the time. Four operations are scheduled this week. What is the probability that at least three are successful?
5. The Reds and Dodgers are equally successful against each other. What is the probability that they split a given four-game series?
6. The Yankees are playing the Pirates in the World Series. The Yankees have a probability .6 of winning any given game. What is the probability that the series goes seven games?
7. A wine company finds that 1% of its bottles go bad in storage. What is the probability of getting a perfect case of six?
8. A cereal company weighs its boxes in groups of five. No box ever

weighs exactly the advertised amount. If more than three boxes are underweight or overweight, the line is shut down and the machines adjusted. Half the boxes are underweight. What is the probability that the line is not shut down?

9. Thirty percent of all Christmas tree light bulbs are blue. They are packed at random into boxes of eight. What percentage of boxes contain exactly two blue bulbs?

10. A student takes a five-question multiple-choice test. Each question has five possible answers. Three or more right answers will pass the test. What is the probability that a student who is guessing completely will pass?

11. In problem 10 what is the probability that a student who can rule out one of the answers to each question before guessing will pass the test?

12. A student taking the test described in problem 10 has a probability .6 of answering any given question correctly. What is her probability of passing?

13. A fisherman has five lines in the water. Each line has a probability .7 of having a fish. What is the probability he has exactly the legal limit of three fish?

14. The probability of rain on any given day is .1. What is the probability of having two or more days of rain in a given week, assuming these are independent events?

4.6 THE BINOMIAL DISTRIBUTION

What Happens when n Gets Large

We return to the nightly gin rummy game of Ian and Martha. In Example 1 of the last section we computed Ian's probability of winning each possible number of games in a five-game series.

WINS	PROBABILITY
5	1/32
4	5/32
3	10/32
2	10/32
1	5/32
0	1/32

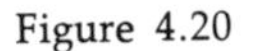
Figure 4.20

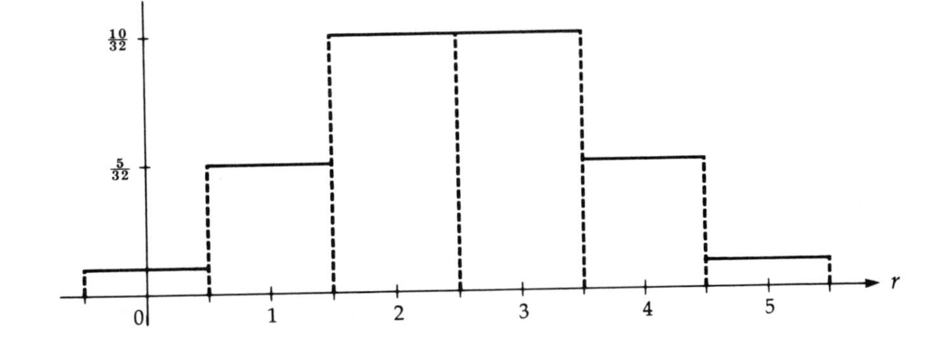

We make a bar graph of these probabilities, using the number of wins r as the horizontal axis (Figure 4.20). An interesting thing about this graph is that the total area under it is 1. This is true because the area of each rectangle equals the probability Ian wins the number of games the rectangle is centered over, and all these probabilities add up to 1. This is a characteristic shared by any continuous probability distribution.

The graph shown is not of a *continuous* distribution, however, since the variable r can only take on a finite number of values, in this case 0, 1, 2, 3, 4, or 5. Such a distribution is called a *discrete* probability distribution.

The particular type of distribution based on repeated trials is called a *binomial distribution*. Our graph shows the binomial distribution with $n = 5$ and $p = .5$.

Suppose now we change n from 5 to 10; that is, we consider what is likely to happen after two weeks of playing. The formula

$$\binom{n}{r} p^r (1 - p)^{n-r}$$

enables us to compute Table 4.5. The graph is shown in Figure 4.21. Perhaps this graph may look familiar. It is reminiscent of the bell-shaped normal curve shown in Figure 4.22. In fact *as n gets larger and larger, the binomial distribution looks more and more like a normal curve.*

The Normal Approximation

For large values of n the normal curve is easier to work with than the binomial distribution. Consider, for example, the following question. Ian and Martha play 100 games of gin rummy. What is the probability that Ian wins 60 or more of them?

The probability that Ian wins exactly 60 of the hundred games is

TABLE 4.5

WINS r	PROBABILITY
10	$\binom{10}{10}\left(\frac{1}{2}\right)^{10}\left(\frac{1}{2}\right)^{0} = 1/1024 = .0010$
9	$\binom{10}{9}\left(\frac{1}{2}\right)^{9}\left(\frac{1}{2}\right)^{1} = 10/1024 = .0098$
8	$\binom{10}{8}\left(\frac{1}{2}\right)^{8}\left(\frac{1}{2}\right)^{2} = 45/1024 = .0439$
7	$\binom{10}{7}\left(\frac{1}{2}\right)^{7}\left(\frac{1}{2}\right)^{3} = 120/1024 = .1172$
6	$\binom{10}{6}\left(\frac{1}{2}\right)^{6}\left(\frac{1}{2}\right)^{4} = 210/1024 = .2051$
5	$\binom{10}{5}\left(\frac{1}{2}\right)^{5}\left(\frac{1}{2}\right)^{5} = 252/1024 = .2461$
4	$\binom{10}{4}\left(\frac{1}{2}\right)^{4}\left(\frac{1}{2}\right)^{6} = 210/1024 = .2051$
3	$\binom{10}{3}\left(\frac{1}{2}\right)^{3}\left(\frac{1}{2}\right)^{7} = 120/1024 = .1172$
2	$\binom{10}{2}\left(\frac{1}{2}\right)^{2}\left(\frac{1}{2}\right)^{8} = 45/1024 = .0439$
1	$\binom{10}{1}\left(\frac{1}{2}\right)^{1}\left(\frac{1}{2}\right)^{9} = 10/1024 = .0098$
0	$\binom{10}{0}\left(\frac{1}{2}\right)^{0}\left(\frac{1}{2}\right)^{10} = 1/1024 = .0010$

$$\binom{100}{60}\left(\frac{1}{2}\right)^{60}\left(\frac{1}{2}\right)^{40} = \frac{100!}{40!60!}\left(\frac{1}{2}\right)^{100}.$$

This is a rather small number and would take a long time to compute, even with a calculator. But to answer the question posed we would have to compute not only it but also Ian's probability of winning exactly 61 games, 62 games, . . . , up to 100 games.

Instead we will assume that the number of games r Ian wins is distributed normally. Since $n = 100$ is a fairly big number, a fairly accurate answer can still be obtained this way. We will need to know the mean m and standard deviation s of this binomial distribution to know the proper normal curve with which to approximate it.

Since $p = 1/2$, Ian expects to win about 1/2 of the 100 games, and $(1/2)(100) = 50$. We might expect that $m = 50$ for this distribution, and indeed that is correct.

More generally, if Ian had a probability p of winning each game and played n games, we would expect him to win $p \cdot n$ of them on the average.

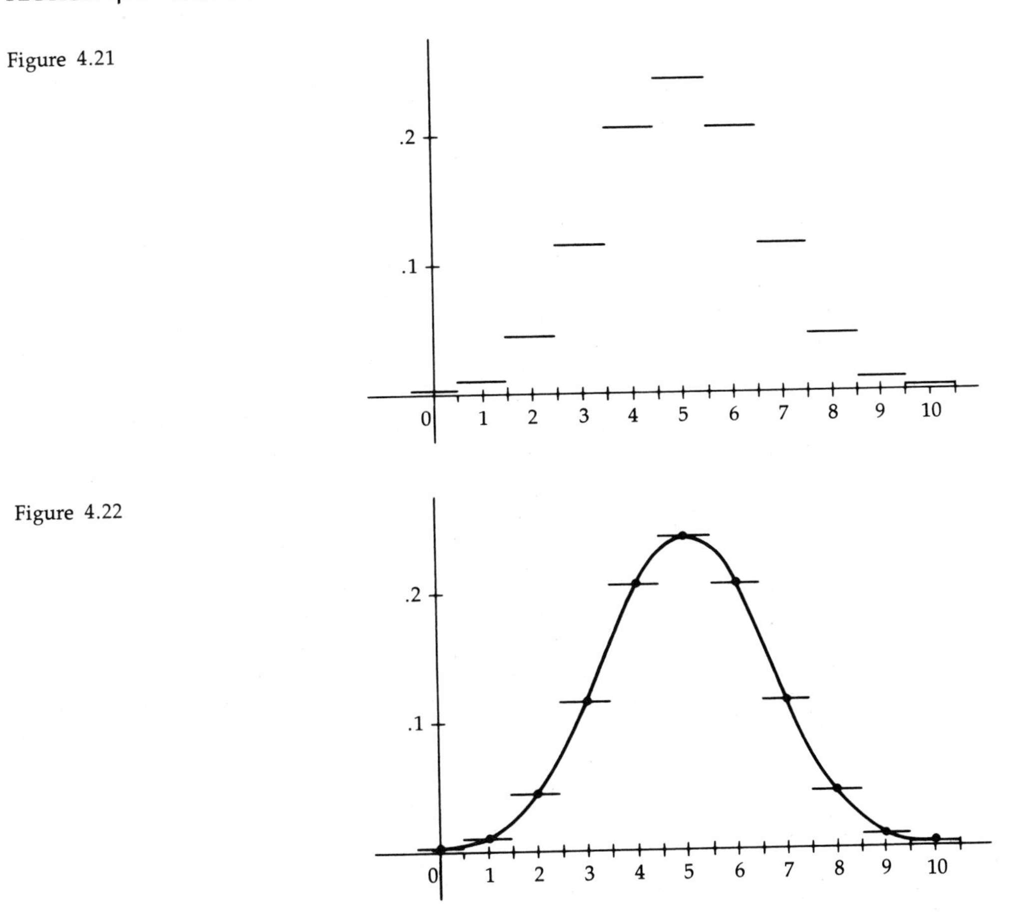

Figure 4.21

Figure 4.22

The mean of the binomial distribution of n trials with probability of success p is $m = p \cdot n$.

The formula for the standard deviation s of the binomial distribution is more difficult to justify. We will simply state that it can be proven that

The standard deviation of the binomial distribution of n trials with probability p of success is $s = \sqrt{n \cdot p \cdot (1 - p)}$.

In our example $n = 100$ and $p = 1/2$. Thus

$$s = \sqrt{100 \cdot \frac{1}{2}\left(1 - \frac{1}{2}\right)} = \sqrt{25} = 5.$$

Now we want $P(r \geq 60)$. We approximate the distribution of r by assuming it is normal with $m = 50$ and $s = 5$.

We can reduce r to a standard normal variable by the formula

$$x = \frac{r - m}{s} = \frac{r - 50}{5}.$$

In particular, when $r = 60$, $x = \dfrac{60 - 50}{5} = 2$. Then

$$\begin{aligned} P(r \geq 60) &= P(x \geq 2) \\ &= P(x > 0) - P(0 < x < 2) \\ &= .5 - .4772 = .0228. \end{aligned}$$

EXAMPLE 1

The Light Bulbs A manufacturer has a contract to sell light bulbs to the government. One bulb in a hundred is defective. Occasionally the government tests a crate of 1000 bulbs. If more than 15 are defective, the company must pay a cash penalty. What is the probability of this happening for a particular test?

Here we have a binomial distribution with $n = 1000$ trials and $p = .01$ (counting finding a defective bulb as a success). The mean of this distribution is $p \cdot n = (.01)(1000) = 10$ and the standard deviation is

$$s = \sqrt{np(1 - p)} = \sqrt{1000(.01)(1 - .01)} = \sqrt{9.9} = 3.15.$$

We want $P(r \geq 15)$. The standard normal variable x satisfies

$$x = \frac{r - m}{s} = \frac{r - 10}{3.15}.$$

When $r = 15$,

$$x = \frac{15 - 10}{3.15} = 1.59.$$

Then

$$\begin{aligned} P(r \geq 15) &= P(x \geq 1.59) \\ &= P(x > 0) - P(0 < x < 1.59) \\ &= .5000 - .4441 = .0559. \end{aligned}$$

EXAMPLE 2

The Hospital A hospital in an isolated community of 20,000 has 25 beds. The probability that any given resident needs hospital care at a particular time is .001. What is the probability that the hospital overflows at any given time?

We have a binomial distribution with $n = 20{,}000$ and $p = .001$. If x_1 is the number of people needing hospital care, then we want $P(x_1 > 25)$.

We use the normal approximation with

$$m = p \cdot n = (.001)(20{,}000) = 20$$

and

$$s = \sqrt{np(1-p)} = \sqrt{20{,}000(.001)(.999)} = \sqrt{19.98} = 4.47.$$

If x has the standard normal distribution, then

$$x = \frac{x_1 - m}{s} = \frac{x_1 - 20}{4.47}.$$

When $x_1 = 25$, then

$$x = \frac{25 - 20}{4.47} = 1.12.$$

$$\begin{aligned} \text{Thus } P(x_1 > 25) = P(x > 1.12) &= P(x > 0) - P(0 < x < 1.12) \\ &= .5 - .3686 \quad \text{(from Table II)} \\ &= .1314. \end{aligned}$$

EXAMPLE 3

The Traffic Jam Every weekday afternoon 80,000 cars travel from Downtown to the North Side by either Vine Street or the Freeway. The probability that a driver chooses the Freeway on any given day is .7. The

traffic moves well unless too many drivers choose the Freeway. In fact, if more than 55,800 drivers take the Freeway, then it becomes overloaded, and traffic slows to a crawl. What is the probability that this happens on any given day?

We have a binomial distribution with $n = 80{,}000$ and $p = .7$. Here

$$m = p \cdot n = (.7)(80{,}000) = 56{,}000$$

and

$$s = \sqrt{np(1-p)} = \sqrt{80{,}000(.7)(.3)} = \sqrt{16800} = 130.$$

We want $P(x_1 > 55{,}800)$, where x_1 is the number of drivers choosing the Freeway.

If we use the normal approximation we have

$$x = \frac{x_1 - 56{,}000}{130},$$

where x has the standard normal distribution. When $x_1 = 55{,}800$,

$$x = \frac{55{,}800 - 56{,}000}{130} = -1.54.$$

Then
$$\begin{aligned} P(x_1 > 55{,}800) = P(x > -1.54) &= P(-1.54 < x < 0) + P(x > 0) \\ &= P(0 < x < 1.54) + .5 \\ &= .4382 + .5 \\ &= .9382. \end{aligned}$$

The probability of a jam is .9382.

Exercises 4.6

1. Calculate the mean and standard deviation for the binomial distribution with $p = 1/2$ and $n = 6400$.
2. Calculate the mean and standard deviation for the binomial distribution with $p = .2$ and $n = 36$.
3. Calculate the mean and standard deviation for the binomial distribution with $p = .3$ and $n = 140$.
4. Calculate the mean and standard deviation for the binomial distribution with $p = .99$ and $n = 1000$.

In each exercise below a situation having a binomial distribution is described. Answer the question by approximating it with a normal distribution.

5. One toaster in 200 is defective. What is the probability that a shipment of 1000 toasters has fewer than 10 defective?

6. The Orioles win 60% of their games. What is the probability that they win 100 games or more in a 162-game season?

7. A 100-item test has five answers given for each multiple-choice question. What is the probability that a guesser gets (a) 30 or more right? (b) 40 or more right?

8. A student taking the test described in the last problem has a probability .3 of getting any one question right. What is her probability of getting 40 or more correct?

9. A student taking the test described in problem 7 has a probability .9 of answering any given question correctly. What is the probability that she gets fewer than 81 right on the test?

10. An experiment uses mice. At least 300 mice must survive until the end of the experiment in order for it to be valid. On the average 20% of the mice die before the experiment is over. If 400 mice start the experiment, what is the probability that it will be valid?

11. In Example 2, what is the probability of having more than 15 empty beds?

12. In Example 3, what is the probability that more than 56,400 drivers take the Freeway?

SOME NECESSARY MATHEMATICS FOR THE CONSUMER

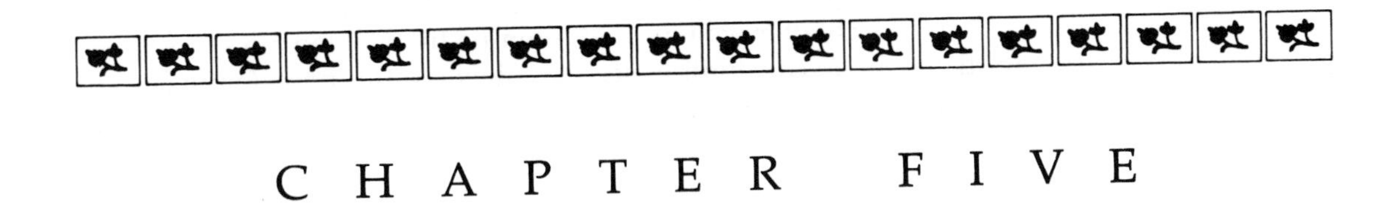

CHAPTER FIVE

The consumer has been much in the news lately: Congress has passed consumer protection bills, and unit pricing has been advocated by many consumer groups. Truth in lending laws have been passed, and lending institutions have been very forward in advertising their rates. These actions have been taken to help us all become more informed consumers. Legislation is not enough, however, since we must be able to use all this new information in an intelligent way.

Consideration of such things as unit pricing, interest rates, compounding periods, cash flows, fixed and variable expenses, depreciation, and insurance costs has been an important aspect of business economics for decades. It is also important for any individual or family that wants to get the most out of the available money.

In this chapter we will first examine the mathematics of interest and then, in a concluding section, delve into some familiar, and hopefully some not so familiar, situations that you as a consumer will meet.

5.1 WHAT IS INTEREST?

We often hear on the radio and television that "interest rates are up" or "the Federal Reserve has just cut the prime lending rate by one-quarter of a per cent." Some banks are advertising that they compound their interest daily. Is this like a compound in chemistry, or a compound fracture? (Well, according to an Oxford dictionary, compound means "not simple" so the answer to the last question is "yes.")

How much difference does a quarter of a per cent make when borrowing money? As with most simple questions, it depends. It depends on for how long the loan is, what the interest rate is, how much is borrowed, *and* the period for compounding. To us, going to a bank for a small, short-term loan, probably one-quarter of a per cent does not make too much difference, but to General Motors, which makes transactions of millions of dollars, it may mean several hundred thousand dollars. If you are buying a house with a $50,000 mortgage for 30 years, it means more than $3000 over the 30 years of the mortgage.

Interest charges can be rather surprising. For example, do you know that if you buy a house with a 30-year mortgage that you will pay more than twice what you borrowed? In fact, at $7\frac{1}{2}\%$ interest you will repay more than $125,000 on a $50,000 mortgage—*$75,000 in interest*. At $9\frac{1}{2}\%$, you will pay twice as much in interest as you borrowed. When you buy a new car with the minimal down payment and finance it for three years, you generally owe more on the car after one year than the car is worth. Does this mean you should not buy a house or a new car? We will say something about these problems later.

What other kind of problems are faced when dealing with interest? *Interest*—not the feeling of curiosity or fascination, but *the charge for a financial loan which is usually a percentage of the amount loaned.* There are surprisingly few real variations, although sometimes the wording or perspective can make the problems seem different.

For example, here are some problems that you will be able to solve by the completion of this chapter (remember, think positively).

1. If Ed deposits his inheritance of $5000 in the bank at $5\frac{1}{2}$% annually, how much will he have in 6 years?
2. Jana decides to save $50 a month for 10 years. How much will she have at the end of that time if she puts it into a savings account paying $5\frac{1}{2}$% monthly?
3. Anja wants to be able to buy a $6000 sports car in 4 years when she gets out of college. She can get a special 9% savings certificate that accumulates interest quarterly. For how much should the savings certificate be made?
4. Gene decides to buy a new car for $5500 and to finance $5000 of it at 8% interest over 3 years (his motorcycle had a $500 trade-in value). How much will his monthly payments be?
5. Peggy wants to put her $100,000 winnings from the racetrack into an 8% annuity that will make a payment to her every three months for the next 20 years. How much will she get each quarter?
6. After a successful weekend in Las Vegas (due no doubt to the extensive knowledge of probability theory he learned in Chapter 3, but perhaps from a kiss from Lady Luck), Stanley decides to put his $30,000 in winnings into a savings account paying 6% interest monthly. He wants to draw out living expenses of $300 a month. How long can he do this?
7. Rebecca decides to buy a new $60,000 house and has $10,000 for closing expenses and down payment. (The minimum down payment is 10% of the cost of the house.) She can get a 30-year mortgage from her bank or from her savings and loan association. The bank will charge $7\frac{1}{4}$% interest and 2 points, whereas the savings and loan charges $7\frac{1}{2}$% but no points. Where should she get her mortgage?
8. Kevin decides to work 30 years and then retire. He decides that when he retires at age 51 in 30 years he can take his savings and invest it in a gold-plated building and loan at 6% and draw on it quarterly for 25 years, at which time, his guru convinces him, the end of the world will occur. Now, he wants to get $3000 every three months after retirement. How much must he invest every six months at 10% semi-annually during the next 30 years so he will be able to achieve his retirement goal?

9. Roberto's job (he got it April 1, 1979) requires him to participate in a retirement fund that only pays 4% annually. His contribution of $300 per quarter is forwarded to the retirement fund. How much will be in the fund for him when he retires on March 31, 2009?

In spite of the apparent variation in the problems, we will find that we only need to consider four different formulas and the tables they generate. Only one of these problems cannot be answered by direct consideration of certain values in an appropriate table.

Simple and Compound Interest

Simple interest can best be explained by an example. Suppose $1000 is put into a savings account that pays 5% simple interest each year for 10 years. What *simple* means is that the interest already earned does not itself earn interest. How much will be in the account at the end of 10 years?

The computation is quite easy. We take 5% of $1000 (that is, (.05)(1000)) and get $50.

This is what our $1000 earns every year for 10 years. The total interest will then be ($50)(10) or $500. We could have done this all at once by applying the formula

(1) principal × rate × time = interest.

Our computation would have been

$$(\$1000)(.05)(10) = \$500.$$

The total in our account at the end of 10 years would be

$$\$1000 + \$500 = \$1500.$$

This was found by the obvious formula principal + interest = amount. This is all there is to say about simple interest. . . . Well, almost. We will point out how foolish it would be to save money this way a little later.

First we will see what happens if the interest already earned is allowed to draw interest during subsequent years. Each time we apply formula (1) the time will be one year. Hence the new amount will be principal + (rate)(principal).

We show the succession of computations for each year:

$$
\begin{array}{llll}
\text{End of year 1:} & 1000 + .05(1000) & = 1000(1 + .05) & = 1000(1.05) \\
\text{End of year 2:} & 1000(1.05) + .05(1000)(1.05) & = 1000(1.05)(1 + .05) & = 1000(1.05)^2 \\
\text{End of year 3:} & 1000(1.05)^2 + .05(1000)(1.05)^2 & = 1000(1.05)^2(1 + .05) & = 1000(1.05)^3 \\
\text{End of year 4:} & 1000(1.05)^3 + .05(1000)(1.05)^3 & = 1000(1.05)^3(1 + .05) & = 1000(1.05)^4 \\
\vdots & \vdots & \vdots & \\
\text{End of year 10:} & 1000(1.05)^9 + .05(1000)(1.05)^9 & = 1000(1.05)^9(1 + .05) & = 1000(1.05)^{10} \\
 & & = 1628.90, &
\end{array}
$$

where we have used our pocket calculator to find the total of \$1628.90.

Thus, by allowing the interest earned each year to also earn interest, at the end of 10 years we have \$128.90 more than we would have otherwise. This process of adding the interest already earned to the principal and treating this sum as the principal for the next interest computation is called *compounding*. The value of compound interest over simple interest should not need further explanation at this stage. (See Figure 5.1.) However, notice that it is on the average \$12.89 per year foolish to invest at simple interest (in our example) rather than compound interest. Since compound interest is such a good thing, how can we determine the amount earned by compounding without going through a long series of computation as we did above? You may have wondered why we put in all the algebraic steps in the above analysis when the first and last expressions are all we need. The reason is that if you examine the algebra a little, you will see that the amount accumulated after n years at 5% interest per year compounded annually is

$$1000(1.05)^n.$$

How can we generalize this? Well, the \$1000 is the amount deposited, which we might call P for principal, and .05 is the interest rate; let's call it i. If we let A be the amount accumulated, the formula becomes

$$A = P(1 + i)^n.$$

We state this as a theorem now, but we will show some more examples next, postponing the proof until later in the section.

THEOREM 1

The Compound Interest Formula The amount A that is accumulated when a single deposit P is left for n periods at interest i *per period* is given by the formula

$$A = P(1 + i)^n.$$

1978=
Make every day count~
at Bloomington Federal
with day-in/day-out Daily Compounded Interest
on every savings dollar you deposit.

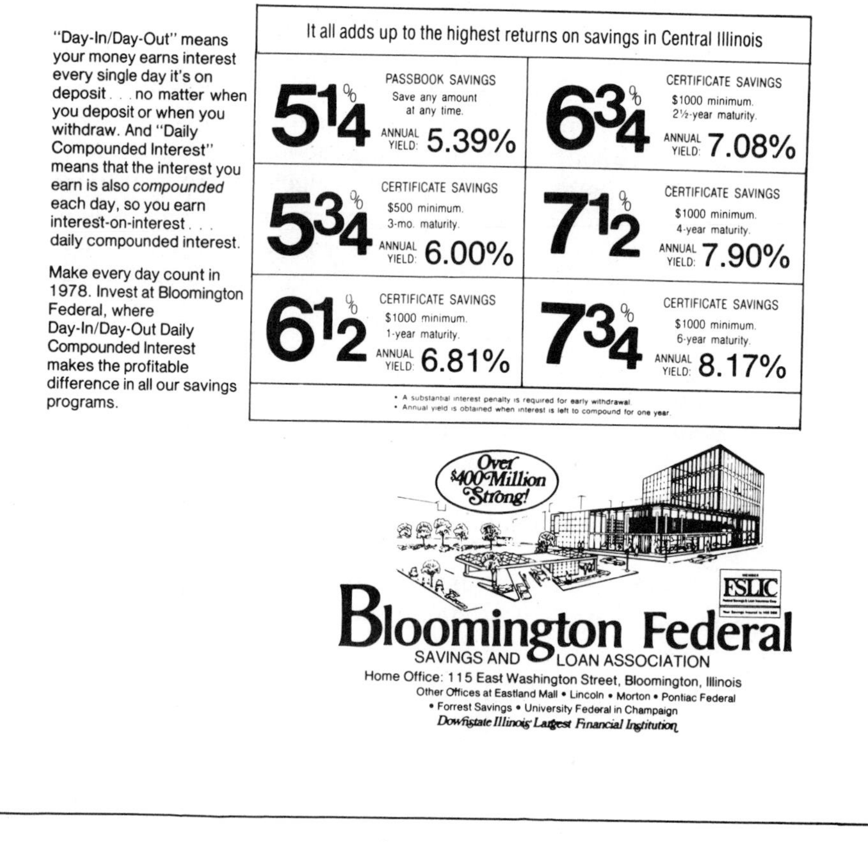

"Day-In/Day-Out" means your money earns interest every single day it's on deposit . . . no matter when you deposit or when you withdraw. And "Daily Compounded Interest" means that the interest you earn is also *compounded* each day, so you earn interest-on-interest . . . daily compounded interest.
Make every day count in 1978. Invest at Bloomington Federal, where Day-In/Day-Out Daily Compounded Interest makes the profitable difference in all our savings programs.
It all adds up to the highest returns on savings in Central Illinois
5¼% PASSBOOK SAVINGS Save any amount at any time. ANNUAL YIELD: 5.39%
6¾% CERTIFICATE SAVINGS $1000 minimum. 2½-year maturity. ANNUAL YIELD: 7.08%
5¾% CERTIFICATE SAVINGS $500 minimum. 3-mo. maturity. ANNUAL YIELD: 6.00%
7½% CERTIFICATE SAVINGS $1000 minimum. 4-year maturity. ANNUAL YIELD: 7.90%
6½% CERTIFICATE SAVINGS $1000 minimum. 1-year maturity. ANNUAL YIELD: 6.81%
7¾% CERTIFICATE SAVINGS $1000 minimum. 6-year maturity. ANNUAL YIELD: 8.17%
• A substantial interest penalty is required for early withdrawal.
• Annual yield is obtained when interest is left to compound for one year.
Over $400 Million Strong!
FSLIC
Bloomington Federal
SAVINGS AND LOAN ASSOCIATION
Home Office: 115 East Washington Street, Bloomington, Illinois
Other Offices at Eastland Mall • Lincoln • Morton • Pontiac Federal
• Forrest Savings • University Federal in Champaign
Downstate Illinois' Largest Financial Institution

Figure 5.1

Courtesy Bloomington Federal Savings and Loan Association.

EXAMPLES

1. Suppose \$200 is deposited for 8 years at 6% interest compounded annually. The amount after 8 years is

$$A = 200(1 + .06)^8 = 200(1.5938) = \$318.76.$$

2. Suppose \$3000 is deposited at 9% compound interest. The amount after 20 years is

$$A = 3000(1 + .09)^{20} = 3000(5.60441) = \$16{,}813.23.$$

Often the term "compounded" is left out when describing interest. Thus "6% quarterly" means "6% compounded quarterly," "6% monthly" means "6% compounded monthly," and "6% daily" means "6% compounded daily." (Usually, but not always, this means 365 times in a year.)

The problem is only slightly complicated by the fact that almost no one compounds annually anymore. The usual periods are quarterly or monthly, although semiannual periods are not unheard of. Some banks and saving and loan associations even advertise daily compounding. How does this change the problem?

For example, suppose we deposit \$500 in a savings account at 6% quarterly for 5 years. The "6% quarterly" means that an annual rate of 6% is paid and interest is compounded quarterly, that is, every 3 months. In one year there will be 4 periods and each period will have an interest rate of $\frac{1}{4}(6\%) = 1\frac{1}{2}\% = .015$. Thus for this problem $P = 500$, $i = .015$, and $n = 20$ (5 years times 4 periods per year). The amount from the formula is thus

$$\begin{aligned} A &= 500(1.015)^{20} \\ &= 500(1.34686) \\ &= \$673.43. \end{aligned}$$

As a last example, suppose \$800 is deposited at 6% monthly. Then after 5 years the accumulated amount will be

$$\begin{aligned} A &= 800(1.005)^{60} \\ &= 800(1.34885) \\ &= \$1079.08. \end{aligned}$$

(Notice $i = .06/12 = .005$ and $n = (12)(5) = 60$.)

Demonstration of the Compound Interest Formula

We will carry out a few steps to show how the process works. Suppose P is the amount deposited at interest rate i per period. We can construct Table 5.1 to show what occurs as time goes on.

If the rate of interest is r per year and the compounding occurs k times in a year for m years, then in the compound interest formula

$$i = r/k$$

and

$$n = k \cdot m.$$

TABLE 5.1

AT END OF PERIOD	DEPOSIT HELD DURING PERIOD	AMOUNT OF INTEREST EARNED DURING PERIOD
1	P	iP
2	$P(1 + i)$	$iP(1 + i)$
3	$P(1 + i)^2$	$iP(1 + i)^2$
.	.	.
.	.	.
.	.	.
n	$P(1 + i)^{n-1}$	$iP(1 + i)^{n-1}$

Present Value and the Tables

Gayl has just won $10,000 in the lottery and wants to put part of her money into a college fund for her sons, Gabe and Greg. She decides that they will need $16,000 between them when they start college in 10 years. How much of her winnings will she need to deposit now at 8% compounded quarterly?

This problem can be solved by viewing the compound interest formula in a different way than we have thus far. The question is, What do we need to deposit so that we will have $16,000 in 10 years, given that our deposit will be earning 8% quarterly interest? Using the formula, we can write (letting $i = .08/4$ and $n = (4)(10)$)

$$16{,}000 = P(1 + .02)^{40}.$$

Then, solving for P, we get

$$P = \frac{16{,}000}{(1.02)^{40}} = \frac{16{,}000}{2.20804} = \$7246.25$$

as the amount Gayl needs to deposit now at 8% quarterly in order to have $16,000 in 10 years.

This shows that we can change the compound interest formula to a present value formula. We simply solve

$$A = P(1 + i)^n$$

for P and obtain

$$P = \frac{A}{(1 + i)^n} = A(1 + i)^{-n}.$$

We state this as

SUM OF COLUMNS 2 AND 3	FACTORED SUM
$P + iP$	$P(1 + i)$
$P(1 + i) + iP(1 + i)$	$P(1 + i)(1 + i) = P(1 + i)^2$
$P(1 + i)^2 + iP(1 + i)^2$	$P(1 + i)^2(1 + i) = P(1 + i)^3$
.	.
.	.
.	.
$P(1 + i)^{n-1} + iP(1 + i)^{n-1}$	$P(1 + i)^{n-1}(1 + i) = P(1 + i)^n$

THEOREM 2

Present Value Formula The present value PV of an amount A desired in the future under the assumption of accumulation of interest at an interest rate i per period compounded for n periods is given by the formula

$$\text{PV} = A \frac{1}{(1 + i)^n} = A(1 + i)^{-n}$$

The only reason to have a new formula is that multiplication is so much easier than division, and tables are readily available which list values of

$$\frac{1}{(1 + i)^n} = (1 + i)^{-n}.$$

In fact, tables for $(1 + i)^n$ are also common. To simplify your life, we have included Table III, which is used for the compound interest formula and Table IV, which is used for the present value formula.

EXAMPLES

1. We deposit \$1500 at 10% quarterly for 11 years. How much will be in our account? For this problem

$$P = 1500,$$
$$i = 2\tfrac{1}{2}\% \qquad (10\% \div 4)$$
$$n = 44 \qquad (4 \times 11).$$

Reading from Table III, we get

$$\begin{aligned} A &= 1500(\text{Table III entry with } i = 2\tfrac{1}{2}\%, n = 44) \\ &= 1500(2.9638) \\ &= \$4445.70. \end{aligned}$$

2. What is the present value of \$4000 if interest is earned at the rate of 8% semiannually for 15 years? Here $A = 4000$, $i = 4\%$, and $n = 30$. Reading from Table IV, we get

$$\begin{aligned}\text{PV} &= 4000(\text{Table IV entry with } i = 4\%, n = 30)\\ &= 4000(.30832)\\ &= \$1233.28.\end{aligned}$$

3. If \$500 is deposited at $5\frac{1}{2}\%$ interest compounded monthly, how much interest will have been earned in 5 years?

This is beyond the scope of our tables, but even a nonprogrammable calculator can be used to solve it in a few minutes.

Any calculator that allows successive multiplication by a constant will do. First we write down the compound interest formula with

$$i = 5\tfrac{1}{2}\% \div 12 = .055/12,$$

$$n = 12 \times 5 = 60,$$

and

$$P = 500.$$

Thus

$$A = 500\,(1 + .055/12)^{60}.$$

So enter .055, divide by 12, and add 1. This is the number to be multiplied by itself 60 times. On our calculator, we press the × button and then the = button 59 times (the first time gives the square) and we get

1.3156984.

Multiplying by 500 gives

$$A = \$657.85,$$

and interest = \$157.85.

However, if you have a calculator with an x^y or y^x button, the calculation is clearly much easier. You simply

enter .055,

divide by 12,

and add 1.

The result should be

$$1.004583333.$$

Then press the x^y or y^x button and press 60. Pressing = should give

$$1.315703746.$$

Finally, multiplying by 500, we get

$$\$657.85,$$

the same answer as above. (If your calculator uses reverse Polish instead of algebraical logic, the order in which the keys must be pressed will be slightly different.)

Exercises 5.1

1. In each of the following compute the simple interest earned by the deposit at the annual rate of interest given over the time period shown, and then the interest earned if compounding takes place as shown.
 (a) \$100, 4%, 10 years, compounded quarterly
 (b) \$50, 12%, 3 years, compounded monthly
 (c) \$1000, 8%, 20 years, compounded semiannually
 (d) \$10,000, 6%, 4 years, compounded quarterly

2. Kathy has deposited \$200 in her savings account, which pays 6% quarterly. How much will she have in 11 years?

3. Which is the better deal, to deposit \$300 at 3% for 15 years compounded semiannually or at $2\frac{1}{2}$% for 18 years with annual compounding? By how much?

> Notice that the *preferred* definition of bimonthly is "once every two months," which is the definition we use. However, occasionally it is used to mean "twice a month," even though in this sense *semimonthly* is the preferred term.

4. Suppose you want \$10,000 in 10 years and you can deposit a sum now in a savings and loan association that pays 8% quarterly. How much should you deposit?

5. In each of the following, find the present value for the interest, time, and future amount given.
 (a) A = \$750, 10% quarterly for 8 years
 (b) A = \$5, 6% bimonthly for 5 years
 (c) A = \$1000, 5% semiannually for 10 years

6. Jean wants to have \$4000 in 6 years when she begins college. Should she deposit her money at 4% quarterly or 5% bimonthly? Why?

7. In 1973, Anthony's "progressive" employer decided that he should be given a cost of living increase in salary of 6% per year. If he was earning $20,000 then, how much was his salary in 1979?

The following problems require the computer or a hand calculator, as they are beyond the scope of our tables.

8. In 1970, John's firm started to give a 10% annual increase in salary. If he earned $15,000 per year then, how much would he be earning 8 years later?

9. In 1966, Randall Trucking started giving its employees a cost of living raise. If the per cent cost of living during each of the following 12 years was as given, how much would a salary of $12,000 have increased to by 1978?

RISE IN CONSUMER PRICE INDEX

1966	3.3%
1967	3.1%
1968	4.7%
1969	6.2%
1970	5.5%
1971	3.4%
1972	3.4%
1973	8.8%
1974	12.2%
1975	7%
1976	4.8%
1977	6.8%

10. Manhattan Island sold for $24 in 1584. If the Indians had deposited that amount at 4% per annum interest, how much would they have now? Supposing it had been compounded semiannually instead of annually, how much difference would there be?

11. Help Ed and Anja out by solving their problems as stated in Section 5.1.

The following problems and discussion refer to Figure 5.1.

12. Show that $5\frac{1}{4}$ percent compounded daily (365 times) gives an *annual yield* of 5.39%, that is, it is the same as 5.39% annual interest rate.

13. Find the amount $1 will earn in 1 year at $5\frac{3}{4}$% compounded daily.

If you use a 10-place calculator with a y^x button for this problem, you should get

1.059180474,

so the answer is 5.918¢. That is, the annual yield of $5\frac{3}{4}\%$ compounded daily should be 5.92% and not 6.00%.

The savings and loan can advertise a higher yield because they are using what is called a 360/365 formula. Let us examine how we have learned to do a problem like this one. Our formula is

$$(1 + i)^n = (1 + .0575/365)^{365}$$

since $P = \$1$, $r = .0575$, $m = 1$, and $k = 365$. However, in the 360/365 method, the yield is computed as

$$(1 + .0575/360)^{365}.$$

In this case the result is

1.060026551,

so the yield would be 6.00% as the ad claims.

14. Find the annual yield on $6\frac{1}{2}\%$ compounded daily and using the 360/365 method. About how much difference is there?

15. Find the annual yield on $6\frac{3}{4}\%$ compounded daily and using the 360/365 method. About how much difference is there?

16. Find the annual yield on $7\frac{1}{2}\%$ compounded daily and using the 360/365 method. About how much difference is there?

Once the annual yield is found, this can be used as the rate of annually compounded interest.

17. How much more does a \$1000, 4-year, $7\frac{1}{2}\%$ savings certificate earn at Bloomington Federal than at an institution giving ordinary daily compound interest?

18. Find the total interest earned on a \$5000, 6-year, $7\frac{3}{4}\%$ savings certificate that earns interest compounded daily. How much less is this than what the corresponding certificate bought at Bloomington Federal would earn?

5.2 SAVINGS AND RETIREMENT PLANS

Suppose instead of saving a lump sum for a period of time, we continually pay into our savings account a fixed amount each period. Such problems were faced by Jana, Kevin, and Roberto in the introduction to this chapter. We show how to solve these problems in this section.

Before we examine the problem, however, it is useful to obtain a preliminary result on geometric series.

Summing a Geometric Series

The derivation of the formula we need requires that we sum what is called a *geometric series.* This is a series of the form

$$a + ar + ar^2 + \cdots + ar^{m-1},$$

where we notice that each term, except the first, is obtained by multiplying the previous term by r, called the *common ratio*. If we let S be this sum, and notice what r times S is, by subtraction we get the following computation:

$$\begin{array}{rrl} & rS = & ar^m + ar^{m-1} + \cdots \quad + ar^2 + ar \\ (-) & S = & \qquad\quad ar^{m-1} + ar^{m-2} + \cdots + ar + a \\ \hline & rS - S = & ar^m \qquad\qquad\qquad\qquad\qquad\qquad - a \end{array}$$

Thus since all the middle terms cancel out, we get

$$rS - S = ar^m - a,$$

or

$$S(r - 1) = ar^m - a,$$

or

$$S = \frac{ar^m - a}{r - 1}.$$

We can remember this by remembering the verse

> The sum of
> a geometric series
> is the
> next unused term minus the first term
> divided by
> the common ratio minus one

An Example

We will show how this formula arises by considering an example before we look at the general situation.

Suppose we decide to deposit $100 every 3 months into a savings account that pays 8% quarterly interest. To be more precise, we will make our first deposit on April 1 and will ask for the balance on our account after we have made our deposit on April 1 two years later. We need to make this assumption about having just made a deposit for a technical reason that we will explain later.

What happens to our money? First notice that we have made *nine* deposits, so we should have more than $900 in our account. Yet if we compute the amount in our account as a result of the first $100 we deposit, there are only 8 compounding periods, each accumulating interest at a 2% rate. By using Table III with $n = 8$ and $i = 2\%$, the amount, call it A_1, including interest from this deposit will then be

$$A_1 = 100\ (1.1717) = \$117.17.$$

The second deposit will only be in the bank for 7 periods, however. Thus A_2 would be

$$\begin{aligned} A_2 &= 100(1 + .02)^7 \\ &= 100(1.1487) \\ &= \$114.87, \end{aligned}$$

using Table III. We show these and the other amounts in Table 5.2 (page 282). (If you use your hand calculator to compute the entries in column 3, you will see a considerable amount of roundoff error.)

This appears to be an extremely tedious way to solve the problem. Let's try to use the generality which is the primary achievement of mathematics to solve the problem in a way that will simplify finding solutions to similar problems.

Notice that each entry in column 5 is just 100 times the number in column 3—admittedly written a little differently. Thus to find the value of the sum, we can just take 100 times the sum

$$(1.02)^0 + (1.02)^1 + (1.02)^2 + (1.02)^3 + \cdots + (1.02)^7 + (1.02)^8.$$

Any nonzero term to the zeroth power is just 1, so our sum is

$$1 + (1.02)^1 + (1.02)^2 + \cdots + (1.02)^7 + (1.02)^8.$$

TABLE 5.2

1 DEPOSIT NUMBER	2 NO. OF PERIODS	3 TABLE III ENTRY (i = 2%)	4 AMOUNT ($100 TIMES COLUMN 3)	5 FORMULA
1	8	1.1717	$117.17	$100(1.02)^8$
2	7	1.1487	114.87	$100(1.02)^7$
3	6	1.1262	112.62	$100(1.02)^6$
4	5	1.1041	110.41	$100(1.02)^5$
5	4	1.0824	108.24	$100(1.02)^4$
6	3	1.0612	106.12	$100(1.02)^3$
7	2	1.0404	104.04	$100(1.02)^2$
8	1	1.0200	102.00	$100(1.02)^1$
9	0	1.0000	100.00	$100(1.02)^0$
Total			$975.47	

What a coincidence! A geometric series with first term $a = 1$, common ratio $r = (1.02)$, and $m = 9$. Our formula from the previous section tells us this sum is

$$S = \frac{(1.02)^9 - 1}{(1.02) - 1} = \frac{(1.02)^9 - 1}{.02}.$$

We can look up $(1.02)^9$ in Table III and get

$$S = \frac{1.1951 - 1}{.02} = \frac{.1951}{.02} = 9.7550.$$

Upon multiplying by $100 we would get $975.50, which only differs by $.03 from the sum obtained in column 4. Why is there any difference at all?

This is easily seen to be roundoff error. If the figures in column 4 of the table are written to the nearest tenth of a cent, the total would be $975.463. Also, if we use our pocket calculator to compute S instead of using Table III, we get

$$S = \frac{(1.02)^9 - 1}{(1.02) - 1} = \frac{0.19509257}{.02} = 9.754628$$

Now when we multiply by $100 we get $975.4628, and both values are the same to the nearest cent.

The General Solution

Suppose that we make n deposits of an amount D with an interest rate of i being earned between deposits. Further assume that compounding occurs when we make our deposits, and that our last deposit was just made, so that no interest has been earned on it. The first deposit has earned interest for $n - 1$ periods, the second for $n - 2$ periods, and so on. The total amount A will then be

$$\begin{aligned} A &= D(1 + i)^{n-1} + D(1 + i)^{n-2} + \cdots + D(1 + i)^2 + D(1 + i) + D \\ &= D\left[(1 + i)^{n-1} + (1 + i)^{n-2} + \cdots + (1 + i)^2 + (1 + i) + 1\right] \\ &= D\,\frac{(1 + i)^n - 1}{(1 + i) - 1} \\ &= D\,\frac{(1 + i)^n - 1}{i}. \end{aligned}$$

Thus we finally arrive at

THEOREM 3

Savings Plan Formula The accumulated amount A in a savings plan that has accrued from n deposits of D each with interest compounded at a rate i between deposits is given by the formula

$$A = D\left[\frac{(1 + i)^n - 1}{i}\right]$$

provided the last deposit was just made. The quantity in the brackets can now be easily computed by using Table III.

EXAMPLE 1

A deposit of $250 is made each quarter for 3 years at 6% interest. Assuming a deposit is made on the first and last day, find the accumulated amount. Here $i = 1\frac{1}{2}\%$ and $n = 13$, since $i = 6\% \div 4$ and $n = (3)(4) + 1$, the last term of 1 representing the deposit that has not earned interest.

$$\begin{aligned} A = 250\,\frac{(1 + .015)^{13} - 1}{.015} &= 250\left(\frac{1.2136 - 1}{.015}\right) = 250\left(\frac{.2136}{.015}\right) \\ &= 250\,(14.24) = \$3{,}560 \end{aligned}$$

Using Table V

The use of the savings plan formula and Table III requires us to divide. Not a completely unreasonable task, but one likely to produce more errors than multiplication, as we noted earlier. To help us we have another table!

Table V in the back of the book gives the values of

$$\frac{(1+i)^n - 1}{i},$$

which is exactly what we want. Don't forget, though, that in using the savings plan formula and Table V you are assuming a deposit that has not earned any interest.

EXAMPLE 2

Nine deposits of \$100 are made at 8% quarterly, one each 3 months for 2 years. How much has accumulated? Here $D = 100$, $i = 2\%$, and $n = 9$, so

$$\begin{aligned} A &= 100(\text{Table V entry with } i = 2\%,\ n = 9) \\ &= 100(9.7546) \\ &= \$975.46. \end{aligned}$$

EXAMPLE 3

If you decide to join the Garden-of-Eden Mutual Fund, which guarantees 6% compounded quarterly, by paying \$200 into it every 3 months, what will be the minimum in your account after 5 years? (This is a purely fictitious problem, as we know of no mutual fund that guarantees *any* percentage.)

In this example we are making 4 deposits in each of 5 years plus the last non-interest bearing deposit, which makes $n = 21$. The interest $i = 6\% \div 4 = 1\frac{1}{2}\% = .015$, while $D = 200$. From the savings plan formula and Table V we get

$$\begin{aligned} A &= 200(\text{Table V entry with } i = 1\tfrac{1}{2}\%,\ n = 21) \\ &= 200(24.4705) \\ &= \$4894.10. \end{aligned}$$

Notice that of this total amount only \$694.10 is interest; the rest is principal.

Naturally we could use this formula to find out what we would

Here is the rarest of the rare-
the premier Mercedes-Benz
touring coupe. The 450SLC.
Come and test drive this limited-edition Mercedes-Benz, engineered
to a sports car philosophy, with the luxury of a fine Sedan.

Mercedes-Benz of North America Inc.

need to save each month (or other time period) to save enough to pay cash for a purchase.

EXAMPLE 4

Suppose you decide to save until you have enough to buy a Mercedes-Benz 450 SLC, which costs \$30,500. With inflation you decide that a new one will cost at least \$35,000 in 10 years. How much must you save at 6% quarterly in order to buy your dream car?

In this case the accumulated amount will be \$35,000, the interest is 1.5%, and $n = 41$. The amount you must deposit quarterly is D, where

$$
\begin{aligned}
35{,}000 &= D(\text{Table V entry with } i = 1.5\%,\ n = 41) \\
&= D(56.0519).
\end{aligned}
$$

So

$$D = 35{,}000/56.0519 = \$624.42.$$

EXAMPLE 5

Perhaps a more realistic example would be furnished by asking how much you need to save each month at 6% monthly interest in order to buy a \$1000 stereo system in 2 years. Here the figures become

$$
\begin{aligned}
1000 &= D(\text{Table V entry with } i = .005,\ n = 25) \\
&= D(26.5591),
\end{aligned}
$$

So

$$D = 1000/26.5591 = \$37.65.$$

Notice that you pay only $941.25 for the system. In Section 5.5 we will see that if you bought the system now for $1000 and paid the usual carrying charge of 1.5% per month interest you would pay $49.92 per month, or $1198.08 total. Is it worth an extra $250 to have the system now?

Exercises 5.2

1. Using Table V find the accumulated amounts in each of the following cases:
(a) D = $150, 4% quarterly for 4 years
(b) D = $300, 6% quarterly for 2 years

2. You deposit $500 in a bank at 5% compounded semiannually and $500 every 6 months after that. How much is in the account after 19 years?

3. Annemarie begins her monthly savings plan on May 1 by depositing $20. She deposits this same amount monthly in her account paying 12% annual interest compounded monthly. How much is in her account on May 1 in 4 years?

4. Carl is hired by the Slick Wick Company on July 1 and must join their retirement fund. He is required to deposit $500 for every 6 months he is employed, but the fund pays 8% compounded semiannually. How much is in his account on July 1 after 12 years? Suppose he had been fired on June 30. How much then?

5. An employee joins the credit union by depositing $50, and he deposits $50 every 3 months. After 9 years, if the credit union pays 6% compounded quarterly, how much is in the account?

6. A new father decides to put $10 every two months into his newborn son's savings account, which pays 6% compounded bimonthly. How much will be in the son's account on his fourth birthday?

7. Having just graduated from college, Steve decides he wants to save up to buy a sports car. He decides he is willing to put $200 every 3 months into his savings account, which pays 8% quarterly. How much will he have in 10 years? How much would he have to put into his account every quarter if he wanted to buy a $25,000 Mercedes-Benz 450SL in 10 years?

8. In problem 6, how much will be in the son's account when he is ready to start college on his nineteenth birthday?

9. Help Jana by solving her problem, Example 2 in Section 5.1.

10. How much is in Carl's account (problem 4) after 18 years? 25 years?

11. How much is in the employee's account (problem 5) after 20 years? After 30 years?

12. Suppose Carl in problem 4 has to deposit $250 every 3 months at the same 8% interest rate but now compounded quarterly. How much would be in his account after 12 years? 18 years? 25 years?

13. How much interest would a deposit of a dollar a year generate in 50 years? 80 years? 100 years? (Take $i = 3\%$ annually.)

14. Kim decides to save for a new living room suite which she estimates will cost $1300. How much will she need to save each month at 6% monthly interest to be able to buy such a suite in 3 years?

15. Nicole wants a new stereo and tape deck, and her father agrees to pay her 10% per month interest on the amount she puts away toward its cost. How much does she need to save each month to have $600 in 18 months?

16. How much would a 20 year old have to save each month at 8% monthly interest in order to have $1,000,000 when he retires at age 65?

17. A new 25-year-old executive decides she wants to retire at 50 with $250,000. How much of her executive salary must she save each month at $8\frac{1}{4}\%$ monthly in order to reach her goal?

18. A 21-year-old school teacher decides that he wants to be able to retire in 30 years with a savings balance of $200,000. Can he reasonably expect to be able to do so if he can get $8\frac{1}{2}\%$ monthly interest on his monthly deposits?

5.3 VARIATIONS AND EXTENDING TABLE V

Since just looking up numbers in tables and doing a little multiplication can get pretty tedious, we include three variations to maintain your interest. (Some students have claimed it's to complicate the situation, but surely we wouldn't do that.)

Two of these variations deal with situations where the number of periods or payments does not fit the prescription given in the savings plan formula, and the third variation investigates what happens when a payment is missed. In this latter case we see that Table III can be used in the solution.

VARIATION 1

Bert joins the payroll savings plan and will have deposited $180 every 6 months into a bank paying 6% compounded semiannually. How much is in the plan after 15 years?

Solution In this case it appears that the first deposit was not made until the end of the first 6 months. This means the first deposit drew interest for only $14\frac{1}{2}$ years or 29 periods. Thus $n = 30$, $i = 3\%$, and $A = 180(47.5754) = \$8563.57$.

VARIATION 2

Jack started saving on January 1, 1960 by depositing $150 on that date and each 3 months thereafter in a bank paying 10% compounded quarterly. How much was in his account on December 31, 1969?

Solution We need to compute the amount that would be in the account on January 1, 1970 assuming a deposit that day and then subtract that deposit. Thus we take

$n = 41$ (4 deposits per year for 10 years plus the fictitious January 1, 1970 deposit),

$i = 2\frac{1}{2}\%$ (10% divided by 4),

$D = \$150$ (the deposit amount)

to obtain

$$A = 150(70.0876) = \$10{,}513.14.$$

Now we subtract the $150.00 not really deposited and get the correct total $10,363.14.

VARIATION 3

A man takes out a savings annuity paying 8% compounded quarterly with quarterly payments of $200, including the first day. Unfortunately, he missed the eleventh payment because he was in the hospital. How much was in the account after 8 years?

Solution We first work the problem assuming no missed payment. Then $n = 4 \cdot 8 + 1 = 33$ and $i = 2\%$, so using Table V, we have $A = 200(46.1116) = \$9222.32$. We can use Table III to find how much of this should be credited to the eleventh payment. Now, the first payment was compounded 32 times, the second payment was compounded 31 times, the third compounded 30 times. Then the eleventh payment must have been compounded 22 times. (Why?) From Table III, with $n = 22$ and $i = 2\%$, we get that the eleventh payment contributed $200(1.5460) = \$309.20$. This amount is to be subtracted. The answer is then

$$9222.32 - 309.20 = \$8913.12.$$

More extensive tables giving more periods and more interest rates can be found in books of mathematical tables. Those of you who have access to a computer can generate your own tables for whatever interest rate i per period you like and for however many periods you would like. However, the formula in the accompanying box shows how our Table V can be extended in length for one of the interest rates we used there. We show one example.

EXTENSION FORMULA FOR TABLE V
For a fixed interest rate i per period, if we let S_n denote the Table V entry for n periods at interest rate i per period, then

$$S_{m+n} = S_m + S_n + iS_mS_n.$$

EXAMPLE

Suppose we deposit \$15 bimonthly in our savings account, which pays 9% per annum compounded bimonthly. How much will we have in our account after 15 years? How much of this will be interest?

Solution First we have to decipher the problem. Bimonthly means every two months, so we make 6 deposits per year or (6)(15) = 90 deposits in 15 years—oops, don't forget that our formula assumes we just made a deposit, so we must take

$$n = 91.$$

Then $i = 9\% \div 6 = 1.5\%$ and $D = \$15$. But our Table V does not go up to $n = 91$. Thus we must think of m and n so that m and n are in our table and

$$m + n = 91.$$

There are lots of possibilities, for example, $m = 45$, $n = 46$, or $m = 43$ and $n = 48$, or $m = 50$ and $n = 41$. We will use the latter numbers. The formula says

$$S_{91} = S_{50} + S_{41} + iS_{50}S_{41}.$$

So using $i = 1.5\%$ and reading from Table V

$$\begin{aligned} S_{91} &= 73.6828 + 56.0619 + (.015)(73.6828)(56.0619) \\ &= 129.7447 + 61.9620 \\ &= 191.7067. \end{aligned}$$

Thus the account has

$$A = 15(191.7067) = \$2875.60$$

in it. Of this amount (15)(91) = \$1365.00 was deposited, so the total interest earned is

$$2875.60 - 1365.00 = \$1510.60.$$

Notice this is rather more than was deposited.

Exercises 5.3

1. When Al joined the Farmers' Union, he was told he would be sent a bill for \$250 for his retirement contribution every 3 months. If the fund pays 8% interest quarterly, how much will Al have in his account on the fifth anniversary of his joining the union?

2. Gerry started a savings account on her twenty-first birthday, April 14, 1978, by depositing \$20. If she deposits \$20 on the fourteenth of every other month into her savings account, which pays 6% bimonthly interest, how much will she have in her account on Valentine's Day in 1984? If she decides to close her account to get married on April 13, 1985, how much will she have?

3. In problem 4 of Section 5.2, suppose Carl was sick and missed the ninth payment. How much would be in the account then on July 1 in 12 years?

4. How much would be in the account of the employee in problem 5 of Section 5.2 if he missed payments 15 and 22?

5. Fred's retirement contribution is \$500 per quarter into a fund paying 10% quarterly interest. How much will be in the account 10 years after he started work if he missed the thirteenth payment due to illness? If he started on June 1, 1969, and missed the first quarter payment in 1973 and the fourth quarter payment in 1974, how much was in his account 10 years after he started work?

6. In problem 3 of Section 5.2, how much is in Annemarie's account after 7 years? If she misses payments number 36 and 57, how much is in her account on April 30 seven years after she started?

7. Suppose Jana in Section 5.1 starts a new account with her \$50 a month with a bank offering 12% monthly interest. By extending Table V (or otherwise) find how much will be in this account 8

years after she started it (on the same day of the month). If she missed payments 50 and 65, how much would she have?

8. Suppose Jana, from problem 7 above, put payment 50 (which she had missed) into the account at the same time as payment 60, and payment 65 got made up when she sent it along with payment 70. Now how much should she have in her account after 8 years?

9. Find how much Roberto (in the introduction) will have when he retires.

10. Professors at a certain college are required to make monthly contributions to a retirement fund paying 6% annually, the interest beginning at the end of the year during which the contributions are made. If Professor Skarf makes a $200 a month contribution, how much will he have in the fund in 30 years? Suppose he had put the money into a savings and loan paying 6% compounded monthly. How much more would he have at the end of 30 years?

5.4 ANNUITIES

We wish to consider now the problem of taking money out of an account on a periodic basis. Since the dictionary defines annuity as "a payment of a fixed sum of money at regular intervals of time," we are clearly discussing annuities. Suppose, as an example, that Hugh decides to retire (he is not able to devote as much time to golf and tennis as he would like) and, in addition to his other income, to set up a trust for himself that will pay him $3000 every 3 months for 12 years. How much does he need to put into the trust so he can receive these payments? The answer will depend, we hope to hear you say, on what the interest rate is. Suppose it is 6% quarterly.

The solution of this problem also requires us to look at a geometric series. However, instead of deriving the formula we simply state the result as

THEOREM 4

Annuity Formula The present value PV of an annuity that pays P dollars at the end of each of n periods during which interest has been paid at a rate of i per period is given by

$$\text{PV} = \frac{P[1 - (1 + i)^{-n}]}{i}.$$

Notice that in this formula it is assumed no payment is paid until the end of a compounding period, so n is the same as the number of compounding periods. This is a little different from the situation studied in Section 5.2, since here the number of compounding periods equals the value of n.

As before, PV stands for present value and represents how much we need to have now in order to make the subsequent payments, assuming the money not paid out is left at the interest i stated. Note that the compounding period must correspond to the payment period. Table VI is just what we need since it computes $[1 - (1 + i)^{-n}]/i$.

For example, in the problem stated in the first paragraph above, $P = \$3000$, $i = 1\frac{1}{2}\%$, and $n = 48$. Thus we have

$$\begin{aligned} \text{PV} &= 3000(\text{Table VI entry with } i = 1\tfrac{1}{2}\%,\ n = 48) \\ &= 3000(34.0426) \\ &= \$102{,}127.80. \end{aligned}$$

Notice that this is appreciably less than the total amount ultimately received over the 12 years, which is $3000(48) = \$144{,}000$. The difference of \$41,872.20 will be generated by compound interest.

As another example, suppose a doctor wishes to set up a fund that will pay for his daughter's college and medical school expenses. He decides that she should have \$5000 a year for 8 years for all expenses including tuition. If he sets up an annuity that pays \$1250 every 3 months and earns 8% quarterly, how much will the annuity cost?

We organize our solution in a manner similar to that used earlier:

$$P = \$1250, \qquad i = 2\%, \qquad n = 32.$$

Using Table VI, we have

$$\begin{aligned} \text{PV} &= 1250(\text{Table VI entry with } i = 2\%,\ n = 32) \\ &= 1250(23.4683) \\ &= \$29{,}335.38 \end{aligned}$$

to the nearest cent.

If we use a calculator, we may compute

$$\text{PV} = 1250[1 - (1 + .02)^{-32}]/.02.$$

Although while actually doing the computation we would not write down the intermediate steps, we do so below so that you can check your computations as you follow the calculations on your calculator. The steps would be

$$
\begin{aligned}
PV &= 1250(1 - .5306333035)/.02 \\
&= 1250(.4693666965)/.02 \\
&= 1250\,(23.46833482) \\
&= 29335.41853.
\end{aligned}
$$

Using 10-place accuracy, the present value would be $29,335.42, a difference of only $.04.

To illustrate another problem that can be solved using the formula, we consider a final example. Two people, both of whom are 30 years old, win $125,000 on a magazine sweepstakes and decide to buy a 40-year annuity that pays only 5% annual interest. How much will the couple receive each year from the annuity and how much more than the $125,000 will they receive?

This requires the use of the annuity formula to find the payment, given the present value. The solution can be found as follows:

$$
\begin{aligned}
125{,}000 &= P(\text{Table VI entry with } i = 5\%,\, n = 40) \\
&= P(17.1591)
\end{aligned}
$$

or

$$
\begin{aligned}
P &= 125{,}000/17.1591 \\
&= \$7284.76.
\end{aligned}
$$

The couple will receive $7284.76 each year for 40 years. How much is interest? We take

$$(40)(7284.76) - 125{,}000$$

to get

$$\$166{,}390.40$$

as the total amount of interest.

With today's interest rates, a more likely approach for our couple, assuming they wanted to spread the income over 40 years, would be to invest in an annuity that pays $6\frac{1}{2}\%$ quarterly interest and receive a payment every three months. We need a computer or calculator for this computation but the formula would read

$$
\begin{aligned}
125{,}000 &= \frac{P[1 - (1 + .065/4)^{-160}]}{(.065/4)} \\
&= P(56.87125244).
\end{aligned}
$$

Thus $\quad P = 125{,}000/56.87125244$

$$= \$2197.95.$$

The annual income would be

$$4(2197.95) = \$8791.80,$$

more than $1500 per year more than before, and the total interest earned will be

$$40(8791.80) - 125{,}000 = \$226{,}672.$$

With your experience at using tables now, you should have no trouble with these exercises. The *annual rent* is simply the total amount paid in one year. In our first example it is (3000)(4) = $12,000, and in the second example it is $5000.

Exercises 5.4

In problems 1 and 2, find the present value and the annual rent of the annuity that has the payments, interest, and time period specified.

1. **(a)** $250 each 6 months for 25 years at 5% semiannually.
(b) $1000 each 3 months for 10 years at 6% quarterly.

2. **(a)** $500 each 4 months for 12 years at 6% compounded three times a year.
(b) $5000 each year for 30 years at 5% annually.

3. Renée decides that she wants an annuity that pays $3000 every 2 months and can get 6% compounded bimonthly. How much must she put into the account in order to give her what she wants during her eight years in college?

4. Jack wants a retirement allowance of $1000 per month for 20 years and he can get 6% compounded monthly. What is the present value of his annuity?

5. Bob decides to place an amount in an account that will pay his daughter $300 per month during her 4 years in college. If he can get 6% monthly, how much must he deposit?

6. Mrs. Lately has $100,000 from her late husband's estate to put into an annuity. She decides, since she is just 60, to buy a 20-year annuity at 8% semiannual interest. What amount will she receive every 6 months? What is the annual rent and the total interest earned over the 20 years?

7. Joan decides that during the next 5 years, while she is starting her restaurant business, all her income from the restaurant will be put back into the business. Consequently, she allocates $60,000 from her savings to buy an annuity that pays 6% monthly for 5 years. How much will each monthly payment be? What is the annual rent and how much interest has she earned?

8. If you won $500,000 in a lottery, how much could you spend each month at 8% monthly interest and still have your money last for 40 years? What is the annual rent?

9. The Illinois State Lottery has a million dollar winner who receives $50,000 per year for 20 years (a total of $1,000,000.) How much must the state put into an annuity at $8\frac{1}{4}\%$ in order to be able to make these payments?

10. Chris has received a $15,000 inheritance and decides to save $10,000 in a blue-ribbon account paying $8\frac{1}{2}\%$ interest (using the other $5000 for a trip to Europe). She is going to buy an annuity in 20 years (when she is 45) and believes she can get $8\frac{1}{2}\%$ quarterly interest so she can get a payment every 3 months. How much will she receive each quarter if she buys a 20-year annuity? What if she buys a 30-year annuity?

11. When Emily was born her father decided to set up a trust for her by investing $5000 into a savings certificate paying $8\frac{1}{4}\%$ interest for 30 years. At that time it automatically converts to a 20-year annuity with monthly payments at 8% monthly interest. How much will she receive each month and what is the annual rent? How much total interest will the $5000 have earned in the 50 years?

5.5 INSTALLMENT BUYING AND RETIREMENT PLANNING

In this section we show first how to use our knowledge to compute installment loan payments. Then we give an example of retirement funding, which requires us to be able to use both Table V and Table VI.

Installment Payments

Amortization is the term used by businesses for the process of paying off a debt by installment payments. Since installment buying is something we are all likely to face, let us consider the following problem. We will look into advantages and disadvantages in the next section.

Kathy has decided to buy a new stereo. The system costs \$783.50 and she can pay \$83.50 down. How much should her monthly payments be if she wants to pay it off in one year and the store charges 1% per month interest? This does not, on the surface, look like anything we have done before.

However, suppose we look at the situation from the point of view of the store. They have essentially loaned \$700 which is to be paid back in 12 installments at an interest rate of 1% per installment. That is, it is like having an annuity with present value \$700, $i = 1\%$, and $n = 12$. Thus the equation

$$700 = P(\text{Table VI entry with } i = 1\%, n = 12)$$

yields

$$P = 700/11.2551$$

so $$P = \$62.19.$$

Notice that Kathy will repay a total of 12(62.19) = \$746.28, or \$46.28 in interest over the year.

Her friend, Bill, decides to buy a \$800 pool table on his bank charge card. He pays $1\frac{1}{2}\%$ interest on the unpaid balance each month and wants to know how much he has to pay each month in order to pay it off in 2 years.

This can be found as before by using the annuities formula. In this case,

$$\text{PV} = \$800, \quad i = 1.5\% \quad \text{and} \quad n = 24,$$

so we have

$$\begin{aligned} 800 &= P(\text{Table VI entry with } i = 1.5\%, n = 24) \\ &= P(20.0304) \end{aligned}$$

or

$$P = 800/20.0304 = \$39.94.$$

Notice that Bill pays \$158.56 interest over the 2 years.

Thus we see that amortization and annuity problems can be solved from the same table. We give one more example before looking at a problem that requires both Table V and Table VI.

As our final example, suppose we buy a new car for $7165 and have $2165 for a down payment. This leaves $5000 to finance over 3 years at 6% annual interest. What will be the monthly payments?

The seller has a $5000 annuity and will receive 36 payments at an interest rate of $\frac{1}{2}$% per period. Thus

$$\begin{aligned} 5000 &= P(\text{Table VI entry with } n = 36, i = \tfrac{1}{2}\%) \\ &= P(32.8714). \end{aligned}$$

By dividing, we observe

$$P = 5000/32.8714 = \$152.11,$$

so our monthly payment will be $152.11. In 3 years we will have paid

$$(36)(152.11) = \$5475.96$$

in all. Our total interest paid will be $475.96.

Retirement Planning

We can combine the use of Tables V and VI to solve problems like the following one on retirement planning.

Having just reached 40, Lin Chiu is concerned about his retirement. Allowing for inflation, he believes he will need $5000 every 3 months, when he retires, for living (and playing) expenses. He wants to receive these funds over a 12-year period because no one in his family has lived beyond 77. How much does he have to put into his account each 6 months, beginning now, for 25 years, to have enough in his account to provide him with the annuity he desires, assuming he can get 6% per annum compounded appropriately in each case?

We first need to know what the present value of the annuity will be when he reaches 65. In this case, the payment is to be $P = \$5000$, the interest $i = 1\frac{1}{2}\%$ for $4 \times 12 = 48$ quarters. Thus the amount needed in his account when he reaches age 65 is given by

$$\begin{aligned} \text{PV} &= 5000(\text{Table VI entry with } i = 1\tfrac{1}{2}\%, n = 48) \\ &= 5000(34.0426) \\ &= \$170{,}213.00 \end{aligned}$$

Now we can view the problem as follows: Lin Chiu wants $170,213 to be in his account in 25 years at 6% semiannually. How much must he pay in each 6 months beginning now? We considered problems like this in

Section 5.2. In this case,

$$170{,}213 = D(\text{Table V entry with } i = 3\%, n = 50)$$
$$= D(112.797).$$

Solving for D, we have

$$D = 170{,}213/112.797$$
$$= \$1509.02$$

Thus Lin Chiu must make 50 payments of $1509.02 over the 25 years. This means that he will have paid only

$$50(1509.02) = \$75{,}451$$

over the 25 years, and this money will have earned nearly $100,000 more in interest before he start to draw on it and nearly $165,000 in interest before it is all gone.

Exercises 5.5

In problems 1–6 find the payment, calculate the total cost of the article, and find how much total interest is paid.

1. You borrow $1,000 to be repaid in 6 installments, 3 months apart. The rate of interest is 6% compounded quarterly.
2. You buy a badly used Rolls Royce which, after some haggling, comes to $4537. Your trade-in counts for $1200. The remainder is to be paid in 12 installments, 1 every 3 months. The rate of interest is 2% for each 3-month period (including charges and insurance).
3. A certain house has a list price of $25,000, of which $5000 must be paid down. The rest is paid in a 20-year mortgage at 5%.
4. In problem 3, calculate the results if the mortgage had been a 30-year mortgage.
5. A man wants to sell his store for $50,000, including fixtures, stock, and goodwill. The buyer offers to pay 10% down and the rest in a 10-year mortgage at 5% compounded annually.
6. A corporation borrows $100,000 for expansion. The payments are amortized over 15 years at 5% compounded annually.
7. Write a computer program that will give the answers to problems 3

and 4, but will assume the payments and compounding occur monthly.

8. Do example 4 in Section 5.1.

9. Do example 5 in Section 5.1.

10. Do example 6 in Section 5.1.

11. Do example 7 in Section 5.1.

12. Dominic decides he wants to retire at age 60 with an annuity that pays $3500 every four months for 15 years. He decides that he can get a 6% triannual policy. He is only 35 and plans to save every 6 months in an account paying 7% semiannual interest. How much must his semiannual payments be? How much will he have paid into his retirement fund and how much will he have earned in interest all together?

13. A 20 year old has gotten his own business going and decides he wants to retire at age 45 with a retirement annuity that pays $8000 semiannually for 25 years at 8% interest. How much must he pay into the account every 6 months at 7% semiannual interest in order to achieve his goal?

14. Referring to Section 5.4, in which a doctor is setting up a fund to pay for his daughter's medical school and college expenses, how much must he save every 3 months at 6% quarterly for 15 years in order to have enough to give her the annuity he decides she will need?

15. How much do you have to save every 3 months at 8% quarterly interest for 10 years in order to be able to pay your son $350 a month from a 4-year annuity paying 6% monthly interest?

16. Consider the finance charges imposed by Stark Brothers' Nurseries on the order blank shown in Figure 5.2 (page 300).
    - **(a)** Find the interest rates on $30 and on $50 where the finance charge is $4.
    - **(b)** Find the interest rates on $80.00 and on $150.
    - **(c)** Discuss the differences found in (a) and (b). Why do you think the nursery is willing to accept the lower rate on the larger amount?

17. Do example 8 in Section 5.1.

5.6 ON BEING A CONSUMER

You are now armed with some knowledge of the mathematics of interest. In this section we will specialize our thinking even more and try to

It's easy to order from Stark Bro's

Before you order make sure you live in a zone where your stock will thrive. The map on page 7 is for this purpose. If your variety needs pollination, also check the chart on page 6 for helpful recommendations. Then you're ready to order.

Order Early

1. Order early to get the stock you want. And indicate Fall or Spring Shipment at the top of the Order Form.
2. Be sure to use order form on page 61 first. Be sure to include your county, regardless of which Order Form you use. If Order Form is missing, use a plain sheet of paper.
3. List the catalog number, how many, size, variety or item and price of each item. Include sales tax and shipping and packing charges — see Order Form.
4. Enclose payment.

POLICY STATEMENT

It is our policy to fill orders exactly as received. If we are sold out of the variety ordered, we reserve the right to select one of similar, equal, or greater value, unless otherwise specified on Order Form. If our selection is not acceptable, prompt adjustment or full refund will be made.

SALES TAX

We are required to collect sales tax for many states. These rates were correct at printing time. If new rates have been adopted since, please use them.

Alabama . . . 4%
California* . 6%
Idaho 3%
Illinois 4%
Kentucky . . . 5%
Michigan . . . 4%
Missouri . . . 3%
N. Carolina 3%†
Oklahoma . . 2%
Penn. 6%
Utah . . . . 4%†

*Includes county tax.
†Also add County Tax where in effect.

Terms of Stark Bro's Monthly Payment Plan

Keep this copy for your records.

Amount to be Financed	FINANCE CHARGE	Monthly Payment
$30.00 to $50.00	$4.00	$ 7.00
$50.00 to $75.00	$5.00	$10.00
$75.00 to $150.00	$7.50	$15.00

Deferred Payment Price is determined by adding Total Cash Price and Finance Charge from table.

Finance Charge begins to accrue at date of shipment.

Pay balance in monthly installments starting **30 days after shipment** of stock. **Number of Payments** can be determined by dividing Total of Payments by Monthly Payment amount (from table).

Upon non-payment of any installment at its maturity, all remaining installments shall become immediately due and payable.

Stark Bro's Nurseries
Louisiana, Missouri 63353

Date ________ 19 ________

Mr.
Mrs.
Miss ________________________________
First Name   Initial   Last Name

Street, Box or Route ________________ County ________

Post Office ________________ State ________ Zip ________

Ship to Person or Address if Other Than Above

☐ **Check if gift order.**

Your Phone No. ________

☐ Check for FALL Shipment
☐ Check for SPRING Shipment

PLEASE USE ORDER FORM ON PAGE 61 FIRST

CATALOG NUMBER	HOW MANY	SIZE	VARIETIES OR ITEMS	PRICES EACH	PRICES TOTAL

• We recommend you purchase STARK TRE-PEP Fertilizer and HOME ORCHARD SPRAY
• For more space, attach sheet of paper listing additional items.

SHIPPING AND PACKING CHART

IF YOUR ORDER TOTALS	UP TO $10	$10 to $15	$15 to $20	$20 to $30	$30 to $60	$60 to $100	$100 & Over
ADD	$1.50	$2.00	$3.00	$4.00	$6.00	$9.00	10%

FOR OFFICE USE ONLY: TYPE / ROUTING / TERMS / ST / CR

4 WAYS TO PAY

1. ☐ **CASH ORDER.** Enclose check or money order. You save C.O.D. charges when you pay cash.
2. ☐ **C.O.D. ORDER.** Enclose 25% down-payment. Customer pays postman balance due plus C.O.D. charges. Minimum C.O.D. order $10.00.
3. ☐ **MONTHLY PAYMENT PLAN.** Send 10% Cash down-payment and fill out Payment Plan below. Minimum amount financed $30.00.
4. **CREDIT CARD.** Charge this purchase
☐ **MASTER CHARGE**
☐ **BANKAMERICARD**

SEE BELOW FOR IMPORTANT INFORMATION

MC Bank No. ☐☐☐☐ Expiration Date ________ Credit Card No. ☐☐☐☐☐☐☐☐☐☐☐☐☐☐☐☐☐

COMMENTS FOR OFFICE USE ONLY

Total Order	$
Tax (see chart at left)	
Shipping & Packing. Add to All Orders. (See chart at left)	
Total Cash Price	
Amount Paid With This Order	
Unpaid Balance of Cash Price, if any. (Check payment method at left)	

STARK BRO'S MONTHLY PAYMENT PLAN

For monthly payment plan, fill out below and include with your order. Please use order form on page 61 first. Keep copy of terms (shown at right) for your records.

Amount to be Financed	FINANCE CHARGE	Monthly Payment
$30.00 to $50.00	$4.00	$ 7.00
$50.00 to $75.00	$5.00	$10.00
$75.00 to $150.00	$7.50	$15.00

(Fill This Out for Monthly Payment Plan Only Not necessary on Cash or C.O.D. Plan)

Unpaid Balance Due on Order After Down Payment (carry forward from bottom line on order) ________ $ ________

Add Finance Charge (from table above) ________ $ ________

Total of payments (add 2 lines above) ________ $ ________

Deferred Payment Price is determined by adding Total Cash Price and Finance Charge from table.
Finance Charge begins to accrue at date of shipment.
Pay balance in monthly installments starting **30 days after shipment** of stock. **Number of Payments** can be determined by dividing Total of Payments by Monthly Payment amount (from table).

Upon non-payment of any installment at its maturity, all remaining installments shall become immediately due and payable.

________ Purchaser's signature

________ Street or R.F.D.

City ________ State ________ Zip ________

________ Spouse's Signature

Figure 5.2

Courtesy Stark Brothers Nurseries.

furnish you with some hints on what to watch out for and what to expect as a consumer.

First we discuss unit pricing and different forms of interest charges made to the consumer. Next we discuss separately buying a car, buying life (and other) insurance, and buying a house. Although there is not much (if any) new mathematics here, we hope you will enjoy the variety of the problems considered. Moreover, we hope that you will be able to buy more for your money after studying this section.

Inflation

We all have heard of inflation. In 1974, the United States suffered double figure inflation and many European countries have had several years with inflation rates higher than 10%.

Inflation is usually measured by the rise in the consumer price index. This is an index that takes into account the price of a representative large sample of items bought by large numbers of people. Included on the list are cars, clothes, food, fuel, housing, and many durable goods (freezers, refrigerators, washers, dryers, and so on), but very expensive items like fur coats, jets, large yachts, and so on are not included.

Naturally, the effect of inflation on any given household can only be measured by that household. The consumer price index gives only an approximation of the actual rise in prices. Many variables affect the rate, including where you live, how you drive, what you eat, and what you wear. A few years ago the cost of eating out was increasing more rapidly than the cost of eating at home. This is no longer true with the plethora of fast food and chain restaurants.

Naturally, what the federal government does affects inflation more than any other single factor. The federal budget is almost unimaginably large; as you can see from the newspaper clipping in Figure 5.3, a trillion is unbelievably large. Even a billion dollars is more than most of us can imagine. We would need to spend $54,757 *every day for*

How much is $1,000,000,000,000?

NEW YORK (AP) — President Carter's budget calls for government spending of more than half a trillion dollars for fiscal 1979.

One trillion dollars has 12 zeros — $1,000,000,009,000. Half a trillion is $500,000,000,000, and the president's budget is $500,174,000,000.

Those are big bucks.

If you're still having difficulty grasping the concept of such an amount, look at it this way:

That much money in dollar bills placed end to end would stretch around the earth more than 1,800 times.

It would reach to the moon and back 250 times.

With that much money, you could make every man, woman and child in Atlanta a millionaire.

You could buy everybody in Minnesota a $115,000 top-of-the-line Rolls-Royce Camargue.

You could have given $6.50 to every human being who has lived and died in the past 600,000 years.

You could give everyone in the world now $120.

Still confusing? Try thinking of it like this: a trillion is 1,000 times a billion. So how much is a billion?

One billion seconds ago, the first atomic bomb had not been exploded.

One billion minutes ago, Christ still walked the earth.

One billion hours ago, people lived in caves.

And one billion dollars ago — in terms of government spending — was yesterday.

Figure 5.3

Daily Pantagraph January 24, 1978. Courtesy Associated Press.

50 years to spend $1 billion, and if you were receiving daily interest at the modest rate of 6% per year, you would need to spend $173,000 every day to use up the full billion dollars in 50 years.

As the table of inflation rates (problem 9 Section 5.1) shows, the actual rate varies from year to year. Nonetheless, an interesting perspective can be provided by supposing the rate is about constant.

Suppose, for example, that the rate is 5%. This probably doesn't seem too bad. After all, a 40¢ cheeseburger will only cost 42¢ next year (5% of 40¢ = 2¢). It looks as if it will take 5 years for it to cost 50¢ and 20 years before the price doubles.

This is an interesting figure—*the years to double*. Just as it should be, it is the number of years it takes for an item to double in price, assuming a fixed rate of interest.

Let us look at our cheeseburger again (if you didn't eat it). We can make the following computation using a 5% inflation rate:

Cost now	$.40
Cost 1 year from now	.42
Cost 2 years from now	.44
Cost 3 years from now	.46
Cost 4 years from now	.48
Cost 5 years from now	.50
Cost 6 years from now	.53
Cost 7 years from now	.56
Cost 8 years from now	.59
Cost 9 years from now	.62
Cost 10 years from now	.65

So the cost will actually have doubled in only 14 years. Why? Because the already increased amount also helps to increase the cost the next year. It is just like figuring compound interest, so we are back using Table III.

EXAMPLE 1

At an inflation rate of 4%, how many years will it take for prices to double? How much will a suit that costs $100 today cost in 10 years?

Solution We look in Table III and find that, in the column headed 4%,

opposite $n = 17$ is 1.9479,

opposite $n = 18$ is 2.0258.

Thus the years to double is about 18. To find the cost in 10 years of a $100 suit, we compute

$\$100(\text{Table III entry with } i = 4\%, n = 10) = 100(1.4802) = \$148.02.$

Thus in 10 years the cost of the suit has increased by 50%.

Very often a manufacturer increases his prices by putting less in a package and charging the same. In these cases the only way to compare the cost is to compute what a single unit would cost. The unit involved will vary from item to item, since it may be ounces, cubic centimeters, liters, grams, quarts, some multiple of one of these, or another measure.

Consideration of cost per unit comes up not only in determining our inflation, but also in deciding which of two competing products or sizes is cheaper. Some grocery stores have the price per unit marked on the shelf to aid the consumer in making a wise choice. The kind of choice you may be faced with is illustrated in the next example.

EXAMPLE 2 Here is your choice for bottles of ketchup. (See Figure 5.4.)

Figure 5.4

	BRAND A	BRAND B
Regular size	12 oz @ $.35	15 oz @ $.43
Large size	38 oz @ $1.09	40 oz @ $1.15

In this case we would find the cost per ounce. The basic formula is

$$\frac{\text{Total cost}}{\text{number of units}} = \text{cost per unit.}$$

This is the cost per ounce:

	BRAND A	BRAND B
Regular size	$.02917	$.02867
Large size	.02868	.02875

Thus a regular size of brand B is the best buy. It sometimes happens, as in this example, that the large size is not the best buy when considering only cost per unit.

With the conversion to metric units there is an even bigger problem. Table 5.3 lists conversion factors that relate certain units. For example, 1 inch is 25.4 millimeters and 1 millimeter is .039 inches.

TABLE 5.3

$\frac{\text{inches}}{.039}$	1	$\frac{\text{millimeters}}{25.4}$
$\frac{\text{feet}}{3.281}$	1	$\frac{\text{meters}}{.305}$
$\frac{\text{miles}}{.625}$	1	$\frac{\text{kilometers}}{1.6}$
$\frac{\text{ounces}}{.035}$	1	$\frac{\text{grams}}{28.35}$
$\frac{\text{pounds}}{.0022}$	1	$\frac{\text{grams}}{453.59}$
$\frac{\text{quarts}}{1.10}$	1	$\frac{\text{liters}}{.909}$
$\frac{\text{U.S. gallons}}{.275}$	1	$\frac{\text{liters}}{3.636}$

The listing

$$\frac{\text{feet}}{3.281} \qquad 1 \qquad \frac{\text{meters}}{.305}$$

in the table means that

$$1 \text{ foot} = .305 \text{ meter}$$

and $3.281 \text{ feet} = 1 \text{ meter}$.

Similarly,

$$1 \text{ quart} = .909 \text{ liters}$$

and $1.10 \text{ quarts} = 1 \text{ liter}$.

EXAMPLE 3

Suppose that gasoline costs 32 cents per liter in England. How much is this per gallon?

Solution We write

$$\frac{32 \text{ cents}}{1 \text{ liter}} \cdot \frac{1 \text{ liter}}{.275 \text{ gallons}}.$$

We cancel the liters

$$\frac{32 \text{ cents}}{\cancel{1 \text{ liter}}} \cdot \frac{\cancel{1 \text{ liter}}}{.275 \text{ gallons}}.$$

and get 32/.275 cents per gallon. This is more than $1.16 per gallon.

EXAMPLE 4

Sugar costs 85 cents for a 250-gram bag and $3.00 for a 2-pound bag. Which is the better buy?

Solution We first have to decide on our unit. We have decided to take 1 pound as the unit because the $3.00 bag costs $1.50 per pound. We make the following computation to find the cost per pound of the 250-gram bag.

$$\frac{\$.85}{250 \cancel{\text{ grams}}} \cdot \frac{453.59 \cancel{\text{ grams}}}{1 \text{ pound}} = \$1.54 \text{ per pound.}$$

Exercises 5.6

1. Inflation is currently at 3%. How many years to double? If a new car costs $4000 now, how much will it cost in 8 years? How many years before it costs $6000?
2. With inflation at a 5% annual rate, how many years to double? What will a $400 color television cost in 6 years? (Note that this does not take into account any reduction in price due to improved technology.) How long before it costs $600?
3. Inflation has been nearly at the 5% rate every 6 months in several European countries. How many years to double? If a car costs 10,000 deutsche marks today, how much will it cost in 5 years?
4. Salad oil costs $.85 for a 16-ounce size and $1.60 for the 30-ounce size. Which is cheaper per ounce?
5. Which is the better buy, shredded wheat in a 10-ounce package for

49¢ or a 15-ounce package for 72¢? Wheatflakes comes in packages of 18 ounces for 87¢ and 12 ounces for 65¢. Which is the better buy? Are the wheatflakes cheaper or more expensive than the shredded wheat?

6. Jam comes in the large economy size 32-ounce jar for $1.59 and the small 12-ounce jar for $.59. Which is the better buy?

7. Dog food comes in 737-gram cans at 30¢ and 430-gram cans at 19¢. Clearly, the larger can is cheaper. (Check to see that this is true.) If the store puts the small can on sale at 18¢, which is the better buy?

8. Peanut butter sells in one store at $.99 for an 18-ounce jar and $1.53 for a 28-ounce jar. John buys the larger jar even though it dries out a little (he habitually leaves the lid off) because he is sure it is cheaper per ounce. Is John right?

9. You can buy six 5.5-ounce cans of the store's brand of tomato juice for 78¢ or six 6-ounce cans of a brand name juice for 84¢. Which is the better buy? (Ignoring any preference you may have.)

10. Which is the better buy, a liter of milk for $.98 or a half gallon of milk for $1.79?

11. Which is the better buy, a half gallon of whiskey at $8.25 or 1.75 liters at $8.00?

12. Which is more, 12 ounces or .35 liters? Is 12 ounces more than $\frac{1}{3}$ liter?

13. If Chicago and St. Louis are 280 miles apart, how many kilometers apart are they?

14. How many kilometers are there between Washington, D.C., and New York City if they are 233 miles apart?

15. Los Angeles and San Francisco are 379 miles apart. How many kilometers is this?

16. If you can go 95 kilometers on 10 liters of gasoline, how many miles per gallon can you go?

17. Suppose that on the road you get 150 kilometers per 10 liters in your Mercedes-Benz. How many miles per gallon do you get?

Forms of Interest Charge

There are several ways in which interest is charged. Most banks state an interest rate that is an "add-on" rate; for example, 8% add-on.

The amount of interest for an *add-on* rate r is found by multiplying the principal times r times the number of years of repayment.

The monthly payments are easily computed as the principal plus interest divided by the total number of months. Thus to borrow $300 for one year at 8% interest (add-on), the total amount repaid would be 300 + .08(300) = $324; so the monthly payments would be 324 ÷ 12 = $27 per month.

Installment loans usually state the interest rate per month, but some give it per year and compound each month. Finance companies usually have a sliding scale that depends on how much you borrow and for how long you borrow it.

The truth in lending legislation requires all interest to be given in terms of *true annual interest*. Monthly rates can be multiplied by 12 to give annual rates. However, add-on interest is harder to calculate. The true annual rate is approximately $1\frac{1}{2}$ times the add-on rate. Here is how it works.

EXAMPLE 1

You go to your local bank and ask for a loan to buy a diamond ring. Your friendly banker offers you a $750 loan at 8% interest (add-on) to be paid back in 17 monthly payments of $46.67 each plus one payment of $46.61. What is the true annual interest rate?

Solution First let us determine how the banker obtained the payment amounts. She took ($750)(8%) to get $60 as the cost for 1 year. But you will be paying back the loan over $1\frac{1}{2}$ years, so she took $(1\frac{1}{2})(60)$ to get $90. This is the interest charge. The computation is now easy.

$$750 + 90 = \$840$$

is the total to be paid, so

$$840/18 = \$46\tfrac{2}{3}$$

is the monthly payment. Now,

$$(18)(46.67) = \$840.06,$$

which would be $.06 too much. Thus there are 17 payments at $46.67 and one at $46.61.

Our problem is really to find i so that

$$750 = 46.6666\left[\frac{1-(1+i)^{-18}}{i}\right].$$

(Note we have taken enough decimal places for the payment to allow a mathematical solution since $(18)(46.6666) = 839.9988$.)

The problem of determining i "is beyond the scope of this course," as our colleagues like to say. But we can find an approximate value by dividing both sides by 46.6666. This gives

$$16.072 = \frac{1 - (1 + i)^{-18}}{i}.$$

Recalling that the right-hand side of this equation is tabulated for certain values of i in Table VI, we could try to find the value 16.072 in the row labeled 18 and then the column heading would give us i.

The row headed by 18 looks like

17.1726 16.3983 15.6726 14.9920 13.7535 13.1897 12.6593 11.6896,

so i must be between 1% and 1.5% and about midway between. In fact, $i = 1.22\%$, so the annual rate is

$$(12)(1.22\%) = 14.64\%.$$

This is over half again as much as the quoted 8% interest. Actually, the true annual interest approaches the add-on interest as the number of payments gets close to one.

Other interest charges are usually given as a monthly rate or an annual rate. To find the annual rate from the monthly rate just multiply by 12.

EXAMPLE 2

When buying a car for $2000, Ken decides to shop around for financing and finds his bank will give him the loan at $6\frac{1}{4}\%$ add-on and his credit union will charge 1% per month. Which is the better deal? And what is the actual monthly rate the bank is charging?

Solution First of all, it is relatively easy to decide which is the better deal. Table VI is used to find the monthly payment for the credit union.

$$\begin{aligned} 2000 &= P(\text{Table VI entry with } i = 1\%, n = 24) \\ &= P(21.2434) \end{aligned}$$

so $$P = 2000/21.2434 = \$94.15.$$

The bank takes

$$(2000)(.0625)(2) = \$250$$

as the amount of interest, so the monthly payment is

$$2250 \div 24 = \$93.75.$$

Thus Ken would save 40¢ a month by borrowing from the bank.

To find the bank's monthly rate, we first take

$$2000 \div 93.75 = 21.3333$$

to obtain the value corresponding to the Table VI entry. Thus we want i so that

$$\frac{1 - (1 + i)^{-24}}{i} = 21.3333$$

A hand calculator can be used to find an approximate value for i by successive trial and error. We know it is slightly less than 1. When we worked on the problem, our guesses went like this:

i	$\frac{1 - (1 + i)^{-n}}{i}$
.0095	21.370
.0096	21.3449
.00965	21.33216
.009647	21.3347
.009645	21.33343

You can see that you can approximate as close as you like by successive computation. The monthly interest rate is very nearly .9645%.

Fortunately you do not have to go through all this computation. Just shop around and let the lending institution do the figuring. And be sure that if one place includes a charge for insurance on the borrower's life, that the other one also does.

Exercises 5.6

18. If you are offered a \$600 loan for a year to be repaid in 12 monthly installments of \$56.55, what approximate true annual interest are you paying?

19. If you borrow \$900 from a friendly loan shark at $2\frac{1}{2}$% per month on the unpaid balance, what is the true annual rate of interest, how much do you pay each month, and how much do you pay in interest assuming you pay the loan off in 12 monthly payments?

20. If you borrow $2000 to buy a used car and pay $96.92 per month for 24 months, approximately what is the true annual rate of interest you would be paying? How much would you pay in interest altogether? If you got this loan at the bank, what was the add-on interest rate that was used?

21. You are trying to decide between a loan for 3 years of $6000 for your new car from a bank at a 5% add-on rate or from your credit union at a $\frac{3}{4}$% per month rate. Which should you borrow from? What is the monthly rate at the bank?

Bank and Other Charge Cards

Credit is an important part of everyone's life. Even if you can pay cash, it is important to borrow to establish credit. You need to establish a credit record and, if you always pay cash, no one (in the credit business) will ever hear about you. Then when you want to make a big purchase such as a house or an automobile, you may have some difficulty.

Bank cards offer an easy and cost-free way to establish credit—provided you always pay off the balance at the end of the month. On those cards, as well as on department store cards, interest is charged only if you do not pay the total balance due.

You should always keep receipts and check them against your statement. Mistakes can and do occur and you should be aware of the new federal law protecting your rights.

> IN CASE OF ERRORS OR INQUIRIES ABOUT YOUR BILL*
> The Federal Truth in Lending Act requires prompt correction of billing mistakes.
> 1. If you want to preserve your rights under the Act, here's what to do if you think your bill is wrong or if you need more information about an item on your bill:
>   a. Do not write on the bill. On a separate sheet of paper write (you may telephone your inquiry but *doing so will not preserve your rights under this law*) the following:
>   i. Your name and account number.
>   ii. A description of the error and an explanation (to the extent you can explain) why you believe it is an error.
>
>   If you only need more information, explain the item you are not sure about, and if you wish, ask for evidence of the charge

*Federal Reserve System, Truth in Lending—Fair Credit Billing Amendments.

such as a copy of the charge slip. Do not send in your copy of a sales slip or other documents unless you have a duplicate copy for your records.

iii. The dollar amount of the suspected error.

iv. Any other information (such as your address) which you think will help the creditor to identify you or the reason for your complaint or inquiry.

b. Send your billing error notice to the address on your bill which is listed after the words: "Please Direct Inquiries To:"

Mail it as soon as you can, but in any case, early enough to reach the creditor within 60 days after the bill was mailed to you. If you have authorized your bank to automatically pay from your checking or savings account any credit card bills from that bank, you can stop or reverse payment on any amount you think is wrong by mailing your notice so the creditor receives it within 16 days after the bill was sent to you. However, you do not have to meet this 16-day deadline to get the creditor to investigate your billing error claim.

2. The creditor must acknowledge all letters pointing out possible errors within 30 days of receipt, unless the creditor is able to correct your bill during that 30 days. Within 90 days after receiving your letter, the creditor must either correct the error or explain why the creditor believes the bill was correct. Once the creditor has explained the bill, the creditor has no further obligation to you even though you still believe that there is an error, except as provided in paragraph 5 below.
3. After the creditor has been notified, neither the creditor nor an attorney nor a collection agency may send you collection letters or take other collection action with respect to the amount in dispute; but periodic statements may be sent to you, and the disputed amount can be applied against your credit limit. You cannot be threatened with damage to your credit rating or sued for the amount in question, nor can the disputed amount be reported to a credit bureau or to other creditors as delinquent until the creditor has answered your inquiry. *However, you remain obligated to pay the parts of your bill not in dispute.*
4. If it is determined that the creditor has made a mistake on your bill, you will not have to pay any finance charges on any disputed amount. If it turns out that the creditor has not made an error, you may have to pay finance charges on the amount in dispute, and you will have to make up any missed minimum or required payments on the disputed amount. Unless you have agreed that your bill was correct, the creditor must send you a written notification of what you owe; and if it is determined that the creditor did make a mistake in billing the disputed amount, you must be given the time to pay which you normally are given to pay undisputed amounts before any more finance charges or late payment charges on the disputed amount can be charged to you.

5. If the creditor's explanation does not satisfy you and you notify the creditor *in writing* within 10 days after you receive his explanation that you still refuse to pay the disputed amount, the creditor may report you to credit bureaus and other creditors and may pursue regular collection procedures. But the creditor must also report that you think you do not owe the money, and the creditor must let you know to whom such reports were made. Once the matter has been settled between you and the creditor, the creditor must notify those to whom the creditor reported you as delinquent of the subsequent resolution.
6. If the creditor does not follow these rules, the creditor is not allowed to collect the first $50 of the disputed amount and finance charges, even if the bill turns out to be correct.
7. If you have a problem with property or services purchased with a credit card, you may have the right not to pay the remaining amount due on them, if you first try in good faith to return them or give the merchant a chance to correct the problem. There are two limitations on this right:
   a. You must have bought them in your home state or if not within your home state within 100 miles of your current mailing address; and
   b. The purchase price must have been more than $50. However, these limitations do not apply if the merchant is owned or operated by the creditor, or if the creditor mailed you the advertisement for the property or services.

Look at the two sample statements shown in Figures 5.5 and 5.6 and note several things.

1. Transactions
2. Payments
3. Finance charge and interest rate
4. Closing date and due date
5. Minimum payment due
6. The annual finance charge, furnished for tax purposes
7. The account numbers
8. One shows credit limit and available credit and the other does not.

You should have noticed that the interest rates are different. State laws differ on what interest rates can be charged. The reason there was a finance charge on the account in Figure 5.5 was the $200 cash advance from the bank. The advance was made on January 7 and the billing date was January 18, so having the $200 for 11 days was equivalent to having $73.33 for the full month. By taking 1.5% of this average daily balance, the finance charge of $1.10 was obtained.

THE ANNUAL FINANCE CHARGE CHARGED TO YOUR ACCOUNT FOR 1978 IS $0.00.

ACCOUNT NUMBER	CREDIT LIMIT	AVAILABLE CREDIT	PAST DUE AMOUNT	BILLING DATE
5211-4036-1713	1,400.00	904.60	.00	01-18-79

TRANSACTION DATE	POSTING DATE	REFERENCE NUMBER	TRANSACTION DESCRIPTION			TRANSACTION AMOUNT
12 13 78	01 01 79	MH6143894	BOOK BAZAAR EASTLAND	BLOOMINGTON	IL	9.40
12 17 78	12 29 78	M 35664012	TIEN TSIN	BLOOMINGTON	IL	22.40
12 19 78	01 17 79	M 00643143	R READS INC	BLOOMINGTON	IL	13.13
12 20 78	01 05 79	ME 0450694	LOWELL JEWELERS	MARION	IL	180.30
12 21 78	01 12 79	ME 1102777	KAY-BEE TOYS 66	BLOOMINGTON	IL	11.52
12 21 78	01 11 79	ME 1058196	FORGET ME NOT FLOWER	BLOOMINGTON	IL	11.55
12 22 78	01 07 79	ME 0638379	1ST NATL BK OF NORMAL	NORMAL	IL	200.00
*	01 17 79	M 00680383	REMNANT HOUSE	BLOOMINGTON	IL	19.75
*	01 04 79	M 36358864	THE FLOWER FARM	CHICAGO	IL	26.25

* TRANSACTION DATE NOT AVAILABLE

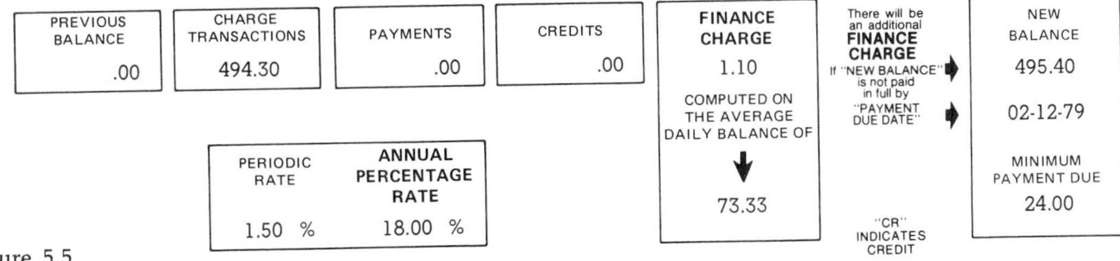

PREVIOUS BALANCE	CHARGE TRANSACTIONS	PAYMENTS	CREDITS	FINANCE CHARGE		NEW BALANCE
.00	494.30	.00	.00	1.10	There will be an additional FINANCE CHARGE if "NEW BALANCE" is not paid in full by "PAYMENT DUE DATE"	495.40
				COMPUTED ON THE AVERAGE DAILY BALANCE OF 73.33		02-12-79
					"CR" INDICATES CREDIT	MINIMUM PAYMENT DUE 24.00

PERIODIC RATE	ANNUAL PERCENTAGE RATE
1.50 %	18.00 %

Figure 5.5

DATE	REFERENCE NUMBER	TRANSACTION DESCRIPTION	PURCHASES AND ADVANCES	PAYMENTS AND CREDITS
* 10-03	4850100327606444	PAYMENT — THANK YOU		28.00

* POSTING DATE · TRANSACTION DATE IS NOT AVAILABLE

MONTHLY ACTIVITY		ACCOUNT NUMBER	CLOSING DATE	TOTAL MINIMUM PAYMENT DUE	
PREVIOUS BALANCE	568.48	230-058-129	10-10-78	PAST DUE	
CASH ADVANCES				CURRENT DUE	27.00
PURCHASES		TOTAL FINANCE CHARGES INCLUDE	ANNUAL PERCENTAGE RATE THIS MONTH		
INSURANCE PREMIUM					
PAYMENTS	28.00	1% MONTHLY PERIODIC RATE ON BALANCE SUBJECT TO FINANCE CHARGE 5.68	11.9	TOTAL DUE	27.00
CREDITS					
DEBIT ADJUSTMENT					
BALANCE SUBJECT TO FINANCE CHARGE	568.48	CASH ADVANCE SET-UP CHARGES			
FINANCE CHARGE	5.68				
LATE CHARGE		IF NO FINANCE CHARGE IS IMPOSED, YOUR CORRESPONDING NOMINAL ANNUAL PERCENTAGE RATE IS 12%		BY PAYMENT DUE DATE	11-04-78
NEW BALANCE	546.16				

PAY "NEW BALANCE" BEFORE "PAYMENT DUE DATE" TO AVOID ANY ADDITIONAL FINANCE CHARGE

Figure 5.6

Notice on the other bill (Figure 5.6) that the finance charge was computed before the payment was deducted. All this is explained on the reverse side of the bill, which we have shown in Figure 5.7. Notice that a cash advance from this bank involves a 4% setup charge.

You should always be aware of what you are doing, especially when borrowing money, which is what you are doing when you charge. Finally, a word about finance companies. *Avoid them like the plague.* Their interest rates vary from 18% to about 24% for smaller loans. Again, each state has its own laws, and you should find out what the laws in your state are. However, only go to a finance company as a last resort. They lend money to high-risk customers and so must charge more. Try your bank or a local savings and loan association, depending on what you want the money for.

Buying a Car

Your first major purchase may well be a car. Most people agree that the best buy is a two-year-old car with under 20,000 miles. There are several reasons for this. The major depreciation has occurred, as Table 5.4 shows.

TABLE 5.4

A NEW CAR HAS THIS PERCENTAGE OF ORIGINAL VALUE LEFT	AFTER THIS NUMBER OF YEARS
70%	1
55%	2
42%	3
33%	4
25%	5

Naturally, this represents an average. Some cars, such as a Mercedes-Benz, depreciate much more slowly. It also depends on the care and use the car has had.

Another reason for buying a two-year-old car is that the car is well broken in but not worn out. If any serious flaws existed when the car was new, they have almost surely surfaced and been repaired.

Some suggestions for buying a used car:

1. Shop first at a new-car dealer's used car lot. Most new-car dealers only keep the best used cars to sell themselves.

NOTICE TO BANK CARD HOLDERS

The following information is provided pursuant to the Federal Truth in Lending Act:

(1) A periodic FINANCE CHARGE will be imposed on next month's statement if the entire "New Balance" is not paid in full by the "Due Date."

(2) Payments must reach the Bank by "Due Date" to insure that they will be credited before "Closing Date."

Thus, by paying in full the entire "New Balance" shown on your periodic statement within twenty-five (25) days after "Closing Date" you will avoid additional periodic FINANCE CHARGES.

FINANCE CHARGES are determined and computed as follows:

(3) Periodic Rate: If any portion of the "New Balance" remains unpaid after "Closing Date" a periodic rate of 1% per month, which is an ANNUAL PERCENTAGE RATE of 12%, is applied to the "Balance Subject to FINANCE CHARGE."

The "Balance Subject to FINANCE CHARGE" is the "New Balance" from the preceding month's statement, which is shown on this month's statement as the "Previous Balance" less credit vouchers (that is, returns or refunds made by merchants) and credit adjustments. Payments are not deducted from "Previous Balance" before periodic rate is applied to "Balance Subject To FINANCE CHARGE."

(4) Cash Advances: On cash advances an additional FINANCE CHARGE (termed a transaction or set-up charge) will be made for each advance in the amount of 4% of the cash advances or $15.00, whichever is less. Cash Advances include all advances made by "Special checks" for example; auto license fees, taxes and insurance premiums, or advances through a bank machine.

On OK Check Guarantee Service transfers an additional FINANCE CHARGE (termed a transaction or set up charge) is made. For each transfer this charge is 2% of the amount transferred or $10.00, whichever is less.

Set up charges for Cash advances and OK Check Guarantee Service transfers are included in the total FINANCE CHARGE and are separately shown on your statement. Where the periodic rate only is applied, the corresponding ANNUAL PERCENTAGE RATE is 12%. If a set up charge is incurred, the ANNUAL PERCENTAGE RATE will be in excess of 12%.

EXTENDED PAYMENT SCHEDULE

IF NEW BALANCE IS:	MINIMUM PAYMENT IS:
UNDER $10	AMOUNT OF NEW BALANCE
$10 TO $200	$10
OVER $200	5%

Late Charge: If the Minimum Amount Due is not received within 15 days after "Due Date," a late charge of $1.00 or 5% of the Minimum Amount Due (whichever is greater) will be made. In addition, you shall be liable for all reasonable attorneys fees and collection costs incurred by the Bank in enforcing collection, whether or not suit is brought.

Insurance Premium: Creditor Life and Disability Insurance Protection is available insuring the balance of your account up to $2500 at an initial cost of 24¢ per $100.00 of the "New Balance." Specific insurance coverage provisions are available upon request from the Bank. Insurance coverage is not required by the Bank as a condition of credit.

Security Interest: The Bank has no security interest securing the payment of your Bank Card account notwithstanding the provisions of any agreement between the Bank and you or any third party.

Figure 5.7

2. Always have a mechanic, either a friend or someone you pay, to look at the engine, test compression, and so on. The engine and drive train, for example, are the most expensive parts to repair.
3. Test *everything*; windows, doors, lights, and so on. Be sure everything works before you sign anything.
4. Read *Consumer Reports* for data on service records of various cars and general tips on buying cars (both used and new).

Buying a new car is another exercise altogether. Here the dealer and particularly his service facilities and policies are more important. Most new cars need adjustments of various sorts, and your dealer must be willing and able to provide them.

Briefly, the rules are:

1. Shop around. Get prices from several dealers on the same or different cars. Find out what the car you want wholesales for and add 10% for the dealer. This should be your maximum. Some dealers might take less.
2. Obtain your financing before you buy the car. See your bank or credit union. Usually, interest rates on new cars are less than on used cars. Only finance through the dealer if you must, since usually the interest charges are higher.
3. Read *Consumer Reports* on how to buy a new car. It also has general information about new cars they have tested.

We conclude this brief section by talking a little about auto insurance. There are four major items of auto insurance, and many lesser ones. We discuss only the four:

1. Bodily injury insurance protects you in case someone is injured in or by your car. There are definite limits which you pay for. For example, a 25/50 policy pays up to $25,000 to any one individual, but at most $50,000 to all individuals injured as a result of one accident. A 100/300 policy is $100,000 per person with a maximum of $300,000 per accident.
2. Property damage insurance pays for damage done to property of others as a result of an accident for which you are responsible. Many states require that bodily injury and property damage insurance be held by every driver. However, the laws vary a great deal from state to state and the penalties are also quite different.
3. Collision insurance pays for repairs to your auto that exceed the deductible (usually $50 or $100) you hold. The deductible applies to each accident, so that if you are hit in the morning on one side and in the afternoon on the other, and you hold

$50 deductible, then you will have to pay $100 toward the cost of repairing the car.

4. Comprehensive insurance pays for damage to your car caused by theft, vandalism, fire, falling trees, and most other natural events. It does not usually cover damage due to floods, war, or acts of terrorism.

The rates for all of these are computed separately and vary so much from area to area and age to age that no generalizations can be made, except perhaps that insurance for those under 25 (especially males) is very high. *Before* buying a car shop around for insurance; be sure you can find some and that you can afford to pay the premium.

Life versus Term Insurance

Speaking of insurance, we should say a few words about life insurance. The two primary types are whole life and term.

Figure 5.8 Courtesy State Farm Insurance Companies

Whole life is also sometimes called straight life or ordinary life. It combines the protection which we associate with all insurance with a guaranteed cash value and usually dividends. The cash value is a way of returning part of your insurance premium. That is, when you buy a whole life policy, part of your money goes toward the cost of the insurance and part goes into what amounts to a savings plan. If you live to retirement, the company either pays you a lump sum or else an annuity from the lump sum. Also you can borrow against this cash value at very favorable interest rates (such as 5% per annum). This is because you are really borrowing your own money.

Term insurance, on the other hand, is very much cheaper, but is strictly an insurance policy for a fixed period of time such as one year, three years, or five years. No cash value and usually no dividends are paid. If you renew the policy, the premium will go up since you are older. The reason it is so much cheaper is precisely because it offers no retirement benefits or cash value.

Many young families buy *decreasing term insurance,* which gives high initial coverage with a gradually decreasing coverage as years go on (and presumably the children get older). The premium, however, remains constant. This often is a good addition to a whole life policy and often takes the form of a guaranteed monthly income for the beneficiary up to a fixed point in time.

Again, our advice is *shop around.* We quote from "A Shopper's Guide to Life Insurance" as prepared by the Pennsylvania Insurance Department, June 1972.

> If you learn nothing else from this Guide, remember that—
> — It pays to shop for life insurance (as well as other kinds of insurance). Costs may vary over 170 percent.
> — You can't tell which is the best buy on straight life insurance by looking at premiums alone. A policy with the lowest premium may actually be the highest cost policy.
> — Use the "average yearly cost" figures in this Guide in deciding which are the lowest cost policies. But remember that cost is only one of many factors to consider in selecting a company.
> — If you already have a straight life policy, it is usually a mistake to drop it to buy coverage from another company.

Since this guide is not available to everyone, and different insurance companies operate in different states, the *interest-adjusted price* of a policy has been devised. The agent selling the policy should be able to furnish the interest-adjusted price. If not, it is probably wise to seek another company.

How much of each kind of insurance should you have? That varies too much to say, but some use five times the annual income as a rule of thumb. The best idea is to talk to several agents. They will discuss with you your situation and give advice. By synthesizing their advice, you should be able to reach a decision. Another source of information is an article entitled "How They Stack Up," in *Forbes Magazine*, March 15, 1975.

Buying a House

The single largest purchase most people make is to buy a house. It is generally better to buy than to rent, especially during inflationary times. Moreover, the interest paid on your mortgage is tax deductible, while none of your rent is.

The three primary quantities of concern to a prospective purchaser of a house are (1) the purchase price, (2) the down payment, and (3) the monthly payment. The first of these depends on many factors including location and size of house and lot. The location factors include recreation facilities nearby, distance from schools, value of other houses in the neighborhood, and the size of the city or town. Building costs are rising at an alarming rate and traditional factors in determining the cost are not always reliable, however.

The best source of reasonable advice is a reputable real estate agent. However, watch the newspapers and talk to friends as well. The real estate agent is legally bound to represent the seller, who, after all, pays the fee.

The down payment depends on what you can afford and may be as low as 5%. However, in this case, the mortgage interest rate will surely be higher. Twenty percent is a common figure for a down payment, but the best mortgage rates are usually reserved for those with one-third or more as a down payment.

Finally, the monthly payment must be considered. You should first compute a rule of thumb maximum for your house payment.

1. Determine your gross monthly income (before any deductions) from all sources.
2. Subtract the monthly payments of any outstanding debts that will not be paid off in 6 months.
3. One-fourth of the result represents your maximum house payment.

A rule of thumb that is sometimes used to determine how expensive a house you can afford is to multiply your annual gross salary by 2.5.

For example, Carlos and Mercedes have a combined monthly income of $2000 before deductions. In addition to a few small department store bills, they pay $165 per month for their new car and $75 for their living and dining room suites. Their maximum monthly house payment is then

$$\tfrac{1}{4}[2000 - (165 + 75)] = \tfrac{1}{4}(2000 - 240) = \tfrac{1}{4}(1760) = \$440.$$

We already know how to compute the house payments on a mortgage since it is just a *very* long term installment loan. However, you will not be surprised to learn that tables have been computed. We reproduce one page from a book of mortgage tables in Table 5.5.

To show how Table 5.5 is used, suppose that you wish to buy a \$59,900 house and you have a \$6000 down payment (about 10%). This leaves \$53,900 to finance with a mortgage. If you can find $8\frac{3}{4}$% financing for 30 years, the table allows you to compute your monthly payments as

$$\begin{aligned}\text{(the number of hundreds)(entry under } 8\tfrac{3}{4} \text{ opposite 30)} &= 539(.7932)\\ &= \$427.53.\end{aligned}$$

Notice that Carlos and Mercedes in our previous example could afford this house. We can also use the table to find the maximum amount Carlos and Mercedes can afford to pay for a house. From the table,

$$\text{(amount financed in 100's)}(.7932) = 440.$$

So the maximum they can finance at $8\frac{3}{4}$% interest for 30 years is

$$440/.7932 = 554.7,$$

or \$55,470. Now, add the money they have available for a down payment, say \$8000 in this example, and we arrive at the maximum purchase price of \$63,470.

There is only one small thing we have forgotten, *closing costs.* In fact, you will see that it is not so small. The largest unexpected item here is something called points. This is the cost of the privilege of borrowing from the savings and loan association or bank. A *point* is a percentage point. One, two, and even three points may be charged, depending on how scarce money is. If lots of money is available to be lent, there may be no points charged. For example, if a savings and loan charges 2 points and you are borrowing \$50,000, you will have to pay the savings and loan association

$$(.02)(50{,}000) = \$1000$$

at the time of closing. This is income-tax deductible as an interest charge, but it has no other redeeming virtues. We show a typical loan settlement statement in Figure 5.9.

Of the charges shown in Figure 5.9, the buyer must pay

TABLE 5.5 MORTGAGE INSTALLMENT TABLE SHOWING THE MONTHLY SUM IN $ TO REDEEM $100 BORROWED

Yrs.	Rate Percent 8	$8\frac{1}{8}$	$8\frac{1}{4}$	$8\frac{3}{8}$	$8\frac{1}{2}$	$8\frac{5}{8}$	$8\frac{3}{4}$	$8\frac{7}{8}$	Yrs.
1	9.0000	9.0104	9.0208	9.0312	9.0417	9.0521	9.0625	9.0729	1
2	4.6731	4.6811	4.6891	4.6971	4.7051	4.7132	4.7212	4.7292	2
3	3.2336	3.2409	3.2482	3.2555	3.2628	3.2701	3.2775	3.2848	3
4	2.5160	2.5230	2.5300	2.5370	2.5441	2.5511	2.5581	2.5652	4
5	2.0871	2.0940	2.1009	2.1078	2.1147	2.1216	2.1286	2.1355	5
6	1.8026	1.8095	1.8163	1.8232	1.8301	1.8369	1.8438	1.8507	6
7	1.6006	1.6075	1.6143	1.6212	1.6281	1.6350	1.6419	1.6488	7
8	1.4501	1.4570	1.4639	1.4708	1.4778	1.4847	1.4917	1.4986	8
9	1.3340	1.3409	1.3479	1.3549	1.3619	1.3689	1.3759	1.3829	9
10	1.2419	1.2489	1.2560	1.2630	1.2701	1.2771	1.2842	1.2914	10
11	1.1673	1.1744	1.1815	1.1886	1.1958	1.2029	1.2101	1.2173	11
12	1.1058	1.1130	1.1202	1.1274	1.1346	1.1419	1.1491	1.1564	12
13	1.0543	1.0616	1.0689	1.0762	1.0835	1.0909	1.0982	1.1056	13
14	1.0108	1.0182	1.0255	1.0329	1.0404	1.0478	1.0553	1.0628	14
15	0.9736	0.9810	0.9885	0.9960	1.0035	1.0110	1.0186	1.0262	15
16	0.9415	0.9490	0.9566	0.9642	0.9718	0.9794	0.9871	0.9948	16
17	0.9136	0.9212	0.9289	0.9366	0.9443	0.9520	0.9598	0.9676	17
18	0.8892	0.8969	0.9047	0.9124	0.9203	0.9281	0.9360	0.9438	18
19	0.8677	0.8755	0.8834	0.8913	0.8992	0.9071	0.9151	0.9231	19
20	0.8488	0.8567	0.8646	0.8726	0.8806	0.8886	0.8967	0.9048	20
21	0.8319	0.8399	0.8480	0.8560	0.8641	0.8723	0.8804	0.8886	21
22	0.8169	0.8250	0.8331	0.8413	0.8495	0.8577	0.8660	0.8742	22
23	0.8035	0.8117	0.8199	0.8282	0.8364	0.8447	0.8531	0.8614	23
24	0.7915	0.7998	0.8081	0.8164	0.8247	0.8331	0.8416	0.8500	24
25	0.7807	0.7890	0.7974	0.8058	0.8143	0.8227	0.8313	0.8398	25
26	0.7709	0.7793	0.7878	0.7963	0.8048	0.8134	0.8220	0.8306	26
27	0.7621	0.7706	0.7791	0.7877	0.7963	0.8050	0.8137	0.8224	27
28	0.7541	0.7627	0.7713	0.7800	0.7887	0.7974	0.8061	0.8149	28
29	0.7468	0.7555	0.7642	0.7729	0.7817	0.7905	0.7994	0.8082	29
30	0.7402	0.7490	0.7578	0.7666	0.7754	0.7843	0.7932	0.8022	30
31	0.7342	0.7430	0.7519	0.7608	0.7697	0.7787	0.7876	0.7967	31
32	0.7288	0.7376	0.7466	0.7555	0.7645	0.7735	0.7826	0.7917	32
33	0.7238	0.7327	0.7417	0.7507	0.7598	0.7689	0.7780	0.7872	33
34	0.7192	0.7282	0.7373	0.7464	0.7555	0.7647	0.7738	0.7831	34
35	0.7150	0.7241	0.7332	0.7424	0.7516	0.7608	0.7700	0.7793	35
36	0.7112	0.7204	0.7295	0.7388	0.7480	0.7573	0.7666	0.7759	36
37	0.7077	0.7169	0.7262	0.7354	0.7447	0.7541	0.7634	0.7728	37
38	0.7045	0.7138	0.7231	0.7324	0.7417	0.7511	0.7606	0.7700	38
39	0.7015	0.7109	0.7202	0.7296	0.7390	0.7485	0.7579	0.7674	39
40	0.6988	0.7082	0.7176	0.7271	0.7365	0.7460	0.7555	0.7651	40
41	0.6963	0.7058	0.7152	0.7247	0.7342	0.7438	0.7533	0.7629	41
42	0.6941	0.7035	0.7130	0.7226	0.7321	0.7417	0.7513	0.7610	42
43	0.6920	0.7015	0.7110	0.7206	0.7302	0.7398	0.7495	0.7592	43
44	0.6900	0.6996	0.7092	0.7188	0.7284	0.7381	0.7478	0.7576	44
45	0.6882	0.6978	0.7075	0.7171	0.7268	0.7365	0.7463	0.7561	45
46	0.6866	0.6962	0.7059	0.7156	0.7253	0.7351	0.7449	0.7547	46
47	0.6851	0.6948	0.7045	0.7142	0.7240	0.7338	0.7436	0.7534	47
48	0.6837	0.6934	0.7031	0.7129	0.7227	0.7326	0.7424	0.7523	48
49	0.6824	0.6921	0.7019	0.7117	0.7216	0.7314	0.7413	0.7512	49
50	0.6812	0.6910	0.7008	0.7107	0.7205	0.7304	0.7403	0.7503	50

Figure 5.9

LOAN SETTLEMENT STATEMENT

To: ______ Date: ______

Property Address: ______

Mailing Address: ______

Loan No.: L-27,531-1 Amount: $ 53,900.00 Rate of interest 9 %; 360 monthly payments are due and payable on the 1st day of each month, beginning October 1, 1978 in the amount of $ 503.00, which is the total of the following items: Payment on loan and interest, $ 434.00; anticipation of taxes, $ 52.00; casualty insurance premium, $ 17.00, accident and health insurance premium, $ -0-; mortgage life insurance premium $ -0-.

1. (A) Amount of loan — $ 53,900.00
   (B) Amount received from you — ______
   (C) Total credits — $ 53,000.00
   (D) Amount due from you — 1,526.01
   Total amount credited to your account — $ 55,426.01

2. We have charged or will charge to your account for expenses incurred in connection with this loan, the following items:
   (A) Preliminary abstracting (receipt attached) — $ ______
   (B) Title search — ______
   (C) Appraisal fee — 50.00
   (D) Initial service charge — 1,078.00
   (E) Credit report — ______
   Total — $ 1,128.00

3. Payments charged to loan:
   (A) To retire existing Mortgage to ______ — $ ______
   (B) Applied to purchase E.&M. Builders, Inc. — 53,562.41
   (C) Revenue Stamps — 60.00
   (D) Interest to 10/1/78 — 471.60
   (E) Tax Deposit ______ months @ ______ — ______
   (F) Insurance — ______
   (G) ______
   (H) ______
   (I) ______
   (J) ______
   Total — $ 54,094.01

4. Recording Fees:
   (A) Recording release — $ ______
   (B) Recording deed — 5.00
   (C) Recording mortgage — 5.00
   (D) Recording affidavit — ______
   (E) ______
   Total — $ 10.00

5. Title Charges:
   (A) Abstracting release — $ ______
   (B) Abstracting deed — ______
   (C) Abstracting mortgage — ______
   (D) Abstracting taxes — ______
   (E) Abstracting affidavit — ______
   (F) Certificate of title — ______
   (G) Owner's Title Policy for $ 59,900 — 174.00
   (H) Mortgage Title Policy — 20.00
   Total — $ 194.00

6. Total deductions for purposes of this loan — 55,426.01
   BALANCE
   Total of loan and other credits — 55,426.01

7. Remarks: ______

By: ______ Attorney

The undersigned acknowledges the receipt of a copy of this loan settlement, and agrees to the correctness thereof, and authorizes and ratifies the disbursement of the funds as stated therein.

August 26, 19 78

______ Borrower

______ Borrower

Appraisal fee (the cost of the lender appraising the house)	$50.00
Loan origination fee (points! 2% of the loan amount of $53,900)	1078.00
Interest from 8/26 to 10/1 @ 13.47 per day (interest on the loan from closing until first payment)	471.60
Title insurance (guarantees that the title is not encumbered)	20.00
Recording fees ($5 for recording each of the deed and mortgage)	10.00
	$1629.60

Thus the total cost to the buyer is the down payment *plus* closing costs. In our example, the total cost is

$$\$6000 + 1629.60 = \$7,629.60.$$

However, as is typical, there is an adjustment to the buyer from the seller in the form of a payment for taxes. In this case, it was for $113.59. Thus the actual cash outlay was only $7516.01. Still, this is 25% more than the $6000 down payment that had been agreed to.

Moreover, there are two additional costs not figured in here. The first is fire and liability insurance (usually in the form of a homeowner's policy). This usually runs between $150 and $300. In our example it costs $204 per year, and the first year must be paid in advance. The subsequent years will be paid by the $17.00 per month that is part of the monthly payment.

The other is the cost of an attorney. We highly recommend that you hire an attorney to represent you at the closing, not because of fear of doing business with a crook (although this is possible), but primarily to watch out for your interests and to ensure that no inadvertent errors are made.

Exercises 5.6

22. If a husband and wife make $2500 gross per month (before deductions) and have only $250 in long-term monthly payments, how large a house payment can they afford?

23. Henry is a bachelor making $20,000 a year gross. He is buying a new car and pays $155 per month on it, but has no other long-term debts. How large a house payment can he afford?

24. What will be the monthly payment on a $55,000 mortgage at $8\frac{7}{8}$% interest for 20 years? for 30 years?

25. What will be the monthly payment on a $40,000 mortgage at $8\frac{3}{4}$% interest for 30 years? for 35 years?

26. If a bank will lend the couple in problem 1 money at $8\frac{3}{4}$%, how much can they afford to borrow on a 30-year mortgage? on a 20-year mortgage?

27. Henry of problem 23 can get a $8\frac{5}{8}$% mortgage from his savings and loan association. What is the maximum amount he can borrow on a 30-year loan? a 20-year loan?

GRAPH THEORY

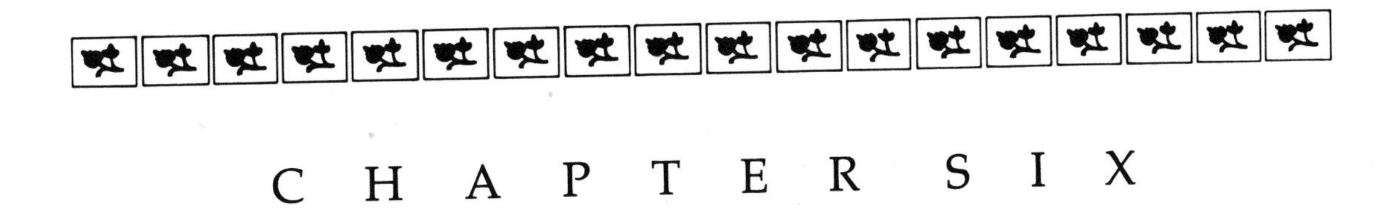

CHAPTER SIX

In this chapter we will see how diverse problems involving scheduling, coloring maps, solving puzzles, finding routes, and planning construction projects can be understood and solved by means of a simple geometric idea—that of a graph.

6.1 SCHEDULING AND COLORING

EXAMPLE 1

The Donohue Picnic Uncle Lester somehow got roped into being in charge of activities at the annual Donohue family picnic. The traditional activities are softball, over-40 softball, hardball, men's volleyball, women's volleyball, mixed volleyball, badminton, the three-legged race, touch football, Frisbee, egg throwing, and horseshoes.

The picnic starts at 4 o'clock, supper takes about an hour, and it gets too dark to play after 9, so Lester decides that he has about 4 hours for the activities. He plans to allow an hour for each one. Of course, more than one activity can go on at the same time, but there are limitations on such bunching up.

Since there is only one baseball diamond, no two of softball, over-40 softball, and hardball can be scheduled at the same time. The same goes for the three-legged race, touch football, Frisbee, and the egg throw, all of which must be held in the same field. Likewise there is only one net, so the three volleyball games and badminton must all be at different times.

There's more. Lester knows that the most athletic of the Donohues will want to play both hardball and touch football, so these cannot be scheduled simultaneously. Lester's own specialties are badminton and Frisbee; he certainly will put them at different times. Finally, the four best horseshoe throwers happen to be enthusiasts, respectively, of hardball, over-40 softball, touch football, and egg throwing. Thus horseshoes must be at a different time from each of these.

Uncle Lester's problem is how to schedule the various activities (there are 12) into the four time slots available so as to satisfy all the conditions listed. Many people confronted with a problem such as Lester's would simply throw up their hands and give up. It sounds very complicated. Lester considered telling everyone he had the gout so that the job would be passed to someone else, but finally he sat down and worked out various schedules until, by trial and error, he found one satisfying all the conditions.

It turns out that Lester's problem is not really a hard one. It only looks hard. In this section we will *not* show how to solve it. What we *will* show is how to boil the problem down to its essential ingredients, into a form, in fact, in which anyone will be able to solve it.

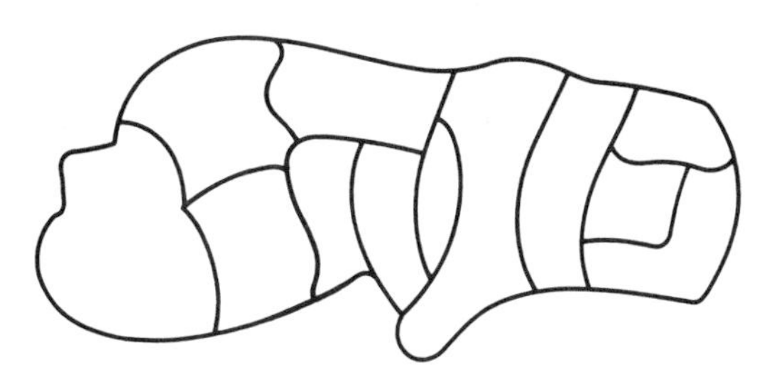
Figure 6.1

EXAMPLE 2

The Map Figure 6.1 shows the legislative districts in a certain county. A political party wants a blown-up version of this map to hang in its campaign headquarters, and hires a sign painter to make it. The painter decides to color each district, with no touching districts having the same color. He finds it quite easy to do this with just four colors.

The reader may want to try to color the map with four colors. Instead of actual colors simply label each district with a 1, 2, 3, or 4, making sure that no two districts that touch get the same number.

This map-coloring problem is exactly the same as Uncle Lester's scheduling problem.

Graphs

The concept that ties together Uncle Lester's scheduling problem and the map-coloring problem is that of a *graph*. A graph is simply a group of points, some of which are joined by lines. Figure 6.2 is an example of one. The points are called *vertices* (a term already introduced in Chapter 1) and the lines are called *edges*. Vertices that are joined by an edge are said to be adjacent; for example, vertices A and B are adjacent in the graph shown in Figure 6.2, but A and D are not.

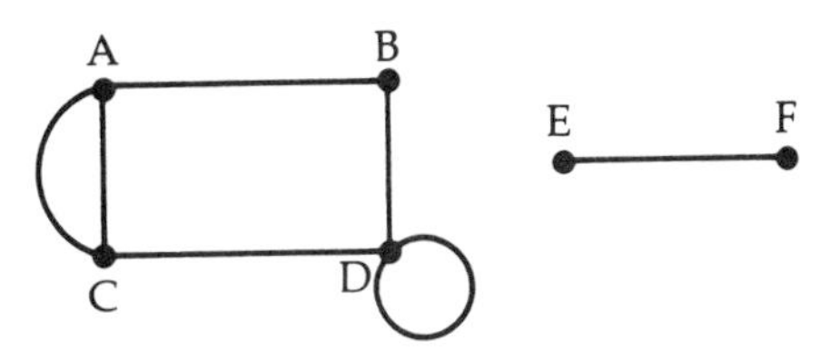

Figure 6.2

A graph may allow us to show the essential relationships in a problem in a very simple way. Consider the map-coloring problem. Let us look at the map again, this time labeling the districts A, B, . . . so we can refer to them. See Figure 6.3.

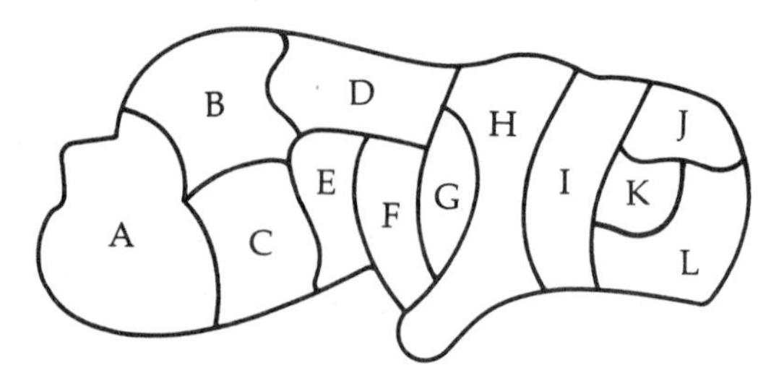

Figure 6.3

Each district has only one characteristic of importance as far as coloring the map is concerned. Whether the district is large or small, compact or skinny, has nothing to do with it. What is important about a district is *which other districts it touches*.

So as to eliminate the inessential attributes of the districts, let us consider each simply as a point, as shown in Figure 6.4. Now let us join with a line those points representing districts with a common border (Figure 6.5).

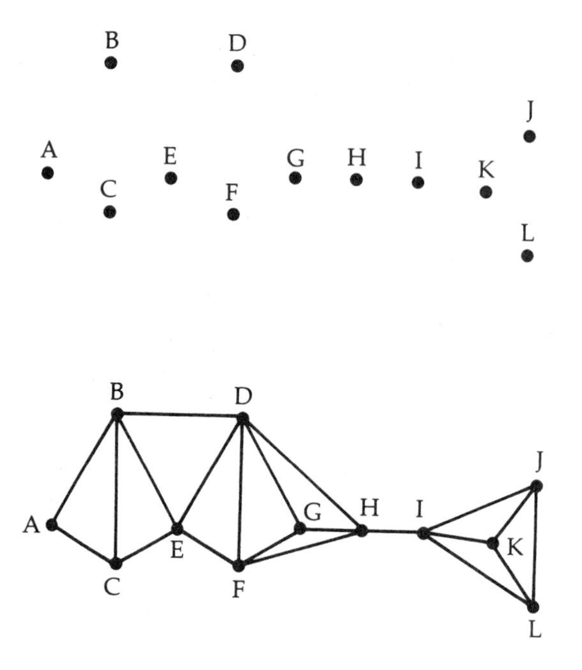

Figure 6.4

Figure 6.5

Figure 6.5, of course, is a graph. The map-coloring problem is equivalent to numbering the vertices of this graph with the numbers from 1 to 4 so that no two adjacent vertices are given the same number. One way of doing this is shown in Figure 6.6.

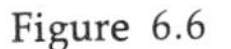
Figure 6.6

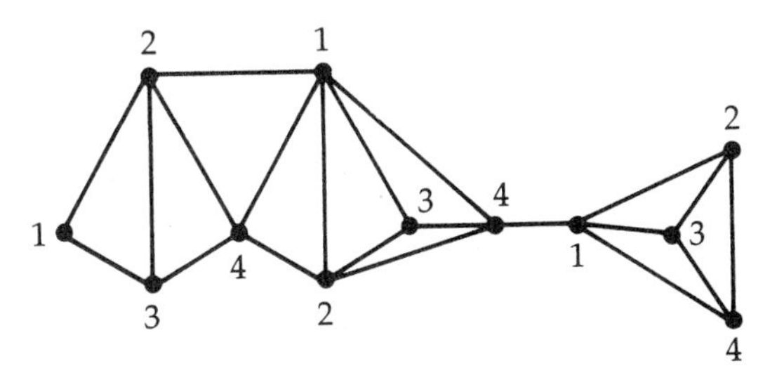

The Scheduling Problem

Now we return to the problem of scheduling sports at the Donohue picnic, introduced in Example 1. Abbreviate the 12 activities as follows:

Softball	SB	Badminton	B
Over-40 softball	O	Three-legged race	3L
Hardball	HB	Touch football	TF
Men's volleyball	MV	Frisbee	F
Women's volleyball	WV	Egg throw	ET
Mixed volleyball	V	Horseshoes	H

Now we will make a graph having 12 vertices labeled as above (see Figure 6.7). We will join with an edge any two vertices corresponding to activities that *cannot* be scheduled at the same time. SB and O are joined, for example, because these represent softball and over-40 softball, which cannot be simultaneous. The rest of the edges were drawn from the conditions given previously in a similar way.

Notice that the edges joining F to TF and 3L to ET cross. There is nothing wrong with this as long as we understand that there is no vertex where they intersect. As long as we agree to put a heavy black dot at each vertex no confusion will be possible because of crossing lines. In

Figure 6.7

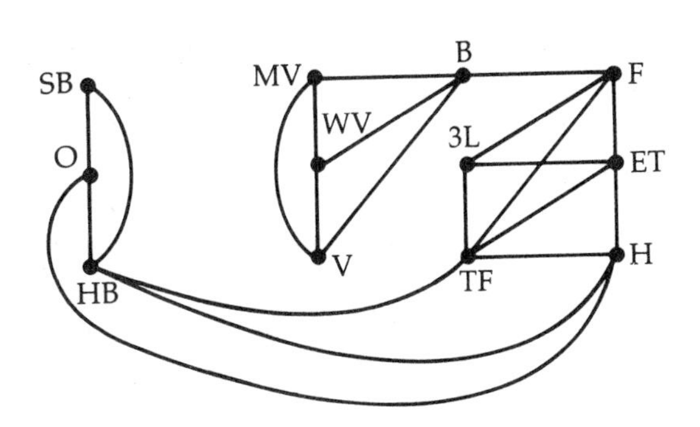

Figure 6.8

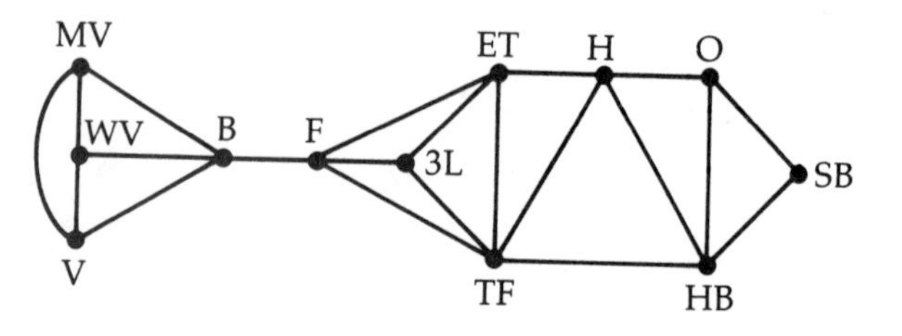

this case, however, we could eliminate this crossing-over (and shorten some of the longer edges) by redrawing the graph as in Figure 6.8.

To solve the scheduling problem we merely have to label each vertex with one of the numbers 1 through 4 (representing the four time periods) in such a way that no adjacent vertices get the same number. This is quite easy; in fact, we have already done it. Turn back to the graph used to color the map of Example 2 (Figure 6.6). If we rotate it 180° we get a version of the graph in Figure 6.8, and the same numbering can be used (see Figure 6.9).

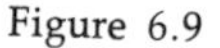
Figure 6.9

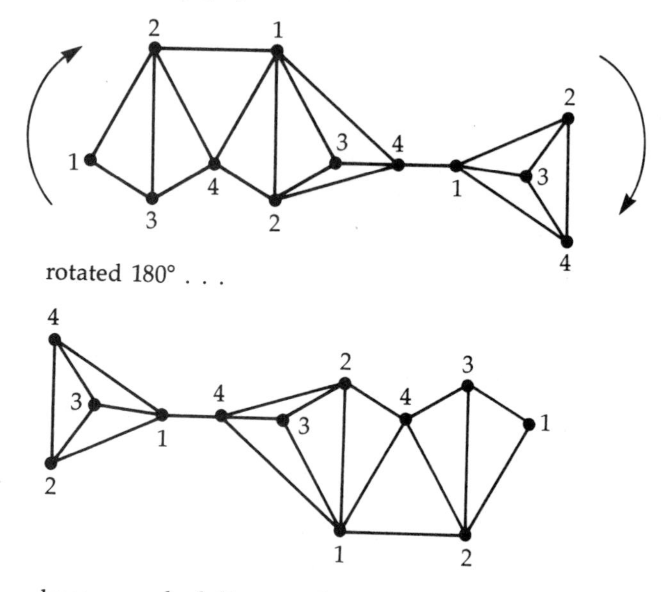

Map coloring graph . . .
rotated 180° . . .
becomes scheduling graph.

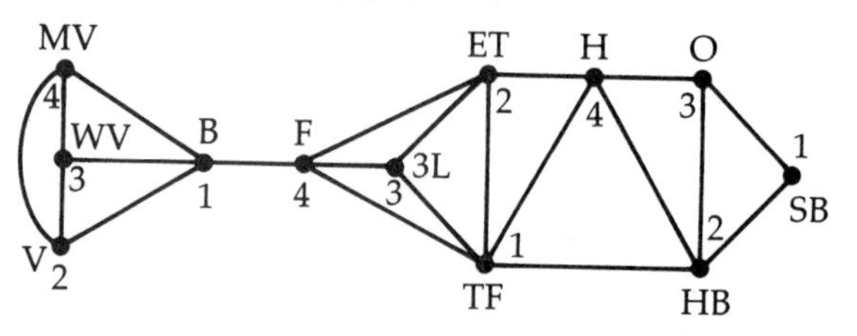

Uncle Lester could have used the following schedule:

PERIOD 1	PERIOD 2	PERIOD 3	PERIOD 4
Badminton	Mixed volleyball	Women's volleyball	Men's volleyball
Touch football	Egg throw	Three-legged race	Frisbee
Softball	Hardball	Over-40 softball	Horseshoes

EXAMPLE 3

The Party In this example we will see how a graph can be used to understand a problem of seating people at a dinner.

The Harrises are planning a dinner party. They want to invite Amos Smith, Betty Smith, Calvin Smith, Antony Jaffee, Beryl Jaffee, Cecile Jaffee, Aaron Belloti, Bess Belloti, and Carter Belloti. The guests will be seated at three tables, but some problems may arise about this. The Smiths will bicker all through dinner if any two of them are seated at the same table, so this must be avoided. The same goes for the Jaffees and for the Bellotis, for that matter. Aaron Belloti will flirt with Cecile Jaffee if they are put at the same table, which would never do, and the same is true of Cecile and Carter Belloti, which is why Cecile should also not sit with Bess Belloti. Is the party possible?

Labeling each vertex with the initials of a prospective dinner guest, and joining vertices standing for people who cannot sit together, we construct the graph shown in Figure 6.10.

Figure 6.10

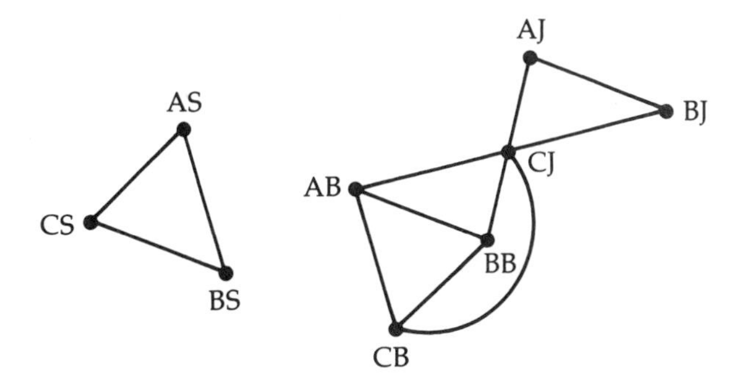

We need to "color" each vertex with one of the numbers 1, 2, or 3 (since there are three tables) in order to solve the Harrises' problem. We quickly see that this is impossible. Vertices AB and BB must get different numbers, say 1 and 2. Then CJ and CB must both be 3. This is forbidden, since they are also joined by an edge. The Harrises must either give up their party or else buy another table.

Exercises 6.1

In problems 1 through 6 make a graph where each region is represented by a point as in Example 2. "Color" each graph with the numbers 1, 2, 3, . . ., n so that adjacent vertices have different numbers, using as small a value of n as possible. *Note*: Two regions in a map are not considered to touch if they meet only at a single point.

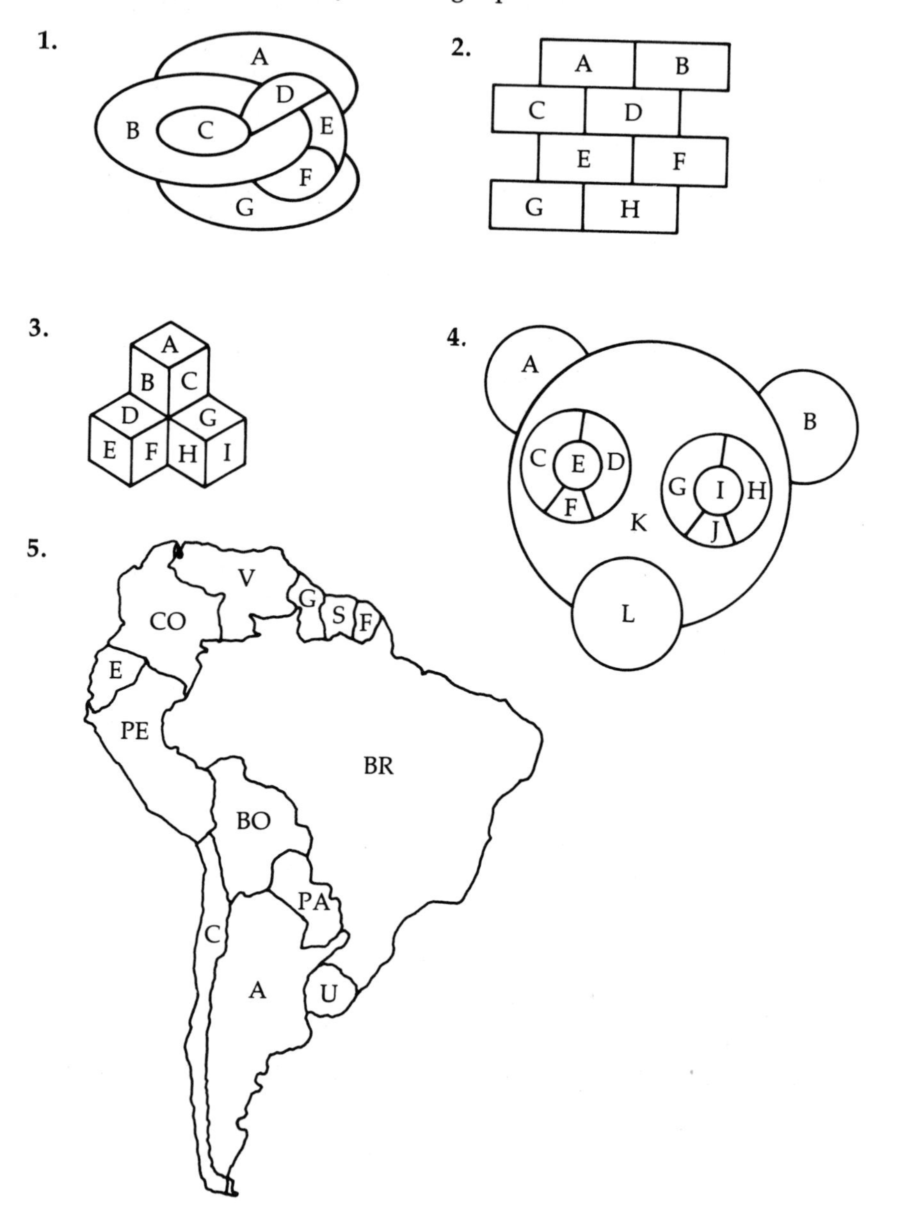

6.

7. Color the graph shown in the least possible number of colors with adjacent vertices different

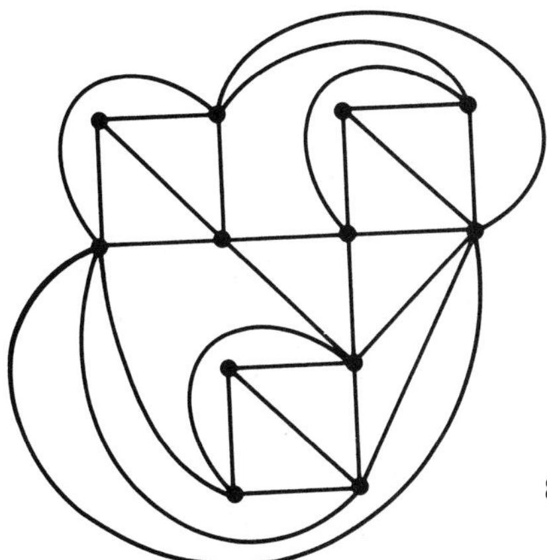

8. Color the graph shown with the least possible number of colors with adjacent vertices different.

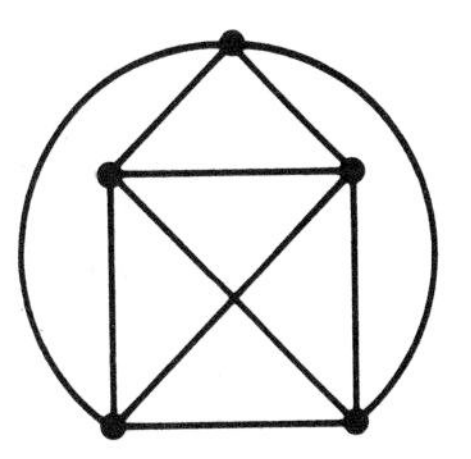

9. Which of the graphs shown are really the same graph, drawn differently?

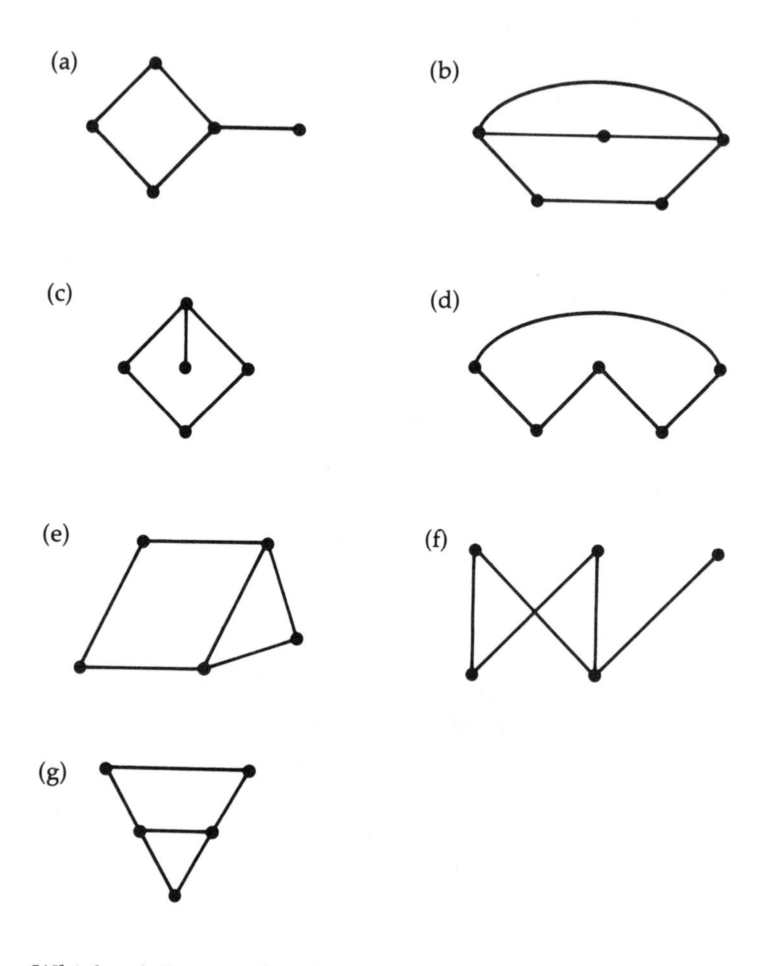

10. Which of the graphs shown are really the same graph, drawn differently?

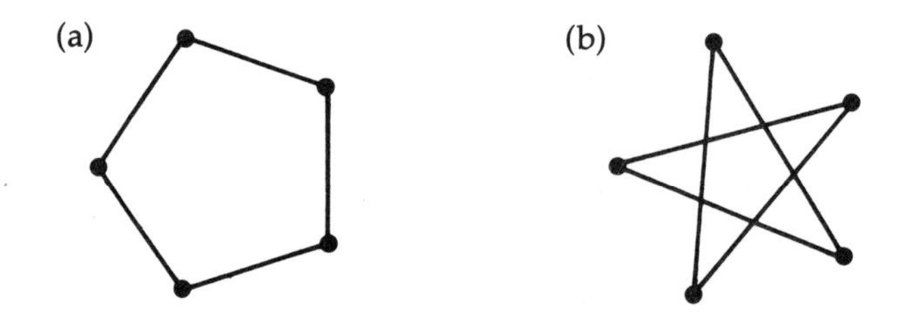

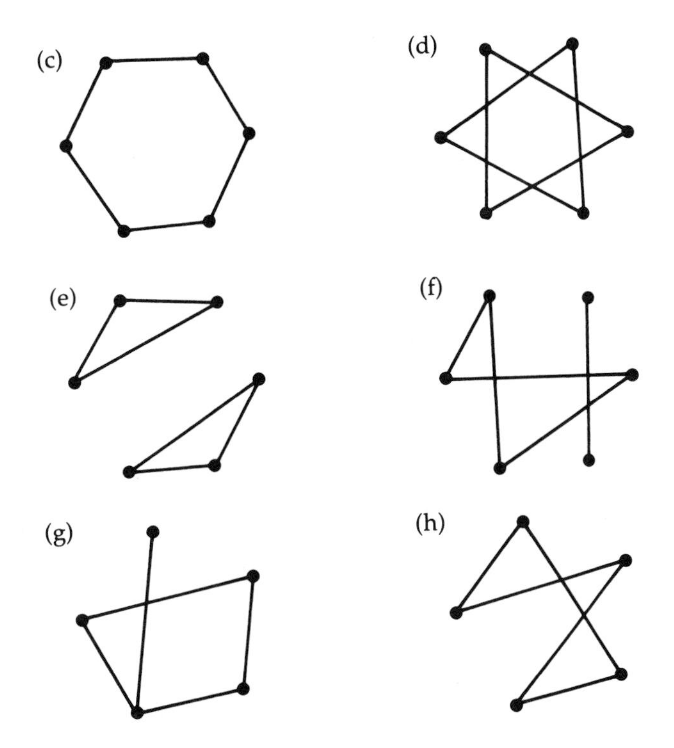

11. Nurse Catalano is trying to schedule the three available time slots at the cat hospital Thursday morning. More than one cat can be treated at the same time, but certain cats would fight if brought in together and so must be given different times. Tabby fights with Alice, Spooksie, and Seymore; Alice fights with Spooksie, Thomson, Mildred, and Foo Foo; Thomson fights with Foo Foo, Winky, and Tiger; and Tiger and Winky fight. Draw a graph, joining the points corresponding to cats that fight. Draw up a schedule if possible.

12. The Russian Commissar of Television has a number of one-hour programs to schedule in the three time slots between 7 and 10 on Friday night. They are *Ballet Highlights, Wheat News, Chess Tips, Dancing Behind the Plow* (an agricultural ballet), *Farmers' Chess Tournament*, and *March of the Pawns* (a chess ballet). No two shows involving ballet, or involving agriculture, or involving chess should be shown at the same time. Make a graph with lines joining points representing shows that cannot be simultaneous. Is the Commissar's task possible, given that there are only two channels?

13. Professor Zich is running a mathematics conference. The following talks have been submitted.

SPEAKER	TOPIC
Professor Abell	Graph Theory in High School
Professor Brown	Computers and Graphs
Professor Clemens	Computers and the Law
Professor Dossey	Statistics and Computers
Professor Edge	Statistics and Tennis
Professor Friedberg	Statistics and Graph Theory
Professor Gilmore	Baseball and Computers
Professor Ha	Computers and Analysis

There are only four time periods for talks, but more than one talk may be given at the same time. Talks with titles involving a single subject (like graphs) should not be scheduled at the same time, however, since a specialist would then have to miss at least one. Brown and Edge should not be scheduled at the same time, since both will draw large audiences. Dossey wants to hear Abell's talk.

Draw a graph where each vertex stands for a speaker and an edge joins each pair of vertices representing speakers who cannot speak simultaneously. Make up a schedule of talks, if possible.

14. At the last minute Zich (in problem 13) learns that Clemens has changed his talk to Graph Theory and Tennis. Draw the graph that is now appropriate and make up a schedule if possible.

6.2 THE FOUR-COLOR THEOREM AND THE SALESMAN PROBLEM

The Four-Color Theorem

One famous map-coloring problem was first posed in 1852. It was found in practice that four colors always seemed to be enough to color any map in such a way that countries with a common border got different colors. That at least four colors are sometimes needed is not hard to see. The map in Figure 6.1 requires four colors, as does the map of South America (see problem 5 of the last section). The simplest map requiring four colors is shown in Figure 6.11. Here each of the four regions touches each of the other three and so four colors are clearly necessary.

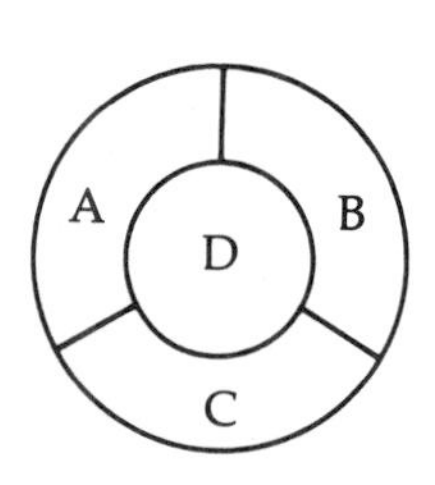

Figure 6.11

The question that caused all the trouble was whether there could possibly be a map requiring five colors. It is not hard to see that it is impossible to draw a map of five regions, each of which touches the other four. (Try it! Remember that countries are not considered to touch

Figure 6.12

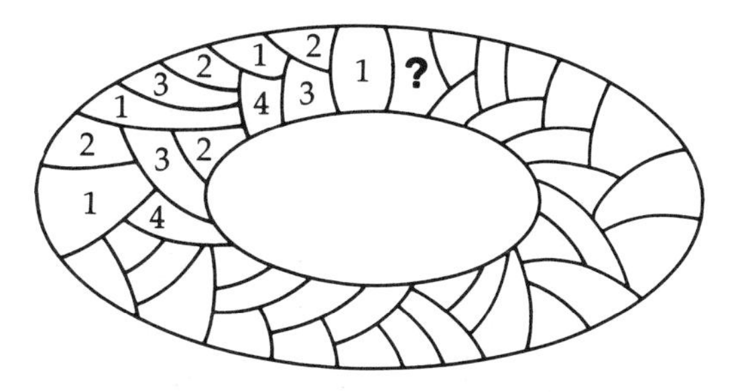

if they meet only at a point.) Nevertheless it was conceivable that some very complicated map might exist so that any attempt to color it with four colors would lead to a contradiction. (See Figure 6.12.)

The *four-color conjecture* was that four colors were always enough. No one was able to prove this for almost 125 years, until 1976, when two mathematicians at the University of Illinois, Kenneth Appel and Wolfgang Haken, proved the conjecture. Their proof entailed the checking of a great many cases with a large computer. Their accomplishment was to break the problem up in such a way that a computer was able to handle it.

Let us point out that although every map can be represented as a graph, not every graph corresponds to a map in a plane. Consider, for example, the graph in Figure 6.13. Here each of the five vertices is

Figure 6.13

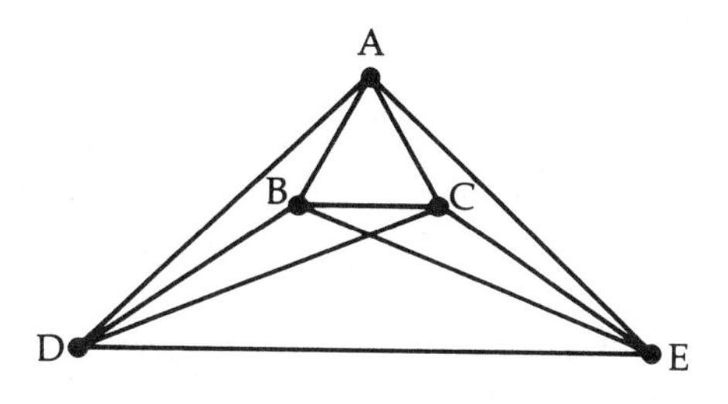

adjacent to the other four, and five colors are needed to color the vertices of the graph. The graph corresponds to no map in a plane, however. In fact it is impossible to redraw the graph so as to eliminate a crossing of edges, as we have between the edges joining B to E and C to D.

EXAMPLE 1

The Tourist A tourist to the island shown wants to visit each country there (see Figure 6.14). Because of the difficulty in crossing borders, she would like to do this as few times as possible, never reentering a country already visited. Is this possible?

Figure 6.14

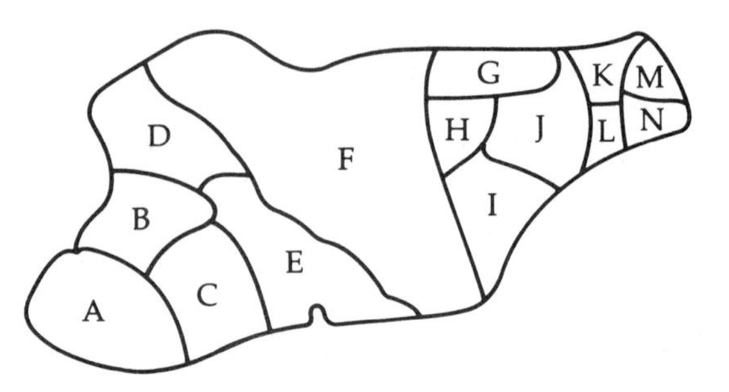

We convert the map to a graph (Figure 6.15), joining the vertices corresponding to countries with a common border.

Figure 6.15

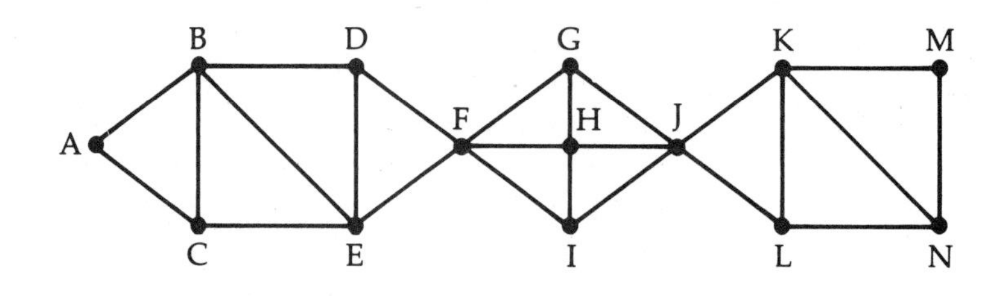

Now the problem is to find a path along the edges of the graph that hits each vertex exactly once. There are apparent bottlenecks at countries F and J; clearly the path must start on one side of each of these and end on the other. Thus the journey must start in one of the two sets A, B, C, D, E and K, L, M, N and end in the other. One solution is A B C E D F G H I J K L N M.

EXAMPLE 2 **The Salesman** A salesman covers the cities identified by letters on the map in Figure 6.16. The lines joining the cities represent roads. He would like to start at his home office H and visit each city exactly once, then return to H. Is this possible?

Figure 6.16

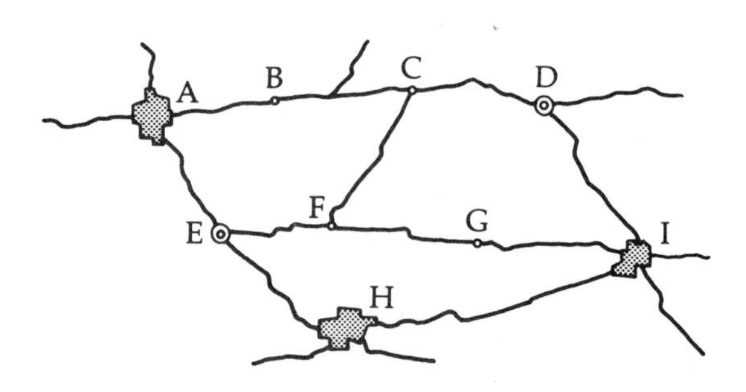

This is a graph-theory problem, pure and simple. Here we are starting with a graph; no transformation is necessary. A little trial and error soon convinces us that no path of the sort called for exists.

Let us denote by AB the road joining A to B, and so on. We could argue that the road HI must be used, for otherwise the salesman would have to pass through E both leaving and returning to H. Then exactly one of the two roads DI and GI must be used. If DI is not used, then the salesman must go through C twice, on his way to and from D. Likewise, if GI is not used, then F must be visited twice. One way or another some city must be visited twice.

Any problem that requires finding a path along the edges of a graph that visits each vertex is often referred to as a *salesman problem*.

EXAMPLE 3

The Old Friends Al would like to visit his old pals from high school but doesn't know where most of them live. He has kept in touch with only Clem. Clem, Dom, and Eddie have kept in touch with each other, as have Dom, and Fred. Fred, Harry, and Guido have kept in touch, as have the pairs George and Ira, Ira and Joel, Harry and Joel, and Joel and Bill. Can Al visit all his old friends, each one exactly once, always getting directions to the next friend from the one he is currently visiting?

Figure 6.17

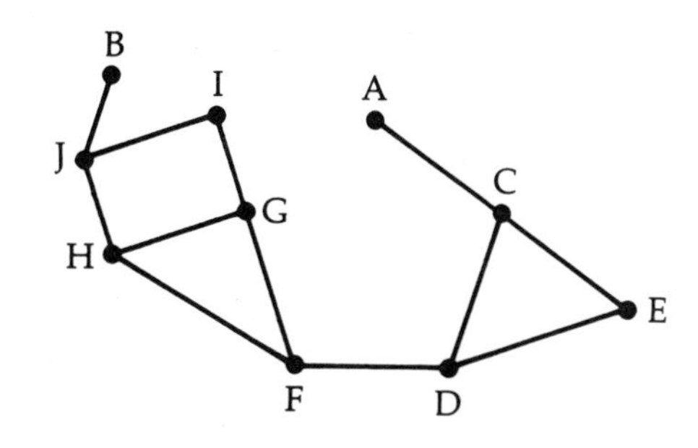

We draw a graph, letting A represent Al, B Bill, and so on, and joining vertices corresponding to friends who have kept in touch. The only path satisfying the conditions is A C E D F H G I J B. See Figure 6.17.

Exercises 6.2

1. Find a path in the graph (page 340) from A back to A again, touching each vertex exactly once.

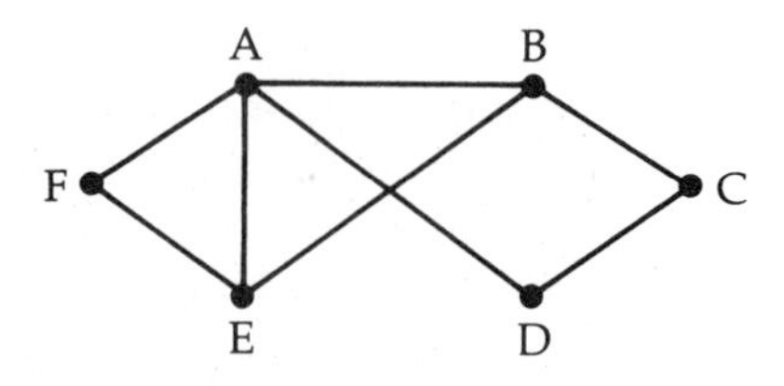

2. Find a path starting at A and visiting each vertex exactly once in the following graph.

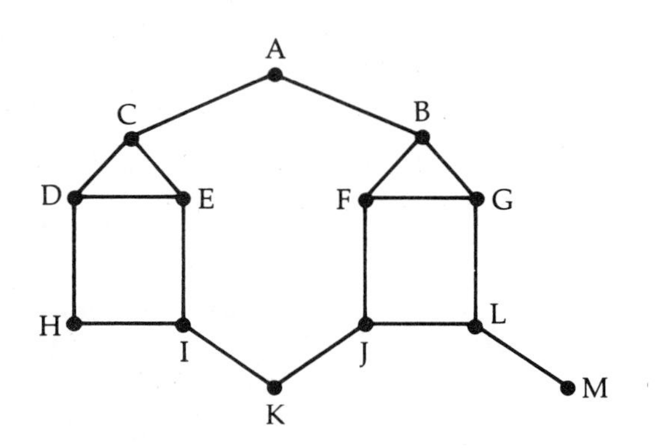

3. Draw a graph corresponding to the following map in the usual way. Find a path starting at A and returning to A visiting every other country exactly once.

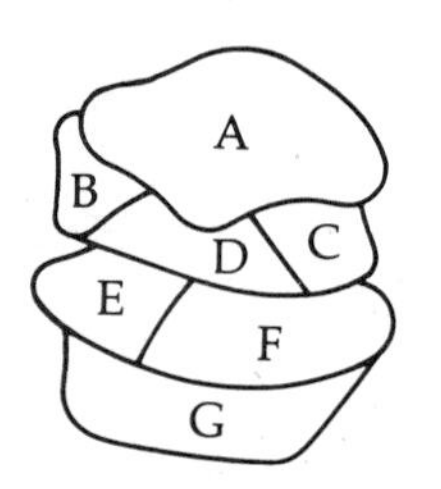

4. Draw a graph representing the map shown in the usual way. Find a path visiting each country exactly once.

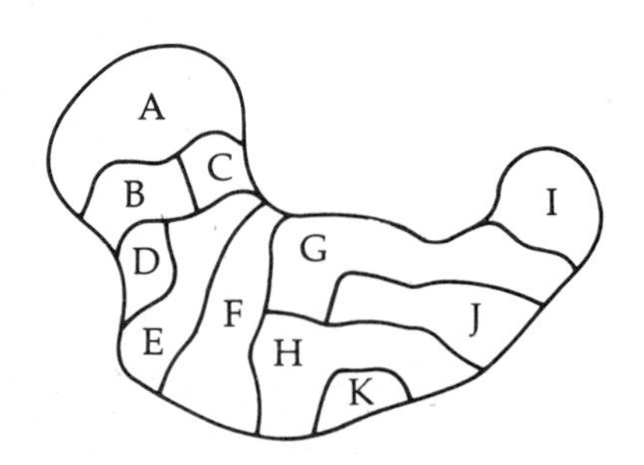

5. A salesman must visit each of the cities shown in the graph, using the roads shown. He would like to visit each city exactly once, starting from and returning to his home office at A. Show how this can be done.

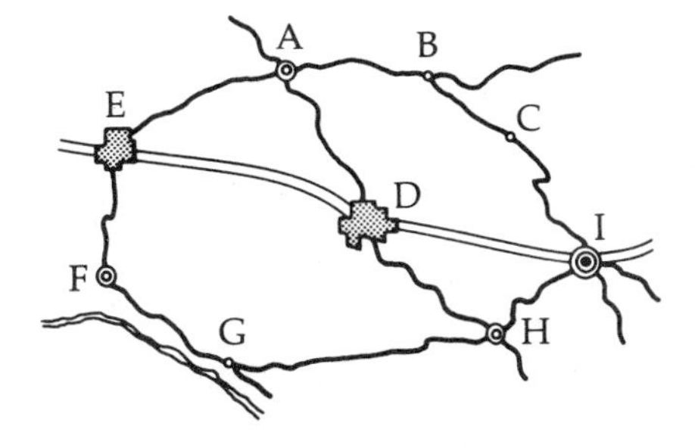

6. When the telephone system was just being introduced, each house in Pottsville was only hooked up to certain other houses. The Andersons, Bilodeaus, and Carters were hooked together, as were the Carters, Jimenezes, and Fleegles, and the Garbers, Hortons, and Drurys. The Enlows were hooked up to the Bilodeaus, Carters, and Fleegles. The Fleegles and Garbers were hooked up, as were the Drurys and Insels. Draw a graph showing these relationships. How could a piece of news be passed from the Andersons to the other families having phones without any family making or receiving more than one call?

7. The figure shows the rooms and doors of a suite of offices. Draw a graph with a vertex for each room, joining vertices if a door connects the corresponding rooms. Can a security guard enter the suite from the hall and visit each room exactly once without returning to the hall until this is done?

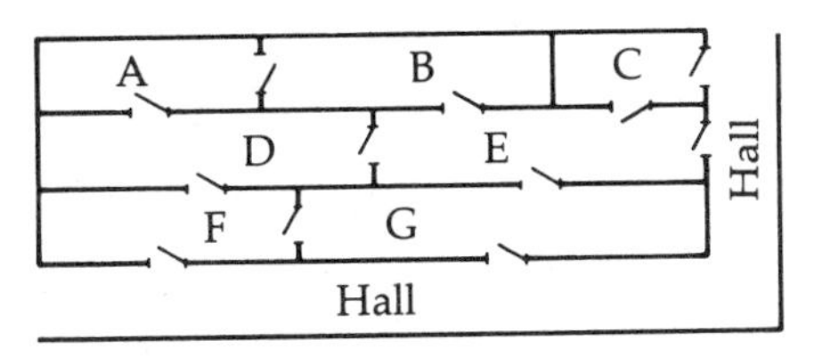

8. The figure shows a maze consisting of 12 compartments. A rat enters the maze at the place marked "In" and runs through the compartments, picking up a morsel of food in each one. The rat finally runs out of compartment K as shown. Draw a graph repre-

senting the maze as in problem 7. Is it possible for the rat to visit each compartment exactly once?

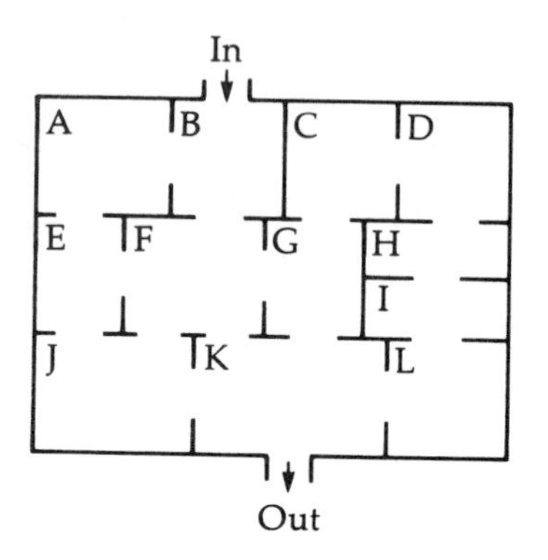

In problems 9 and 10 find, if possible, a path visiting each town in the maps shown exactly once, starting at A and ending at B.

9.

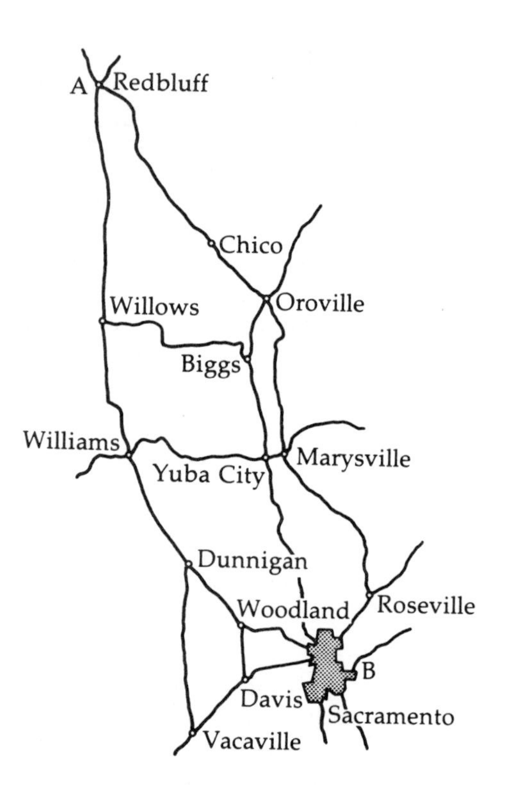

10.

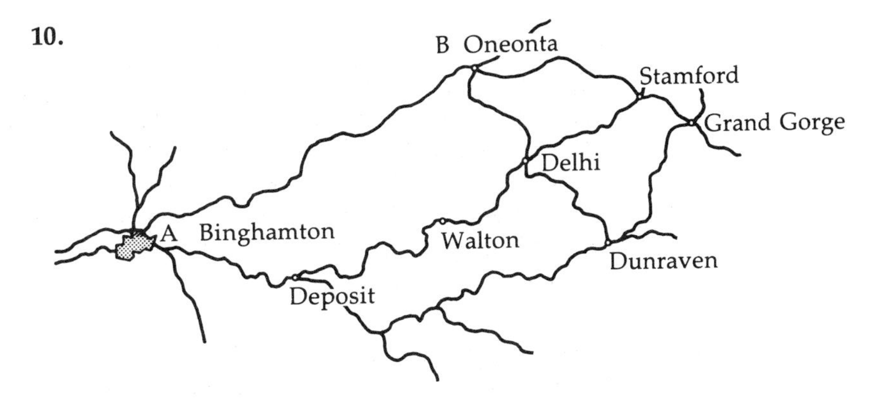

Even though by the four-color theorem four colors suffice to color any map on a plane (flat surface) or sphere (such as the earth), on other surfaces a different number may be needed. On a *torus* (the doughnut shape shown), for example, *seven* colors may be needed.

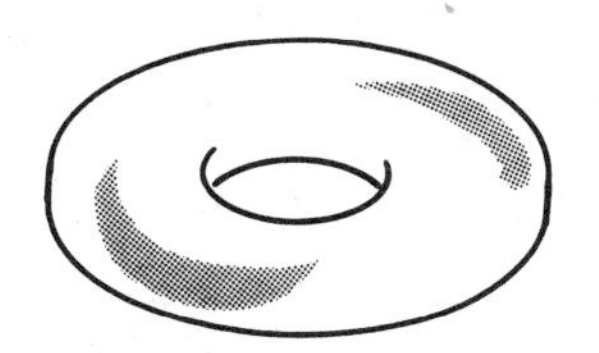

To see why the torus presents different coloring problems from the plane, notice that in the plane there is no way to join A and B below by an edge without crossing circle $\mathscr{C}$.

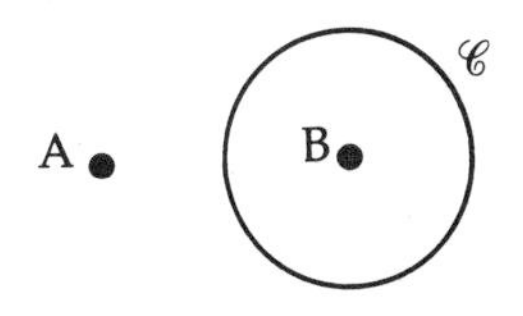

If $\mathscr{C}$ is drawn as shown on a torus, however, then A and B can be joined without crossing $\mathscr{C}$ no matter how they are placed.

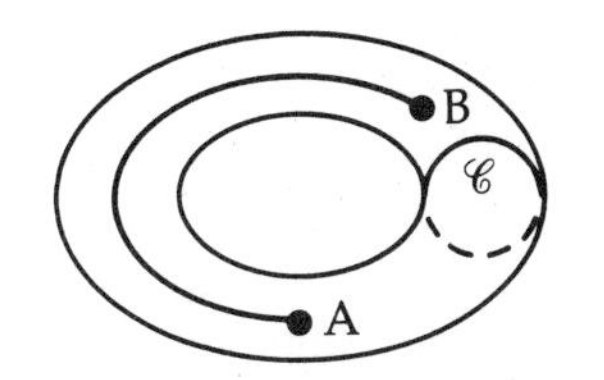

11. Show that the graph in Figure 6.13 can be drawn on a torus without any edges touching.

12. Draw a map on a torus requiring five colors.

It can be shown that every map on a torus can be colored with seven colors. Moreover, maps on a torus requiring seven colors exist.

13. Draw a map on a torus consisting of seven regions, each of which touches the other six.

6.3 THE MAILMAN PROBLEM

In this section we will consider problems that require finding a path on a graph tracing each edge exactly once. This attention to *edges* rather than *vertices* is what distinguishes the *mailman problem* from the *salesman problem* of the last section.

EXAMPLE 1

Delivering the Mail The mailman must deliver to both sides of the streets *inside* his territory (Figure 6.18), between B and E, for example, but only to one side of the streets along the boundary, such as from A to

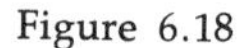
Figure 6.18

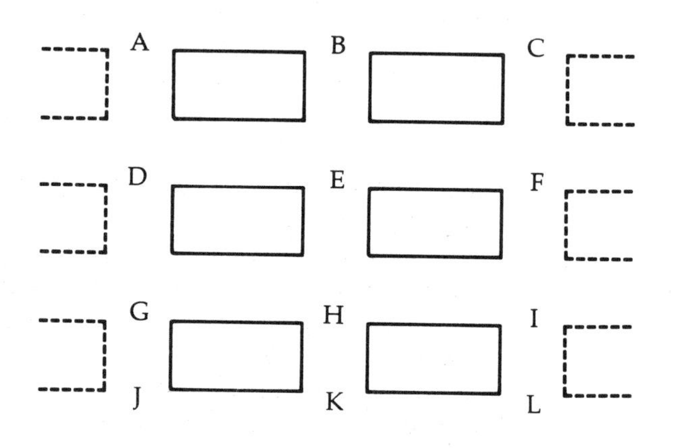

B. He would like to be able to drive his truck to some spot and deliver mail to all the streets that are his responsibility without having to walk along a side of a street he has already delivered to. Ideally, he would end up where he parked the truck. Is this possible?

First let us replace our street map with a graph. The intersections of the streets will be the vertices of the graph. (Doing this involves the assumption that we can neglect the work of crossing the street at an

Figure 6.19

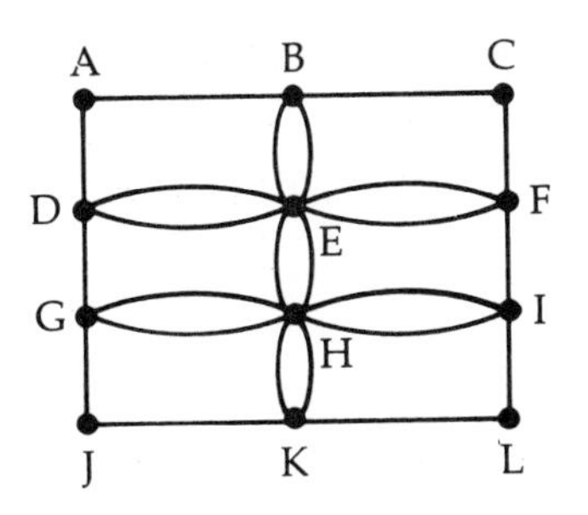

intersection.) Since the mailman must deliver to both sides of certain streets we will join the vertices representing the ends of these streets by *two* edges. See Figure 6.19.

Here we have a problem different from those of the last section. Instead of trying to visit each *vertex* exactly once, we want a path following each *edge* exactly once. A little fooling around provides many solutions. One is A B C F E B E D E F I H E H G H I L K H K J G D A. The reader should trace this out on the graph.

Vertices with an Odd Number of Edges

An old puzzle is to trace the figure shown in Figure 6.20 without retracing any line or lifting the pencil from the paper. This is the same sort of problem as the last. One solution is shown in Figure 6.21.

Figure 6.20

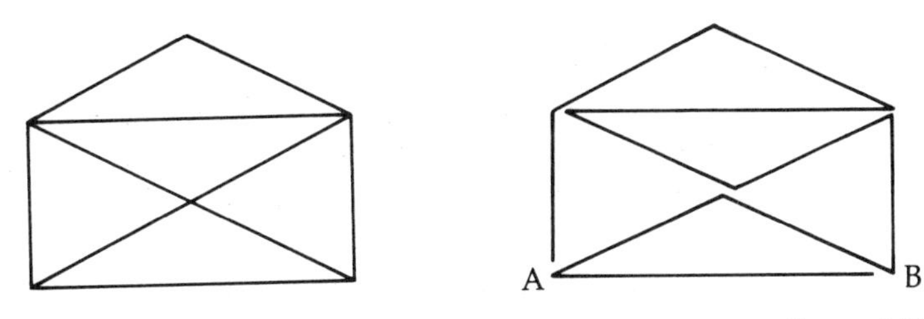

Figure 6.21

One can find other solutions, but all of them seem to begin and end at the two points A and B. A little thought shows why this must be so. There are three edges at the vertex A. If A were not the beginning or end of the path, then A would have to have an *even* number of vertices attached to it, for each time an edge came in another would have to go out. The same goes for B.

We see that *if a path is to be found in a graph tracing each edge exactly once, then any vertex having an odd number of edges emanating from it must be either the beginning or end of the path.*

EXAMPLE 2

The Seven Bridges of Königsberg The Pregel River at Königsberg, Prussia, contained two islands accessible by seven bridges, as shown in Figure 6.22. The residents of Königsberg liked to stroll along the bridges

Figure 6.22

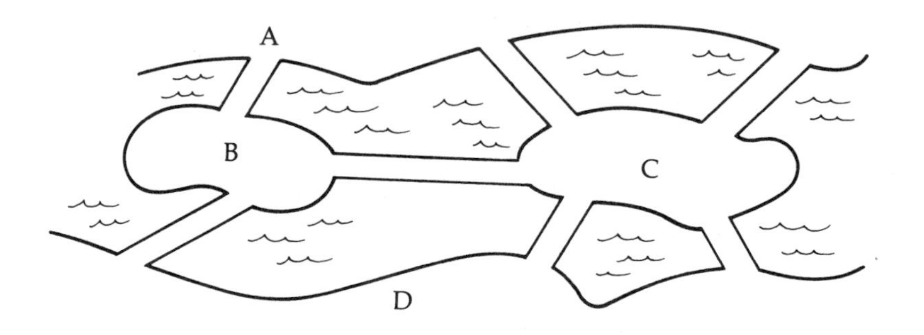

on warm summer evenings. The question arose whether one could cross each bridge exactly once on such a stroll.

It is probably almost automatic by now that we transform the map into a graph, the bridges becoming edges and the islands and sides of the river vertices (see Figure 6.23).

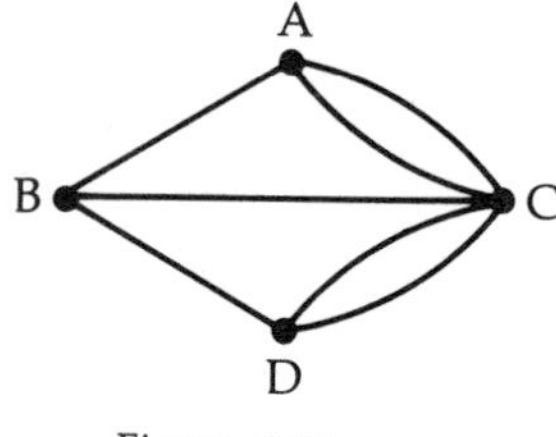

Figure 6.23

We note that A, B, and D have three edges attached to them, while C has five. Since each of A, B, C, and D has an *odd* number of vertices attached to it, each of these must be the beginning or end of a path tracing each edge exactly once. But we can have only one beginning and one end; all four points cannot occupy these roles. We see that the proposed tour of the bridges of Königsberg is impossible.

A Generalization

The following condition for the solvability of the mailman problem should be clear by now.

For a graph to have a path tracing each of its edges exactly once, the graph must have either exactly zero or two vertices with an odd number of edges attached to them. If there are two such vertices, the path must begin at one of them and end at the other. If there are zero such vertices the path must begin and end at the same vertex.

Euler's Theorem

Königsberg really existed (it is now part of Russia and has been renamed Kaliningrad), as did its seven bridges, and the Königsberg Bridge problem was first solved by the great Swiss mathematician Leonhard Euler in 1735. In fact Euler at the same time solved every similar path problem with the following theorem.

THEOREM *A connected graph has a path tracing each edge exactly once if and only if there are either exactly* zero *or* two *vertices with an odd number of edges attached to them.*

It is the "if" part of the theorem that is new to us; that is, if there are exactly zero or two vertices attached to an odd number of edges, *then there is a solution.*

A word should be said about the word "connected" in the above theorem. This just means that there is *some* path along the edges of the graph from any vertex of the graph to any other vertex. Saying that the graph is connected rules out a graph like Figure 6.24, for which it is

Figure 6.24

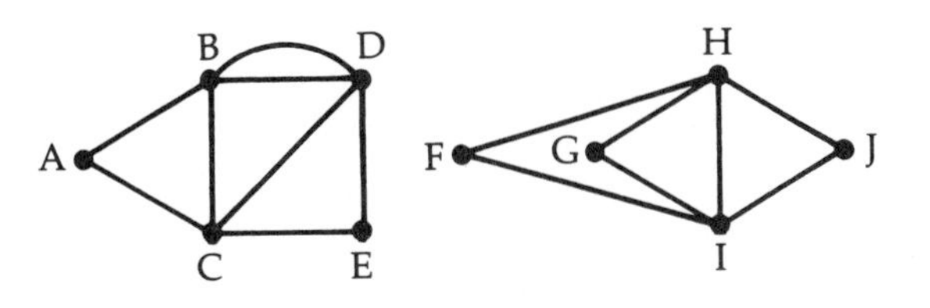

clear no path exists even though no vertex is attached to an odd number of edges. Notice that there is no path from E to F.

LEONHARD EULER
Euler was born in Basel, Switzerland, in 1707, but spent most of his life in other countries, first Russia, then Prussia (he spent 25 years in Berlin at the invitation of Frederick the Great), then St. Petersburg, Russia, again.

Always an incredible worker, Euler lost the sight of one eye through strain in 1735. Soon after returning to Russia in 1766 he developed a cataract in his other eye and spent the last years of his life totally blind. Even this did not impair his unexcelled productivity. He continued to publish many papers, doing the mathematics in his head.

Among Euler's writings was a famous calculus text. He died in St. Petersburg in 1783.

Exercises 6.3

1. Find, if possible, a path tracing each edge exactly once.

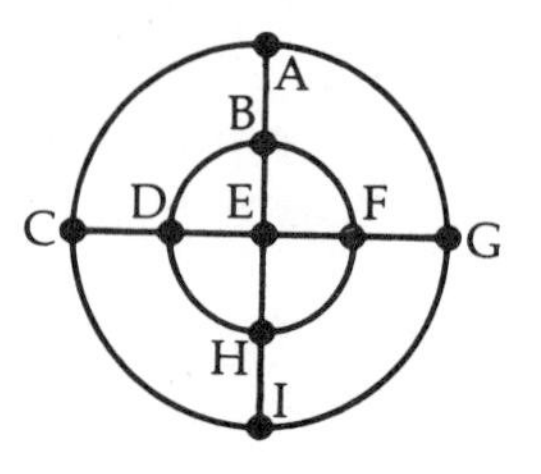

2. Find, if possible, a path tracing each edge exactly once.

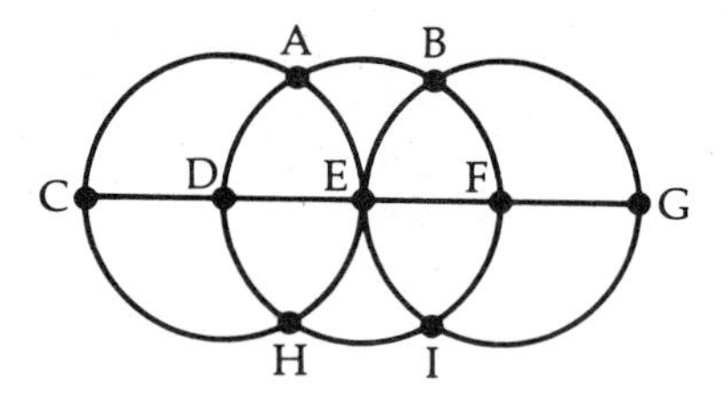

3. An ice cream vendor drives his truck on the same streets the mailman delivers to. (See Example 1.) He only wants to go down each street *once,* however. (Otherwise the parents get angry.) Draw a graph corresponding to this situation. Can the ice cream vendor go down each street exactly once, starting and ending at the same place? Can he do it if he starts and ends at different places?

4. A letter carrier has the territory shown.

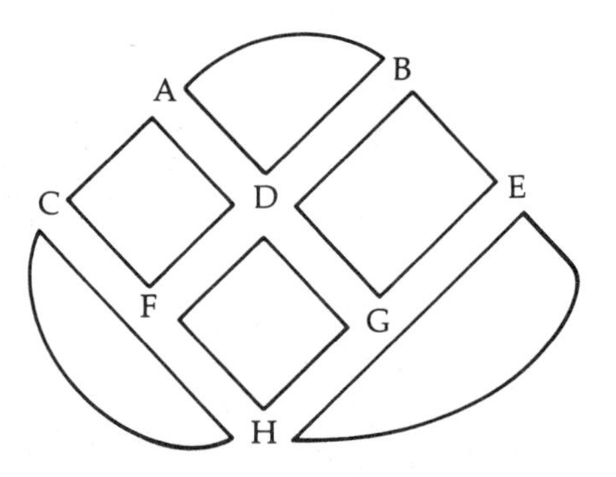

As in Example 1, he need only deliver to one side of the streets on the boundary of this territory. Draw a graph corresponding to this map, and find, if possible, a path tracing each edge exactly once.

5. A suite of offices is shown.

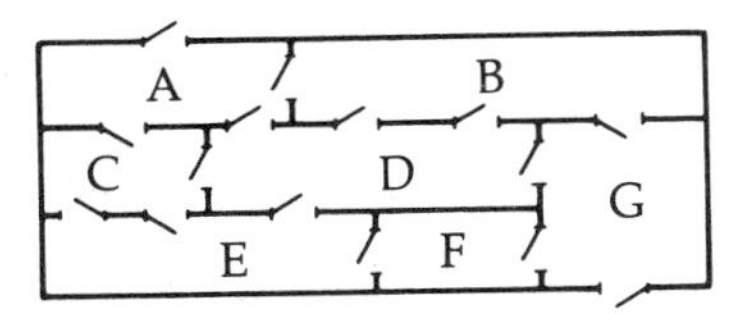

Draw a graph with vertices representing offices and edges representing doors between offices. A security guard wants to enter office A from the outside and go through each door exactly once, leaving the suite through office G. Is this possible?

6. A political scientist wants to study the customs procedures for each pair of countries sharing a border on the island shown.

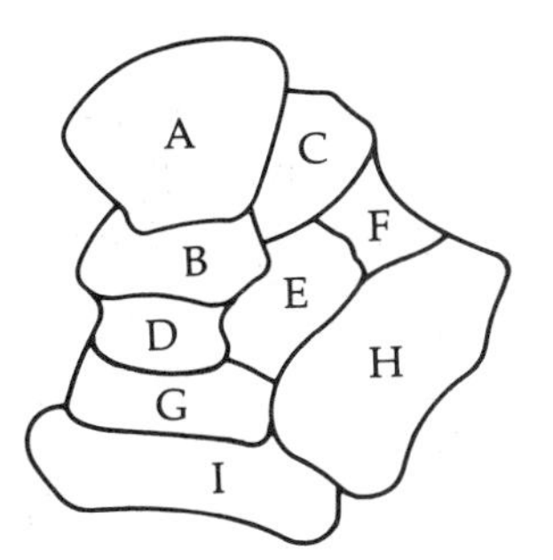

She would like to cross each border exactly once. Draw a graph corresponding to the map and find a way to do this, if possible.

7. Can the security guard in problem 7 of Section 6.2 start in the hall and go through each inside door of the office suite in that problem exactly once and then return to the hall without going through any more inside doors? If so, how?

8. Is it possible to arrange a scenic drive traveling each road joining the cities shown in problem 10, Section 6.2, exactly once? It is not necessary to start or end in any particular city. If so, how?

9. Is it possible to take a scenic drive across each road between cities on the map shown (page 350), traveling on each road exactly once? The trip may start and end in any of the cities. If so, how?

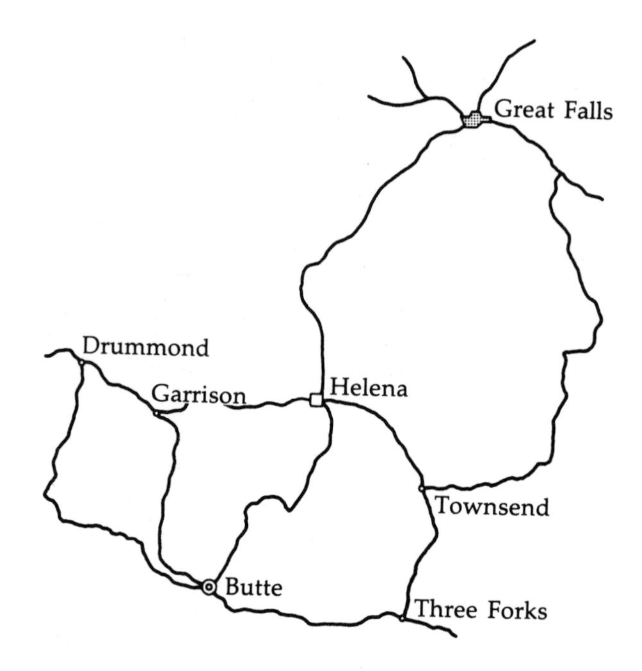

10. A smuggler makes her living taking valuables across the borders of the countries shown. To avoid detection, she never wants to return to a border already crossed. Can she cross every border exactly once? If so, how? Draw a graph representing the map.

11. Can a security guard go through each door in the suite of rooms shown exactly once, entering and leaving as shown? If so, how? Draw the appropriate graph.

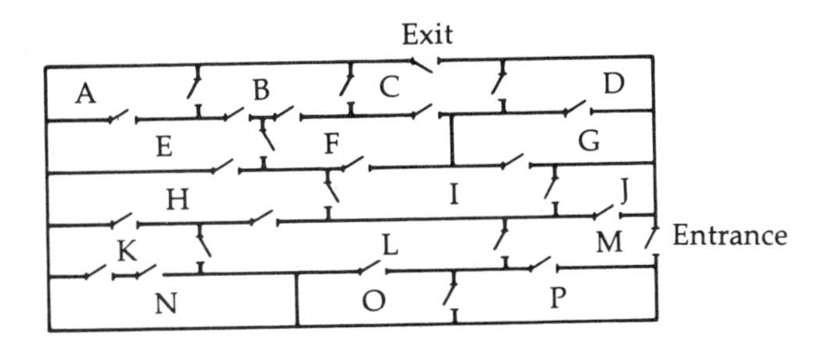

12. Aaron, Bob, Chester, Don, Eric, Frank, and Graham comprise a squash league in which each of them is to play a match at the local court against each of the others. They have pooled their money to buy a ball, which must be used for each match in turn. One of the players in any given match must play in the next match so that the ball will be at hand. Is this possible? If so, how? Draw the graph.

6.4 DIRECTED GRAPHS

A *directed graph* is simply a graph in which each edge carries an implied direction along with it, much like a one-way street.

EXAMPLE 1

Emptying Mailboxes At each intersection of the territory of the mailman introduced in Section 6.3 there is a mailbox. A Postal Service employee goes around in a truck several times a day and empties these mailboxes. She would like to do this in such a way that no intersection is visited more than once; to do otherwise would not be efficient. She has one problem that did not affect the mailman delivering the mail on foot: The streets involved are one-way streets.

The little squares in the map in Figure 6.25 represent the mail-

Figure 6.25

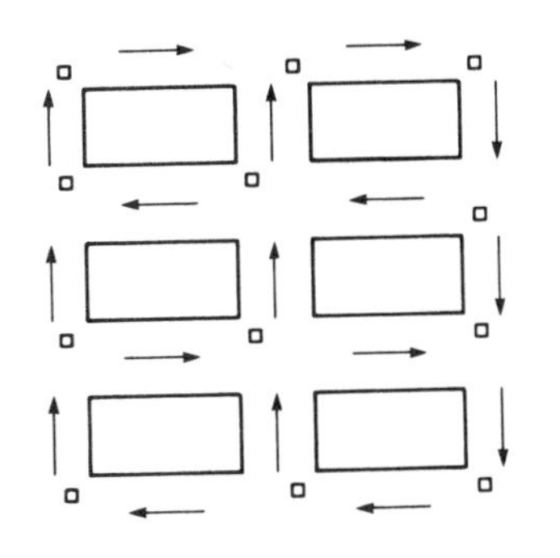

boxes, and the one-way pattern is indicated by arrows. We also use arrows to indicate the permissible directions in our graph (Figure 6.26). One path our mailbox emptier could take is A B C F E D G H I J K L.

Figure 6.26

Here we have an example of a *directed graph*, one where each edge carries an implied direction along with it.

EXAMPLE 2

The Police Officer A police officer is to cover the streets in his squad-car. Certain of them are one-way, as shown in Figure 6.27. (An arrow only implies that the street is one-way for one block.) He would like to ride down each street exactly once and then be back at his starting point. Is this possible? The corresponding graph is shown in Figure 6.28.

One solution is A D E G D C B F G H B A E H F C A.

Figure 6.27

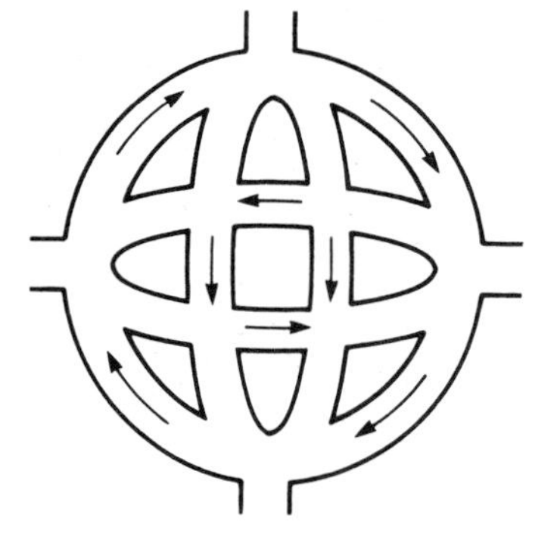

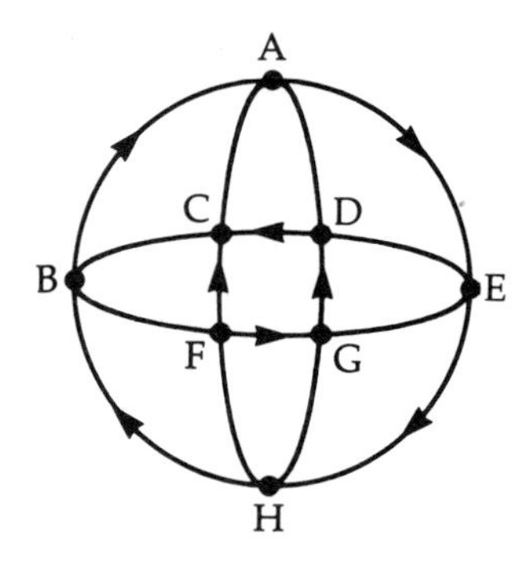

Figure 6.28

A General Principle

A delivery person covers the same territory as the police officer in Example 2, but she needs to go down each two-way street each way so as to make her deliveries on the proper side. (She covers the one-way streets by lane-hopping from side to side.) Can she do this without repeating?

The appropriate graph is shown in Figure 6.29. Here we have used two directed edges for each two-way street. Notice that there are

Figure 6.29

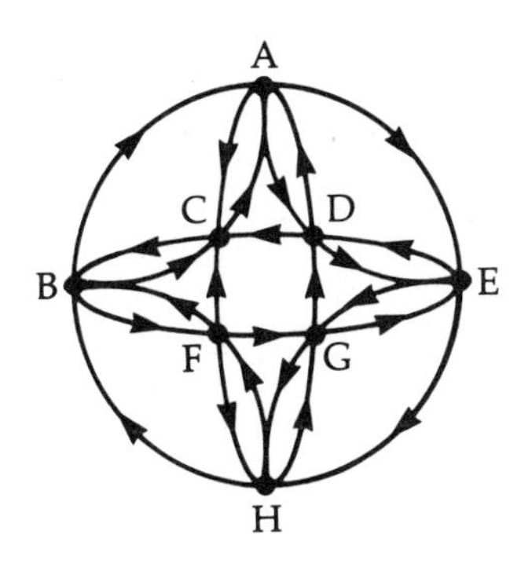

four edges directed into point C and only two directed out. The third time the delivery person enters intersection C she will have no new path out.

We have worked out a general principle:

In order that a path should exist tracing all the edges of a graph, all of whose edges are directed, the number of edges directed *into* each vertex must equal the number of edges directed *away* from that vertex, unless the vertex is the beginning or end of the path, but not both. At the beginning of the path there must be an extra edge directed away from the vertex, and at the end there must be an extra edge directed in.

EXAMPLE 3

The Football Booster John Doaks has made the rash statement that his favorite football team, Illinois, was better than Oklahoma in 1970. That year Illinois beat Oregon, Syracuse, and Purdue; while Oklahoma was beaten by Oregon State, Texas, and Kansas State.

In order to back up his claim John consults the records to find that in 1970 Syracuse beat Navy, Army, and Penn State; Purdue beat

Stanford and Iowa; Kansas State was beaten by Kentucky and Kansas; and Oregon State was beaten by UCLA, USC, and Stanford. Will John's claim hold water?

We make a directed graph of the teams mentioned, with an arrow from A to B in case A beat B (Figure 6.30).

Figure 6.30

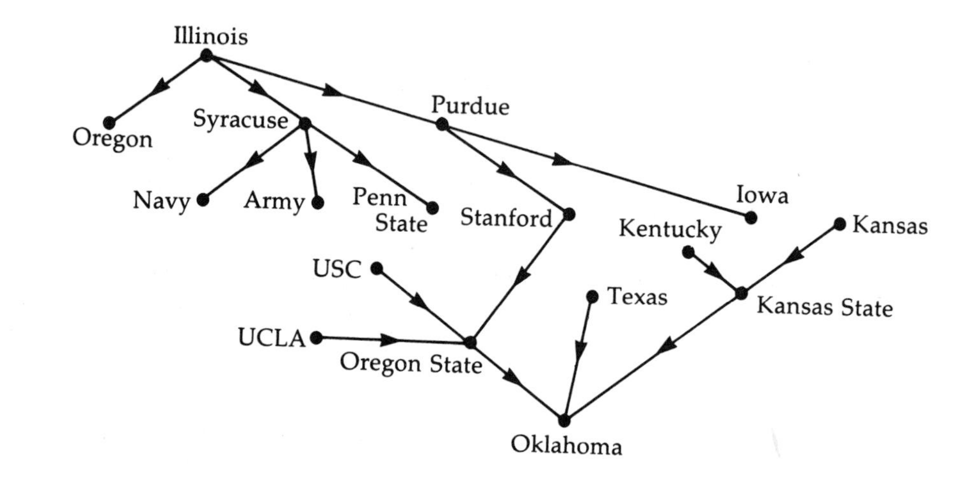

We find the path Illinois beat Purdue beat Stanford beat Oregon State beat Oklahoma.

Exercises 6.4

1. A delivery truck must deliver to the mailman's territory. (See Example 1 of Section 6.3.) All the streets are two-way. The driver need only deliver to one side of the streets on the boundary of the territory but to both sides of the streets inside the territory. The truck can only park on the right side of the street and the driver is not allowed to carry his deliveries across the street. Draw the appropriate directed graph. Can you find a path covering each side of a street in the territory exactly once?

2. A tourist wants to tour all the countries on the island shown, each one exactly once. It is easy to cross certain borders in certain directions, namely, from A to C, D to C, E to D, C to E, F to E, F to B, G to F, B to G, C to B, and either way between A and B. Draw a directed graph showing the easy crosses. Can the trip be arranged with only easy crosses, starting at A?

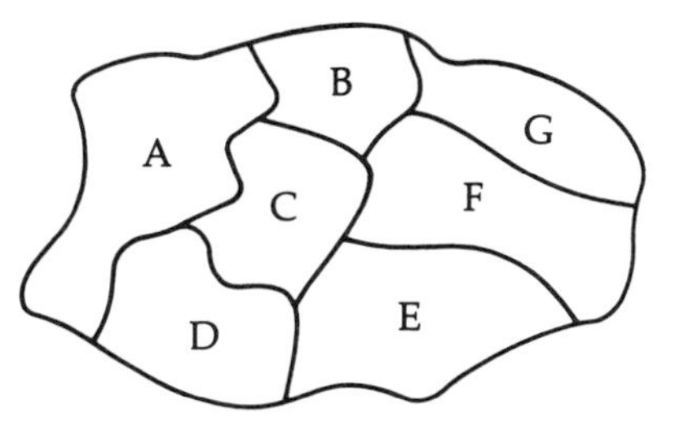

3. A police department has a rather primitive communication system. Certain police cars can call certain others, but not always vice versa. The table shows which cars can call which other cars.

CAR	CAN CALL CAR
1	2
2	3, 5
3	1, 6
4	2
5	2, 3, 6
6	4, 5

It is decided to test the system by giving a message to one car and having it relayed through the system as fast as possible. Each possible call is to be made exactly once in this test. A car may relay the message more than once, but each time it is called it makes at most one call. Draw a directed graph for this system. Is the test possible? If so, how?

4. A knows the phone number of E; B knows the number of A and C; C knows the number of E and D; D knows the number of E and F; E knows the number of A and B; F knows the number of A and B. Draw a directed graph showing these relationships. How can a message be passed from A to F?

5. In tennis Al beat Bob and Cal; Bob beat Don and Emil; Emil beat Gus; Fred beat Al and Cal; and Gus beat Don and Fred. Draw a directed graph. Could an argument be made, as in Example 3 of this section, that Al is better than Fred?

6. Andy, Sam, Tom, and Susan frequently play bridge together. In between hands they talk about their lives. Andy makes the most money, Tom and Susan about the same, and Sam the least. Susan

has the nicest car and Tom the second nicest; Andy and Sam have junkers. Sam is a professional violinist, Tom and Andy are fair whistlers, while Susan can't carry a tune. Draw a directed graph with an arrow going from A to B if A has any reason to feel superior to B. Does everyone have grounds for feeling superior to everyone else?

7. Harry Creech lived in a cottage on the grounds of a vacant mansion, for which he was the caretaker. On the night of the 13th, at midnight, he thought he saw a light in the mansion, and walked up to investigate. A well-dressed gentleman lay inside the door, stabbed through the heart. Creech returned to his cottage and called the police, who came immediately. It had rained the previous morning, and everyone was careful not to disturb the tracks in the wet earth around the mansion, which are shown in the drawing. The dashed lines represent all the tracks found going to and

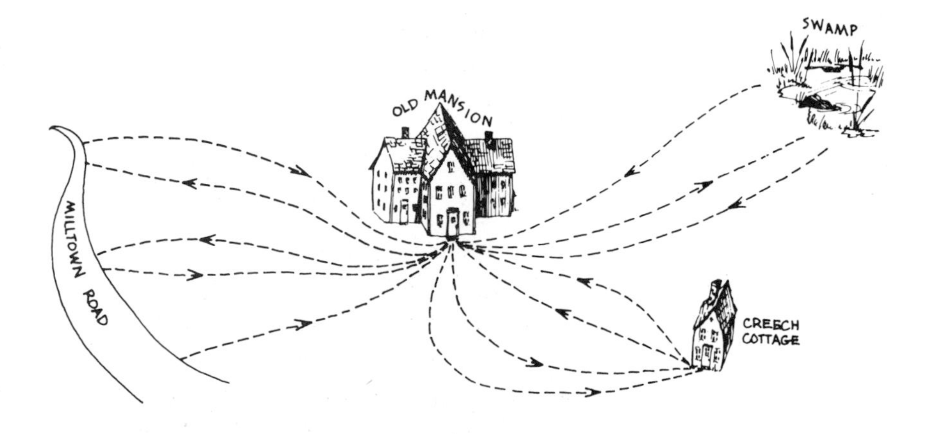

from the mansion, with their directions indicated by arrows. Of course it was possible to infer the direction of the tracks from the heel and toe marks. Creech told Inspector Parker that he had inspected every room of the mansion the afternoon of the 13th and could swear no one was there. "I hope you don't suspect me in this thing, Inspector," said Creech. "I don't even *know* the bloke what was done in."

"Not at all, my good man," returned Inspector Parker. "As a matter of fact we'll soon have the murderer in custody; I know where he is right now."

Where is the murderer and how does Inspector Parker know?

8. A taxi driver taking an innocent fare from the airport to his hotel

decides to pass over every possible street in between in order to run up the meter. He is afraid to repeat a street because then his passenger might catch on. The streets are all one-way, as shown in the figure. Is this possible? If so, how?

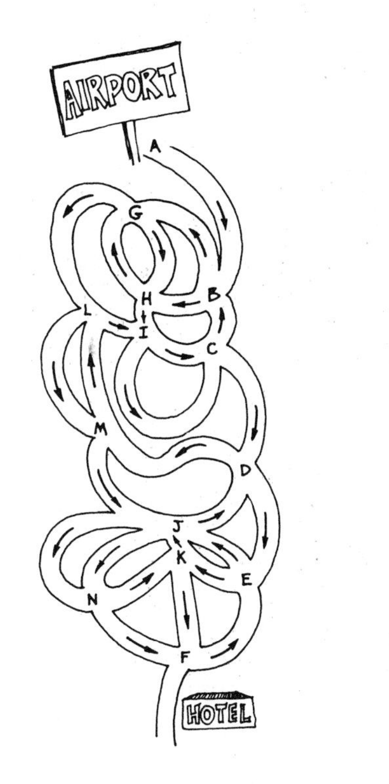

Problem 8

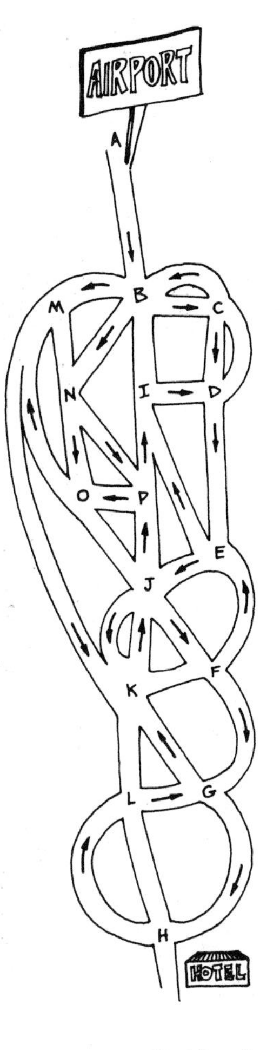

Problem 9

9. The dishonest taxi driver of problem 8 loses his job because of complaints and so moves to a new town. The pattern of streets between the airport and the hotel in the new town is shown in the figure. Can he drive on each street between the airport and hotel in this town without repeating? If so, how?

10. Inspector Parker has the job of finding Mr. Blain, who wandered off into the fen after being kicked in the head by a cow. The figure shows the police map of Blain's tracks. The northeast corner of the fen is especially rough and it was not possible to determine which way Blain was headed on some portions of his path, namely between C and D (twice), D and F, D and I, and I and F. Can you fill in the arrows at these spots?

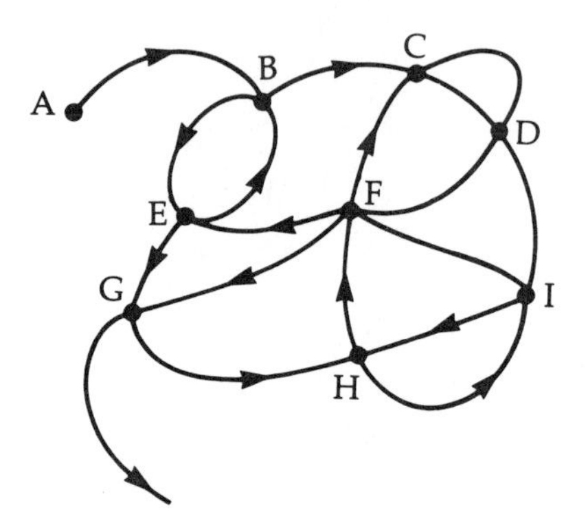

11. The figure shows the path traced by a mouse after eating three ounces of an experimental breakfast cereal. The direction of the mouse at each part of its path was supposed to have been recorded by a lab assistant, but the mouse went so fast that only a few arrows were inserted. Can you supply the rest?

12. In various Superbowls the Packers beat the Chiefs and Raiders; the Jets beat the Colts; the Chiefs, Dolphins, Steelers, and Raiders beat the Vikings; the Cowboys beat the Dolphins and Broncos but lost to the Colts and Steelers; and the Redskins lost to the Dolphins. Which teams could be argued to be better than the Broncos, using the idea of Example 3?

 6.5 TRANSITIVE DIRECTED GRAPHS

EXAMPLE 1

The Spy An international spy has been given the mission of finding out which country has the biggest bomb. Through his various contacts he discovers that England's bomb is bigger than France's; France's is bigger than Italy's and than Germany's; England's is bigger than Germany's; and both Italy's and Germany's are bigger than Sweden's. In order to understand this information better, he makes a directed graph, putting an arrow from country A to country B if A's bomb is bigger than B's. See Figure 6.31.

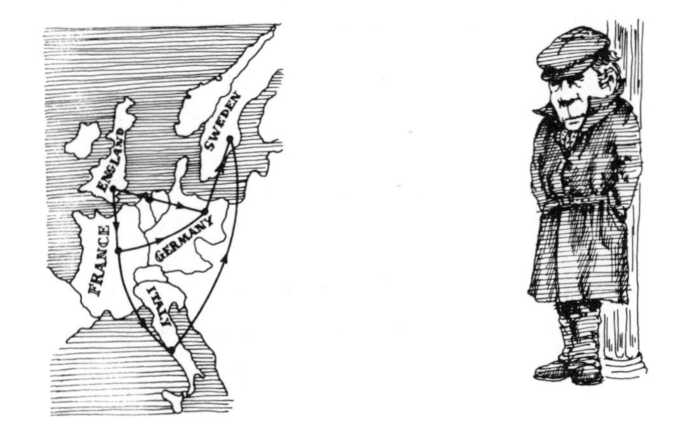

Figure 6.31

The spy notices that certain other directed edges can be drawn even though he has no direct information about them. Since England's bomb is bigger than France's and France's is bigger than Italy's, England's must certainly also be bigger than Italy's. Likewise England's must be bigger than Sweden's, as must France's. He adds these edges to his graph (Figure 6.32).

On the other hand, this merely complicates the graph without adding any information that is not obvious. In fact, the edge England → in the original graph was implied by the edges England → Germany in the original graph was implied by the edges England → France and France → Germany. For the sake of clarity it might be better merely to present the graph as shown in Figure 6.33, allowing the directed edges implied by those displayed to be understood.

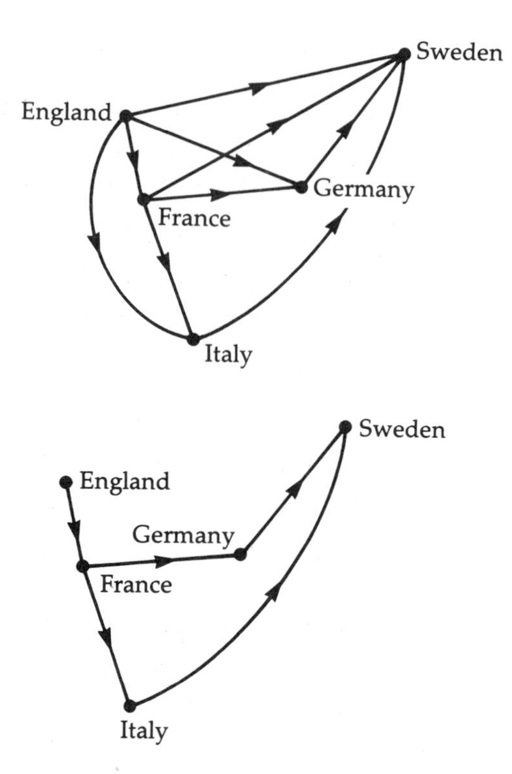

Figure 6.32

Figure 6.33

Transitivity

The relation "has a bigger bomb than" is *transitive* in the sense that if A has a bigger bomb than B and B has a bigger bomb than C, then A has a bigger bomb than C. A directed graph is said to be *transitive* if whenever there are directed edges from A to B and from B to C, then there is a directed edge from A to C. If we know a graph is transitive we may simplify things by drawing only those edges not implied by other edges and the transitivity property. For example, consider the graph shown in Figure 6.34. Suppose we all agree that this is a transitive graph (perhaps based on the real-life situation it represents). Then the

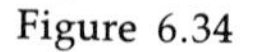
Figure 6.34

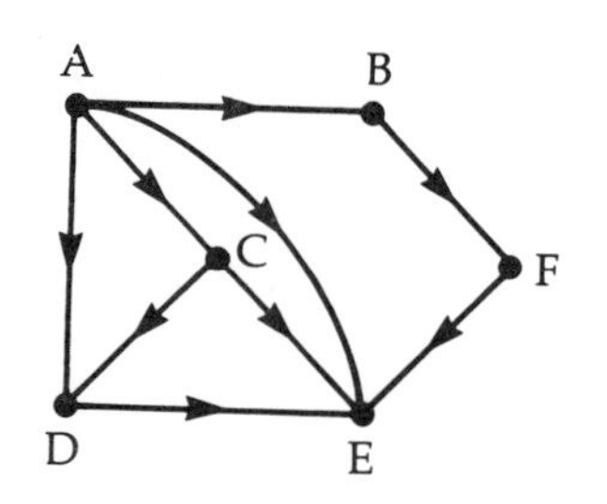

Figure 6.35

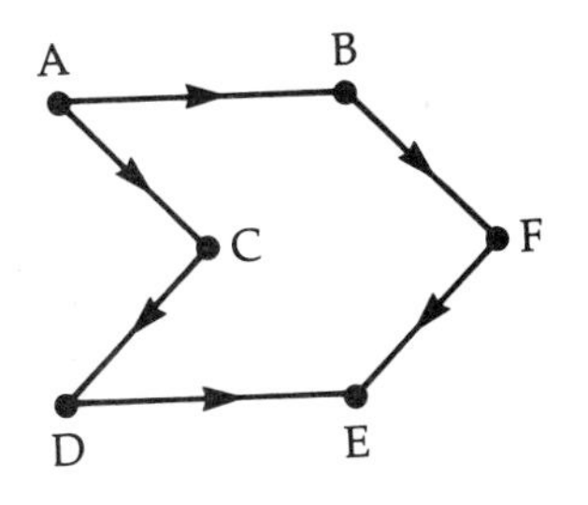

edge A → D is implied by A → C and C → D, and may be removed. Likewise A → E is implied by A → C and C → E; remove it. Finally C → E is implied by C → D and D → E. We are left with the graph shown in Figure 6.35.

EXAMPLE 2

The Letter An efficiency expert analyzes the steps necessary to write and send a letter. She finds they are to: get paper, get pencils, write the letter, get an envelope, get a stamp, stamp the envelope, address the envelope, put the letter in the envelope, and mail the letter. Certain of these steps must precede certain other steps. For example, one must find a stamp before one can stamp the envelope. The expert decided to write A → B if step A must precede step B. She found that

Get paper → Write letter, Put letter in envelope, Mail letter

Get pencils → Write letter, Address envelope, Put letter in envelope, Mail letter

Write letter → Put letter in envelope, Mail letter

Get envelope → Stamp envelope, Address envelope, Put letter in envelope, Mail letter

Get stamp → Stamp envelope, Mail letter

Address envelope → Mail letter

Stamp envelope → Mail letter

Put letter in envelope → Mail letter

The expert started to make a directed graph showing these relationships, but realized it would be very complicated. She also realized the graph would be transitive, since if step A must be done before step B and B before C, clearly A must be done before C. She decided to leave out those directed edges implied by other edges and transitivity. It is possible to do this *before drawing the graph* by a careful analysis of how the activities are related. For example since

Get paper → Write letter

and

Write letter → Put letter in envelope,

the edge

Get paper → Put letter in envelope

could be left out.

Crossing out all other implied edges left her with

Get paper → Write letter, ~~Put letter in envelope, Mail letter~~

Get pencils → Write letter, Address envelope, ~~Put letter in envelope, Mail letter~~

Write letter → Put letter in envelope, ~~Mail letter~~

Get envelope → Stamp envelope, Address envelope, Put letter in envelope, ~~Mail letter~~

Get stamp → Stamp envelope, ~~Mail letter~~

Address envelope → Mail letter

Stamp envelope → Mail letter

Put letter in envelope → Mail letter

The graph of this is shown in Figure 6.36.

Figure 6.36

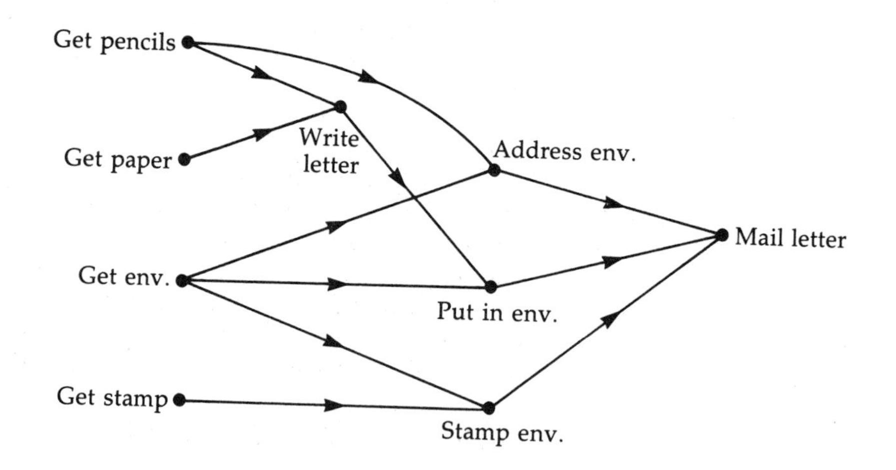

EXAMPLE 3

The Team The members of the foundry basketball team need to be able to get messages to each other in case they are challenged to a game on short notice. Antonio knows how to get in touch with Ben, Chuck

Figure 6.37

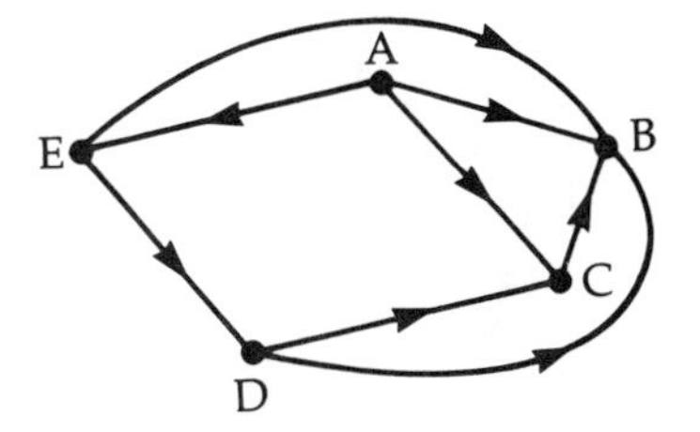

and Eliot; Chuck can contact Ben; Dan can get in touch with Ben and Chuck; and Eliot can contact Ben and Dan. Of course, a message can be relayed from one person to another so that if we draw a directed edge from one person to another whenever the first can get a message to the second, then our graph will be transitive. The graph is shown in Figure 6.37.

Some of these edges are superfluous, given that the graph is transitive. Since D → C and C → B, we can erase D → B. Likewise A → B is implied by A → C and C → B. Also E → D, D → C, and C → B together imply E → B, which can be removed. In the same way A → E → D → C, so A → C is not needed. We are left with the graph shown in Figure 6.38. To get a message to the team one should clearly start with Antonio.

Figure 6.38

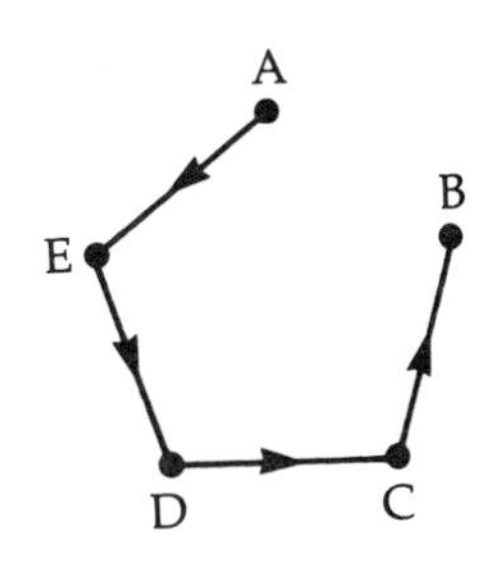

Exercises 6.5

In the first six problems redraw the graphs removing superfluous edges, assuming transitivity.

1

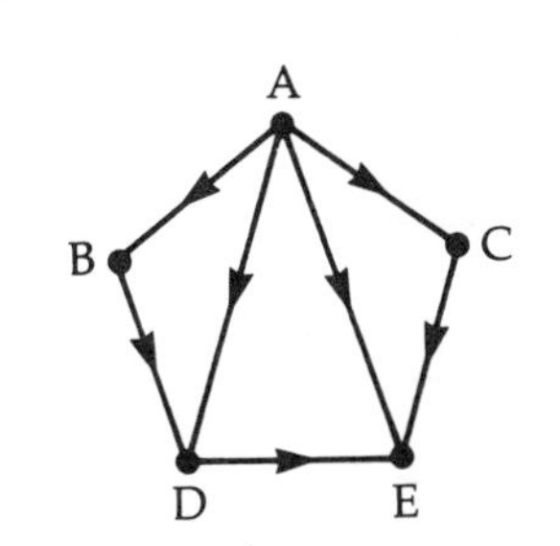

2.

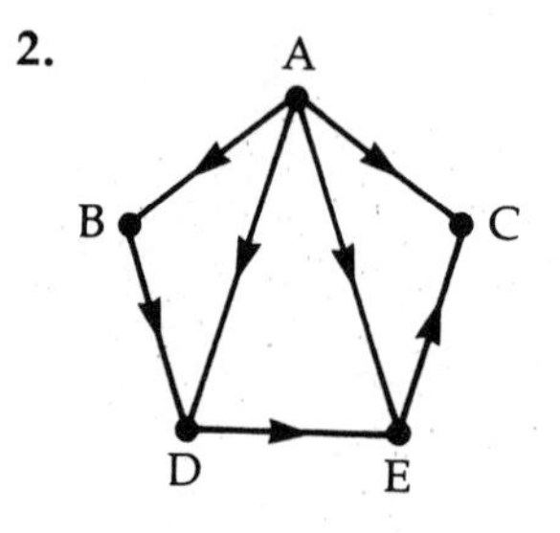

3.

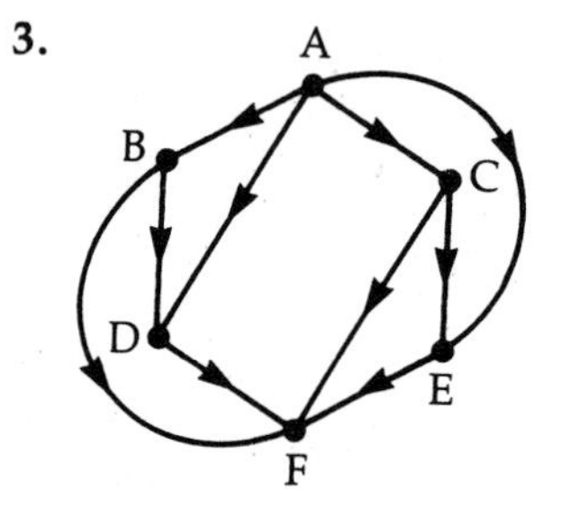

4.

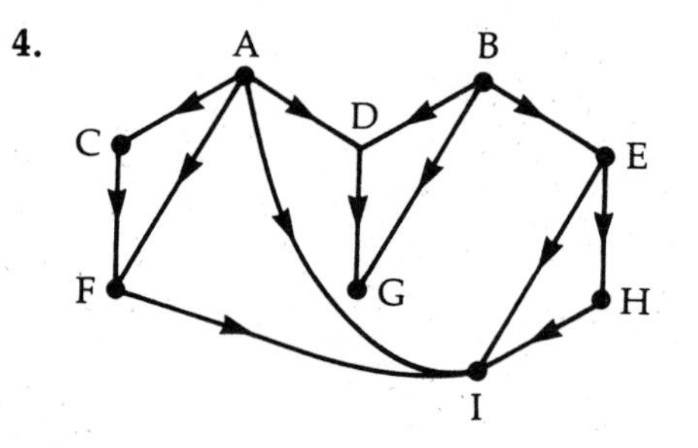

5.

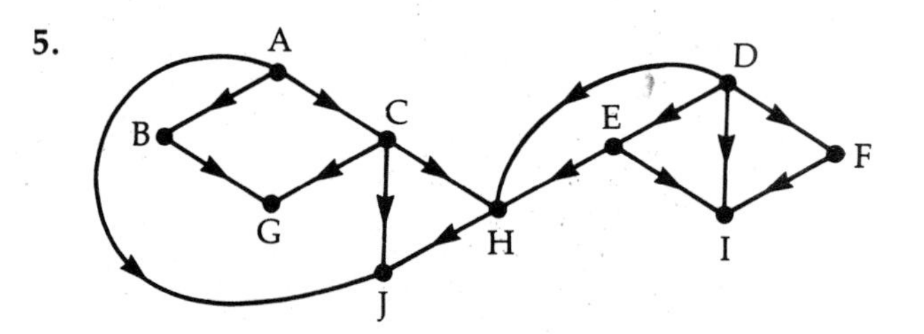

6.

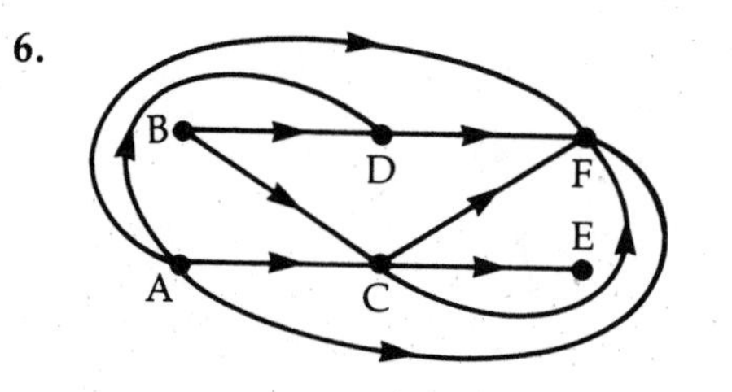

7. The Earth is further from the sun than Mercury or Venus. Saturn and Jupiter are both further out than Earth, Mars, Venus, or Mercury. Neptune and Uranus are further out than Saturn, Mars or Venus, and Pluto is further out than Neptune. Saturn is further out than Jupiter. Express these statements as a graph, where $A \rightarrow B$ if planet A is further from the sun than B, with edges implied by transitivity removed.

8. Consider the following steps involved in taking a bath:

A fill tub

B get undressed

C find soap

D get in tub

E wash

F get out of tub

G dry off

H dress

Construct a directed graph with no edges implied by transitivity where step $X \rightarrow$ step Y if step X must precede step Y.

9. The accompanying table gives the population (in millions) and area (in square miles) of Alabama, Alaska, Arizona, Arkansas, California, Colorado, and Connecticut.

STATE	POPULATION	AREA
Alabama	3.6	51,600
Alaska	0.4	586,400
Arizona	2.3	113,900
Arkansas	2.1	53,100
California	21.5	158,700
Colorado	2.6	104,200
Connecticut	3.1	5,000

Write State $A \rightarrow$ State B if either of these numbers is larger for state A than state B. Draw a directed graph of the situation, leaving out edges implied by transitivity. Is transitivity a reasonable condition for the graph to satisfy?

10. A person going on a plane trip needs to

A call a travel agent and order a ticket

P pick up the ticket at the agent

S pack a suitcase

G go to the airport

C check in the suitcase at the airport lobby

T present the ticket at the boarding gate

B board the plane

Make a graph of this, drawing an arrow from any action to any action that must come after it, but leaving out edges implied by transitivity.

11. A man is wearing: (A) underpants, (B) an undershirt, (C) socks, (D) a shirt, (E) pants, (F) shoes, (G) a tie, (H) a vest, (I) a sport coat, (J) an overcoat, and (K) a hat. Certain of these items must be put on before certain others. Make a directed graph, drawing an arrow from any item of clothing that must be put on before any other item to the second item, but leaving out edges implied by transitivity. Put in an edge only if it is physically impossible to don the second item before the first. For example, although it would be awkward to put on the tie after the overcoat, it is not impossible. Do not allow contortionists' tricks, like putting on the shirt after the vest, however.

12. Repeat problem 11, this time drawing an arrow from one item to another if it would be merely awkward or odd to don the second item first. Leave out edges implied by transitivity.

13. Consider the following: (N) North America, (C) Canada, (A) Akron, (S) St. Louis, (B) Boston, (P) Paraguay, (R) Rome, (U) United States, (O) Ohio, (M) Missouri, (W) Western Hemisphere, (T) Toronto, and (E) Europe. Make a directed graph where $X \to Y$ if X contains Y, omitting edges implied by transitivity.

14. The senior class at Tipp City High School has only six members. Three of these, Alan, Benita, and Clara, are running for class president. School by-laws do not allow the candidates to vote, so the only voters are Xaviera, Yetta, and Zack. Xaviera likes Benita best and Alan least of the three candidates. Yetta likes Clara best and Benita least, while Zack likes Alan best and Clara least. Make a graph with a vertex for each candidate, drawing an arrow from a candidate to another if the first candidate would beat the second in a two-person election. Is this a transitive graph?

6.6 PERT

In this section we will explain a method that allows a complicated project, perhaps involving many individual tasks performed by a number of people or groups, to be done in the shortest possible time. The method is called PERT.

PERT Used to Write a Letter

Recall that in Example 2 of Section 6.5 we considered an efficiency expert who was studying the steps involved in writing a letter. She came up with a transitive graph (with implied edges omitted) that showed which steps had to precede which others. We repeat that graph in Figure 6.39 with one extra step added preceding all the others, namely "start."

Figure 6.39

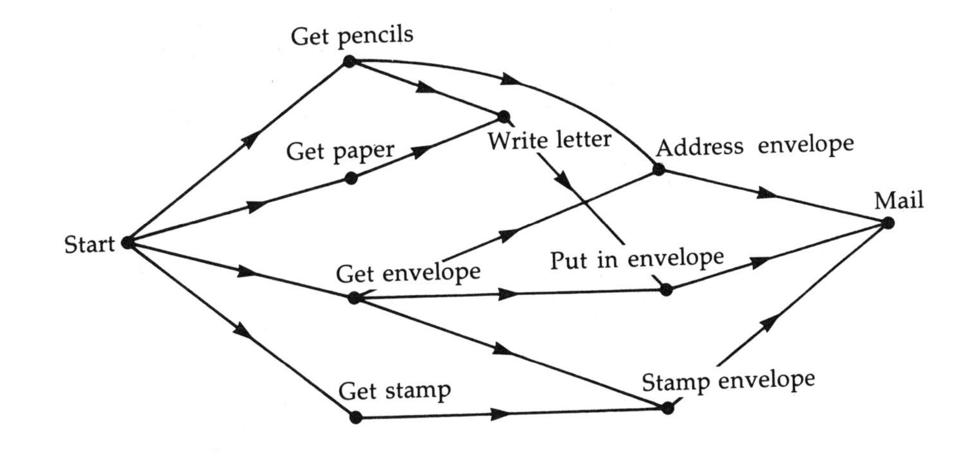

Now the efficiency expert determines how long each of these steps usually takes, coming up with the following table:

STEPS	TIME IN MINUTES
Start	0
Get pencils	2
Get paper	3
Get envelope	4
Get stamp	2
Write letter	15
Put in envelope	1
Stamp envelope	1
Address envelope	2
Mail letter	10

All these times add up to 40 minutes, but this does not mean it has to take 40 minutes to get out a letter. Some of these steps can be done simultaneously. While one person is writing the letter another can be addressing the envelope, for example. There is a limit to how much time can be saved this way, however. The letter cannot be put in the envelope until it is written no matter how many people are working on the project. Let us see how much time can really be saved.

We repeat our graph, this time replacing each vertex with a circle containing the time the corresponding step requires (Figure 6.40).

Figure 6.40

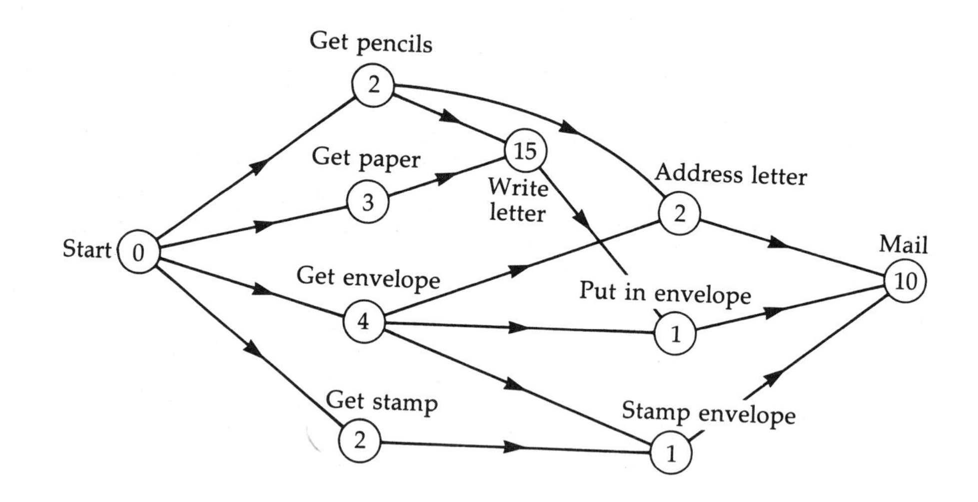

Now let us write after each vertex a number telling the least amount of time, *from the start of the whole operation,* until the corresponding step can be completed. We start at the left of the graph. It takes no time to start, so we put a 0 by that vertex. Each of the steps Get pencils, Get paper, Get envelope, and Get stamp can be done right away, but each will take a certain amount of time. We enter these times as shown in Figure 6.41.

Now consider the vertex Write letter. Figure 6.40 shows writing the letter cannot even begin until Get pencils and Get paper have been done. These take 2 and 3 minutes, respectively. Even if they are done simultaneously, perhaps by two different people, it will be at least 3 minutes before the writing can start, and so at least $3 + 15 = 18$ minutes before it can finish. We enter an 18 by the vertex Write letter (Figure 6.42).

In the same way we enter $5 = 4 + 1$ by Stamp envelope because it will be 4 minutes before the envelope stamping can even begin (Figure 6.43).

In general *we write by each vertex the sum of the number inside its*

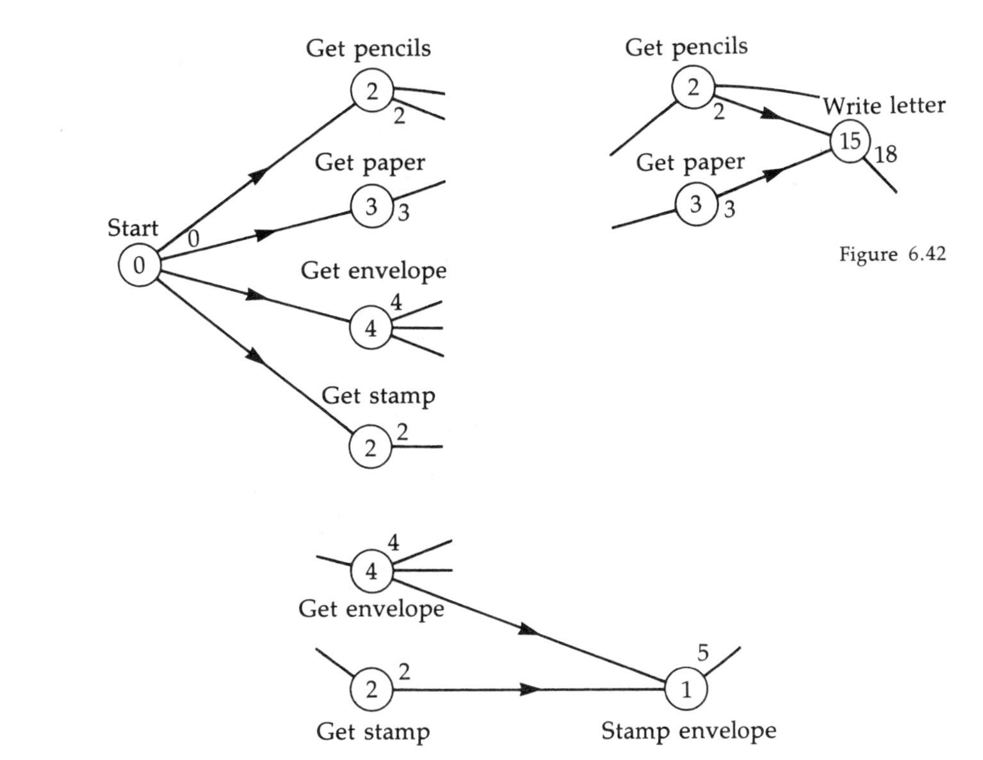

Figure 6.41

Figure 6.42

Figure 6.43

circle and the largest of all numbers written next to vertices leading into that vertex.

When this process is applied to the whole graph, we have Figure 6.44. We see that the project will take at least 29 minutes no matter how it is organized.

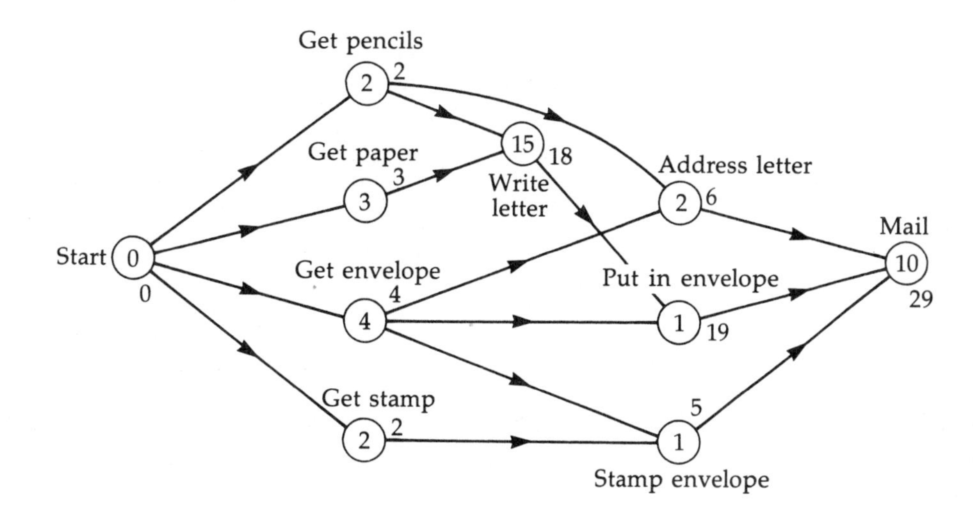

Figure 6.44

The Critical Path

The analysis we just applied to the process of getting out a letter is called *PERT*, which stands for *Program Evaluation and Review Technique*. PERT was developed in the 1950s by the U.S. Navy for the planning of the Polaris submarine. Its use has become widespread in large government and private industrial projects since then.

The PERT analysis of the letter project tells us more than merely that a minimum of 29 minutes is needed to complete it. Let us go back to the graph we developed in the last section (see Figure 6.45).

Figure 6.45

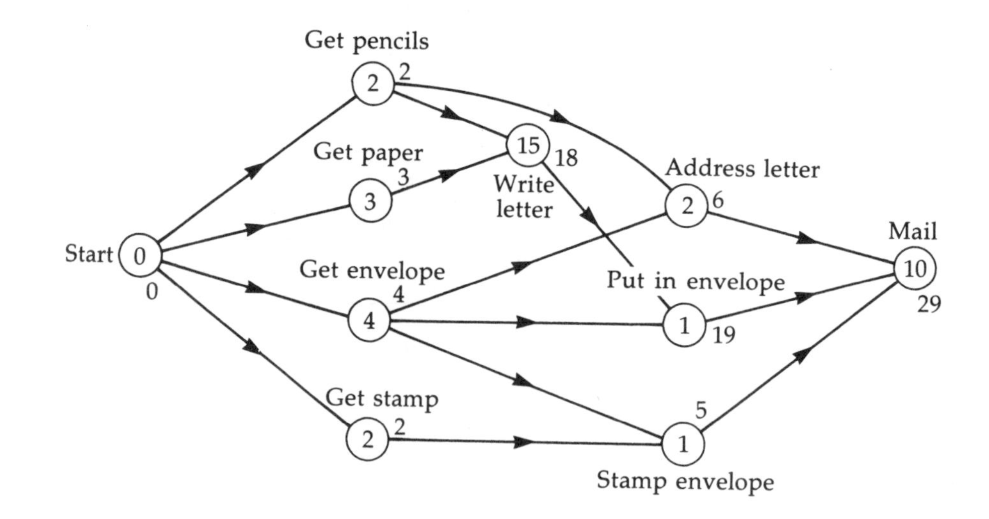

What is really taking up the time? Let us work backwards, starting from the right side of the graph. The reason the letter cannot be mailed for 29 minutes is that it takes 19 minutes until it can be put in the envelope. This is true because it takes 18 minutes until the letter gets written. This is because it takes 3 minutes to get the paper. We are tracing a path backwards from Mail letter, choosing the edge leading into each vertex that has the largest number next to the preceding vertex. It is the sequence shown in Figure 6.46 that takes up the 29 minutes; the other steps can be done with time to spare.

This path is called the *critical path*. This path is of special interest for two reasons. First of all, any delay in taking the steps along this path

Figure 6.46

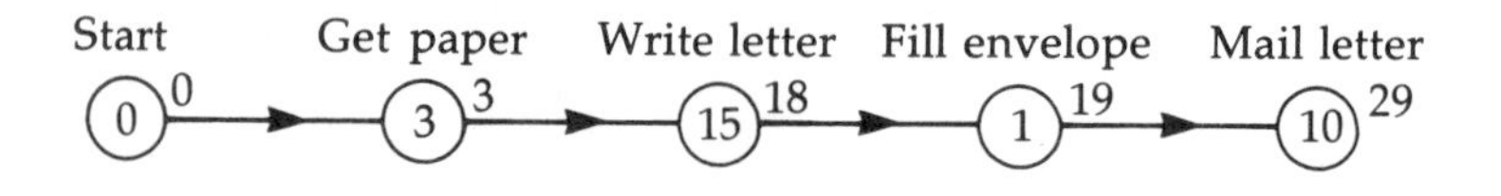

will delay the whole project. If the letter writer dawdles, for example, and takes 17 minutes to write the letter instead of 15, the whole operation will take 2 minutes longer. This is not true of steps not on the critical path. If it took 15 minutes to get the stamp instead of 2 minutes, for example, the total time could still be held to 29 minutes.

The other important characteristic of the critical path is that the steps along it are exactly where time can be saved. For example it takes 3 minutes to get the paper and 4 minutes to get an envelope. Without the PERT analysis we might figure putting the envelopes in a more convenient place would be a good way to expedite the process. This is not true, however, since Get envelope is not on the critical path. On the other hand Get paper is; and any time we can cut from this step will shorten the whole operation.

EXAMPLE

Building a House Let us apply PERT to the construction of a house. The main steps to be done are to dig a hole (2 days), lay the foundation (5 days), build the walls (2 days), build the roof (2 days), cover the roof with plywood (1 day), shingle the roof (2 days), wire for electricity (3 days), install the plumbing (4 days), brick the outside of the house (6 days), put up drywall (4 days) and paint the interior (3 days). We list each activity and also those which it must immediately precede (that is, relations implied by transitivity have been left out):

ACTIVITY	FOLLOWING ACTIVITY
Hole	Foundation
Foundation	Walls
Walls	Roof, Electricity, Plumbing, Brick
Roof	Cover roof
Cover roof	Shingle
Shingle	Drywall
Electricity	Drywall
Plumbing	Drywall
Brick	Drywall
Drywall	Paint

We construct the graph in Figure 6.47. Evidently the house can be built in 22 days.

Working backwards, we find the critical path marked with heavy lines, namely: Hole Foundation Walls Brick Drywall Paint.

Figure 6.47

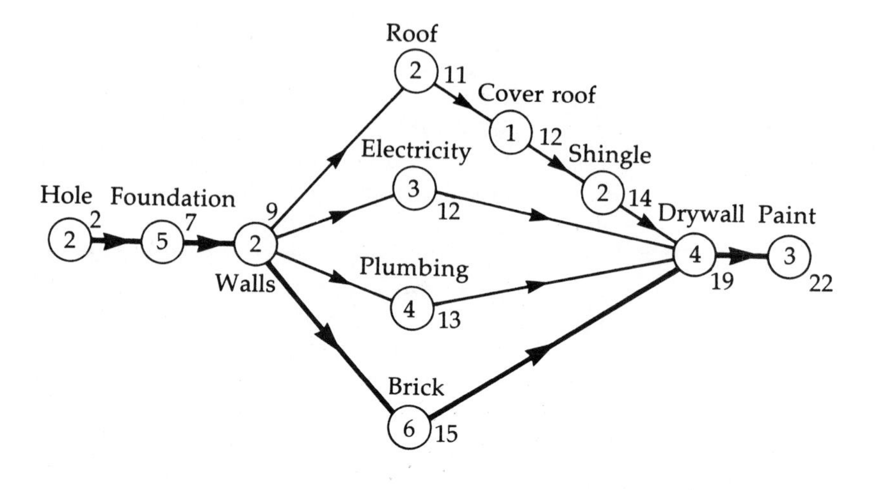

Exercises 6.6

In the first 10 problems the graph gives the time in days necessary to do the steps in a project. How long will it take to do the whole project? Find a critical path.

1.

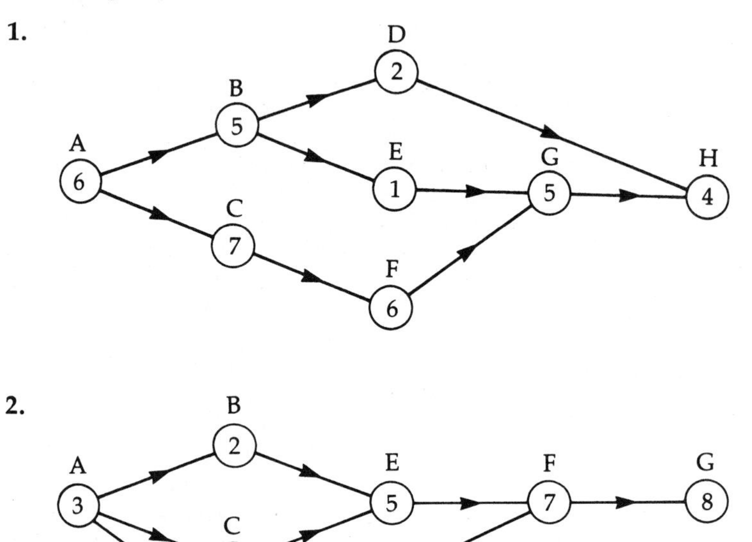

2.

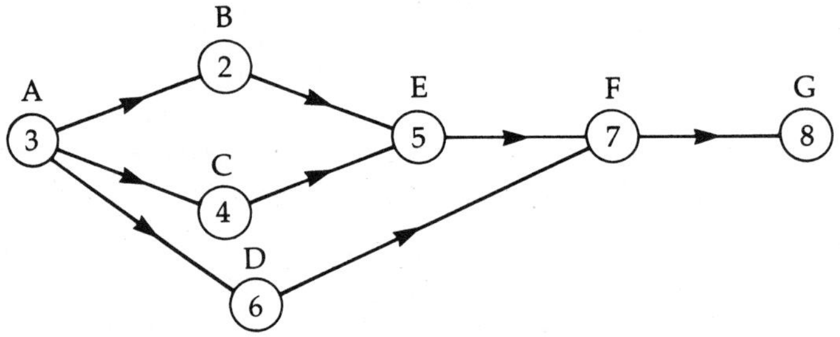

A 3, B 2, C 4, D 6, E 5, F 7, G 8

3.

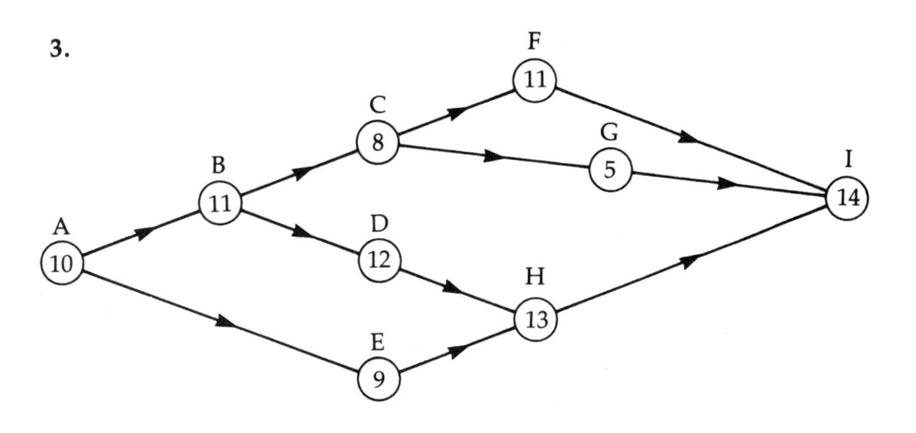
F
11
C
8
G
5
B
11
I
14
A
10
D
12
H
13
E
9

4.

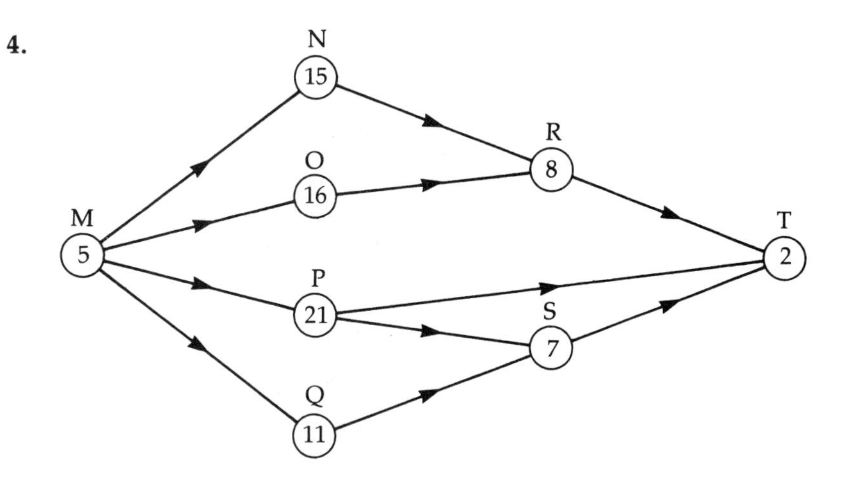
N
15
R
8
O
16
M
5
T
2
P
21
S
7
Q
11

5.

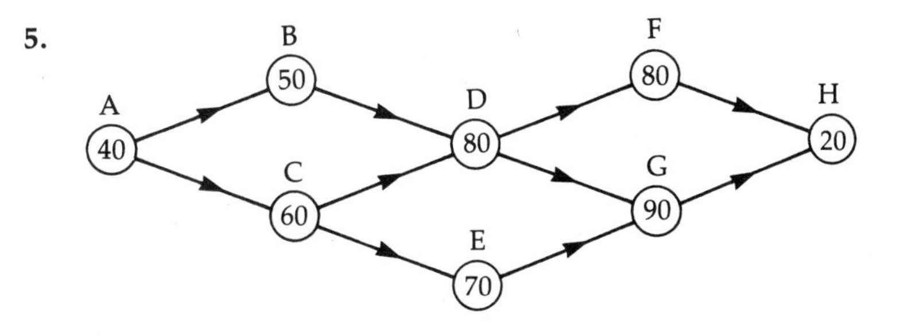
B
50
F
80
A
40
D
80
H
20
C
60
G
90
E
70

6.

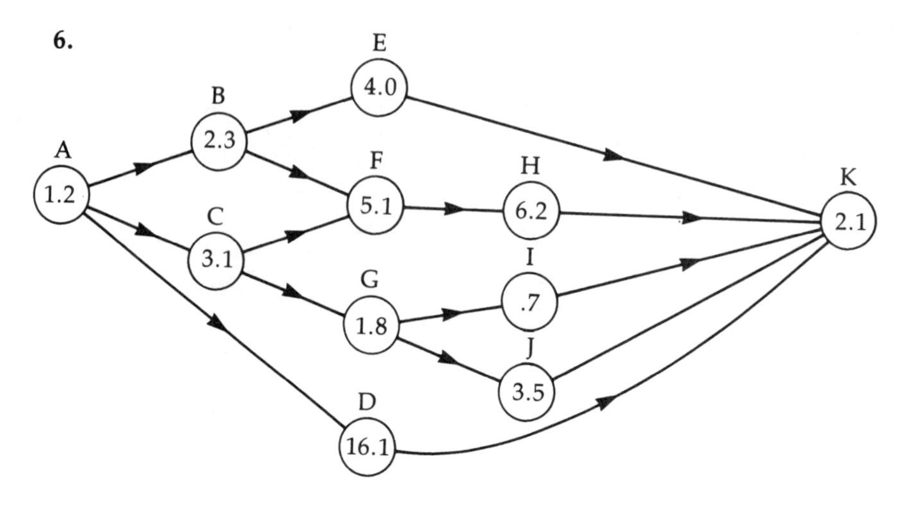

E
4.0
B
2.3
A
1.2
F
5.1
H
6.2
K
2.1
C
3.1
I
.7
G
1.8
J
3.5
D
16.1

7.

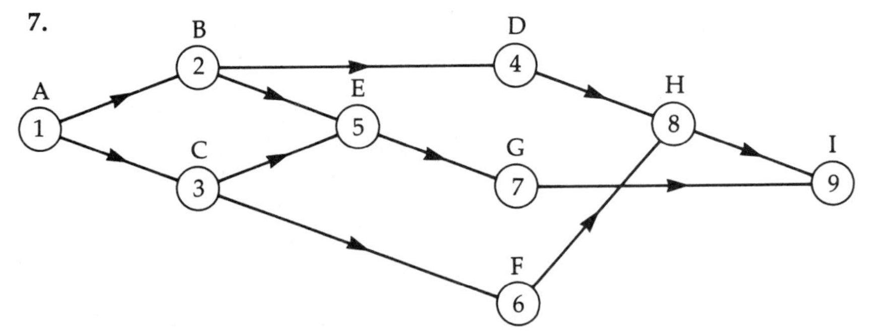

B
2
D
4
A
1
E
5
H
8
C
3
G
7
I
9
F
6

8.

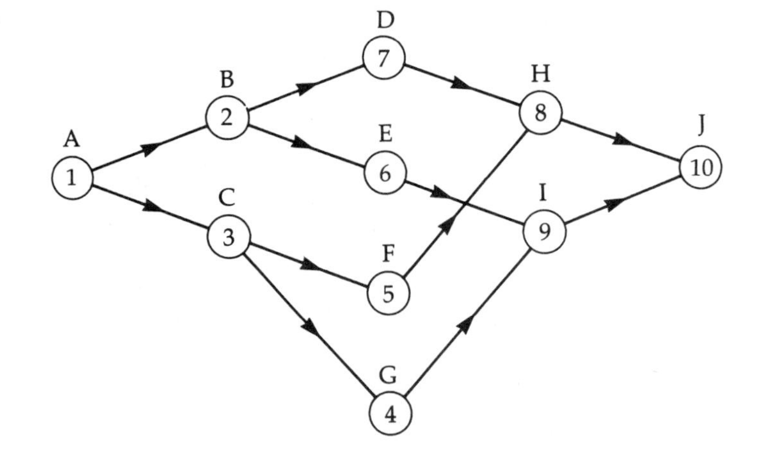

D
7
B
2
H
8
A
1
E
6
J
10
C
3
I
9
F
5
G
4

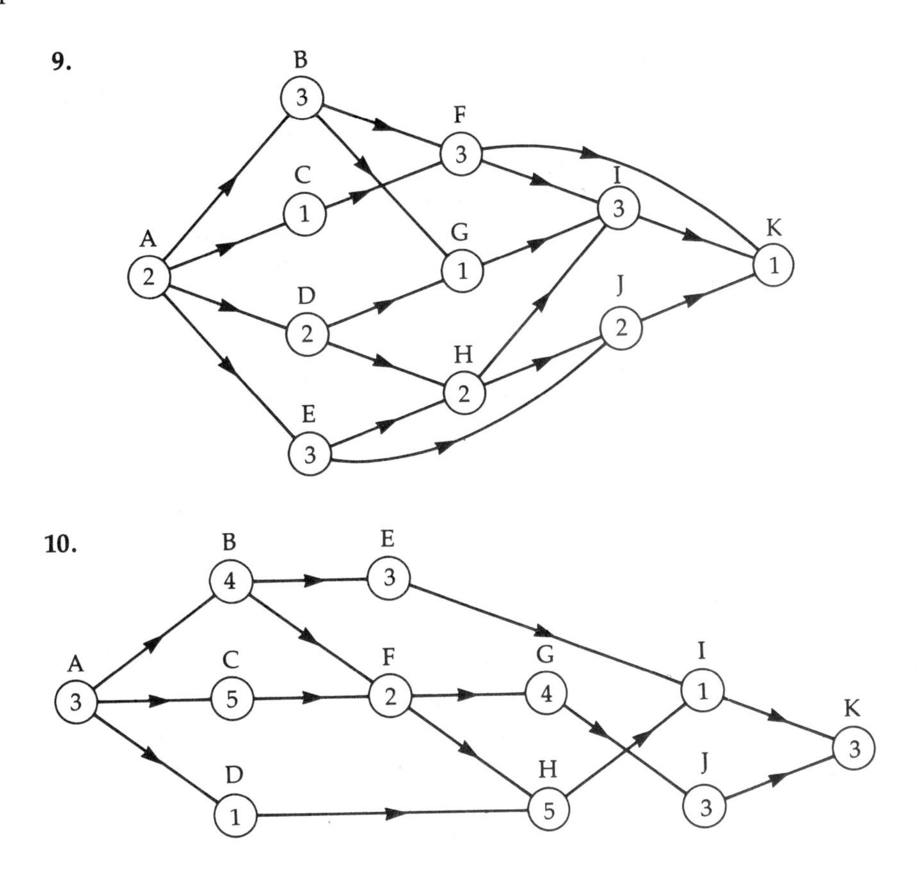

11. The accompanying table gives the steps necessary to apply PERT to a real-world situation, the time it might take to do each step, and the steps that must come immediately after each step.

	STEP	TIME IN MINUTES	FOLLOWING STEPS
A	decide what the steps are	5	B, C
B	time each step	120	E
C	decide which steps precede which others	10	D, E
D	eliminate relationships implied by transitivity	5	E
E	draw graph, including times	10	F
F	put numbers by vertices	5	G
G	find critical path	3	

Draw the graph for this task and find how long it takes to complete step G. Find the critical path.

12. In order to grow tomatoes one must (A) decide what variety to grow (40 minutes), (B) prepare the ground (20 minutes), (C) buy a tomato plant (30 minutes), (D) plant the tomato plant (15 minutes), (E) water the newly planted tomato (5 minutes), (F) let nature take its course (45 days), and (G) pick your first tomato (1 minute). Draw a graph for this process. What is the shortest time it can take? Find a critical path.

13. The table gives the times in minutes of various operations involved in making a cake and the operations each step must immediately precede.

	STEP	TIME	FOLLOWING STEPS
A	start	0	B
B	get ingredients	5	C, E, G
C	break eggs	1	D
D	beat eggs	10	F
E	scald milk	10	F
F	mix batter	12	H
G	make icing	15	L
H	pour cake in pan	1	J
I	preheat oven	20	J
J	bake cake	30	K
K	cool cake	30	L
L	ice cake	15	

Draw a PERT graph for making a cake. How long must it take? Find the critical path.

14. The table gives the time in days for the various steps in making a motion picture and the steps that must immediately follow each step. Draw a PERT graph for making a movie. How long must it take? Find the critical path.

	STEP	TIME	FOLLOWING STEPS
A	get financing	30	B, D, E, F
B	buy script	10	C, G, H
C	hire cast	15	I, J
D	rent studio	5	G, H
E	hire technicians	10	I, J
F	hire director	5	I, J
G	construct outdoor sets	20	I
H	construct indoor sets	15	J
I	shoot outdoors	20	K
J	shoot indoors	20	K
K	process film	15	L
L	edit film	20	

MATRICES AND MARKOV CHAINS

CHAPTER SEVEN

 7.1 INTRODUCTION AND PROBLEM EXAMPLES

It is difficult to give an idea of the vast extent of modern mathematics. The word "extent" is not the right one: I mean extent crowded with beautiful detail—not an extent of mere uniformity such as an objectless plain, but of a tract of beautiful country seen at first in the distance, but which will bear to be rambled through and studied in every detail of hillside and valley, stream, rock, wood, and flower. But, as for every thing else, so for a mathematical theory—beauty can be perceived but not explained.

These words were written in 1883 by Arthur Cayley, one of the three most prolific mathematicians of all time (his *Collected Mathematical Papers*, comprising 966 papers, are contained in 13 large quarto volumes of about 600 pages each). The basic tool that we introduce in this chapter, namely the theory of matrices, was one of Cayley's greatest inventions and originated in 1858 in a setting quite different from any we shall introduce here. We hope that you can see some beauty in mathematics (and especially will see some beauty in what follows), but in keeping with our commitment to applications let us consider some problems first.

There are many problems in physics and chemistry, such as problems relating to quantum mechanics and the structure of crystals, which can be answered using matrix theory, and matrices are even used in constructing secret codes. However, we will use examples from business, genetics, and political science to stimulate our interest in learning about these as yet undefined quantities.

Arthur Cayley (1821–1895), son of an English merchant, loved novels (which he read in French, German, Italian, and classical Greek) as well as mathematics. Unwilling to take holy orders to retain his position at Trinity College, Cambridge, he left in 1846 to take up law, which he followed as a profession until 1863, although he continued to publish extensively in mathematics. In that year Cambridge established a new professorship in mathematics (the Sadlerian) and Cayley promptly accepted the offered post. Among his many contributions should be mentioned the concept of n-dimensional space and its geometry and the means to unify Euclidean and non-Euclidean geometry into a single comprehensive theory.

EXAMPLE 1

John Thread runs the Screwy Manufacturing Company, which manufactures 17 different kinds of screws. There are seven wood screws of lengths $\frac{1}{4}$, $\frac{3}{8}$, $\frac{1}{2}$, $\frac{5}{8}$, $\frac{3}{4}$, $\frac{7}{8}$ and 1 inch; six sheet metal screws of lengths 0.5, 1, 1.5, 2, 2.5, and 3 centimeters, all with round heads; and four flat-headed screws of lengths 1, 2, 3, and 4 centimeters. Short screws (of length less than $\frac{3}{4}$ inch) are packaged with five to a package, while longer screws come three to a package. In addition there are three combination packages called English, Metric, and Super made up as follows:

English: Four of each of the seven screws measured in inches.

Metric: Three of each of the six short metric screws (2 centimeters or less) and two of each of the four longer metric screws.

Super: One screw of each of the 17 lengths and two additional for each screw of length less than $\frac{3}{4}$ inch.

On a typical day, John receives orders for 2000 packages of each length of small screw, 1500 packages of each length of longer screw, 1000 English, 2000 Metric, and 500 Super. How many of each type of screw does he need to manufacture to fill this day's orders?

EXAMPLE 2

To pass the time while climbing the Matterhorn, your guide, Jacques, gives you this problem. There are three mountains, call them M, G, W, on which you can act as a guide. The first week you climb M twice and W once for a total height of 12.7 kilometers, the second week you go up each once totaling 12.5 kilometers, but the third week a total height of 20.9 kilometers is reached by climbing M once and the other two twice. Now, how high are each of the mountains? (Jacques knows the answer because the mountains are the Mönch, the Gross-Wannenhorn and the Weisshorn.)

EXAMPLE 3

Gamblers Ruin Having just won $20,000 in the Illinois State Lottery, Jane decides to go to Las Vegas to increase her capital. She finds the Friendly Gambling House, which allows her to choose heads when tossing a coin that is biased to come up heads with probability 5/8. She decides to risk $10,000 at $10 per toss of the coin, and to keep playing until she has lost it all, or has broken the bank, which has $990,000 in it. What is the probability that Jane breaks the bank? If she can only stay to play 1000 times, what is the probability that she will have more when she stops than when she started? If she finds the Even-Break Saloon, which uses an honest coin, but only has a bank with $10,000 in it, what are her chances of breaking the bank before going broke? Is there any way to estimate about how long it will take before Jane or the bank is broke?

EXAMPLE 4

Ralph and Carl are tired of matching pennies, and Carl suggests that they play the following game. Ralph takes a red 5 and a black 2, and Carl takes a red 7 and a black 3 from a deck of cards. At a given signal they simultaneously expose one of their cards; if the cards match in color, Ralph wins the (positive) difference between the numbers on the cards; if the cards are of different colors, Ralph loses the (positive) difference between the numbers on the cards played. After losing regularly for a while, Ralph suggests they alter the game by exchanging Carl's red 7 for a red 3. Carl agrees to play this new game, but suggests that they agree to choose their red and black cards with equal probability. Is the honest Ralph protected from an unscrupulous Carl if he agrees to this added condition? Can Ralph protect himself in any way when playing this new game? How much should Ralph expect to lose if he plays as well as he can?

EXAMPLE 5

In genetics, certain characteristics are said to be dominant and others recessive. For example, having brown eyes is a dominant characteristic and having blue eyes is a recessive characteristic. In its simplest form the assumption is that the color of the eyes is governed by a pair of genes, each of which may be of one of two types, say B or b. An individual with a BB pair will have brown eyes and be called *dominant,* one with a Bb (which is the same genetically as bB) pair will have brown eyes but be called *hybrid,* and an individual with a bb pair will have blue eyes and be called *recessive.* In mating, each parent provides one of his or her types of genes to each offspring and the *basic assumption of genetics* is that these genes are selected at random, that is, independently of each other and with equal probabilities. Thus for example, the child of two dominant (recessive) parents must be dominant (respectively, recessive), and the child of one dominant and one recessive parent must be hybrid. Imagine a family tree in which one of the original parents is a hybrid and the other of unknown eye color. Each offspring subsequently mates with a hybrid. Can you say anything about the distribution of eye color genetic characteristics of the tenth or even fifth generation? Will there be any blue-eyed children at all? Does it matter whether or not the unknown ancestor had blue eyes or brown eyes?

EXAMPLE 6

Suppose that in the previous example each offspring has a blue-eyed mate. Does that matter? In particular, if a brown-eyed dominant individual enters an all blue-eyed community, will more and more people in the community have brown eyes as time goes on?

EXAMPLE 7

Suppose that in a certain city census figures show that each year 6% of the people in the city (proper) move to the suburbs and 2% of the people in the suburbs move inside the city limits. Assuming that the total number of people in the city and the suburbs together remains the same as it is today, 5 million, will the populations of the city and suburb ever settle down, and if so what will be the ultimate population in the city?

EXAMPLE 8

In studying a certain precinct, a political science student has found that in the last 50 years the outcome of elections in succeeding years has gone from a Labour victory in one year to a Tory victory the next year 15 times and from a Tory victory to a Labour victory 20 times. Can she predict what the chances for a Tory victory are in say 10 years? What would she need to know in order to improve her prediction?

Exercises 7.1

1. Which two mathematicians were more prolific than Cayley?
2. Play the games in Example 4 at least 10 times and keep track of the totals paid. Which seems more fair?*
3. Discuss Example 3. What do you think Jane should do with her money?
4. Given that 1 centimeter = 0.3937 inches, in Example 1 how many of the metric screws will be packaged in sets of five to a package?
5. Make a table with 17 rows and 20 columns. Label each row by a screw length and each column by a screw length and the headings English, Metric, and Super. Then fill in each column with the number of each type of screw which will be in the package whose name heads the column.
6. Discuss Examples 5 and 6. What do you think will happen in the long run?

7.2 WHAT IS A MATRIX?

A *matrix* is, very simply, any rectangular array of numbers. Matrices with only one row or only one column are called *vectors*. We give some examples:

*This game and others like it can be solved by using matrices and are analyzed in discussions of a new branch of mathematics called game theory. We do not answer the questions raised in Example 4 in this text, but refer the reader to one of the many texts on finite mathematics; for example, *Introduction to Finite Mathematics*, by J. Kemeny, J. Snell, and G. Thompson (Prentice-Hall, Englewood Cliffs, N.J., 1974).

$$\begin{pmatrix} 3 & -36 \\ 17 & 1.6 \\ -8 & \frac{1}{2} \\ \sqrt{19} & 1.39 \\ 0.2 & -.207 \\ -5 & 87 \end{pmatrix} \qquad (3 \quad \pi \quad -7) \qquad \begin{pmatrix} 3 & 4 & 5 \\ -6 & 7 & -8 \\ \pi & 13 & \frac{3}{2} \end{pmatrix} \qquad \begin{pmatrix} 17 \\ \sqrt{8} \\ -3 \\ 49 \end{pmatrix}$$

$$\begin{pmatrix} 1.3 & 47 & -8 & 10^5 \\ -.037 & 6/5 & -\sqrt{7} & \sqrt{8 + 3(10)^{1/4}} \end{pmatrix}$$

How can anything so simple have relevance to the examples listed in the last section? This, of course, is part of the beauty of the situation: a simple idea that has applications to many problems. Let us begin by examining one such application.

Simple Explosion

Most manufacturers and suppliers today stock a variety of merchandise. When accepting and filling orders it is essential to be able to learn quickly exactly how much of each type of merchandise is needed. Example 1 in the previous section furnishes an instance of what we have in mind. We begin, however, with a simpler example.

EXAMPLE 1

A bandage company sells medium and large cans of bandage assortments. The medium-size assortment contains 2 large compresses, 10 medium plaster strips, and 20 small plaster strips, while the large size contains 5 large compresses, 25 medium plaster strips, and 30 small plaster strips. How many bandages are required to fill an order for 125 large cans and 75 medium cans?

Common Sense Solution No special knowledge is needed to solve this problem. For example, we could compute as follows:

COMPRESSES

$2 \times 75 = 150 =$ No. of large compresses in 75 medium cans

$5 \times 125 = \underline{625} =$ No. of large compresses in 125 large cans

Sum $= 775 =$ No. of large compresses needed for the order

MEDIUM STRIPS

$10 \times 75 = 750 =$ No. of medium strips in 75 medium cans

$25 \times 125 = \underline{3125} =$ No. of medium strips in 125 large cans

Sum $= 3875 =$ No. of medium strips needed for the order

SMALL STRIPS

$20 \times 75 = 1500 =$ No. of small strips in 75 medium cans

$30 \times 125 = \underline{3750} =$ No. of small strips in 125 large cans

Sum $= 5250 =$ No. of small strips needed for the order

The answer is then 775 compresses, 3875 medium strips, and 5250 small strips.

Matrix Solution For this solution, we first build what is called the *inventory-file matrix,* which expresses in matrix form the number of each type of bandage contained in the two cans:

	MEDIUM CAN	LARGE CAN
COMPRESSES	2	5
MEDIUM STRIPS	10	25
LARGE STRIPS	20	30

We then consider the *production vector:*

$$\begin{pmatrix} 75 \\ 125 \end{pmatrix} \begin{matrix} \text{MEDIUM CANS} \\ \text{LARGE CANS} \end{matrix}$$

which has the entries in the same order as the top of the inventory-file matrix. We now combine these two "matrices" in the following manner:

$$\begin{pmatrix} 2 & 5 \\ 10 & 25 \\ 20 & 30 \end{pmatrix} \begin{pmatrix} 75 \\ 125 \end{pmatrix} = \begin{pmatrix} (2 \times 75) + (5 \times 125) \\ (10 \times 75) + (25 \times 125) \\ (20 \times 75) + (30 \times 125) \end{pmatrix} = \begin{pmatrix} 150 + 625 \\ 750 + 3125 \\ 1500 + 3750 \end{pmatrix} = \begin{pmatrix} 775 \\ 3875 \\ 5250 \end{pmatrix}.$$

The *result vector* gives the totals in the same order as in the inventory-file matrix, that is, 775 compresses, 3875 medium strips, and 5250 small strips.

What we have done is to combine a three-row by two-column matrix with a two-row by one-column matrix to obtain a three-row by one-column matrix. Let us look at this process of combination a little more closely before we do another example.

The general process is called row-by-column multiplication because each row in the first matrix is "multiplied" by the column matrix. For example, to obtain the second entry 3875 in the answer, we combined the second row with the column matrix as follows:

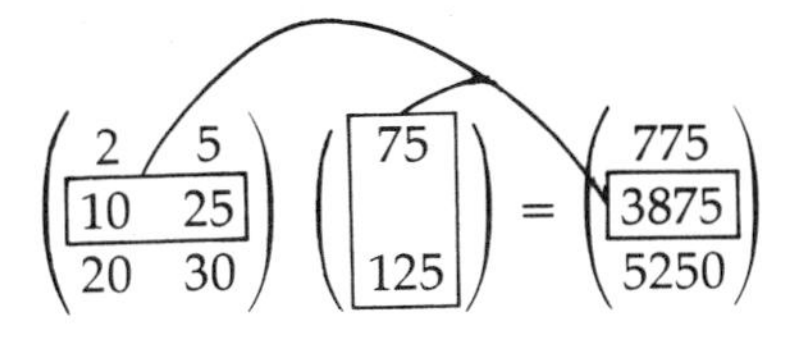

This combination process is called *matrix multiplication* even though it involves both multiplication and addition of the numbers involved. Let us look at this process more closely.

To find the first entry in the answer matrix, we multiply the first row of the inventory file matrix by the production vector:

$$\begin{pmatrix} \boxed{2 \quad 5} \\ 10 \quad 25 \\ 20 \quad 30 \end{pmatrix} \begin{pmatrix} 75 \\ \\ 125 \end{pmatrix} \quad \left.\begin{matrix} 2 \times 75 = 150 \\ + \\ 5 \times 125 = 625 \end{matrix}\right\} = 775.$$

The third entry was found similarly by multiplying the first entry in the row by the first entry in the column, the second entry in the row by the second entry in the column, and so on, and then adding up all the products:

$$\begin{pmatrix} 2 \quad 5 \\ 10 \quad 25 \\ \boxed{20 \quad 30} \end{pmatrix} \begin{pmatrix} 75 \\ \\ 125 \end{pmatrix} \quad (20 \times 75) + (30 \times 125) = 1500 + 3750 = 5250.$$

We consider a few more examples just showing how this "multiplication" process works.

EXAMPLE 2

$$\begin{pmatrix} 2 & 3 \\ 8 & 6 \\ 4 & 5 \end{pmatrix} \begin{pmatrix} 2 \\ 5 \end{pmatrix} = \begin{pmatrix} (2 \times 2) + (3 \times 5) \\ (8 \times 2) + (6 \times 5) \\ (4 \times 2) + (5 \times 5) \end{pmatrix} = \begin{pmatrix} 4 + 15 \\ 16 + 30 \\ 8 + 25 \end{pmatrix} = \begin{pmatrix} 19 \\ 46 \\ 33 \end{pmatrix}.$$

EXAMPLE 3

In this example we will show a computational aid for working these problems. It helps to keep track of which numbers are multiplied and the position of the answer. We do this by showing successive stages in the work.

$$\begin{pmatrix} 2 & 4 & 6 \\ 1 & 3 & 5 \\ 2 & 3 & 7 \end{pmatrix} \cdot \begin{pmatrix} 2 \\ 5 \\ 3 \end{pmatrix} = \begin{pmatrix} 2 & 4 & 6 \\ 1 & 3 & 5 \\ 2 & 3 & 7 \end{pmatrix} \cdot \begin{pmatrix} 2 \\ 5 \\ 3 \end{pmatrix} \begin{matrix} \times 2 \\ \times 4 \\ \times 6 \end{matrix}$$

STAGE 1 STAGE 2

$$\begin{pmatrix} 2 & 4 & 6 \\ 1 & 3 & 5 \\ 2 & 3 & 7 \end{pmatrix} \cdot \begin{pmatrix} 2 \\ 5 \\ 3 \end{pmatrix} \begin{matrix} \times 2 \\ \times 4 \\ \times 6 \end{matrix} = \begin{pmatrix} 4 + 20 + 18 \\ \\ \\ \end{pmatrix} = \begin{pmatrix} 42 \\ \\ \\ \end{pmatrix}$$

STAGE 3

$$\begin{pmatrix} 2 & 4 & 6 \\ 1 & 3 & 5 \\ 2 & 3 & 7 \end{pmatrix} \cdot \begin{pmatrix} 2 \\ 5 \\ 3 \end{pmatrix} \begin{matrix} \times \not{2}1 \\ \times \not{4}3 \\ \times \not{6}5 \end{matrix} = \begin{pmatrix} 4 + 20 + 18 \\ 2 + 15 + 15 \\ \end{pmatrix} = \begin{pmatrix} 42 \\ 32 \\ \end{pmatrix}$$

STAGE 4

$$\begin{pmatrix} 2 & 4 & 6 \\ 1 & 3 & 5 \\ 2 & 3 & 7 \end{pmatrix} \cdot \begin{pmatrix} 2 \\ 5 \\ 3 \end{pmatrix} \begin{matrix} \times \not{2}\not{1}2 \\ \times \not{4}\not{3}3 \\ \times \not{6}\not{5}7 \end{matrix} = \begin{pmatrix} 4 + 20 + 18 \\ 2 + 15 + 15 \\ 4 + 15 + 21 \end{pmatrix} = \begin{pmatrix} 42 \\ 32 \\ 40 \end{pmatrix}$$

STAGE 5

We conclude with an example, similar to Example 1, of a *simple explosion* problem (so called because we "explode" the order into its component parts by using a matrix).

EXAMPLE 4

The Goody Candy Company makes three sizes of boxes of assorted candy, the Snack Sack, the Go Getter and the Farm Family. The Snack Sack holds 2 caramels, 4 soft creams, and 6 mint chewies, the Go Getter holds 2 caramels, 6 soft creams, and 5 mint chewies, and the Farm Family holds 8 caramels, 7 soft creams, and 10 mint chewies. If the Corner Drug Store orders 6 Snack Sacks, 5 Go Getters, and 8 Farm Families, how many of each kind of candy is needed by Shipping?

Inventory-file matrix:

	S.S.	G.G.	F.F.
CARAMELS	2	2	8
SOFT CREAMS	4	6	7
MINT CHEWIES	6	5	10

Production vector: $\begin{pmatrix} 6 \\ 5 \\ 8 \end{pmatrix}$

Result vector:

$$\begin{pmatrix} 2 & 2 & 8 \\ 4 & 6 & 7 \\ 6 & 5 & 10 \end{pmatrix} \cdot \begin{pmatrix} 6 \\ 5 \\ 8 \end{pmatrix} \begin{matrix} \times \not{2}\not{4}6 \\ \times \not{2}\not{6}5 \\ \times \not{8}\not{7}10 \end{matrix} = \begin{pmatrix} 12 + 10 + 64 \\ 24 + 30 + 56 \\ 36 + 25 + 80 \end{pmatrix} = \begin{pmatrix} 86 \\ 110 \\ 141 \end{pmatrix}$$

Shipping needs 86 caramels, 110 soft creams, and 141 mint chewies (the factory speciality!).

Exercises 7.2

1. Combine the following matrices by means of the process described above:

(a) $\begin{pmatrix} 2 & 3 \\ 4 & 1 \end{pmatrix} \begin{pmatrix} 3 \\ 2 \end{pmatrix}$

(b) $\begin{pmatrix} 5 & 13 \\ 7 & 8 \\ 5 & 3 \end{pmatrix} \begin{pmatrix} 9 \\ 4 \end{pmatrix}$

(c) $\begin{pmatrix} 1 & 3 & 2 \\ 6 & 10 & 8 \\ 5 & 9 & 11 \end{pmatrix} \begin{pmatrix} 6 \\ 5 \\ 4 \end{pmatrix}$

(d) $\begin{pmatrix} 1 & 4 & 2 \\ 8 & -3 & 6 \\ 5 & 9 & 10 \\ -4 & 3 & 2 \end{pmatrix} \begin{pmatrix} 2 \\ 3 \\ 4 \end{pmatrix}$

In each of problems 2–7, (a) write the inventory-file matrix, and (b) find how many of each part is needed to fill the order.

2. A library supply house must send 5 index cards and 3 posters for each large hardbound book it sells and 3 index cards and 1 poster for each softbound book. The order is for 5 hardbound books and 12 softbound books.

3. A toy company sells two sizes of badminton sets. The large set contains 4 racquets and 5 birdies and the small set contains 2 racquets and 3 birdies. The order is for 15 large and 20 small sets.

4. A lamp manufacturer makes three sizes of lamps. The desk lamp requires 4 feet of wire cord, one switch, and one length of tubing; the table lamp requires 5 feet of cord, one switch, and two lengths of tubing; and the floor lamp requires 7 feet of cord, 3 switches, and 4 lengths of tubing. The order is for 6 table, 5 desk, and 8 floor lamps.

5. A manufacturer of badminton equipment sells her product in two different size packages. In one she has two racquets, one net, and three shuttlecocks, and in the other four racquets, one net, and five shuttlecocks. The order is for 27 of the smaller and 39 of the larger packages.

6. A manufacturer makes a tiny, small, regular, and large size ordinary blade screwdriver and a tiny, regular, and large size Phillips-head screwdriver. They are sold in four different size packages, as follows:

	PHILLIPS	BLADE
Small set	1 tiny	1 tiny, 1 small
Regular set	1 tiny, 1 regular	1 small, 1 regular
Handyman set	1 of each	1 of each
Builder's set	1 tiny, 1 regular, 1 large	1 tiny, 1 small, 2 regular, 2 large

The order is for 22 Small, 25 Regular, 37 Handyman, and 14 Builder's sets.

7. A glass manufacturer makes three sets of assorted glasses, called Home, Bar, and Office. The Home set has 4 highball, 6 water, and 4 wine glasses; the Bar set has 8 highball, 4 water, 6 wine glasses, and 2 whiskey jiggers; and the Office set has 10 highball, 2 water, 8 wine glasses, and 4 whiskey jiggers. The order is for 14 Home, 8 Bar, and 3 Office sets.

8. The cycle repair kit contains 2 spark plugs, 4 wrenches, 2 screwdrivers, 1 pliers, and 1 carburetor kit; the auto repair kit contains 8 spark plugs, 6 wrenches, 3 screwdrivers, 2 pliers, and 2 carburetor kits; and the truck repair kit contains 12 spark plugs, 10 wrenches, 4 screwdrivers, a pry bar, 3 pliers, and a diesel kit. How many rows and columns are in the inventory-file matrix? How many of each part is needed to fill an order for 40 cycle, 20 auto, and 10 truck repair kits?

9. Solve Example 1 in Section 7.1.

7.3 OPERATIONS ON MATRICES

We saw in the last section that it was convenient to define a rather complicated form of "multiplication" on certain kinds of matrices. Mathematicians are looking for opportunities of this sort. In particular, what often happens is that quantities of some sort arise—either naturally or in someone's imagination. A "natural" question is then to ask whether one can define operations of addition and multiplication under which these quantities act like real numbers. Additional questions usually follow. The first, assuming it is possible to define addition and multiplication, would be whether the resulting "arithmetic" seems useful.

> "God made the integers; all the rest is the work of man." Leopold Kronecker (1823–1891)

Others take several forms depending on what the answers to the previous questions were. Roughly speaking, we want to approximate ordinary arithmetic as much as possible and still have the resulting processes and answers have some usefulness. This is what has been done with the quantities we are considering: matrices. The following definitions have all stood the test of time—over 100 years—and because of the wealth of applications of matrices, most of which we cannot go into, they must in some sense be "right." These criteria of similarity to established theories and applicability are used by most mathematicians when working out a new theory.

Terminology

Before we go on, it is convenient to have some terminology and notation. For those using the computer, the function notation is included as well as the subscript notation.

A matrix, being a rectangular array of numbers, has rows and columns. We start numbering them both from the upper left-hand corner:

	COLUMN ONE	COLUMN TWO	COLUMN THREE
ROW ONE	6	−3	5
ROW TWO	1	4	π
ROW THREE	$\sqrt{2}$	8	2
ROW FOUR	4	−7	$\sqrt{10}$

Row two is $(1 \quad 4 \quad \pi)$ and column three is

$$\begin{pmatrix} 5 \\ \pi \\ 2 \\ \sqrt{10} \end{pmatrix}$$

The *dimension* of this matrix is 4×3 (read four-by-three) since it has 4 rows and 3 columns. We say a matrix has dimension $m \times n$, or is an $m \times n$ matrix, when it has m rows and n columns.

Each row or column by itself is a *vector*. A vector is then a special sort of matrix, one which is either just a single row or just a single column. The following are all vectors:

$$(1, -12, 2.4, 3) \qquad \begin{pmatrix} -13 \\ 68 \\ 372 \end{pmatrix} \qquad \begin{pmatrix} 2\sqrt{8} \\ -6.7 \\ \frac{17}{239} \\ \sqrt{83} \\ -19 \end{pmatrix}$$

$$(-\sqrt{5}, \tfrac{6}{5}, \sqrt{\tfrac{5}{3}}, 19, 147, 3^{50})$$

Notice that the row vectors shown here have the entries separated by commas. It is conventional to use commas to separate entries in row vectors, but to use spaces to separate entries in matrices. This leads to a potential notational conflict since a $1 \times n$ matrix is a row vector. We use whichever notation seems most appropriate at the time.

The number in the third row and the second column (8 in our earlier example) is called the (3, 2)th or 3, 2th entry. So in general the entry in the ith row and jth column is called the i, jth entry.

Finally, if we call the matrix A, then by a_{ij} we will mean the i, jth entry in the matrix A. So, for

$$A = \begin{pmatrix} 6 & -3 & 5 \\ 1 & 4 & \pi \\ \sqrt{2} & 8 & 2 \\ 4 & -7 & \sqrt{10} \end{pmatrix},$$

we have

$$a_{12} = -3, \qquad a_{23} = \pi, \qquad a_{42} = -7, \qquad a_{43} = \sqrt{10}.$$

We also write a_{ij} as $a(i, j)$ or even $A(i, j)$, especially in computer programs.

Rather than trying to learn all these definitions before going on, let us learn them as we go along. The more they are used, the more familiar they will become.

One final word before we get down to business. Did you notice that in the first part of this section we only mentioned addition and multiplication? "How about subtraction and division?" you might ask. In some sense subtraction and division are "inverse" operations to addition and multiplication. In particular, addition must be available before we define subtraction, and multiplication must be defined before division is possible. However, as we will see, multiplication can be defined without a meaningful division process being possible. Enough meandering, let's get down to business.

Addition of Matrices

This turns out to be fairly easy. We define the sum of two matrices to be the matrix obtained by adding corresponding elements. This is best illustrated by the following examples:

1. $\begin{pmatrix} 3 & 2 \\ 1 & 4 \end{pmatrix} + \begin{pmatrix} 5 & 4 \\ 6 & 8 \end{pmatrix} = \begin{pmatrix} 3+5 & 2+4 \\ 1+6 & 4+8 \end{pmatrix} = \begin{pmatrix} 8 & 6 \\ 7 & 12 \end{pmatrix}$

2. $\begin{pmatrix} 3 & 2 \\ 1 & 4 \end{pmatrix} + \begin{pmatrix} 0 & -3 \\ 2 & -2 \end{pmatrix} = \begin{pmatrix} 3 & -1 \\ 3 & 2 \end{pmatrix}$

3. $\begin{pmatrix} 1 & -2 & 3 \\ 3 & 1 & 4 \\ 7 & -5 & \sqrt{2} \\ 3 & -4 & 6 \end{pmatrix} + \begin{pmatrix} 2 & 4 & -5 \\ 0 & 2 & -1 \\ -4 & 2 & 5 \\ 2 & 2 & -3 \end{pmatrix} = \begin{pmatrix} 3 & 2 & -2 \\ 3 & 3 & 3 \\ 3 & -3 & \sqrt{2}+5 \\ 5 & -2 & 3 \end{pmatrix}$

4. $\begin{pmatrix} 1 & 3 \\ 2 & 4 \\ 1 & 5 \end{pmatrix} + \begin{pmatrix} 2 & 1 \\ 3 & 4 \end{pmatrix} = \begin{pmatrix} 3 & 4 \\ 5 & 8 \\ ? & ? \end{pmatrix} = \text{nonsense}$

Oops, there is nothing to add the 1 and 5 from the first matrix to in the second matrix. This illustrates a principle that is not stated but which is implicit in the definition.

Addition is only defined for matrices of the same dimension.

(Recall "dimension" is another word for "size" in this context.) The formal definition would be written this way:

If A and B are two $m \times n$ matrices, then $A + B = C$ where the entries in the $m \times n$ matrix C are found by taking $c_{ij} = a_{ij} + b_{ij}$.

In functional notation $C(i, j) = A(i, j) + B(i, j)$.

An example of how this might be used is furnished by considering a manufacturer of step ladders who has two plants. Both plants make steel and aluminum ladders in 4-foot and 6-foot lengths. They make different amounts since one plant is newer and closer to the steel supply. The total daily output of both plants together can be found by adding their production matrices together as follows.

	PLANT ONE		PLANT TWO	
	ALUMINUM	STEEL	ALUMINUM	STEEL
4 FT	140	110	185	230
6 FT	125	95	155	180

TOTAL PRODUCTION

$$\begin{pmatrix} 140 & 110 \\ 125 & 95 \end{pmatrix} + \begin{pmatrix} 185 & 230 \\ 155 & 180 \end{pmatrix} = \begin{pmatrix} 325 & 340 \\ 280 & 275 \end{pmatrix}$$

After learning to add in the first grade, you next went to subtraction. In most books nowadays, subtraction is defined in terms of addition. What do we mean by that? For example,

What is $7 - 4$?

It is that number C so that $7 = C + 4$.

Now we think back over our addition tables and recall that

$7 = 3 + 4$.

Hence we conclude $7 - 4 = 3$.

We could do the same thing with matrices (and indeed a conscientious mathematician should). If we do, we are led naturally to the result that subtraction is done entrywise exactly like addition is. Thus

5. $$\begin{pmatrix} 8 & 6 \\ 7 & 12 \end{pmatrix} - \begin{pmatrix} 5 & 4 \\ 6 & 8 \end{pmatrix} = \begin{pmatrix} 8-5 & 6-4 \\ 7-6 & 12-8 \end{pmatrix} = \begin{pmatrix} 3 & 2 \\ 1 & 4 \end{pmatrix},$$

6. $$\begin{pmatrix} 3 & -1 \\ 3 & 2 \end{pmatrix} - \begin{pmatrix} 0 & -3 \\ 2 & -2 \end{pmatrix} = \begin{pmatrix} 3-0 & -1-(-3) \\ 3-2 & 2-(-2) \end{pmatrix} = \begin{pmatrix} 3 & 2 \\ 1 & 4 \end{pmatrix},$$

just as we would get by comparing with examples 1 and 2 above.

A further few words for those of us who are interested in the structure of matrices. There are matrices that correspond to the number zero and to negatives in the sense that they have similar properties. A *zero matrix* **O** is defined to have every entry a zero. There is of course a different one for each size of matrix. Here are four examples:

$$(0 \quad 0 \quad 0) \quad \begin{pmatrix} 0 & 0 & 0 \\ 0 & 0 & 0 \\ 0 & 0 & 0 \end{pmatrix} \begin{pmatrix} 0 & 0 & 0 \\ 0 & 0 & 0 \\ 0 & 0 & 0 \\ 0 & 0 & 0 \end{pmatrix} \begin{pmatrix} 0 \\ 0 \\ 0 \\ 0 \end{pmatrix}$$

The *negative* of a matrix A is called $-A$ and is obtained by taking the negative of each entry of A. Two examples follow:

$$-\begin{pmatrix} 7 & 3 & 2 \\ \sqrt{5} & -4 & 3 \\ 7 & 6 & -5 \end{pmatrix} = \begin{pmatrix} -7 & -3 & -2 \\ -\sqrt{5} & 4 & -3 \\ -7 & -6 & 5 \end{pmatrix}, \quad -\begin{pmatrix} 8 \\ -6 \\ 4 \\ -2 \end{pmatrix} = \begin{pmatrix} -8 \\ 6 \\ -4 \\ 2 \end{pmatrix}.$$

Notice that the sum of a matrix and its negative is a zero matrix.

$$\begin{pmatrix} 8 \\ -6 \\ 4 \\ -2 \end{pmatrix} + \begin{pmatrix} -8 \\ 6 \\ -4 \\ 2 \end{pmatrix} = \begin{pmatrix} 0 \\ 0 \\ 0 \\ 0 \end{pmatrix}, \quad \begin{pmatrix} 3 & -\sqrt{5} \\ -1.6 & 17.4 \end{pmatrix} + \begin{pmatrix} -3 & \sqrt{5} \\ 1.6 & -17.4 \end{pmatrix} = \begin{pmatrix} 0 & 0 \\ 0 & 0 \end{pmatrix}.$$

The last two things we notice is that addition of matrices is commutative and associative. Recall that the validity of these laws for real numbers implies that the order in which addition takes place is immaterial. Of course, there are two kinds of order here. For example, $7 + 5 = 5 + 7$, since both equal 12, is an example of the commutative law, while $7 + (5 + 3) = (7 + 5) + 3$ is an example of the associative law. The analogous equations should seem obvious for matrices too. (Those sceptics among you can test some examples and look for a counterexample.)

Thus, if A, B, C are any three $m \times n$ matrices, then

$$A + B = B + A,$$
$$(A + B) + C = A + (B + C).$$

So far so good. Addition for matrices seems to be just like addition of numbers. Now how about multiplication?

Multiplication of Matrices

What happens if we define multiplication as we did addition, that is, entrywise? Then

$$\begin{pmatrix} 2 & 1 \\ 3 & 2 \end{pmatrix} \times \begin{pmatrix} 2 & 2 \\ 1 & 4 \end{pmatrix} = \begin{pmatrix} 2 \times 2 & 1 \times 2 \\ 3 \times 1 & 2 \times 4 \end{pmatrix} = \begin{pmatrix} 4 & 2 \\ 3 & 8 \end{pmatrix}$$

This seems fairly simple—and it is. Unfortunately, it doesn't seem to be of much use! Instead, the multiplication which we used in Section 7.2 is much more useful since it allows for applications to many problems. Recall we could multiply a matrix times a column vector provided the number of columns in the matrix was the same as the number of rows (or entries, in this case) in the column vector. For example, a 2×3 matrix times a 3×1 column vector gives a 2×1 column vector as follows:

$$\begin{pmatrix} 1 & 3 & -2 \\ 2 & 4 & 5 \end{pmatrix} \begin{pmatrix} 2 \\ -1 \\ 3 \end{pmatrix} = \begin{pmatrix} 1 \times 2 + 3 \times (-1) + (-2) \times 3 \\ 2 \times 2 + 4 \times (-1) + 5 \times 3 \end{pmatrix}$$

$$= \begin{pmatrix} 2 + (-3) + (-6) \\ 4 + (-4) + 15 \end{pmatrix} = \begin{pmatrix} -7 \\ 15 \end{pmatrix}.$$

We can extend this to the multiplication of matrices where the number of columns in the first matrix is the same as the number of rows in the second. We simply perform the multiplication on each column of the second matrix separately. An illustration will show what we mean. How do we perform the multiplication

$$\begin{pmatrix} 1 & 3 & -2 \\ 2 & 4 & 5 \end{pmatrix} \begin{pmatrix} 2 & 1 & 0 \\ -1 & 2 & -2 \\ 3 & 4 & 1 \end{pmatrix} ?$$

We perform each multiplication

$$\begin{pmatrix} 1 & 3 & -2 \\ 2 & 4 & 5 \end{pmatrix} \begin{pmatrix} 2 \\ -1 \\ 3 \end{pmatrix}, \begin{pmatrix} 1 & 3 & -2 \\ 2 & 4 & 5 \end{pmatrix} \begin{pmatrix} 1 \\ 2 \\ 4 \end{pmatrix}, \begin{pmatrix} 1 & 3 & -2 \\ 2 & 4 & 5 \end{pmatrix} \begin{pmatrix} 0 \\ -2 \\ 1 \end{pmatrix},$$

and put the results together. When we do this, the result is the following:

$$\begin{pmatrix} 1 & 3 & -2 \\ 2 & 4 & 5 \end{pmatrix} \begin{pmatrix} 2 & 1 & 0 \\ -1 & 2 & -2 \\ 3 & 4 & 1 \end{pmatrix} =$$

$$\begin{pmatrix} 1 \times 2 + 3 \times (-1) + (-2) \times 3 & 1 \times 1 + 3 \times 2 + (-2) \times 4 & 1 \times 0 + 3 \times (-2) + (-2) \times 1 \\ 2 \times 2 + 4 \times (-1) + 5 \times 3 & 2 \times 1 + 4 \times 2 + 5 \times 4 & 2 \times 0 + 4 \times (-2) + 5 \times 1 \end{pmatrix}$$

$$= \begin{pmatrix} 2 - 3 - 6 & 1 + 6 - 8 & 0 - 6 - 2 \\ 4 - 4 + 15 & 2 + 8 + 20 & 0 - 8 + 5 \end{pmatrix} = \begin{pmatrix} -7 & -1 & -8 \\ 15 & 30 & -3 \end{pmatrix}$$

Two additional examples follow.

7. $$\begin{pmatrix} 2 & -1 \\ 3 & -2 \end{pmatrix} \cdot \begin{pmatrix} 4 & 2 \\ 3 & 4 \end{pmatrix} = \begin{pmatrix} 2 \times 4 + (-1) \times 3 & 2 \times 2 + (-1) \times 4 \\ 3 \times 4 + (-2) \times 3 & 3 \times 2 + (-2) \times 4 \end{pmatrix}$$

$$= \begin{pmatrix} 8 + (-3) & 4 + (-4) \\ 12 + (-6) & 6 + (-8) \end{pmatrix}$$

$$= \begin{pmatrix} 5 & 0 \\ 6 & -2 \end{pmatrix}$$

8. $$\underset{2 \times 3}{\begin{pmatrix} 1 & 3 & -2 \\ 2 & 4 & 5 \end{pmatrix}} \underset{3 \times 4}{\begin{pmatrix} 2 & 3 & 4 & 3 \\ -1 & 0 & -3 & -2 \\ 3 & -2 & 2 & 1 \end{pmatrix}} = \underset{2 \times 4}{\begin{pmatrix} -7 & 7 & -9 & -5 \\ 15 & -4 & 6 & 3 \end{pmatrix}}$$

Let us look at example 8 more closely. First, notice that we are multiplying a 2×3 matrix with a 3×4 matrix and getting a 2×4 matrix as an answer. Thus the number of rows in the answer is the same as the number of rows in the first factor and the number of columns in the answer is the same as the number of columns in the second factor. The entry in the ith row and the jth column of the answer is found by "multiplying" the ith row of the first factor by the jth column of the second factor. To put it another way, if A_i denotes the ith row of A and ${}_jB$ denotes the jth column of B, then $A \cdot B = C$, where c_{ij} is the product of A_i and ${}_jB$. Pictorially,

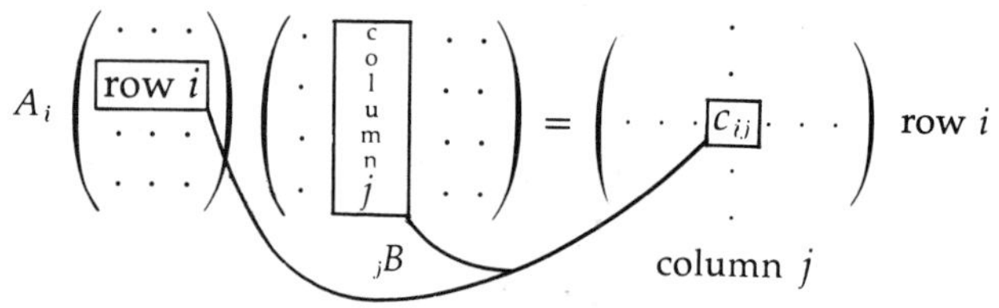

In describing the process of matrix multiplication, it is necessary to be able to talk about a single column of the matrix B. We have introduced the notation ${}_jB$ on an ad hoc basis. It is not standard notation and we do not use it elsewhere.

Let's see how this worked in example 8. Thus we are supposing

$$A = \begin{pmatrix} 1 & 3 & -2 \\ 2 & 4 & 5 \end{pmatrix}, \quad B = \begin{pmatrix} 2 & 3 & 4 & 3 \\ -1 & 0 & -3 & -2 \\ 3 & -2 & 2 & 1 \end{pmatrix}, \quad C = \begin{pmatrix} -7 & 7 & -9 & -5 \\ 15 & -4 & 6 & 3 \end{pmatrix}.$$

How is the 7 in C obtained? Well, $7 = c_{12}$ so we take

$$A_1 = (1 \quad 3 \quad -2) \text{ times } {}_2B = \begin{pmatrix} 3 \\ 0 \\ -2 \end{pmatrix}.$$

Then

$$A_1 \cdot {}_2B = (1 \quad 3 \quad -2) \cdot \begin{pmatrix} 3 \\ 0 \\ -2 \end{pmatrix} = 1 \cdot 3 + 3 \cdot 0 + (-2)(-2) = 3 + 0 + 4 = 7.$$

Also,

$$-5 = c_{14} = A_1 \cdot {}_4B = (1 \quad 3 \quad -2) \begin{pmatrix} 3 \\ -2 \\ 1 \end{pmatrix} = 3 + (-6) + (-2) = 3 - 8 = -5,$$

$$15 = c_{21} = A_2 \cdot {}_1B = (2 \quad 4 \quad 5) \begin{pmatrix} 2 \\ -1 \\ 3 \end{pmatrix} = 4 + (-4) + 15 = 15,$$

$$6 = c_{23} = A_2 \cdot {}_3B = (2 \quad 4 \quad 5) \begin{pmatrix} 4 \\ -3 \\ 2 \end{pmatrix} = 8 + (-12) + 10 = 6.$$

Notice that you can do arithmetic in your head or do just a little scratch paperwork like this:

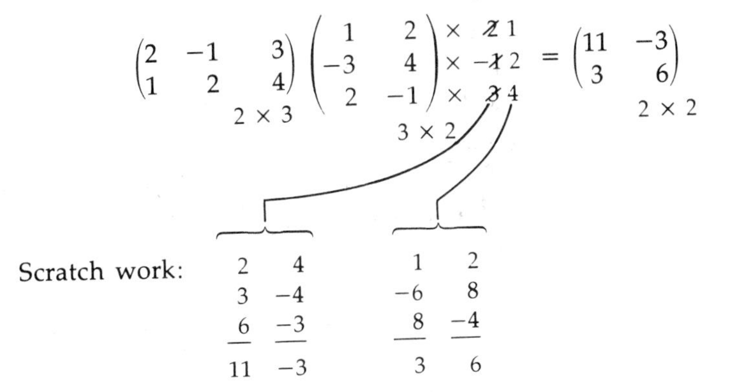

Remark The fact that the number of columns in the first factor must equal the number of rows in the second factor in order for this "row by column" multiplication to be possible restricts the sizes of the matrices that can be multiplied. In symbols,

$$A_{n\times m} \cdot B_{m\times k} = C_{n\times k},$$

where by this we mean that the result of multiplying an $n \times m$ matrix A with an $m \times k$ matrix B is an $n \times k$ matrix C. (We call this the "middle-m-exclusion principle," since the m gets lost.) Pictorially the dimensions must match up as follows:

$$\begin{pmatrix} \cdot & \cdot & \cdot & \cdot & \cdot \\ \cdot & \cdot & \cdot & \cdot & \cdot \\ \cdot & \cdot & \cdot & \cdot & \cdot \end{pmatrix} \begin{pmatrix} \cdot & \cdot & \cdot & \cdot & \cdot & \cdot & \cdot & \cdot \\ \cdot & \cdot & \cdot & \cdot & \cdot & \cdot & \cdot & \cdot \\ \cdot & \cdot & \cdot & \cdot & \cdot & \cdot & \cdot & \cdot \\ \cdot & \cdot & \cdot & \cdot & \cdot & \cdot & \cdot & \cdot \\ \cdot & \cdot & \cdot & \cdot & \cdot & \cdot & \cdot & \cdot \end{pmatrix} = \begin{pmatrix} \cdot & \cdot & \cdot & \cdot & \cdot & \cdot & \cdot & \cdot \\ \cdot & \cdot & \cdot & \cdot & \cdot & \cdot & \cdot & \cdot \\ \cdot & \cdot & \cdot & \cdot & \cdot & \cdot & \cdot & \cdot \end{pmatrix}$$

Before going on, you should practice by doing several of the following.

Exercises 7.3

For problems 1–7 let

$$A = \begin{pmatrix} 2 & 3 \\ 5 & 4 \\ 1 & 6 \end{pmatrix}, \quad B = \begin{pmatrix} 1 & 4 \\ 6 & -1 \\ 2 & 2 \end{pmatrix}, \quad C = \begin{pmatrix} 1 & -1 \\ -1 & 5 \\ -1 & 4 \end{pmatrix}, \quad D = \begin{pmatrix} 3 & 7 \\ 11 & 3 \\ 3 & 8 \end{pmatrix},$$

$$E = \begin{pmatrix} 1 & -2 \\ 3 & 4 \end{pmatrix}, \quad F = \begin{pmatrix} 0 & 2 \\ -3 & -3 \end{pmatrix}, \quad G = \begin{pmatrix} 1 & 1 \\ 1 & 1 \end{pmatrix}, \quad H = \begin{pmatrix} -2 & 1 \\ -4 & -5 \end{pmatrix},$$

$$I = \begin{pmatrix} 1 & 0 \\ 0 & 1 \end{pmatrix}.$$

1. **(a)** Find $A + B$, $B - C$, $D + C$, $C - D$.
   (b) Find $A + C$, $B - D$, $D - B$.
2. **(a)** Find $E + F$, $E + H$, $G + H$, $E - H$.
   (b) Find $E + (G + H)$, $(E + G) + H$, $(E + H) + G$.
3. **(a)** Are A, B, C, and D related? How?
   (b) What is $A + E$?
4. **(a)** Find $A \cdot E$, $(B + C) \cdot E$, $C \cdot F$, $F \cdot G$.
   (b) Find $F \cdot I$, $(E + F) \cdot H$.
5. **(a)** Find $\text{A} \cdot (E + F)$, $\text{G} \cdot I$, $I \cdot H$, $H \cdot I$.
   (b) Find $\text{C} \cdot G$, $F \cdot H$, $B \cdot F$.
6. **(a)** Find $A \cdot G$, $A \cdot I$.
   (b) Find $(D + C) \cdot E$, $A \cdot (E + G + H)$.
7. **(a)** Can you find $A \cdot B$ or $E \cdot A$?
   (b) Does $E \cdot F = F \cdot E$?

For problems 8–11 let

$$J = \begin{pmatrix} 1 & 2 & 0 \\ 3 & 0 & 2 \\ -1 & 3 & 2 \end{pmatrix}, \quad K = \begin{pmatrix} 1 & -1 & 1 \\ 0 & 2 & 3 \\ 0 & 0 & 3 \end{pmatrix}, \quad L = \begin{pmatrix} 1 & 0 & 3 \\ -1 & 2 & 0 \\ -2 & 3 & -1 \end{pmatrix}$$

$$M = \begin{pmatrix} -2 & 9 & -6 \\ -1 & 5 & -3 \\ 1 & -3 & 2 \end{pmatrix}, \quad I = \begin{pmatrix} 1 & 0 & 0 \\ 0 & 1 & 0 \\ 0 & 0 & 1 \end{pmatrix}.$$

8. **(a)** Find $J \cdot L$, $K \cdot L$, $L \cdot K$.
(b) Find K^2 and K^3.

9. **(a)** Find $J \cdot M$, $L \cdot M$, $M \cdot L$.
(b) Find L^2, L^3.

10. **(a)** Find $(JL)M$ and $J(LM)$.
(b) If **O** is the 3×3 zero matrix, what is $J \cdot \mathbf{O}$? $\mathbf{O} \cdot J$?

11. **(a)** Find $J + K$ and $(J + K) \cdot L$. Is $(J + K) \cdot L = J \cdot L + K \cdot L$?
(b) Show that sometimes matrices commute under multiplication and other times they do not.
(c) Does $I \cdot X = X \cdot I$ for all matrices X? If not, for which ones is it true?

12. A mattress manufacturer makes soft, firm, and extra firm mattresses in both double and queen size and has warehouses in Dallas, Denver, and Los Angeles. Each month the company sends 120 soft, 240 firm, and 180 extra firm queen-size mattresses and two-thirds that number of double-bed-size mattresses of each firmness to each warehouse.
(a) Organize the number of mattresses each warehouse gets into a 2×3 matrix.
(b) In October, the Denver and Los Angeles warehouses asked for additional mattresses (besides the usual shipment) as shown below:

	DENVER			LOS ANGELES		
	SOFT	FIRM	EX. FIRM	SOFT	FIRM	EX. FIRM
Q	20	30	45	25	45	30
D	15	50	70	20	30	25

Find the matrix representing the shipment to each of Denver and Los Angeles. What was the total shipment to all warehouses of each kind of mattress?

7.4 FURTHER PROPERTIES

We first look at powers of matrices and products when one factor is a vector. Problems like these are important in the following sections on Markov chains. Next we ask some questions typical of those a mathematician interested in abstract structure would ask. These latter results are not used elsewhere in the book and so may be omitted.

Powers and Products with a Vector

If A is a matrix, when will

$$A \cdot A$$

be defined? If A is an $m \times n$ matrix, then the remark at the end of the last section shows that the number of columns in A must equal the number of rows in A.

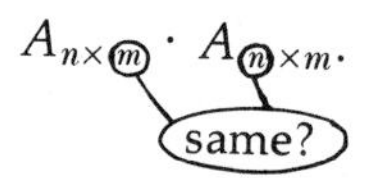

Thus A *must* have the same number of rows and columns. Such a matrix is called a *square* matrix.

If A is a square matrix, say A is an $n \times n$ matrix, then

$$A_{n\times n} \cdot A_{n\times n} = C_{n\times n} \text{ ,}$$

so $C = A^2$ is also an $n \times n$ matrix. This shows that $A^3 = A \cdot A^2$ is also a square $n \times n$ matrix. Surely an example is in order after all this abstraction.

Suppose

$$A = \begin{pmatrix} 1 & 3 \\ 2 & -1 \end{pmatrix};$$

then

$$A^2 = A \cdot A = \begin{pmatrix} 1 & 3 \\ 2 & -1 \end{pmatrix}\begin{pmatrix} 1 & 3 \\ 2 & -1 \end{pmatrix} = \begin{pmatrix} 1+6 & 3-3 \\ 2-2 & 6+1 \end{pmatrix} = \begin{pmatrix} 7 & 0 \\ 0 & 7 \end{pmatrix},$$

$$A^3 = A \cdot A^2 = \begin{pmatrix} 1 & 3 \\ 2 & -1 \end{pmatrix}\begin{pmatrix} 7 & 0 \\ 0 & 7 \end{pmatrix} = \begin{pmatrix} 7 & 21 \\ 14 & -7 \end{pmatrix}$$

$$= A^2 \cdot A = \begin{pmatrix} 7 & 0 \\ 0 & 7 \end{pmatrix}\begin{pmatrix} 1 & 3 \\ 2 & -1 \end{pmatrix} = \begin{pmatrix} 7 & 21 \\ 14 & -7 \end{pmatrix},$$

$$A^4 = A \cdot A^3 = \begin{pmatrix} 1 & 3 \\ 2 & -1 \end{pmatrix} \begin{pmatrix} 7 & 21 \\ 14 & -7 \end{pmatrix} = \begin{pmatrix} 49 & 0 \\ 0 & 49 \end{pmatrix}.$$

What do you think happens to higher powers of A? We will pursue this in the exercises.

Consider next a matrix

$$B = \begin{pmatrix} 0 & 2 & 1 \\ 3 & 2 & 0 \\ 0 & 0 & 2 \end{pmatrix},$$

where $$B^2 = \begin{pmatrix} 0 & 2 & 1 \\ 3 & 2 & 0 \\ 0 & 0 & 2 \end{pmatrix} \begin{pmatrix} 0 & 2 & 1 \\ 3 & 2 & 0 \\ 0 & 0 & 2 \end{pmatrix} = \begin{pmatrix} 6 & 4 & 2 \\ 6 & 10 & 3 \\ 0 & 0 & 4 \end{pmatrix},$$

and $$B^3 = \begin{pmatrix} 12 & 20 & 10 \\ 30 & 32 & 12 \\ 0 & 0 & 8 \end{pmatrix}.$$

Can you guess what the last row in B^4 and B^5 will look like? Compute them to check. Notice that you can find the last row without finding all the entries,

$$B^4 = B \cdot B^3 = \begin{pmatrix} 0 & 2 & 1 \\ 3 & 2 & 0 \\ 0 & 0 & 2 \end{pmatrix} \begin{pmatrix} 12 & 20 & 10 \\ 30 & 32 & 12 \\ 0 & 0 & 8 \end{pmatrix} = \begin{pmatrix} \cdot & \cdot & \cdot \\ \cdot & \cdot & \cdot \\ 0 & 0 & 16 \end{pmatrix},$$

by multiplying the third row $(0 \quad 0 \quad 2)$ by the matrix B^3.

This leads us to consider what happens when we multiply a vector times a matrix. If the vector is the special kind that has zero in every entry except one, you can easily see what happens. Notice the result is also a vector.

$$(0 \quad 0 \quad 3 \quad 0) \begin{pmatrix} 1 & 3 & 0 \\ -2 & 6 & 3 \\ 3 & -2 & 5 \\ 1 & 0 & 0 \end{pmatrix} = (9 \quad -6 \quad 15),$$

$$(0 \quad 1 \quad 0 \quad 0) \begin{pmatrix} 1 & 6 \\ 2 & -7 \\ -3 & 5 \\ 5 & -2 \end{pmatrix} = (2 \quad -7).$$

How about column vectors?

$$\begin{pmatrix} -1 & 6 & 2 & 3 \\ 2 & 6 & 3 & 5 \\ 0 & 7 & 6 & -5 \end{pmatrix} \begin{pmatrix} 0 \\ 0 \\ 1 \\ 0 \end{pmatrix} = \begin{pmatrix} 2 \\ 3 \\ 6 \end{pmatrix},$$

$$\begin{pmatrix} 1 & -2 & 1 \\ 0 & -5 & -1 \\ 56 & 17 & 2 \\ 23 & 9 & 4 \end{pmatrix} \begin{pmatrix} 0 \\ 0 \\ 3 \end{pmatrix} = \begin{pmatrix} 3 \\ -3 \\ 6 \\ 12 \end{pmatrix}.$$

Notice that for the row vectors, the result of multiplying gave three times the third row and one times the second row. For the column vectors, the multiplication picks out the third column in the first case and three times the third column in the second case. Can you formulate a general principle? Problem 3 in the exercises gives further examples and problem 4 looks at the general case.

Further Properties of Multiplication

Now that we know how to multiply matrices, we will consider briefly to what extent matrices act arithmetically like real numbers. Do they satisfy the associative and commutative laws for multiplication? The answers are yes and no, respectively.

Associative Law For any matrices A, B, and C for which $A \cdot B$ and $B \cdot C$ are defined, we have

$$(A \cdot B) \cdot C = A \cdot (B \cdot C).$$

We saw examples of this in the exercises earlier.

Commutative Law The commutative law certainly does not hold in general, as we saw in the exercises to Section 7.3. In fact, it is not easy to find two matrices that do commute (although again we encountered some in the exercises). For example,

$$\begin{pmatrix} 2 & 1 \\ -1 & 2 \end{pmatrix} \begin{pmatrix} 1 & 1 \\ 1 & 1 \end{pmatrix} = \begin{pmatrix} 3 & 3 \\ 1 & 1 \end{pmatrix},$$

while

$$\begin{pmatrix} 1 & 1 \\ 1 & 1 \end{pmatrix} \begin{pmatrix} 2 & 1 \\ -1 & 2 \end{pmatrix} = \begin{pmatrix} 1 & 3 \\ 1 & 3 \end{pmatrix}$$

shows that in general $A \cdot B = B \cdot A$ is not true.

Notice that the question of commutativity doesn't make sense unless A and B are both square matrices (that is, they have the same number of rows as columns) because usually one of $A \cdot B$ and $B \cdot A$ would be undefined. Can you think of matrices A and B that are not square and yet $A \cdot B$ and $B \cdot A$ are both defined? (See problem 5 in the exercises.)

If we restrict ourselves to square matrices, at least two reasonable questions come to mind. (1) What is the relation between addition and multiplication? (2) What about division? The first question is answered easily.

Distributive Law For any $n \times n$ matrices A, B, C,

$$A \cdot (B + C) = A \cdot B + A \cdot C,$$

$$(A + B) \cdot C = A \cdot C + B \cdot C.$$

Proofs of this and the associative laws can be found in books on linear algebra, but are not very instructive.

Division is a more interesting question: As long as two numbers are nonzero, we can divide one by the other. For example,

$$11 \div 8 = 1.375.$$

Why? Well, because $(1.375) \cdot (8) = 11.000$. That is, $a \div b = c$ provided $a = c \cdot b$. Division by zero is excluded because $c \cdot 0 = 0$ for any c, so $a \div 0 = c$ means $a = c \cdot 0 = 0$ so a would have to be zero but c could be anything. Mathematicians get out of this uncomfortable situation by not allowing division by zero.

Getting back to matrices, if A and B are not zero, can we find a C so that $A \div B = C$ in the sense that $A = C \cdot B$? The answer is no in general, as the following discussion shows. Let

$$B = \begin{pmatrix} 1 & 1 \\ 1 & 1 \end{pmatrix}, \qquad A = \begin{pmatrix} 1 & 2 \\ 3 & 4 \end{pmatrix}.$$

Could there be a matrix

$$C = \begin{pmatrix} a & b \\ c & d \end{pmatrix}$$

so that $A = C \cdot B$? If so, then

$$C \cdot B = \begin{pmatrix} a & b \\ c & d \end{pmatrix} \cdot \begin{pmatrix} 1 & 1 \\ 1 & 1 \end{pmatrix} = \begin{pmatrix} a + b & a + b \\ c + d & c + d \end{pmatrix} = \begin{pmatrix} 1 & 2 \\ 3 & 4 \end{pmatrix} = A.$$

$$\text{Thus} \qquad a + b = 1, \qquad c + d = 3,$$
$$a + b = 2, \qquad c + d = 4.$$

These equations are clearly contradictory, so no such numbers a, b, c, d and thus no such matrix exists.

Thus we see that division of matrices is not generally possible. A very interesting study is to consider for what matrices it *is* possible. However, we shall not do it since we have already learned more about matrices than we need for what follows.

Exercises 7.4

1. Let

$$A = \begin{pmatrix} 2 & 0 & 0 \\ 1 & 1 & 1 \\ 0 & -1 & 1 \end{pmatrix}, \quad B = \begin{pmatrix} \frac{1}{2} & \frac{1}{2} \\ 0 & 1 \end{pmatrix}, \quad C = \begin{pmatrix} \frac{1}{2} & 0 & \frac{1}{2} \\ 0 & 1 & 0 \\ \frac{1}{2} & 0 & \frac{1}{2} \end{pmatrix}.$$

(a) Find A^2, A^3, A^4, A^5, A^6. Can you predict any of the entries in any higher power?

(b) Compute B^2, B^4, B^8. Notice one entry getting close to zero?

(c) Compute C^2 and C^3. What about higher powers of C?

2. Let $A = \begin{pmatrix} 1 & 3 \\ 2 & -1 \end{pmatrix}$. Recall we computed A^2, A^3, and A^4 earlier in this section. Compute $A^6 = A^2 \cdot A^4$ and $A^8 = A^4 \cdot A^4$. Can you generalize these answers to predict what A^{2k} will look like?

3. Let $H = \begin{pmatrix} 2 & -3 & 1 \\ -1 & 0 & 2 \\ 3 & 1 & -1 \\ 1 & 2 & -2 \end{pmatrix}, \quad J = \begin{pmatrix} \frac{1}{2} & \frac{1}{2} & 0 \\ 0 & \frac{1}{2} & \frac{1}{2} \\ \frac{1}{3} & \frac{1}{3} & \frac{1}{3} \end{pmatrix},$

$$w = (2, 0, 0, 0), \qquad u = (0, 0, 1, 0), \qquad v = (0, 0, 0, -1),$$

$$x = \begin{pmatrix} 0 \\ 3 \\ 0 \end{pmatrix}, \quad y = \begin{pmatrix} 0 \\ 0 \\ 1 \end{pmatrix}, \quad z = (\tfrac{2}{9}, \tfrac{4}{9}, \tfrac{3}{9})$$

(a) Compute wH, Jx, Hy, zJ.

(b) Compute yz and zy.

(c) Compute uH and vH.

4. **(a)** If u is a vector with n entries which are zero except for a 1 in the ith position and A is an $n \times m$ matrix, what is uA?

(b) If v is a vector with n entries which are zero except for a k in the ith position, and A is an $n \times m$ matrix, describe what vA looks like.

5. Let

$$A = \begin{pmatrix} 2 & -4 & 1 \\ 0 & 3 & -1 \end{pmatrix} \quad \text{and} \quad B = \begin{pmatrix} -4 & 1 \\ 1 & 3 \\ 0 & 2 \end{pmatrix}.$$

(a) Compute $A \cdot B$ and $B \cdot A$.
(b) Even though both matrices in (a) are defined, why are they not equal?
(c) Besides square matrices, for what other size matrices are the products $A \cdot B$ and $B \cdot A$ defined? In these cases is it ever possible for $A \cdot B$ to equal $B \cdot A$?

6. Find two 3×3 matrices that do not commute under multiplication.

7. Show in the following example that division is possible. The method we suggest is similar to that used in the text. For example, to find C so that

$$\begin{pmatrix} 1 & 2 \\ 3 & 4 \end{pmatrix} \div \begin{pmatrix} 1 & 1 \\ 2 & 3 \end{pmatrix} = C,$$

let

$$C = \begin{pmatrix} a & b \\ c & d \end{pmatrix}$$

and set

$$\begin{pmatrix} a & b \\ c & d \end{pmatrix} \cdot \begin{pmatrix} 1 & 1 \\ 2 & 3 \end{pmatrix} = \begin{pmatrix} 1 & 2 \\ 3 & 4 \end{pmatrix}$$

to obtain the system of equations

$$a + 2b = 1 \qquad c + 2d = 3$$

$$a + 3b = 2 \qquad c + 3d = 4$$

and deduce that $a = -1$, $b = c = d = 1$. So

$$C = \begin{pmatrix} -1 & 1 \\ 1 & 1 \end{pmatrix}.$$

(a) Find $\begin{pmatrix} 1 & 0 \\ 3 & 1 \end{pmatrix} \div \begin{pmatrix} 1 & 1 \\ 2 & 3 \end{pmatrix}$.

(b) Find $\begin{pmatrix} -1 & 1 \\ 1 & 1 \end{pmatrix} \div \begin{pmatrix} 1 & 2 \\ 3 & 4 \end{pmatrix}$.

(c) Find $\begin{pmatrix} 1 & 2 \\ 3 & 4 \end{pmatrix} \div \begin{pmatrix} 3 & -1 \\ -2 & 1 \end{pmatrix}$.

8. For any square matrix A is $A^2 \cdot A = A \cdot A^2$? Why?

9. Suppose that D and B are 3×3 and $B \cdot D = I$, where I is the identity matrix

$$I = \begin{pmatrix} 1 & 0 & 0 \\ 0 & 1 & 0 \\ 0 & 0 & 1 \end{pmatrix}.$$

Show that $C = A \div B$ can always be found. As an example take

$$B = \begin{pmatrix} 1 & 2 & 0 \\ 3 & 5 & 1 \\ 0 & -2 & 1 \end{pmatrix}, \quad D = \begin{pmatrix} 7 & -2 & 2 \\ -3 & 1 & -1 \\ -6 & 2 & -1 \end{pmatrix} \text{ and } A = \begin{pmatrix} 2 & 3 & 1 \\ 4 & 0 & -2 \\ -1 & 2 & 1 \end{pmatrix}.$$

Find C. (C can be found by multiplying two matrices, one of which is A.)

10. By considering L and M from problem 9, Section 7.3, and B and D from problem 9 above, what would you conjecture the value of $S \cdot T$ to be, if you were given that $T \cdot S = I$, the identity matrix, and that T and S are both square?

7.5 MARKOV CHAINS

In Chapter 3 we studied processes where the outcome could only be determined with a certain probability. Moreover, in these situations we often assumed that the events were independent. For example, when tossing a coin, what happened one time has no effect on what will happen the next time. Suppose that we consider a more complicated process, one in which the previous outcome does affect in some way what will happen next. An example will illustrate what we mean.

EXAMPLE 1

Willie the Weasel is a notorious second-story man. He is either free, in jail, or under interrogation. If free on one day, the next day he is caught and jailed with probability .3 and brought in for interrogation with probability .1. If he is in jail on one day, with probability .2 he escapes the next day and is free, and with probability .3 he is taken to be interrogated. Finally, if he is being questioned one day, he will be questioned again the next with probability .2, and he will be sent to jail or set free with equal probability.

We first illustrate the problem with a *transition diagram* (Figure 7.1), which is a graph in which F, I, and J stand for the states of being free, under interrogation, and jailed and the edges are labeled with the probabilities.

Figure 7.1

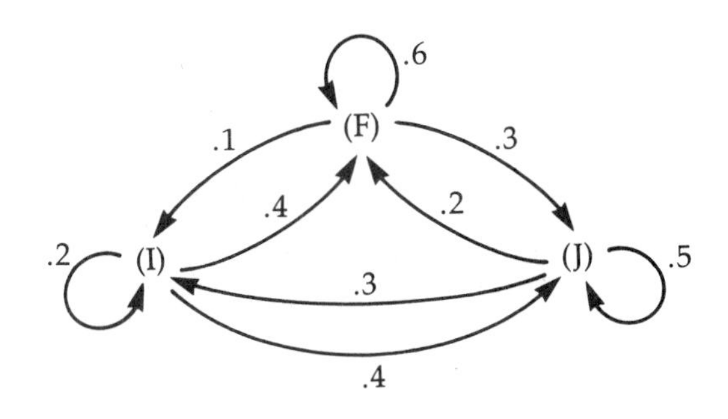

We can also show the transitions by means of a matrix. To do this, we label the rows and columns, in the same order, with the "states," and let the entry in the ith row and jth column be the probability of going from the state labeling row i to the state labeling row j. Again, an example should make it clear. The *transition matrix* for Willie would look like this:

		SECOND STATE		
		FREE	INTERROGATION	JAIL
FIRST STATE	FREE	.6	.1	.3
	INTERROGATION	.4	.2	.4
	JAIL	.2	.3	.5

Notice that each row adds to 1. This is because the next day Willie is assumed to be free, in jail, or being interrogated.

Before going on to show how useful this representation is, let us look at one more example and give you a chance to construct some transition matrices.

EXAMPLE 2 In a particular nuclear reaction, an initial element N is changed by an electron beam into a second element M, which in turn decays into a substance O. However, during the reaction creating O, a certain percentage of M is changed back to N. (The element O then also decays into N.) To be specific, we will view a time unit as being one minute. In each minute, 25% of the amount of N in the electron beam is changed to M; 50% of the amount of M is changed to O and 25% is changed to N; and 50% of the amount of O is changed back to N.

Figure 7.2

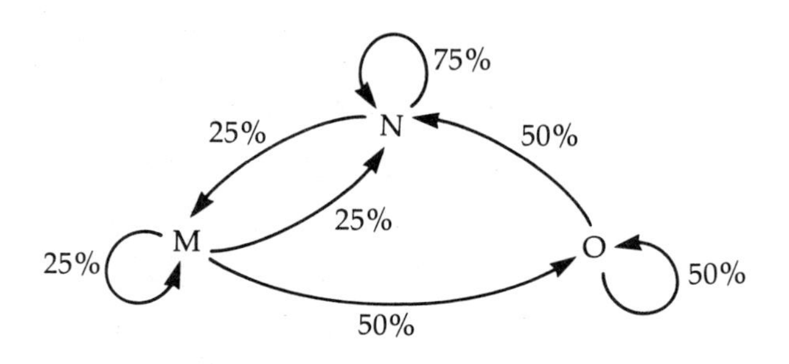

These conditions are shown graphically in Figure 7.2. The transition matrix would be

		SECOND STATE		
		N	M	O
FIRST STATE	N	.75	.25	0
	M	.25	.25	.50
	O	.50	0	.50

Exercises 7.5

1. Given the transition graphs shown, form the corresponding transition matrices.

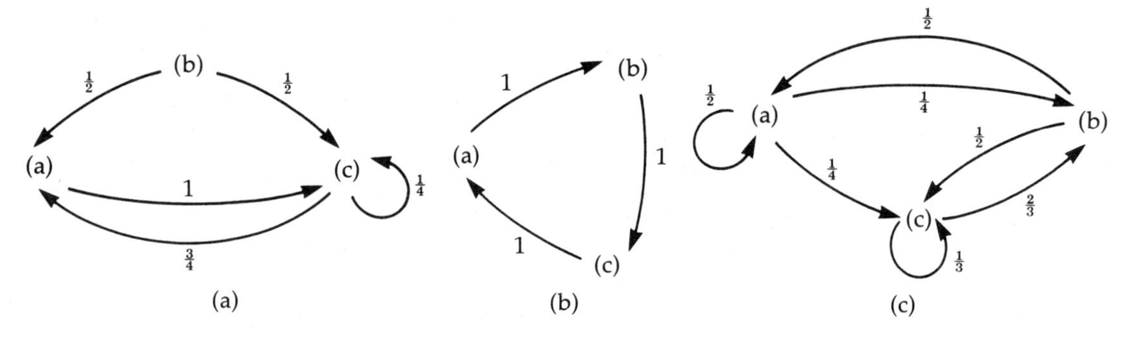

2. Draw a transition graph that corresponds to the following transition matrices:

$$\begin{pmatrix} \frac{1}{2} & 0 & \frac{1}{2} \\ 0 & 0 & 1 \\ \frac{1}{4} & \frac{1}{4} & \frac{1}{2} \end{pmatrix} \quad \begin{pmatrix} 0 & 1 & 0 \\ 0 & 1 & 0 \\ \frac{1}{2} & 0 & \frac{1}{2} \end{pmatrix} \quad \begin{pmatrix} 0 & 1 \\ 1 & 0 \end{pmatrix} \quad \begin{pmatrix} 0 & 1 & 0 & 0 \\ 1 & 0 & 0 & 0 \\ 0 & \frac{1}{3} & \frac{1}{3} & \frac{1}{3} \\ 0 & 0 & \frac{1}{2} & \frac{1}{2} \end{pmatrix}$$

For each of the following, draw the transition graph and give the transition matrix that describes the changes.

3. Larry and Earl play squash regularly. If Earl won the match last time, Larry has a 40% chance of winning next time, while if Larry won last time, he has a 50% chance of winning next time.

4. In a local village election, if the Democrats won in the last election, they have a 60% chance of winning next time and a 40% chance of losing to the Republicans. If the Republicans won last time, then they will win again 70% of the time and the Democrats will win the other 30%.

5. Subway trains are either on time or late (never early). If a train is on time, the next train will also be on time 70% of the time. However, if a train is late, then the next train will be late 80% of the time.

6. Students in the College of Arts and Sciences declare a major in social science, humanities (which includes mathematics), or the natural and physical sciences. A student majoring in social science has a 30% chance of changing to humanities and a 30% chance of changing to sciences. A humanities student only changes 40% of the time and is equally likely to choose science or social science. A science student changes to humanities 10% of the time and to social science 30% of the time.

7. On any given day a manic-depressive is either high, normal, or low. If she is high, she stays high or goes low with equal likelihood and returns to normal 60% of the time. When normal she remains normal half the time, goes high 20%, and low 30% of the time. If low, she never goes high and remains low 60% of the time, returning to normal the other 40%.

8. The corn yield in Illinois each year can be described as poor, average, or good. A good year is followed by another 60% of the time, and a poor or average year each 20% of the time. A poor year follows a poor year 50% of the time, and an average year follows an average year 80% of the time. Whenever a change occurs, the other two outcomes occur with equal probability.

9. According to tradition, California never has two rainy days in a row. After it rains it is cloudy or clear with equal chance. If it is clear or cloudy on a day, it will be the same on the next day with

probability $\frac{1}{2}$; if there is a change from clear or cloudy, then half the time it is a change to rain.

10. In Utopia, people are executives, plumbers, or civil servants. Children of executives are distributed among the jobs with 40% executives, 40% civil servants, and 20% plumbers. In the case of children of civil servants, 60% are civil servants and the rest are divided evenly between executives and plumbers. Finally, of children of plumbers, 40% follow in their parents' footsteps, while the rest are twice as likely to be civil servants as executives.

7.6 RESULT AFTER n TRANSITIONS

Let us return to Example 1 of Section 7.5 and suppose Willie is free. What is the probability that he will be free in two days? We can examine this situation by looking at what is called a Kemeny tree. Recall the transition matrix is

	FREE	INTERROGATION	JAIL
FREE	.6	.1	.3
INTERROGATION	.4	.2	.4
JAIL	.2	.3	.5

Thus we have

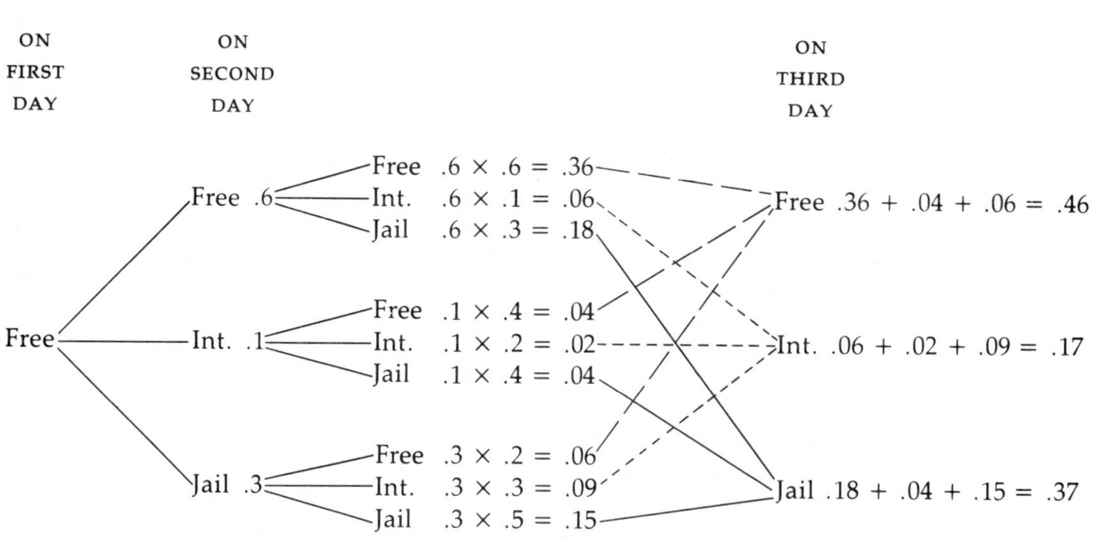

We read along each branch to see how the probabilities change from day to day. For example, since Willie starts out free, he is free on the

second day with probability .6. Then he is free on the second day and under interrogation the third day with probability .6 × .1 = .06 (since we multiply probabilities in this case). But to find the probability he is under interrogation the third day without regard to what happened on the second day, we must add together the three probabilities opposite Int. on the third day. This gives a probability of .17. Similarly, Willie has a 46% chance of being free on the third day and a 37% chance of being in jail. Thus the "current situation" after two days can be represented by the vector

$$(.46, \quad .17, \quad .37).$$

Now, these numbers were obtained by a multiplication and addition process which should be familiar. In fact, notice that

$$(.6 \quad .1 \quad .3)\begin{pmatrix} .6 & .1 & .3 \\ .4 & .2 & .4 \\ .2 & .3 & .5 \end{pmatrix}$$

$$= (.6 \times .6 + .1 \times .4 + .3 \times .2, \quad .6 \times .1 + .1 \times .2 + .3 \times .3, \quad .6 \times .3 + .1 \times .4 + .3 \times .5)$$

$$= (.36 + .04 + .06, \quad .06 + .02 + .09, \quad .18 + .04 + .15)$$

$$= (.46, \quad .17, \quad .37).$$

This represents the situation after two days, so we should get a representation for the situation after three days by computing

$$(.46, .17, .37)\begin{pmatrix} .6 & .1 & .3 \\ .4 & .2 & .4 \\ .2 & .3 & .5 \end{pmatrix}$$

$$= (.276 + .068 + .074 \quad .046 + .034 + .111, \quad .138 + .068 + .185)$$

$$= (.418, .191, .391).$$

This means, of course, that after three days, Willie has a 41.8% chance of being free, a 19.1% chance of being in interrogation, and a 39.1% chance of being in jail.

Notice that if we let

$$A = \begin{pmatrix} .6 & .1 & .3 \\ .4 & .2 & .4 \\ .2 & .3 & .5 \end{pmatrix},$$

then

$$(1, 0, 0)\,A = (1, 0, 0)\begin{pmatrix} .6 & .1 & .3 \\ .4 & .2 & .4 \\ .2 & .3 & .5 \end{pmatrix} = (.6, .1, .3).$$

Moreover, each of the vectors we have obtained can be written in terms of powers of A like this:

After one day: $(.6, .1, .3) = (1, 0, 0)\,A$

After two days: $(.46, .17, .37) = (.6, .1, .3)\,A = [(1, 0, 0)A]\,A = (1, 0, 0)\,A^2$

After three days: $(.418, .191, .391) = (.46, .17, .37)A = [(1, 0, 0)A^2]A = (1, 0, 0)\,A^3$

We would expect that the situation would be represented by

After n days: $(1, 0, 0)\,A^n$

As you expect from the previous discussion by the *current situation vector* we mean a vector whose entries represent the probabilities of being in the respective states. Thus if the current situation vector is $(.25 \quad .15 \quad .60)$, then there is a 25% chance that the first state (whatever it is) holds, a 15% chance that the second state holds, and a 60% chance that the third state holds. We now state our working *principle*.

PRINCIPLE

If v is the current situation vector and A is the transition matrix, then the situation after n transitions will be represented by vA^n.

Notice that once we compute A^n, we can compare the result of having started with a different situation by only making one matrix multiplication. We show the first four powers of the matrix A from Example 1.

$$A = \begin{pmatrix} .6 & .1 & .3 \\ .4 & .2 & .4 \\ .2 & .3 & .5 \end{pmatrix}, \quad A^2 = \begin{pmatrix} .46 & .17 & .37 \\ .40 & .20 & .40 \\ .34 & .23 & .43 \end{pmatrix},$$

$$A^3 = \begin{pmatrix} .418 & .191 & .391 \\ .400 & .200 & .400 \\ .382 & .209 & .409 \end{pmatrix}, \quad A^4 = \begin{pmatrix} .4054 & .1973 & .3973 \\ .4000 & .2000 & .4000 \\ .3946 & .2027 & .4027 \end{pmatrix}.$$

From A^4 we can read that if Willie starts out in jail, he has a 39.46% chance of being free in four days. (We take $(0 \quad 0 \quad 1)\,A^4 = (.3946 \quad .2027 \quad .4027)$.) Notice that each row is very nearly the same.

EXAMPLE

Example 2, Section 7.5, Revisited Recall that the transition matrix

$$B = \begin{pmatrix} .75 & .25 & 0 \\ .25 & .25 & .50 \\ .50 & 0 & .50 \end{pmatrix} = \begin{pmatrix} \frac{3}{4} & \frac{1}{4} & 0 \\ \frac{1}{4} & \frac{1}{4} & \frac{1}{2} \\ \frac{1}{2} & 0 & \frac{1}{2} \end{pmatrix}$$

represents the transitions between the elements N, M, and O during an electron-beam bombardment. We compute to get

$$B^2 = \begin{pmatrix} \frac{5}{8} & \frac{2}{8} & \frac{1}{8} \\ \frac{4}{8} & \frac{1}{8} & \frac{3}{8} \\ \frac{5}{8} & \frac{1}{8} & \frac{2}{8} \end{pmatrix}, \quad B^3 = \begin{pmatrix} \frac{19}{32} & \frac{7}{32} & \frac{6}{32} \\ \frac{19}{32} & \frac{5}{32} & \frac{8}{32} \\ \frac{20}{32} & \frac{6}{32} & \frac{6}{32} \end{pmatrix},$$

$$B^4 = \begin{pmatrix} \frac{76}{128} & \frac{26}{128} & \frac{26}{128} \\ \frac{78}{128} & \frac{24}{128} & \frac{26}{128} \\ \frac{78}{128} & \frac{26}{128} & \frac{24}{128} \end{pmatrix} = \begin{pmatrix} .593750 & .203125 & .203125 \\ .609375 & .187500 & .203125 \\ .609375 & .203125 & .187500 \end{pmatrix}.$$

Thus if we start with one unit of element N, in four minutes 59.375% will be N, 20.3125% will be M, and the same percentage will be O.

Notice that each row is about (.6 .2 .2). What happens in our experiment if we start with an amount that is 60% N, 20% M, and 20% O. After one minute (.6, .2, .2) B would give the distribution of amounts of N, M, and O. But

$$(.6, .2, .2) \begin{pmatrix} .75 & .25 & 0 \\ .25 & .25 & .50 \\ .50 & 0 & .50 \end{pmatrix} = (.6, .2, .2),$$

so the distribution is unchanged. We will have more to say about this phenomenon after you have done some problems.

Exercises 7.6

1. What happens if the current situation vector (.4, .2, .4) is applied to the transition matrix A from Example 1 of Section 7.5?

2. Compute the first four powers of

$$\begin{pmatrix} 0 & 1 \\ 1 & 0 \end{pmatrix}.$$

Can you say anything about higher powers? How about successive powers of

$$\begin{pmatrix} \frac{1}{2} & 0 & \frac{1}{2} \\ 0 & 1 & 0 \\ 0 & \frac{1}{2} & \frac{1}{2} \end{pmatrix}?$$

Each of the following is related to the corresponding problem in Section 7.5.

3. If Larry has just won a squash match from Earl, what are his chances of winning the fourth match from now?

4. The Democrats have just won the local election. Is the chance of their winning the third election from now larger or smaller than 45%?

5. A friend tells me that he arrived late on the third train before the next one. What is the chance that the next one will be on time?

6. Assuming changes in major are allowed each semester (twice a year) and the distribution of majors is currently social science 40%, humanities 40%, and science 20%, what will be the distribution next year after two opportunities to change? How about after three chances? What do you think the distribution of majors will be in the long run? If each area has one-third of the students in it, what will be the distribution after the next opportunity to change major?

7. If our patient has been normal today, what are her chances of being high the day after tomorrow?

8. The last two years have been good years for corn yield. What are the chances for a good yield the year after next? What are the chances for a good or average yield in three years? What are the long run prospects for a good yield?

9. If it rained yesterday in California, what are the chances for a clear day tomorrow? What do you think the chances for a clear day next week are?

10. What are the chances that the grandchild of an executive is also an executive? What is the chance that a plumber's grandchild will be an executive? How about his or her greatgrandchild?

7.7 THE MAIN THEOREM: A 2 × 2 EXAMPLE

We have seen several examples in which it appears that high powers of a matrix have all the rows the same. Moreover, there seem to be some distributions that are left unchanged by the transition matrix. (Individuals may be moved around, but the overall distribution is the same.) In this section we will examine the situation more closely by working out a 2 × 2 example in detail. We will note the relationships that occur in this example and will write the general theorem in the next section.

EXAMPLE 1

Student Help We have noticed that our weekly quizzes cause a curious pattern of seeking help among the students in the class. In particular, we notice that about $\frac{1}{3}$ of the students who seek help before the quiz one week also seek help the next week, while $\frac{5}{6}$ of the students who do not seek help one week *do* seek it the following week.

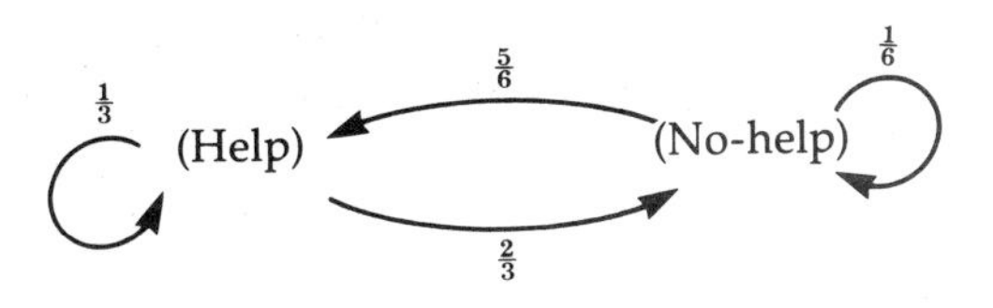

Figure 7.3

If we draw a transition diagram and label our states help and no-help, we would get the diagram shown in Figure 7.3. Notice we are using the fact that either a student gets help or not to get the other percentages. The transition matrix is then

$$\begin{array}{c} \\ \text{HELP} \\ \text{NO-HELP} \end{array} \begin{array}{c} \text{HELP}\quad\text{NO-HELP} \\ \begin{pmatrix} \frac{1}{3} & \frac{2}{3} \\ \frac{5}{6} & \frac{1}{6} \end{pmatrix} \end{array}$$

Our problem is that we do not know how many of our 90 students we should expect to seek help each week, and we would like to know what to expect.

We let

$$A = \begin{pmatrix} \frac{1}{3} & \frac{2}{3} \\ \frac{5}{6} & \frac{1}{6} \end{pmatrix}$$

and compute powers of A. But instead of finding A^2, A^3, A^4, and so on, suppose we compute A^2, A^4, A^8, A^{16}, which we can do by successively squaring. We do this below.

$$A^2 = \begin{pmatrix} \frac{1}{3} & \frac{2}{3} \\ \frac{5}{6} & \frac{1}{6} \end{pmatrix} \begin{pmatrix} \frac{1}{3} & \frac{2}{3} \\ \frac{5}{6} & \frac{1}{6} \end{pmatrix} = \begin{pmatrix} \frac{1}{9} + \frac{10}{18} & \frac{2}{9} + \frac{2}{18} \\ \frac{5}{18} + \frac{5}{36} & \frac{10}{18} + \frac{1}{36} \end{pmatrix} = \begin{pmatrix} \frac{12}{18} & \frac{6}{18} \\ \frac{15}{36} & \frac{21}{36} \end{pmatrix} = \begin{pmatrix} \frac{2}{3} & \frac{1}{3} \\ \frac{5}{12} & \frac{7}{12} \end{pmatrix},$$

$$A^4 = A^2 \cdot A^2 = \begin{pmatrix} \frac{2}{3} & \frac{1}{3} \\ \frac{5}{12} & \frac{7}{12} \end{pmatrix} \begin{pmatrix} \frac{2}{3} & \frac{1}{3} \\ \frac{5}{12} & \frac{7}{12} \end{pmatrix} = \begin{pmatrix} \frac{4}{9} + \frac{5}{36} & \frac{2}{9} + \frac{7}{36} \\ \frac{10}{36} + \frac{35}{144} & \frac{5}{36} + \frac{49}{144} \end{pmatrix} = \begin{pmatrix} \frac{7}{12} & \frac{5}{12} \\ \frac{25}{48} & \frac{23}{48} \end{pmatrix},$$

$$A^8 = A^4 \cdot A^4 = \begin{pmatrix} \frac{7}{12} & \frac{5}{12} \\ \frac{25}{48} & \frac{23}{48} \end{pmatrix} \begin{pmatrix} \frac{7}{12} & \frac{5}{12} \\ \frac{25}{48} & \frac{23}{48} \end{pmatrix} = \begin{pmatrix} \frac{321}{576} & \frac{255}{576} \\ \frac{107}{192} & \frac{85}{192} \end{pmatrix} = \begin{pmatrix} .55729 & .44271 \\ .553385 & .446615 \end{pmatrix},$$

$$A^{16} = A^8 \cdot A^8 = \begin{pmatrix} .5573 & .4427 \\ .5534 & .4466 \end{pmatrix}^2 = \begin{pmatrix} .555561 & .444439 \\ .555546 & .444454 \end{pmatrix}.$$

Realizing that $\frac{5}{9} = .555555 \ldots$ and $\frac{4}{9} = .444444 \ldots$, it appears that the higher the power of A we compute, the more nearly the result looks like

$$\begin{pmatrix} .55555 \ldots & .44444 \ldots \\ .55555 \ldots & .44444 \ldots \end{pmatrix}.$$

That is, A^n seems to be approaching

$$W = \begin{pmatrix} \frac{5}{9} & \frac{4}{9} \\ \frac{5}{9} & \frac{4}{9} \end{pmatrix}.$$

If we let $w = (\frac{5}{9}, \frac{4}{9})$, then every row of W is the same as w. Also notice that

$$wA = (\tfrac{5}{9}, \tfrac{4}{9}) \begin{pmatrix} \frac{1}{3} & \frac{2}{3} \\ \frac{5}{6} & \frac{1}{6} \end{pmatrix} = (\tfrac{5}{27} + \tfrac{20}{54}, \quad \tfrac{10}{27} + \tfrac{4}{54})$$

$$= (\tfrac{30}{54}, \tfrac{24}{54}) = (\tfrac{5}{9}, \tfrac{4}{9}),$$

that is,

$$w \cdot A = w.$$

Let us return now to our original problem. We wanted to know how many of our 90 students we could expect to seek help each week. If x is the percentage who sought help this week, then

$$(x, 1 - x)$$

is the current situation vector. We saw in the last section that the situation after n weeks would be given by

$$(x, 1 - x)A^n.$$

Now we argue this way: If n is very large, then A^n is approximately W, so

$$
\begin{aligned}
(x, 1 - x)A^n &= (x, 1 - x)W \\
&= (x, 1 - x)\begin{pmatrix} \frac{5}{9} & \frac{4}{9} \\ \frac{5}{9} & \frac{4}{9} \end{pmatrix} \\
&= (\tfrac{5}{9}x + \tfrac{5}{9}(1 - x), \quad \tfrac{4}{9}x + \tfrac{4}{9}(1 - x)) \\
&= (\tfrac{5}{9}, \tfrac{4}{9})
\end{aligned}
$$

since the terms containing x cancel out. This shows that *no matter what the current situation is, in the long run the situation tends to become that described by* w. We illustrate this principle by taking (.1, .9) as our current situation and looking at the situation change from week to week.

First week: $(.1, .9)A = (.1, .9)\begin{pmatrix} .333 & .667 \\ .833 & .167 \end{pmatrix} = (.7833, .2167)$

Second week: $(.7833, .2167)A = (.4417, .5583)$

Third week: $(.4417, .5583)A = (.6125, .3875)$

Fourth week: $(.6125, .3875)A = (.527, .473)$

Eighth week: $(.1, .9)A^8 = (.1, .9)\begin{pmatrix} .55729 & .44271 \\ .553385 & .466615 \end{pmatrix} = (.5538, .4462)$

Thus by the eighth week at least (and actually by the sixth week) 55% of the students are coming in for help and this percentage is not changing significantly.

Thus no matter how many students we start out helping, in the long run $\frac{5}{9}$ will be coming in for help, that is, 50 out of the 90 students.

Recapitulating, we found

---

1. Powers A^n approach W, a matrix with all rows the same.
2. If w is one of the equal rows of W, then $wA = w$.
3. No matter what the current situation vector is, in the long run the situation will become that described by w.

EXAMPLE 2 Given the matrix

$$B = \begin{pmatrix} \frac{1}{2} & \frac{1}{2} \\ \frac{1}{3} & \frac{2}{3} \end{pmatrix},$$

find the matrix W which B^n approaches and the vector w for which $wB = B$.

We simply compute powers of B until we can guess the vector w.

(We change to decimals when the fractions become too complicated.)

$$B^2 = \begin{pmatrix} \frac{1}{2} & \frac{1}{2} \\ \frac{1}{3} & \frac{2}{3} \end{pmatrix} \begin{pmatrix} \frac{1}{2} & \frac{1}{2} \\ \frac{1}{3} & \frac{2}{3} \end{pmatrix} = \begin{pmatrix} \frac{1}{4} + \frac{1}{6} & \frac{1}{4} + \frac{2}{6} \\ \frac{1}{6} + \frac{2}{9} & \frac{1}{6} + \frac{4}{9} \end{pmatrix} = \begin{pmatrix} \frac{5}{12} & \frac{7}{12} \\ \frac{7}{18} & \frac{11}{18} \end{pmatrix},$$

$$B^4 = \begin{pmatrix} \frac{5}{12} & \frac{7}{12} \\ \frac{7}{18} & \frac{11}{18} \end{pmatrix} \begin{pmatrix} \frac{5}{12} & \frac{7}{12} \\ \frac{7}{18} & \frac{11}{18} \end{pmatrix} = \begin{pmatrix} .40046 & .59954 \\ .39969 & .60031 \end{pmatrix},$$

$$B^5 = \begin{pmatrix} \frac{1}{2} & \frac{1}{2} \\ \frac{1}{3} & \frac{2}{3} \end{pmatrix} \begin{pmatrix} .40046 & .59954 \\ .39969 & .60031 \end{pmatrix} = \begin{pmatrix} .400075 & .599925 \\ .39995 & .60005 \end{pmatrix}.$$

It looks like $w = (.4, .6) = (\frac{2}{5}, \frac{3}{5})$, so we compute wB and obtain

$$wB = (\tfrac{2}{5}, \tfrac{3}{5}) \begin{pmatrix} \frac{1}{2} & \frac{1}{2} \\ \frac{1}{3} & \frac{2}{3} \end{pmatrix} = (\tfrac{1}{5} + \tfrac{1}{5}, \ \tfrac{1}{5} + \tfrac{2}{5}) = (\tfrac{2}{5}, \tfrac{3}{5}) = w.$$

Thus powers of B must be approaching

$$\begin{pmatrix} .4 & .6 \\ .4 & .6 \end{pmatrix},$$

as we thought.

In the next section we introduce terminology so that we can state the main theorem. We also look at larger examples. But first we think you should try to take powers of some 2×2 matrices and guess what the limiting matrix is.

Exercises 7.7

In problems 1 to 8, compute powers of the given matrix until you can guess the limiting matrix W. Check your guess by taking a row w of W and checking that $wT = w$ for the given matrix T. (Notice that $\frac{1}{9} = .1111 \ldots$ and $\frac{1}{11} = .090909 \ldots$, so for example, $.4444 \ldots = \frac{4}{9}$ and $.272727 \ldots = \frac{3}{11}$.)

1. $T = \begin{pmatrix} \frac{1}{3} & \frac{2}{3} \\ \frac{2}{3} & \frac{1}{3} \end{pmatrix}$

2. $T = \begin{pmatrix} \frac{1}{2} & \frac{1}{2} \\ \frac{1}{6} & \frac{5}{6} \end{pmatrix}$

3. $T = \begin{pmatrix} \frac{2}{3} & \frac{1}{3} \\ \frac{1}{6} & \frac{5}{6} \end{pmatrix}$

4. $T = \begin{pmatrix} \frac{1}{4} & \frac{3}{4} \\ \frac{3}{8} & \frac{5}{8} \end{pmatrix}$

5. $T = \begin{pmatrix} \frac{1}{2} & \frac{1}{2} \\ \frac{1}{7} & \frac{6}{7} \end{pmatrix}$

6. $T = \begin{pmatrix} \frac{1}{8} & \frac{7}{8} \\ \frac{1}{4} & \frac{3}{4} \end{pmatrix}$

7. $T = \begin{pmatrix} \frac{1}{9} & \frac{8}{9} \\ \frac{1}{3} & \frac{2}{3} \end{pmatrix}$

8. $T = \begin{pmatrix} \frac{3}{4} & \frac{1}{4} \\ \frac{2}{3} & \frac{1}{3} \end{pmatrix}$

9. A certain fisherman is seldom lucky at catching fish two days in a row. In fact, if he caught a fish today, he only has a 20% chance of catching a fish when he goes out tomorrow. However, if he did not catch a fish today, he has a $\frac{2}{3}$ chance of catching one tomorrow. In the long run, what is his chance of catching a fish on some day in the future?

10. If Kevin remembers to feed his dog Spot one day, he forgets to do it the next day 60% of the time. However, the day after he forgets to feed Spot, he remembers to feed him 90% of the time (his mother reminds him). On the average, how often does Kevin remember to feed his dog?

11. The Redbirds and the Bluebirds are crosstown rivals. It seems that after the Redbirds win a game, the Bluebirds win the next game $\frac{3}{4}$ of the time, while after the Bluebirds win, the Redbirds win next only $\frac{3}{8}$ of the time. Out of 120 games, about how many would you expect the Bluebirds to win?

12. My calculator seems to malfunction regularly. After it works correctly once, it malfunctions the next time $\frac{1}{3}$ of the time. After it malfunctions, it works the next time $\frac{5}{6}$ of the time. Out of 98 attempts, how many would you expect to be successful?

7.8 THE MAIN THEOREM: THE GENERAL CASE

Now that we have looked at several examples with 2×2 matrices we will make some definitions that will simplify our discussion and we will state the main theorem. Besides another 2×2 example we will illustrate the theorem with some larger matrices. In subsequent sections we will show how to find the limiting matrix without computing powers of the given transition matrix. As is typical for mathematicians, we will try to define some terms precisely so that we agree on what we are talking about.

We have been considering situations involving a process of change. In fact, our examples have had a finite number of possible states, there has been a change from one state to another during some fixed time period, and the probability of going from the ith state to the jth state depends only on i and j and not on how long the process has been going on. Such a process is called a *Markov chain*.

If we let the states be $E_1, E_2, \ldots, E_n$ and the probability of going from E_i to E_j be p_{ij}, then we can represent the Markov chain by the $n \times n$ matrix

$$T = \begin{pmatrix} p_{11} & p_{12} & \cdots & p_{1n} \\ p_{21} & p_{22} & \cdots & p_{2n} \\ \cdot & \cdot & & \cdot \\ \cdot & \cdot & & \cdot \\ \cdot & \cdot & & \cdot \\ p_{n1} & p_{n2} & \cdots & p_{nn} \end{pmatrix}$$

Notice, since the process must be in some state after being in state E_1, say, that the sum of the entries in row one must be 1. This means that with probability 1 the process will be in some state. Similarly for each other row. This motivates the next definitions.

DEFINITION A row vector is called a *probability vector* if it has nonnegative entries that add to 1.

EXAMPLES

$$(\tfrac{1}{2}, \tfrac{1}{2}),\ (0, 1),\ (\tfrac{1}{3}, \tfrac{1}{4}, \tfrac{5}{12}),\ (\tfrac{1}{3}, \tfrac{2}{3}, 0),\ (\tfrac{1}{5}, \tfrac{2}{5}, 0, 0, \tfrac{2}{5})$$

are all probability vectors, but

$$(-\tfrac{1}{2}, \tfrac{1}{3}, \tfrac{1}{2}, \tfrac{2}{3}),\ (0, \tfrac{1}{2}, \tfrac{1}{2}, \tfrac{1}{3}),\ (-\tfrac{1}{2}, 1, \tfrac{1}{2}),\ (0, 1, 1)$$

are not.

DEFINITION A *transition matrix* is a square matrix in which each row is a probability vector; that is, each entry is nonnegative and the entries in any row add to 1.

$$\begin{pmatrix} \frac{1}{2} & \frac{1}{2} \\ 0 & 1 \end{pmatrix}, \quad \begin{pmatrix} \frac{1}{3} & 0 & \frac{2}{3} \\ 0 & \frac{1}{2} & \frac{1}{2} \\ \frac{1}{4} & \frac{3}{4} & 0 \end{pmatrix}, \quad \begin{pmatrix} .66 & .26 & .08 \\ .03 & .76 & .21 \\ .15 & .55 & .30 \end{pmatrix}$$

are transition matrices, but the following are not:

$$\begin{pmatrix} \frac{1}{2} & \frac{2}{3} \\ 0 & 1 \end{pmatrix}, \quad \begin{pmatrix} .667 & .333 \\ .111 & .888 \end{pmatrix}, \quad \begin{pmatrix} \frac{1}{3} & 0 & \frac{2}{3} \\ -\frac{1}{4} & \frac{1}{2} & \frac{3}{4} \\ 0 & \frac{1}{2} & \frac{1}{4} \end{pmatrix}, \quad \begin{pmatrix} .33 & .33 & .33 \\ .50 & .25 & .25 \\ .67 & .33 & 0 \end{pmatrix}.$$

DEFINITION The probability vector w is a *fixed point* of the matrix T when $w = wT$.

EXAMPLE 1 Consider the transition matrix

$$T = \begin{pmatrix} \frac{1}{2} & \frac{1}{2} \\ \frac{1}{4} & \frac{3}{4} \end{pmatrix} = \begin{pmatrix} .50 & .50 \\ .25 & .75 \end{pmatrix}.$$

If $w = (\frac{1}{3}, \frac{2}{3})$, then

$$wT = (\tfrac{1}{3}, \tfrac{2}{3}) \begin{pmatrix} \frac{1}{2} & \frac{1}{2} \\ \frac{1}{4} & \frac{3}{4} \end{pmatrix} = (\tfrac{1}{6} + \tfrac{2}{12}, \tfrac{1}{6} + \tfrac{6}{12})$$

$$= (\tfrac{4}{12}, \tfrac{8}{12})$$

$$= (\tfrac{1}{3}, \tfrac{2}{3}) = w,$$

so that w is a fixed point of T. If we should start with w as the initial probabilities of being in the two states, then the probabilities will never change,

$$w \cdot T^2 = (wT) \cdot T = w \cdot T = w, \quad \text{and so on.}$$

We say that such a process is in *equilibrium*. What happens to powers of T?

We compute powers of T:

$$T^2 = \begin{pmatrix} .3750 & .6250 \\ .3125 & .6875 \end{pmatrix},$$

$$T^3 = \begin{pmatrix} .343750 & .656250 \\ .328125 & .671875 \end{pmatrix},$$

$$T^4 = \begin{pmatrix} .33593750 & .66406250 \\ .33203125 & .66796875 \end{pmatrix}.$$

It looks as if the matrix T^n is becoming more and more like the matrix

$$W = \begin{pmatrix} .333 & .667 \\ .333 & .667 \end{pmatrix} = \begin{pmatrix} \frac{1}{3} & \frac{2}{3} \\ \frac{1}{3} & \frac{2}{3} \end{pmatrix}$$

as n gets larger. In mathematical terminology we say T^n *approaches* W to mean each entry in T^n is getting closer and closer to the corresponding entry in W as n gets larger and larger.

Notice that the rows of W are all alike, and are in fact a fixed point of T. This is the typical situation which we state, along with some other facts, in a theorem. Before doing this we need one more definition.

DEFINITION A transition matrix is called *regular* when some power of it has only positive entries.

EXAMPLES Most of the transition matrices we have encountered have been regular. Notice that those in problem 2 of Section 7.5 are not. Also

$$C = \begin{pmatrix} \frac{1}{2} & \frac{1}{2} & 0 \\ 0 & \frac{1}{2} & \frac{1}{2} \\ 0 & 0 & 1 \end{pmatrix}$$

is not regular since the last row of C^n will always be $(0 \quad 0 \quad 1)$. However,

$$D = \begin{pmatrix} \frac{1}{2} & \frac{1}{2} & 0 & 0 & 0 \\ \frac{1}{3} & 0 & \frac{2}{3} & 0 & 0 \\ 0 & \frac{1}{4} & 0 & \frac{3}{4} & 0 \\ 0 & 0 & \frac{1}{3} & 0 & \frac{2}{3} \\ 0 & 0 & 0 & \frac{1}{2} & \frac{1}{2} \end{pmatrix}$$

is regular. To see this we do not need to compute all the entries in successive powers of D but only be sure they are positive. Consequently, we replace the positive entries in D by X and similarly mark the positive entries in powers of D like this:

$$D = \begin{pmatrix} X & X & 0 & 0 & 0 \\ X & 0 & X & 0 & 0 \\ 0 & X & 0 & X & 0 \\ 0 & 0 & X & 0 & X \\ 0 & 0 & 0 & X & X \end{pmatrix}, \quad D^2 = \begin{pmatrix} X & X & X & 0 & 0 \\ X & X & 0 & X & 0 \\ X & 0 & X & 0 & X \\ 0 & X & 0 & X & X \\ 0 & 0 & X & X & X \end{pmatrix},$$

$$D^3 = \begin{pmatrix} X & X & X & X & 0 \\ X & X & X & 0 & X \\ X & X & 0 & X & X \\ X & 0 & X & X & X \\ 0 & X & X & X & X \end{pmatrix},$$

Now clearly D^4 will have all positive entries. Similarly, if

$$E = \begin{pmatrix} 0 & \frac{1}{2} & \frac{1}{2} \\ 1 & 0 & 0 \\ 1 & 0 & 0 \end{pmatrix},$$

then

$$E^2 = \begin{pmatrix} X & 0 & 0 \\ 0 & X & X \\ 0 & X & X \end{pmatrix}, \quad E^3 = \begin{pmatrix} 0 & X & X \\ X & 0 & 0 \\ X & 0 & 0 \end{pmatrix}, \quad E^4 = \begin{pmatrix} X & 0 & 0 \\ 0 & X & X \\ 0 & X & X \end{pmatrix}$$

and it is clear that E is not regular.

Finally, we can state the fundamental theorem:

THEOREM

If T is a regular transition matrix, then there is a matrix W each of whose rows is the same probability vector w and

1. The powers T^n approach W
2. Each entry in w is positive
3. The vector w is the unique fixed point probability vector of T
4. For any probability vector p, $p \cdot T^n$ approaches w.

(For a proof of these results see *Finite Mathematical Structures* by Kemeny, Mirkil, Snell, and Thompson (Prentice-Hall, Englewood Cliffs, New Jersey, 1959).)

Before continuing with some examples, we want to spell out an important consequence of this theorem. By part 4 of the theorem, no matter what probability vector p we start with, pT^n approaches W, provided T is regular. Thus in a large number of steps, the probability that the process is in the jth state will be very nearly w_j, the jth component of w.

Now recall that when considering independent trials, the probability p of a given outcome, say a, can also be interpreted as giving the fraction of time a will occur when repeating the trial a large number of times. That is, if a coin comes up heads with probability $\frac{1}{3}$, then in 1000 trials we expect about $\frac{1}{3}$ of them to result in heads. In a similar fashion, for regular Markov chains, the entries w_j play the same role. That is, in a large number of transitions the chain will be in the jth state about w_j of the time.

Let us return to Example 1 and see what the theorem tells us.

EXAMPLE 1

Revisited Here the matrix

$$T = \begin{pmatrix} \frac{1}{2} & \frac{1}{2} \\ \frac{1}{4} & \frac{3}{4} \end{pmatrix}$$

is clearly regular. We saw how powers of T approached

$$W = \begin{pmatrix} \frac{1}{3} & \frac{2}{3} \\ \frac{1}{3} & \frac{2}{3} \end{pmatrix}.$$

Then $w = (\frac{1}{3}, \frac{2}{3})$ and we have all parts of the theorem except 3 and 4. To show 3 we consider any other probability vector $x = (q, \quad 1 - q)$ where $0 < q < 1$. Then, if x is a fixed point of T,

$$x = x \cdot T,$$

so

$$\begin{aligned} (q, \quad 1 - q) &= (q, \quad 1 - q) \cdot \begin{pmatrix} \frac{1}{2} & \frac{1}{2} \\ \frac{1}{4} & \frac{3}{4} \end{pmatrix} \\ &= (q/2 + \tfrac{1}{4} - q/4, q/2 + \tfrac{3}{4} - 3q/4) \\ &= (q/4 + \tfrac{1}{4}, \quad \tfrac{3}{4} - q/4). \end{aligned}$$

Equating corresponding terms we get

$$\begin{aligned} q &= q/4 + \tfrac{1}{4}, \\ 1 - q &= \tfrac{3}{4} - q/4, \end{aligned}$$

or

$$\begin{aligned} 3q/4 &= \tfrac{1}{4}, \\ \tfrac{1}{4} &= \tfrac{3}{4}q, \end{aligned}$$

and both yield $q = \frac{1}{3}$. Thus $x = (\frac{1}{3}, \frac{2}{3}) = w$, so w is the only fixed point of T.

Finally, to illustrate part 4, if $p = (.9, .1)$ then

$$pT = (.9 \quad .1) \begin{pmatrix} .5 & .5 \\ .25 & .75 \end{pmatrix} = (.475, .525),$$

$$pT^2 = (pT) \cdot T = (.475, .525) \begin{pmatrix} .5 & .5 \\ .25 & .75 \end{pmatrix} = (.36875, .63125),$$

$$\begin{aligned} pT^3 &= (.36875, \quad .63125) \begin{pmatrix} .5 & .5 \\ .25 & .75 \end{pmatrix} \\ &= (.9, .1) \begin{pmatrix} .343750 & .656250 \\ .328125 & .671875 \end{pmatrix} \quad \text{(by previous Example 1)} \\ &= (.3421875, .6578125), \end{aligned}$$

and you can see that pT^n is approaching $(\frac{1}{3}, \frac{2}{3})$.

EXAMPLE 2 In this example, let

$$T = \begin{pmatrix} \frac{1}{2} & \frac{1}{2} & 0 \\ 0 & \frac{1}{2} & \frac{1}{2} \\ \frac{1}{3} & \frac{1}{3} & \frac{1}{3} \end{pmatrix}.$$

Since T^2 has all nonzero entries, T is regular. We look for w by letting

$$w = (x, y, z) \qquad \text{with } x + y + z = 1$$

and suppose $wT = w$. Then

$$(x, y, z) \begin{pmatrix} \frac{1}{2} & \frac{1}{2} & 0 \\ 0 & \frac{1}{2} & \frac{1}{2} \\ \frac{1}{3} & \frac{1}{3} & \frac{1}{3} \end{pmatrix} = (x, y, z)$$

So we get a system of four equations in three unknowns:

$$\begin{aligned} \tfrac{1}{2}x \phantom{{}+\tfrac{1}{2}y} + \tfrac{1}{3}z &= x, \\ \tfrac{1}{2}x + \tfrac{1}{2}y + \tfrac{1}{3}z &= y, \\ \tfrac{1}{2}y + \tfrac{1}{3}z &= z, \\ x + y + z &= 1, \end{aligned}$$

the first three from multiplying the previous equation and equating entries and the last one from the fact that w is a probability vector. The first and third give

$$\begin{aligned} \tfrac{1}{3}z &= \tfrac{1}{2}x, \\ \tfrac{1}{2}y &= \tfrac{2}{3}z, \end{aligned}$$

or

$$\begin{aligned} 2z &= 3x, \\ 3y &= 4z, \end{aligned}$$

Hence

$$3y = 4z = 6x \quad \text{or} \quad y = 2x, \qquad z = \tfrac{3}{2}x.$$

Substituting into the last equation we get

$$1 = x + y + z = x + 2x + \tfrac{3}{2}x = \tfrac{9}{2}x$$

so

$$x = \tfrac{2}{9}$$
$$y = \tfrac{4}{9}$$
$$z = \tfrac{3}{9} \qquad (\text{or } \tfrac{1}{3}).$$

The theorem then asserts that T^n approaches

$$W = \begin{pmatrix} \frac{2}{9} & \frac{4}{9} & \frac{3}{9} \\ \frac{2}{9} & \frac{4}{9} & \frac{3}{9} \\ \frac{2}{9} & \frac{4}{9} & \frac{3}{9} \end{pmatrix}$$

and the fixed point of T is $w = (\frac{2}{9}, \frac{4}{9}, \frac{3}{9}) = (.222, .444, .333)$.

EXAMPLE 3

The weather on our island paradise is not the best. If it is clear one day, it is equally likely to be clear or cloudy on the next, but it never rains. If it is cloudy, the next day it is rainy or again cloudy, each equally likely. If it rains one day, then it can be clear or cloudy or rainy the next day with equal likelihood. About what percentage of the days is it clear and what percentage is it cloudy? In a month, how many rainy days do you expect?

Solution We draw a transition graph in Figure 7.4 and then translate it to a transition matrix

	CLEAR	CLOUDY	RAINY
CLEAR	$\frac{1}{2}$	$\frac{1}{2}$	0
CLOUDY	0	$\frac{1}{2}$	$\frac{1}{2}$
RAINY	$\frac{1}{3}$	$\frac{1}{3}$	$\frac{1}{3}$

Notice that this is the same as T from Example 2. Thus $w = (\frac{2}{9}, \frac{4}{9}, \frac{3}{9})$ is the fixed point for this Markov chain. We expect it to be clear $\frac{2}{9}$ of the time, cloudy for $\frac{4}{9}$, and rainy for $\frac{1}{3}$. In a typical month, since 30 (the number of

Figure 7.4

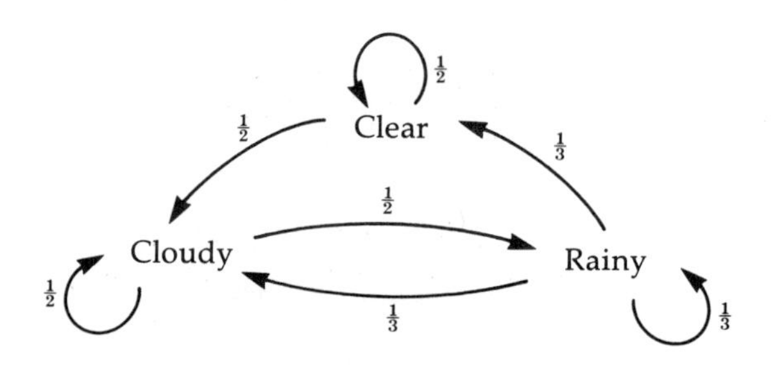

days in a month) is fairly large, we would expect it to be rainy on 10 days.

Exercises 7.8

1. Find the fixed point for the matrix

$$\begin{pmatrix} \frac{1}{2} & 0 & \frac{1}{2} \\ 1 & 0 & 0 \\ \frac{1}{4} & \frac{1}{4} & \frac{1}{2} \end{pmatrix}.$$

   (Did you check to see if it is regular?)

2. Show that (0, 1, 0) is a fixed point for the matrix

$$A = \begin{pmatrix} 0 & 1 & 0 \\ 0 & 1 & 0 \\ \frac{1}{2} & 0 & \frac{1}{2} \end{pmatrix}.$$

   Is A a regular matrix?

Exercises 3–10 refer to the exercises with the same number in Section 7.5.

3. In the long run, say out of 100 games, how many will Earl win? What is the fixed point for this transition matrix?
4. What percentage of the elections can the Democrats expect to win in the next 30 years?
5. What percentage of the time are trains late, in the long run?
6. In the long run, how many students do you expect to be declared for each major if there is a constant collegiate enrollment of 12,000 students?
7. During a typical 30-day month, how often will our patient be normal and how often will she be low?
8. During the next 50 years, how often will the Illinois corn yield be average and how often good?
9. How often does it rain in California (according to our data)?
10. In Utopia, what percentage of the adults do you expect to be in each of the professions plumbers, executives, and civil servants?

11. By considering the data given in problem 9 of Section 7.5, determine how often it is clear (that is, what percentage of the time).

12. Solve the problem in Example 7 of Section 7.1.

The theorem in this section shows that there are two basic ways to find the fixed point of a regular transition matrix. The first would be as we have in this section, by solving a system of equations. The second would be to compute powers of the matrix until the rows of two successive powers are the same.

As a theoretical matter, this latter situation may never occur. However, as a practical matter any computational device will have to round off eventually and, for large enough n, there will come a time when the computation of T^{n+1} will be the same as the computation of T^n. As a practical matter, w will be any row of one of these matrices.

If you have access to a computer and the matrices are not too large, this is probably a most practical method.

We present some problems here for solution by this method. In the next section we consider how to find a fixed point by the first method, that of solving a system of equations.

Find the fixed point of each of the following matrices (if it exists).

13. $$\begin{pmatrix} \frac{1}{2} & \frac{1}{2} & 0 & 0 & 0 \\ \frac{1}{3} & 0 & \frac{2}{3} & 0 & 0 \\ 0 & \frac{1}{2} & 0 & \frac{1}{2} & 0 \\ 0 & 0 & \frac{2}{3} & 0 & \frac{1}{3} \\ 0 & 0 & 0 & \frac{1}{2} & \frac{1}{2} \end{pmatrix}$$

14. $$\begin{pmatrix} \frac{1}{3} & \frac{2}{3} & 0 & 0 & 0 & 0 \\ \frac{1}{3} & 0 & \frac{2}{3} & 0 & 0 & 0 \\ 0 & \frac{1}{3} & 0 & \frac{2}{3} & 0 & 0 \\ 0 & 0 & \frac{1}{3} & 0 & \frac{2}{3} & 0 \\ 0 & 0 & 0 & \frac{1}{3} & 0 & \frac{2}{3} \\ 0 & 0 & 0 & 0 & \frac{1}{3} & \frac{2}{3} \end{pmatrix}$$

15. $$\begin{pmatrix} 0 & 1 & 0 & 0 & 0 \\ \frac{1}{2} & 0 & \frac{1}{2} & 0 & 0 \\ 0 & \frac{1}{2} & 0 & \frac{1}{2} & 0 \\ 0 & 0 & \frac{1}{2} & 0 & \frac{1}{2} \\ 0 & 0 & 0 & 1 & 0 \end{pmatrix}$$

16. View your social security number as *abc-de-fghi* and let

$$T = \begin{pmatrix} .abc & 1 - .abc & 0 \\ 0 & .def & 1 - .def \\ .ghi & 1 - .ghi & 0 \end{pmatrix}.$$

Find T^n for large n and deduce the fixed point of your matrix.

17. In a certain Northwest city the day after a snowy day it is rainy 50% of the time and cloudy 50% of the time. It never snows the day after a rainy day, but it is equally likely to be rainy, clear, or cloudy. The day after a cloudy day can be any of snowy, rainy, clear, or cloudy all with probability $\frac{1}{4}$. Finally, after a clear day it never rains, but it is clear, cloudy, or snowy with probabilities $\frac{1}{6}$, $\frac{1}{3}$, and $\frac{1}{2}$ respectively. In the long run, how many clear days are there in a month? What is the fixed point of the transition matrix associated with this Markov chain?

18. Find the fixed point of the matrix

$$\begin{pmatrix} \frac{1}{2} & \frac{3}{10} & \frac{1}{5} & 0 \\ 0 & 0 & \frac{1}{2} & \frac{1}{2} \\ \frac{1}{3} & 0 & \frac{1}{3} & \frac{1}{3} \\ \frac{1}{4} & \frac{3}{4} & 0 & 0 \end{pmatrix}$$

7.9 SOLVING SYSTEMS OF EQUATIONS

In this section we show how to use matrices to solve the systems of equations that arise in using the equation $wT = w$ to find the fixed point of a regular matrix T. Readers are urged to read Chapter V, Section 5 of *Introduction to Finite Mathematics*, 3rd ed., by J. Kemeny, J. Snell, and G. Thompson (Prentice-Hall, Englewood Cliffs, New Jersey, 1974) for a more complete account of the theory of systems of linear equations. We will only consider the case of m equations in m unknowns.

For the special case of three equations in three unknowns, there is a simple geometric realization (Figure 7.5). Each equation represents a plane in Euclidean 3-space and a solution is a point on the intersection of the three planes—if there is an intersection.

We will only illustrate the case when there is a unique solution since in the application to Markov chains we know this is the case. Recall that in Chapter 1 we solved systems of two equations in two unknowns. The technique here is similar to one of the methods there. We will first work out an example in detail, analyzing the different

Figure 7.5

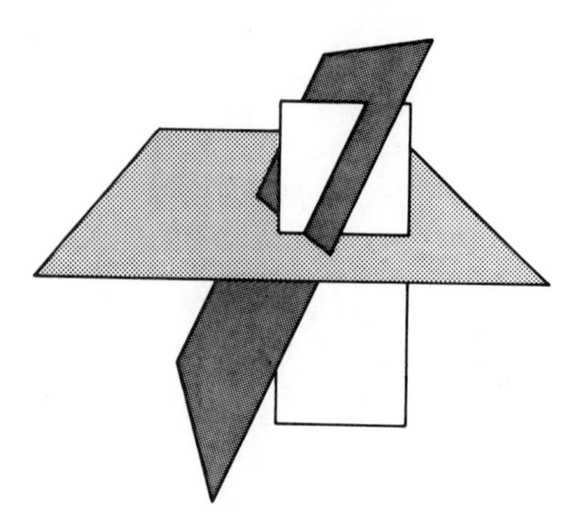

steps. Then we will show how to do the problem using matrices to simplify the method.

EXAMPLE 1

Solve the system of equations

$$\begin{aligned} 2x + 2y + 4z &= 2, \\ x + 2y &= -4, \\ 2x + 3y + 5z &= 3. \end{aligned}$$

First make the top equation have a coefficient of x equal to 1, then by subtracting multiples of the first equation from the others eliminate the x terms from the other equations:

$$\begin{aligned} x + 2y &= -4, \quad \text{(Interchange Eqs. 1 and 2)} \\ 2x + 2y + 4z &= 2, \\ 2x + 3y + 5z &= 3; \end{aligned}$$

$$\begin{aligned} x + 2y &= -4, \\ -2y + 4z &= 10, \quad \text{(Multiply Eq. 1 by } -2 \text{ and add to Eq. 2)} \\ -y + 5z &= 11. \quad \text{(Multiply Eq. 1 by } -2 \text{ and add to Eq. 3)} \end{aligned}$$

Now we multiply Equation 2 by $-\frac{1}{2}$ to obtain 1 as a coefficient and then by adding to Equation 3 and adding -2 times it to Equation 1 we get

$$\begin{aligned} x + 2y \qquad &= -4, \\ y - 2z &= -5, \\ -\ y + 5z &= 11, \end{aligned}$$

or

$$\begin{aligned} x \quad + 4z &= 6, \\ y - 2z &= -5, \\ 3z &= 6. \end{aligned}$$

Finally, multiplying the last equation by $\frac{1}{3}$ and using this to eliminate z from the other equations, we get

$$\begin{aligned} x \qquad &= -2, \\ y \quad &= -1, \\ z &= 2. \end{aligned}$$

Notice there were three types of changes:

1. Interchange two equations.
2. Multiply an equation by a nonzero constant.
3. Multiply one equation by a constant and add it to another equation.

We now simplify our problem by changing the equations into a matrix, simply by erasing the variables and the equal signs. We get the matrix

$$\begin{pmatrix} 2 & 2 & 4 & 2 \\ 1 & 2 & 0 & -4 \\ 2 & 3 & 5 & 3 \end{pmatrix}.$$

Notice that a zero is placed where a variable in an equation is missing. Now we "operate" on this matrix by using the *elementary row operations.*

1. Interchange two rows.
2. Multiply each entry in a row by a nonzero constant.
3. Multiply one row by a constant and add it to another row (viewing the rows as vectors).

The aim is to obtain a matrix of the form

$$\begin{pmatrix} 1 & 0 & 0 & a \\ 0 & 1 & 0 & b \\ 0 & 0 & 1 & c \end{pmatrix}$$

from which we get $x = a$, $y = b$, and $z = c$.

In order to show how one matrix is obtained from another by means of these operations, we will use the following abbreviations:

1. $\boldsymbol{i} \leftrightarrow \boldsymbol{j}$ means interchange row i and row j.
2. $k\boldsymbol{i}$ means multiply row i by k.
3. $k\boldsymbol{j} + \boldsymbol{i}$ means multiply row j by k and add the result to row i.

The sequence of matrices that correspond to the changes made in the system of equations during the process of solution would look like this.

$$\begin{pmatrix} 2 & 2 & 4 & 2 \\ 1 & 2 & 0 & -4 \\ 2 & 3 & 5 & 3 \end{pmatrix} \xrightarrow[\mathbf{1} \leftrightarrow \mathbf{2}]{} \begin{pmatrix} 1 & 2 & 0 & -4 \\ 2 & 2 & 4 & 2 \\ 2 & 3 & 5 & 3 \end{pmatrix} \xrightarrow[\substack{-2\mathbf{1} + \mathbf{2} \\ -2\mathbf{1} + \mathbf{3}}]{} \begin{pmatrix} 1 & 2 & 0 & -4 \\ 0 & -2 & 4 & 10 \\ 0 & -1 & 5 & 11 \end{pmatrix}$$

$$\xrightarrow[-\frac{1}{2}\mathbf{2}]{} \begin{pmatrix} 1 & 2 & 0 & -4 \\ 0 & 1 & -2 & -5 \\ 0 & -1 & 5 & 11 \end{pmatrix} \xrightarrow[\substack{-2\mathbf{2} + \mathbf{1} \\ \mathbf{2} + \mathbf{3}}]{} \begin{pmatrix} 1 & 0 & 4 & 6 \\ 0 & 1 & -2 & -5 \\ 0 & 0 & 3 & 6 \end{pmatrix} \xrightarrow[\frac{1}{3}\mathbf{3}]{} \begin{pmatrix} 1 & 0 & 4 & 6 \\ 0 & 1 & -2 & 5 \\ 0 & 0 & 1 & 2 \end{pmatrix}$$

$$\xrightarrow[\substack{2\mathbf{3} + \mathbf{2} \\ -4\mathbf{3} + \mathbf{1}}]{} \begin{pmatrix} 1 & 0 & 0 & -2 \\ 0 & 1 & 0 & -1 \\ 0 & 0 & 1 & 2 \end{pmatrix}$$

As you can see the main saving is in not having to write down x, y, z and $=$ all the time. Also, the arithmetic seems to be easier to keep track of. (By the way, you should check that $x = -2, y = -1, z = 2$ actually is a solution of the original system.)

EXAMPLE 2 Solve the system of equations

$$\begin{aligned} x - y - z &= 1, \\ 2x \qquad - 3z &= 1, \\ 2x - 3y + z &= 10. \end{aligned}$$

Our work looks like this.

$$\begin{pmatrix} 1 & -1 & -1 & 1 \\ 2 & 0 & -3 & 1 \\ 2 & -3 & 1 & 10 \end{pmatrix} \xrightarrow[\substack{(-2)1 + 2 \\ (-2)1 + 3}]{} \begin{pmatrix} 1 & -1 & -1 & 1 \\ 0 & 2 & -1 & -1 \\ 0 & -1 & 3 & 8 \end{pmatrix} \xrightarrow[2 \leftrightarrow 3]{} \begin{pmatrix} 1 & -1 & -1 & 1 \\ 0 & -1 & 3 & 8 \\ 0 & 2 & -1 & -1 \end{pmatrix}$$

$$\xrightarrow[(-1)\,2]{} \begin{pmatrix} 1 & -1 & -1 & 1 \\ 0 & 1 & -3 & -8 \\ 0 & 2 & -1 & -1 \end{pmatrix} \xrightarrow[\substack{2 + 1 \\ -22 + 3}]{} \begin{pmatrix} 1 & 0 & -4 & -7 \\ 0 & 1 & -3 & -8 \\ 0 & 0 & 5 & 15 \end{pmatrix} \xrightarrow[\frac{1}{5}3]{} \begin{pmatrix} 1 & 0 & -4 & -7 \\ 0 & 1 & -3 & -8 \\ 0 & 0 & 1 & 3 \end{pmatrix}$$

$$\xrightarrow[\substack{33 + 2 \\ 43 + 1}]{} \begin{pmatrix} 1 & 0 & 0 & 5 \\ 0 & 1 & 0 & 1 \\ 0 & 0 & 1 & 3 \end{pmatrix}.$$

Therefore,

$$x = 5,$$
$$y = 1,$$
$$z = 3.$$

Finally, we use this method to find the fixed point of a transition matrix.

EXAMPLE 3 Find the fixed point of the transition matrix.

$$T = \begin{pmatrix} .2 & .6 & .2 \\ .4 & .1 & .5 \\ .4 & 0 & .6 \end{pmatrix}.$$

We first must convert to equations using $wT = w$; thus

$$(x, y, z) \begin{pmatrix} .2 & .6 & .2 \\ .4 & .1 & .5 \\ .4 & 0 & .6 \end{pmatrix} = (x, y, z)$$

becomes

$$.2x + .4y + .4z = x,$$
$$.6x + .1y \qquad = y,$$
$$.2x + .5y + .6z = z,$$

so

$$\begin{aligned} -.8x + .4y + .4z &= 0, \\ .6x - .9y \qquad &= 0, \\ .2x + .5y - .4z &= 0. \end{aligned}$$

Notice that the sum of any two of these equations is the negative of the third equation. This will *always* be true (unless you have made a mistake) for transition matrices. Thus take two of these equations and the equation which asserts that w is a probability vector, and our system is

$$\begin{aligned} x + \quad y + \quad z &= 1, \\ -.8x + .4y + .4z &= 0, \\ .6x - .9y \qquad &= 0. \end{aligned}$$

Thus we consider

$$\begin{pmatrix} 1 & 1 & 1 & 1 \\ -.8 & .4 & .4 & 0 \\ .6 & -.9 & 0 & 0 \end{pmatrix} \xrightarrow[\substack{102 \\ 103}]{} \begin{pmatrix} 1 & 1 & 1 & 1 \\ -8 & 4 & 4 & 0 \\ 6 & -9 & 0 & 0 \end{pmatrix}$$

$$\xrightarrow[\substack{81 + 2 \\ -61 + 3}]{} \begin{pmatrix} 1 & 1 & 1 & 1 \\ 0 & 12 & 12 & 8 \\ 0 & -15 & -6 & -6 \end{pmatrix} \xrightarrow[\frac{1}{12}2]{} \begin{pmatrix} 1 & 1 & 1 & 1 \\ 0 & 1 & 1 & \frac{2}{3} \\ 0 & -15 & -6 & -6 \end{pmatrix}$$

$$\xrightarrow[\substack{-12 + 1 \\ 152 + 3}]{} \begin{pmatrix} 1 & 0 & 0 & \frac{1}{3} \\ 0 & 1 & 1 & \frac{2}{3} \\ 0 & 0 & 9 & 4 \end{pmatrix} \xrightarrow[\frac{1}{9}3]{} \begin{pmatrix} 1 & 0 & 0 & \frac{1}{3} \\ 0 & 1 & 1 & \frac{2}{3} \\ 0 & 0 & 1 & \frac{4}{9} \end{pmatrix}$$

$$\xrightarrow[-3 + 2]{} \begin{pmatrix} 1 & 0 & 0 & \frac{1}{3} \\ 0 & 1 & 0 & \frac{2}{9} \\ 0 & 0 & 1 & \frac{4}{9} \end{pmatrix}$$

so

$$w = (\tfrac{1}{3}, \tfrac{2}{9}, \tfrac{4}{9}).$$

There is still some arithmetic and working with fractions. You cannot avoid fractions using this method. The approach used in problems 13–18 in the exercises of Section 7.8 should be used if you must avoid fractions.

EXAMPLE 4 As a last example, we will redo Example 2 of Section 7.8 by this method. The problem is to find the fixed vector of

$$T = \begin{pmatrix} \frac{1}{2} & \frac{1}{2} & 0 \\ 0 & \frac{1}{2} & \frac{1}{2} \\ \frac{1}{3} & \frac{1}{3} & \frac{1}{3} \end{pmatrix}$$

From $wT = w$, we get

$$\begin{aligned} \tfrac{1}{2}x \qquad\quad + \tfrac{1}{3}z &= x, \\ \tfrac{1}{2}x + \tfrac{1}{2}y + \tfrac{1}{3}z &= y, \\ \tfrac{1}{2}y + \tfrac{1}{3}z &= z, \end{aligned}$$

or

$$\begin{aligned} -\tfrac{1}{2}x \qquad\quad + \tfrac{1}{3}z &= 0, \\ \tfrac{1}{2}x - \tfrac{1}{2}y + \tfrac{1}{3}z &= 0, \\ \tfrac{1}{2}y - \tfrac{2}{3}z &= 0. \end{aligned}$$

We take the first two and $x + y + z = 1$, and so consider the matrix.

$$\begin{pmatrix} 1 & 1 & 1 & 1 \\ -\frac{1}{2} & 0 & \frac{1}{3} & 0 \\ \frac{1}{2} & -\frac{1}{2} & \frac{1}{3} & 0 \end{pmatrix} \xrightarrow[\substack{62 \\ 63}]{} \begin{pmatrix} 1 & 1 & 1 & 1 \\ -3 & 0 & 2 & 0 \\ 3 & -3 & 2 & 0 \end{pmatrix} \xrightarrow[\substack{31 + 2 \\ -31 + 3}]{\qquad\qquad} \begin{pmatrix} 1 & 1 & 1 & 1 \\ 0 & 3 & 5 & 3 \\ 0 & -6 & -1 & -3 \end{pmatrix}$$

$$\xrightarrow[\frac{1}{3}2]{} \begin{pmatrix} 1 & 1 & 1 & 1 \\ 0 & 1 & \frac{5}{3} & 1 \\ 0 & -6 & -1 & -3 \end{pmatrix} \xrightarrow[\substack{-2 + 1 \\ 62 + 3}]{\qquad\qquad} \begin{pmatrix} 1 & 0 & -\frac{2}{3} & 0 \\ 0 & 1 & \frac{5}{3} & 1 \\ 0 & 0 & 9 & 3 \end{pmatrix} \xrightarrow[\frac{1}{9}3]{} \begin{pmatrix} 1 & 0 & -\frac{2}{3} & 0 \\ 0 & 1 & \frac{5}{3} & 1 \\ 0 & 0 & 1 & \frac{1}{3} \end{pmatrix}$$

$$\xrightarrow[\substack{-\frac{5}{3}3 + 2 \\ \frac{2}{3}3 + 1}]{\qquad\qquad} \begin{pmatrix} 1 & 0 & 0 & \frac{2}{9} \\ 0 & 1 & 0 & \frac{4}{9} \\ 0 & 0 & 1 & \frac{1}{3} \end{pmatrix}$$

Therefore

$$w = (\tfrac{2}{9}, \tfrac{4}{9}, \tfrac{1}{3}) \text{ as before.}$$

Exercises 7.9

1. Solve the following systems of equations by using matrices.

(a)
$$\begin{aligned} x + y - 4z &= -4 \\ 2x + 3y - 2z &= -4 \\ x - 2y + z &= 7 \end{aligned}$$

(b)
$$\begin{aligned} 3x - 2y - 4z &= 3 \\ -2x + 6y &= 2 \\ x - 4y + 2z &= 1 \end{aligned}$$

(c)
$$\begin{aligned} 4x + 3y + 2z &= 12 \\ 2x + y - 4z &= 2 \\ x - 7z &= -3 \end{aligned}$$

(d)
$$\begin{aligned} 2x + 4y - z &= 3 \\ x - 3y + 3z &= 6 \\ 3x - 3y + 2z &= 7 \end{aligned}$$

2. Solve the system of equations

$$\begin{aligned} x + 3z - w &= 3, \\ 2x - 3y - z + 2w &= -5, \\ x + 2y + 2z - 3w &= 5, \\ 3x - y + 7z - 3w &= 5. \end{aligned}$$

3. Find the fixed point of the transition matrix.

$$\begin{pmatrix} \frac{1}{2} & 0 & \frac{1}{2} \\ \frac{1}{2} & \frac{1}{2} & 0 \\ \frac{1}{4} & \frac{1}{4} & \frac{1}{2} \end{pmatrix}$$

4. Find the fixed point, by using matrices, of the transition matrix in problem 16, Section 7.8.

5. Solve the problem in Example 2 of Section 7.1.

6–10. Find the fixed point for the transition matrix in the problem from Section 7.5 with the same number.

11. Find the fixed points for the regular transition matrices of problems 13–15 in Section 7.8.

12. Find the fixed point for the matrix of problem 17 in Section 7.8.

13. Find the fixed point for the matrix of problem 18 in Section 7.8.

14. Solve the problem in Example 5 of Section 7.1.

7.10 TWO EXTENSIONS

In this section we look at (1) possible extensions, an illustration of which is given in Example 8, Section 7.1, and (2) absorbing Markov chains.

Analyzing A Long Chain

Recall that a Markov chain is a changing process which at any time is in one of a finite number of states and that the probability of going from state i to state j depends only on i and j and not on how long the process has been going. In processes that occur naturally, such as the type of weather on a given day, we are free to choose what we will call a *state*. In past examples we have taken the weather on a single day as our state. However, observation shows that it is better to look at two-day periods. We give an example to illustrate what we mean.

EXAMPLE 1

We assume a South Sea island on which it is either sunny or rainy. Over a long period of time, the percentage of sunny days and rainy days will not be as instructive as the likelihood of change.

We consequently look at two-day periods, Letting S denote sunny and R rainy, a typical week (Sunday through Sunday) might be

S S R S S S R S.

Our states would be SS, SR, RS, and RR and the sequence of states for the week would be

SS to SR to RS to SS to SS to SR to RS.

Notice that the state RR did not occur since there were not two rainy days in a row. Also notice that the first letter of a state must be the same as the last letter of the previous state since they both represent the weather on the same day.

S S S S S S S
R R R R S S S
S S S R R R R
R S S S S S S
S R S

Now, we have been keeping track of the weather on the island paradise for the last month and it has been as given in the accompanying box. We have also counted the number of times the weather changed from SS to SS and SS to SR, and so forth.

NUMBER OF CHANGES

	SS	SR	RS	RR
SS	14	3	0	0
SR	0	0	1	2
RS	2	0	0	0
RR	0	0	2	5

Now we want to predict the weather based on this presumably typical month. First notice that if we just look at the weather today and ask about tomorrow, we would have states S and R, and this table of changes:

FROM	TO	
	S	R
S	17	3
R	3	7

This would give the transition matrix

$$\begin{pmatrix} \frac{17}{20} & \frac{3}{20} \\ \frac{3}{10} & \frac{7}{10} \end{pmatrix}$$

with fixed point $(\frac{2}{3}, \frac{1}{3})$. So we could say that it is raining about $\frac{1}{3}$ of the time. In fact, it doesn't matter what today is, in the long run it will be rainy tomorrow with probability $\frac{1}{3}$. This is not very helpful, really.

Suppose we look at the four-state Markov chain with transition matrix

$$\begin{array}{c} \\ \text{SS} \\ \text{SR} \\ \text{RS} \\ \text{RR} \end{array} \begin{array}{c} \begin{array}{cccc} \text{SS} & \text{SR} & \text{RS} & \text{RR} \end{array} \\ \begin{pmatrix} \frac{14}{17} & \frac{3}{17} & 0 & 0 \\ 0 & 0 & \frac{1}{3} & \frac{2}{3} \\ 1 & 0 & 0 & 0 \\ 0 & 0 & \frac{2}{7} & \frac{5}{7} \end{pmatrix} \end{array}$$

which we get from the boxed table. The cube of this matrix has all positive entries, so the matrix is regular. Moreover, its fixed point is

$$(\tfrac{17}{30}, \tfrac{3}{30}, \tfrac{3}{30}, \tfrac{7}{30}).$$

Now we see several things emerging. If we have had two sunny days in a row, the probability of another sunny day is $\frac{14}{17}$ or 82%. In like manner, if it is rainy today, it will clear up and be sunny tomorrow with only probability .3. Our ability to predict is considerably improved.

EXAMPLE 2 Let us consider Example 8 from Section 7.1. On the basis of the information given, we could make this table:

	TORY	LABOUR
TORY	?	20
LABOUR	15	?

However, we don't know how many times either party was able to win twice in a row. Consequently, we cannot predict anything. Moreover, it is almost certain that the given data is wrong. (If you think about it, the number of changes each way must be nearly the same). If our political science student had the sequence of election results we could give a prediction based on an analysis like that in the previous example. The exercises let you do that.

Exercises 7.10

1. The last-half century or so of results in a particular district showed these results, where T denotes Tory and L denotes Labour.

   T L L T T T L T L L T L L L T L T T T L

   T T T L L T L L L T T L L T L L L L L T T

   L T L L T L T L L T

   (a) Find the fixed point for the transition matrix of the Markov chain with two states T and L where you assume the above data are typical.

   (b) Do the same for the four state Markov chain with states TT, TL, LT, and LL.

2. Hugh and Larry have a long-time tennis rivalry and in looking back at the sets they have played it is noted that wins occurred in the following sequence:

   L H H L L L H L H H

   L L L H H L H L L H

   L H L L L H H L H L

   H L H L H H L H L H

   L H L L H L H L H L L H

   (a) In the long run, what percentage of the games will Larry win? Does $\frac{28}{52}$ represent a good approximation to his expected percentage of wins? (Use a two-state Markov chain.)

   (b) Now analyze the four-state Markov chain (with states LL, LH, HL, HH) and determine what the chances are that if Larry has won the last two sets that he will win the next set as well.

Absorbing Markov Chains

To introduce this section we will first consider, as is our wont, an example.

EXAMPLE 1

Imagine a four-state Markov chain a_1, a_2, a_3, a_4 in which a_1 and a_4 are *absorbing*, that is, it is impossible to leave the state once there. Suppose also that from a_3 it is equally likely to stay at a_3 or go to a_4, while from a_2 it is possible to go to a_1 or a_3 each with probability $\frac{1}{4}$ and impossible to go directly to a_4.

Then the transition matrix looks like this:

$$T = \begin{pmatrix} 1 & 0 & 0 & 0 \\ \frac{1}{4} & \frac{1}{2} & \frac{1}{4} & 0 \\ 0 & 0 & \frac{1}{2} & \frac{1}{2} \\ 0 & 0 & 0 & 1 \end{pmatrix}.$$

Notice that powers of T have the shape:

$$\begin{pmatrix} X & 0 & 0 & 0 \\ X & X & X & X \\ 0 & 0 & X & X \\ 0 & 0 & 0 & X \end{pmatrix}$$

so that T is not regular. In fact, T is called absorbing. Notice what happens to high powers of T

$$T^6 = \begin{pmatrix} 1 & 0 & 0 & 0 \\ \frac{63}{128} & \frac{2}{128} & \frac{6}{128} & \frac{57}{128} \\ 0 & 0 & \frac{1}{64} & \frac{63}{64} \\ 0 & 0 & 0 & 1 \end{pmatrix}, \quad T^{10} = \begin{pmatrix} 1 & 0 & 0 & 0 \\ .4995 & .0010 & .0049 & .4946 \\ 0 & 0 & .0010 & .9990 \\ 0 & 0 & 0 & 1 \end{pmatrix}.$$

It looks like T^n is approaching

$$\begin{pmatrix} 1 & 0 & 0 & 0 \\ \frac{1}{2} & 0 & 0 & \frac{1}{2} \\ 0 & 0 & 0 & 1 \\ 0 & 0 & 0 & 1 \end{pmatrix}$$

and in fact it is. We would interpret this as saying that if we start in stage a_2, there is a 50-50 chance of being absorbed by state a_1 or by state a_4. However, if we start in a_3 we will surely be absorbed into state a_4. This is an example of an *absorbing Markov chain.* A Markov chain is *absorbing* when it has at least one absorbing state and it is possible to get from any state to an absorbing state (not necessarily in one step).

A standard example of an absorbing Markov chain is the study of a particle in motion along a line that is assumed to have absorbing boundaries. We ask you to analyze one such problem below.

There is a rather interesting theory for absorbing Markov chains which allows the absorbing probabilities to be computed and even gives the expected number of transitions to absorption. (See Kemeny et al., *Finite Mathematical Structure.*)

We simply give a few exercises which show absorbing Markov chains arise in real world situations.

Exercises 7.10

3. A particle moves on a line segment of length 4 so that each time it moves it moves one unit right with probability $\frac{2}{5}$ and one unit left with probability $\frac{3}{5}$. Name the states 0, 1, 2, 3 and 4 and further assume that if the particle ever reaches an end point that it stays there. What is the probability that a particle starting at 1 will be absorbed at 4?

4. A gambler has \$300 to bet and decides to bet \$100 on each throw of a single die. He wins if the die comes up 1 or 2 and loses otherwise. He decides to quit when he has \$500 (or if he is broke, of course). What is the probability that he quits with \$500? (The six states are the number of hundreds of dollars he has, and 0 and 500 are absorbing.)

5. In problem 4, suppose the gambler bets on heads when an honest coin is tossed. How much better are his chances?

6. Suppose you toss a coin until two tails or three heads appear. View this as a Markov process with states TT, T, H, HH, and HHH, with TT and HHH as absorbing. What happens to high powers of the transition matrix? Can you interpret what this means?

7. In light of the information obtained by doing problem 4 above, analyze Example 3 of Section 7.1. How large will the transition matrix need to be to use matrix methods to solve this problem? Is it feasible to do so?

8. Solve the problem in Example 6 of Section 7.1.

APPENDIX

In this appendix we more formally describe the statements introduced in Chapter 2 and give a few additional statements which some students may be interested in learning. Although this presentation is more terse, we do give a few examples.

We do not discuss the use of string variables, the comparison of string expressions, subroutines, or the use of files for referencing or storing. In other words, we omit more than we include, but recall that we intend only to give you a taste of these languages. We want to whet your appetite for more. Students who are interested in the full power of a language should consult a manual for that language or take a regular course in which the language is taught.

BASIC

We present the material in this portion of the Appendix according to the following outline.

Constants and Variables
  Simple, subscripted, numeric, string
Special Statements and Functions
  REM, DIM, BASE
  INT, ROF, SGN, SQR, SIN, COS, TAN
  LEN, ABS, EXP, LGT, LOG
Input and Output
  DATA, RESTORE, NODATA, READ, INPUT, PRINT, TAB
Matrix Operations
  Arithmetic, functions, read, input, print
Diagnostics

As mentioned earlier, we have been selective in expanding the list of statements over that presented in Chapter 2. For example, we do not discuss the IMAGE statement, which allows complete control over the way the output from a PRINT statement appears much the way FORMAT statements in FORTRAN do. However, learning the statements described here will greatly increase the power and flexibility of the programs that you can write.

Constants and Variables

We discussed numeric constants and variables in Chapter 2. A *string* is a collection of letters, numerals, and symbols. It is usually enclosed in quotation marks and then called "quoted text." Unquoted strings are only allowed in DATA statements and as input data. We have used quoted strings extensively in Chapter 2. Two special rules are worth citing.

1. The length of a string is 0 to 78 characters.
2. An unquoted string constant cannot begin with a digit, plus (+), minus (−), comma (,), period (.), blank, or quote.

String variables, like numeric variables, can be simple or subscripted. *Simple string variables* are named by two-character identifiers. The first character must be alphabetic and the second must be a dollar sign ($). *Subscripted string variables* are written as simple string variables followed by a maximum of three subscripts enclosed in parentheses. The conditions described for numeric variable subscripts in Chapter 2 also hold for string variable subscripts.

EXAMPLES

String constants: ABC, "I AM HERE", A7B4, XYZ

Simple string variables: A$, X$, N$

Subscripted string variables: A$(13), B$(2,L+4), P$(N+2, 175, B(2))

Special Statements and Functions

1. The REM statement is a nonexecutable statement and has no effect on the program execution. It is used to insert explanatory remarks or comments into a program for the benefit of the programmer and is not of any value to someone who just runs the program. The comment or explanation that follows REM must not exceed 72 characters. We describe this formally by

Format

REM *st*

st Any explanation or comment of not more than 72 characters

2. The DIM statement establishes the dimensions of an array.

Format

DIM $v_1(nm_1, \ldots, nm_3), \ldots, v_n(nm_1, \ldots, nm_3)$

$v_1 - v_n$	Numeric or string variables
$nm_1 - nm_3$	One to three integers, separated by commas, representing the maximum value of each subscript

If an array will have more than 10 values for any subscript, then a DIM statement is required. However, to save space in the working memory of the computer, the DIM statement may be used to dimension an array with upper subscript limits less than 10. The lower boundary of a subscript is normally 1, but can be 0, depending on the computer system or if the BASE statement (defined next) is used. DIM statements may occur anywhere in a program.

EXAMPLES

```
60  DIM C(4, 15)
```

This statement reserves space for a 4 × 15 array of 60 elements. The subscripted variable C(2, 9) references the element in the second row and ninth column of the array.

```
110  DIM Y8(45), D$(5, 20), C(6, 6)
```

This statement reserves space for

Y8	a one-dimensional numeric array with 45 elements
D$	a two-dimensional string array of $5 \cdot 20 = 100$ elements
C	a two-dimensional numeric array of $6 \cdot 6 = 36$ elements

3. The BASE statement is optional but specifically defines the beginning of all arrays. There can be only one BASE statement in any program and it must precede all DIM statements or references to arrays.

Format

BASE *m*

m must be 0 or 1

EXAMPLE

```
10  BASE 0
20  DIM B(1, 2, 3), S$(2, 10)
```

Since line 10 says all arrays begin at 0, the three-dimensional numeric array B has $2 \cdot 3 \cdot 4 = 24$ elements in it and the two-dimensional string array S$ has $3 \cdot 11 = 33$ elements in it.

4. Special Functions We define eleven standard mathematical functions and one string function. In these functions x stands for a numeric expression of any complexity and may include other functions.

ABS(x)	Gives the absolute value of x.
EXP(x)	Gives the value of e to the power x.
LGT(x)	Gives the base 10 logarithm of x for $x > 0$, and otherwise the program stops with an execution error.
LOG(x)	Gives the natural logarithm of x for $x > 0$, and otherwise an execution error causes the program to stop.
INT(x)	Gives the largest integer not exceeding x.
ROF(x)	Gives the value of the integer nearest to x.
SGN(x)	Gives 1 if x is positive, 0 if x is 0, and -1 if x is negative.
SQR(x)	Gives the square root of x if x is positive, otherwise an execution error causes the program to stop.
SIN(x)	Gives the sine of the angle x expressed in radians.
COS(x)	Gives the cosine of the angle x expressed in radians.
TAN(x)	Gives the tangent of the angle x expressed in radians.
LEN(*sc*)	Gives the length of *sc*, which may be a string constant, expression, or variable. For example, if B$ has value CABOOSE, then LEN(B$) is 7.

Some of these are illustrated in the program with printout shown here.

```
10  LET A = -3.4
20  LET B = 12.25
30  LET C$ = "2YOUNG2BE4GOTTEN"
40  PRINT INT(A); ABS(A); ROF(A); SGN(A)
50  PRINT INT(B); ABS(B); ROF(A); SQR(B)
60  PRINT INT(SQR(B)); LEN(C$); SQR(LEN(C$))
```

```
RNH
-4  3.4  -3  -1
 12     12.25     12     3.5
 3     16     4
```

Input and Output

1. The DATA statements are nonexecutable and may appear anywhere in the program. They are all considered to be next to each other, and the data are placed in sequential order in one big data block (referred to as an *internal data block*). The data can be used only by READ statements.

Format

DATA $c_1, c_2, \ldots, c_n$

$c_1 - c_n$ numeric or string constants

Both quoted and unquoted strings are allowed, but recall the restrictions on the first symbol in unquoted strings.

2. The RESTORE statement is most useful when the same data can be used over and over. An internal data block has a pointer associated with it that indicates the next constant to be read. The RESTORE statement sends the pointer back to the first constant in the block.

3. The NODATA statement is used to determine if all the data have been read.

Format

NODATA *ln*

ln a line number

If the pointer in the internal data block is at the end of the data, the program transfers to the statement with line number *ln*.

4. The READ statement is used to read information furnished by DATA statements.

Format

READ $v_1, v_2, \ldots, v_n$

$v_1 - v_n$ numeric, string, or subscripted variable identifier

As each item of data is read and assigned to the next variable in the READ statement, the pointer in the internal data block is advanced to the next item of data. If the program tries to read strings into numeric variables or numbers into strings, the diagnostic "BAD DATA IN READ" is issued. If the end of the data block is reached during the execution of the READ statement, the diagnostic "END OF DATA AT line number" is given. In both cases the program is terminated.

5. The INPUT statement provides for the user to enter data during execution of the program from the terminal.

Format

INPUT $v_1, v_2, \ldots, v_n$

$v_1 - v_n$ numeric, string, or subscripted variable identifier

Each time an INPUT statement is executed, a question mark is displayed at the current print position of the terminal line. The user must then enter data in such a way that (i) numeric constants are separated by commas or by spaces, (ii) string constants are separated by commas, (iii) there is a one-to-one correspondence between data items and variables, (iv) numbers must be entered for numeric variables, and (v) quoted or unquoted strings must be entered for string variables. Extra blank spaces in the data are ignored. A carriage return marks the end of data to be entered. If too little data is entered, a diagnostic message "NOT ENOUGH DATA, TYPE IN MORE" is issued, and the user should continue to enter data until all the variables have been provided for. If too much data or unacceptable data is entered, a diagnostic message will be issued and the entire data list must be retyped. Using a PRINT statement along with the INPUT statement avoids confusion as to how many and what type of data items to enter (see program AREA1 in Section 2.8).

6. The PRINT statement is our only statement for obtaining output.

Format

PRINT $e_1de_2d\ .\ .\ .\ de_nd$

$e_1 - e_n$	expression, variable, or constant (numeric or string)
d	either a comma (,) or a semicolon (;). The final d may be omitted.

If there is no final comma or semicolon, the next printing begins at the start of a new line. Each print line is normally separated into five *zones* of 15 spaces each. A comma causes the next printing to take place in the next zone, which would be the first zone of the next line when the last zone of a line is filled. A semicolon has no spacing effect. Because numbers are printed with a blank or minus sign in front and a blank behind, two positive numbers would be separated by two spaces. However, strings are printed consecutively without any preceding or intervening blanks when they are separated by a semicolon.

7. The TAB function is used in a PRINT statement to provide spacing.

Format

TAB(*ne*)

ne	a numeric constant, variable, or expression indicating the next print position number counted beginning with zero

The semicolon should be used with the TAB function. The TAB function can be used only in a PRINT statement and should have the semicolon used as separator.

EXAMPLE We illustrate 1–7 in this program.

```
 10  NODATA 60
 20  READ A, B$, C(2)
 30  DATA 1.3, STARS, 19, 2.6, "STRIPES", 29
 40  PRINT A; B$; "="; C(2)
 50  GOTO 10
 60  LET X1 = 123
 70  LET Y1 = 678
 80  LET A$ = "CLOSE"
 90  LET B$ = "FAR"
100  PRINT "01234567890"
110  PRINT X1; Y1
115  PRINT X1, Y1
120  PRINT A$; B$
130  PRINT A$, B$
140  PRINT TAB(5); Y1; TAB(9); A$; TAB(18); B$
150  PRINT TAB(5); Y1; A$; TAB(18); B$
160  END
RNH

 1.3 STARS = 19
 2.6 STRIPES = 29
01234567890
 123  678
 123            678
CLOSEFAR
CLOSE           FAR
      678 CLOSE     FAR
      678 CLOSE     FAR
```

Matrix Operations

It is possible to construct programs to perform matrix operations with the statements you have available. In fact, you should do so for the practice. However, BASIC also provides statements explicitly for matrix operations.

We briefly describe some of these statements here. We suggest you experiment in order to test the limits of use and to increase your facility with the computer.

All matrices will be viewed as two-dimensional arrays. As such, the DIM statement can be used to specify their size (explicit dimensioning). In addition, in some cases the size can be given within a matrix statement (implicit dimensioning). However, only matrices up to 10 × 10 can be specified by implicit dimensioning. The following accomplish the same thing:

EXPLICIT	IMPLICIT
DIM A(3,4),B(9,7)	MAT READ A(3,4),B(9,7)
.	
.	
.	
MAT READ A,B	

1. **Matrix Arithmetic** There are four matrix operations which can be described as follows:

MAT $m_1 = m_2 + m_3$	Addition
MAT $m_1 = m_2 - m_3$	Subtraction
MAT $m_1 = m_2 * m_3$	Multiplication
MAT $m_1 = (ne) * m_3$	Scalar multiplication by a numerical expression

Here m_1, m_2, and m_3 are any matrix identifiers (numeric subscripted variable). Also m_1 must be a different identifier from both m_2 and m_3 in matrix multiplication. Thus MAT $m_1 = m_1 * m_3$ is not allowed, but MAT $m_1 = m_1 + m_3$ is allowed.

2. **Input and Output** The MAT READ, MAT INPUT, and MAT PRINT statements are illustrated at the end of this paragraph. The MAT READ statement is filled from the internal data block row by row. Similarly, MAT INPUT requires the user to type each row separately and to press the carriage return after each row has been entered. Finally, MAT PRINT causes matrices to be printed out in row order using the same rules (relative to comma and semicolon separators) as the normal PRINT statement. A blank line is automatically generated after each row.

EXAMPLE

```
PROGRAM     MATEX

05  BASE 1
10  DIM A1(3,3),A2(3,3),A3(3,3),A4(3,3)
```

```
 20  MAT READ A(3,3)
 30  MAT A1 = A + A
 40  MAT A2 = A * A1
 50  MAT A3 = A2 - A1
 60  MAT INPUT B(2,3)
 70  DIM A4(2,3)
 80  MAT A4 = (9**1.5) * B
 90  DATA 1,0,2,-3,2,1,2,0,-2
100  MAT PRINT A;A1;A2;A3;B;A4
110  END
READY.
RNH

?  1,2,3
?  4,5
  NOT ENOUGH DATA, TYPE IN MORE          AT     60
?  6,7
  TOO MUCH DATA   -RETYPE AT     60
?  4,5,6

  1    0    2

 -3    2    1

  2    0   -2


  2    0    4

 -6    4    2

  4    0   -4


 10    0   -4

-14    8  -12

 -4    0   16


  8    0   -8

 -8    4  -14

 -8    0   20
```

```
1   2   3

4   5   6


27.                 54.                 81.
108.                135.                162.
```

If statement 100 had looked like

```
100  MAT PRINT A;A1;A2;A3;B;A4;
```

then the matrix A4 would have been spaced via the semicolon spacing instead of the zone spacing. The last two matrices would have appeared on the output like this:

```
1   2   3

4   5   6


27.   54.   81.

108.   135.   162.
```

We mention without explanation the inversion and transposition functions, and the matrix-generating statements.

MAT m = INV(a)	m becomes the inverse of a which cannot be larger than 50 × 50
MAT m = TRN(a)	m becomes the transpose of a
MAT m = ZER[(ne_1 [,ne_2])]	generates a matrix of all zeros
MAT m = CON[(ne_1 [,ne_2])]	generates a matrix of ones
MAT m = IDN[(ne_1 [,ne_2])]	generates an identity matrix

Items enclosed in brackets are optional, where ne_1 and ne_2 are numeric expressions giving the number of rows and columns.

Diagnostics

We have already mentioned some of the diagnostic messages that the system gives when an error is made. Some of the other more common ones are

END NOT LAST	END statement placed prior to the last statement in a BASIC program.
FOR WITHOUT NEXT	FOR statement has no balancing NEXT statement.
ILLEGAL CHARACTER	Unrecognizable character.
ILLEGAL LINE REF	Incorrectly written line number, or line number referenced >99999.
ILLEGAL LINE NUMBER	Line number >99999.
ILLEGAL STATEMENT	Statement does not begin with a recognizable word or is written incorrectly.
MISSING LINE NUMBER	Statement written without a line number.
NEXT WITHOUT FOR	NEXT statement has no balancing FOR statement.
READ WITHOUT DATA	Program containing a READ statement has no DATA statements.
BAD DATA IN READ	String read, numeric expected; or vice versa.
END OF DATA	READ statement executed when internal data block is exhausted.
ILLEGAL FILE NAME	File name does not correspond to alphanumeric rule.
MATRIX DIMENSION ERROR	Dimension inconsistency in one of the MAT statements or is greater than 50 × 50 for INV function.

FORTRAN

We present the material in this portion of the Appendix according to the following outline.

Constants and Variables
  Logical, literal, subscripted
Specification Statements and Built-in Functions
  DIMENSION, IMPLICIT, type
  SQRT, EXP, ALOG, ALOG10, IABS, ABS, INT, AINT, SIN, COS,
    TAN
Input and Output
  Implied DO loops
  FORMAT, READ, WRITE
Looping and Branching
  IF, DO

Constants and Variables

1. There are exactly two logical constants and they are .TRUE. and .FALSE.. A logical variable can be any valid symbolic name that is declared in a LOGICAL declaration statement. For example, the statement

```
LOGICAL TOM, B, ILL, A(5)
```

declares each of TOM, B, and ILL to be logical variables and A to be a one-dimensional logical array.

There are three logical operations that are used in FORTRAN, just as they are in usual logical statements. These operations

```
.AND.,   .OR.,   .NOT.
```

can be employed to make logical statements for use in the logical IF statement.

Logical assignment statements take the general form

logical variable = logical expression.

EXAMPLES

```
TOM = B .OR. ILL
B = X.EQ.Y .OR. A(1)
ILL = .NOT. B .AND. A(2)
```

where X and Y are real or integer variables.

2. *Literal* or *Hollerith* constants are delimited in two ways. Either the string of characters can be enclosed in apostrophes or the string can be preceded by wH where w is the number of characters in the string.

EXAMPLES

```
'COMPUTE'
'RAINY CLOUDY SUNNY'
18H GO TO THE MOVIES.
'ISN''T'
5HISN'T
```

Notice that if apostrophes are used to delimit the literal, then a single apostrophe within the literal is represented by two apostrophes.

3. Recall that variables whose names begin with I, J, K, L, M, and N are integers and those beginning with other letters are real (or floating point). It is possible to override this implicit rule by specifying the type of variable using INTEGER, REAL, or LOGICAL. For example,

```
INTEGER STEPS,   QUANT,   ALT,   PRIME
```

declares that in this program, STEPS, QUANT, ALT and PRIME will all be integer-valued variables. Similarly,

```
REAL LIFE,   BAKER,   INDEX
```

provides that LIFE, BAKER, and INDEX will all be real variables in this program. We say more about this later under specification statements.

4. Subscripted variables or arrays are specified in several ways. We describe a few in the next section. A subscripted variable is a simple variable followed by from one to seven numbers separated by commas and enclosed in parentheses. (See the discussion in the BASIC section for further insight into subscripted variables.)

   For example, ABLE(15, 19) would denote the entry in the fifteenth row and the nineteenth column of an array with name ABLE. X(13) would stand for the thirteenth value of a one-dimensional array called X.

Subscripted variables are very useful in DO loops and also, of course, two-dimensional arrays can be viewed as matrices. We say more about them in the next section.

Finally, the subscripts may contain arithmetic expressions, function references, and other subscripted variables as in

```
BAH(I,J*K/14,CL(19,3)*(4+3*T(8)))
```

Specification Statements and Built-in Functions

1. The most common way to specify the size of a subscripted variable (or array) is by means of a DIMENSION statement. It takes this form:

DIMENSION $a_1(k_1), a_2(k_2), \ldots, a_n(k_n)$

where $a_1, \ldots, a_n$ are variable names, and $k_1, \ldots, k_n$ are each composed of from one to seven unsigned integer constants, separated by commas.

EXAMPLES

```
DIMENSION  A(20),  BLOB(15,20),  VOL(5,10,15),  INDEX(4,4)
```

A — has 20 subscripts A(1), A(2), . . . , A(20).

BLOB — is a 15-row by 20-column array with BLOB(3,17) denoting the element in the third row and seventeenth column.

VOL — represents a three-dimensional array with $5 \cdot 10 \cdot 15 = 750$ entries. Thus VOL(4,8,12) is the entry in the fourth row and eighth column of the twelfth layer of the array.

INDEX — is a 4×4 matrix.

In fact, the values of the coordinates in the arguments of the variables represent the largest values allowed. It is not necessary to use the whole array. We illustrate this later.

2. The IMPLICIT statement allows the programmer to specify the type (that is, integer, real, or logical) of variables according to the initial letter of their names. If used, it must be the first statement in a program. An illustration should make this clear.

```
IMPLICIT INTEGER   (A-D,L-N),REAL(E-H,O-X,Z),LOGICAL(I,J,Y)
```

In this program all variables beginning with letters A through D and L through N would be integer variables, those with initial letters E through H, O through X, and Z would denote real variables, and those with initial letters I, J, and Y would denote logical variables.

3. The *explicit specification statements* INTEGER, REAL, LOGICAL, enable the user to specify the type of a variable or array according to its particular name, to specify the size of an array, and to assign initial data to variables and arrays. Any explicit specification statement overrides the IMPLICIT statement. Some examples should clarify the situation.

```
INTEGER   ORGAN / 83 / ,VALUE
```

This statement declares that ORGAN and VALUE are of integer type and specifies 83 as the initial value of ORGAN.

```
REAL   JEST(3)/ .5,9.,1.1/ ,NONO(6,6)/ 36*0.0/
```

This statement declares that JEST is a one-dimensional variable with three arguments and JEST(1), JEST(2), and JEST(3) are initialized at .5, 9., and 1.1, respectively. Moreover, NONO is declared to be a 6×6 array with all 36 values initialized at zero. Thus dimensioning is done just as in the DIMENSION statement and initial values are placed between slashes and separated by commas. If k consecutive initial values are the same constant c, then $/ \ k * c \ /$ can be used to give the first k entries in the array the initial value c.

```
INTEGER   ABLE(6,6)/ 30*1,6*2/
```

gives the 36-element array ABLE the initial value 1 in the first thirty places and 2 in the last six places.

4. Built-in Functions Notice that the integer functions start with I.

FUNCTION NAME WITH ARGUMENT	DEFINITION
SQRT(a)	$\sqrt{a}$, square root
EXP(a)	e^a, exponential function
ALOG(a)	$\ln(a)$, natural logarithm
ALOG10(a)	$\log_{10}(a)$, common logarithm
IABS(n)	$\lvert n \rvert$, absolute value of the integer n
ABS(a)	$\lvert a \rvert$, absolute value of the real a
INT(n)	just gives $\operatorname{sgn}(n)[\lvert n \rvert]$
AINT(a)	gives the integer part of a, $\operatorname{sgn}(a)[\lvert a \rvert]$
SIN(a)	sine of a when a is in radians
COS(a)	cosine of a when a is in radians
TAN(a)	tangent of a when a is in radians

Input and Output

We discuss some additional aspects of format-free input and output. In addition, we introduce some of the immediately useful features of the FORMAT statement. If your computer does not admit the WATFIV language, you may need to use READ and WRITE as described below.

In this section we introduce the FORMAT statement by giving examples using the PRINT statement. Near the end, we specify READ and WRITE and briefly describe their uses with FORMAT.

When used with the FORMAT statement, the PRINT statement takes this form:

PRINT *b*, *list*

where *b* is the statement number (XXXXX in the description of the

FORMAT statement given next) of the FORMAT statement describing the manner in which the values of the variables from *list* are to be printed.

The general form of a FORMAT statement is as follows.

XXXXX FORMAT ($f_1,f_2, \ldots, f_n$)

where $f_1,f_2, \ldots, f_n$ are format codes.

We will describe three number data codes and two literal data codes.

*a*I*w*	Describes integer data fields
*a*F*w.d*	Describes real data fields
*a*E*w.d*	Describes real data fields
*w*H	Transmits literal data
'Literal'	Transmits literal data

Here *a* is an optional unsigned integer constant used to denote the number of times the format code is to be used. If *a* is omitted, the code is used only once. Also, *w* is an unsigned positive integer constant that specifies the number of characters in the field, and *d* specifies the number of places to be printed after the decimal point.

You should have noticed that the sequence of statements

```
X=12.345
PRINT,X
```

will produce

```
0.1234500E 02
```

The F format code can be used to make this more readable.

```
      X=12.345
      PRINT 999,X
999   FORMAT ('O',F7.3)
```

causes the printer to doublespace and then print

```
 12.345
```

There are two things which need explaining here. The first data code is 'O' and is called a control character. There are five control characters, the first four of which cause printing to start on a new line.

' '	Single space
'O'	Double space
'–'	Triple space
'1'	Skip to the top of a new page
'+'	Do not skip to a new line

The first code in the FORMAT statement for a PRINT statement must be a control character.

The second code, F7.3, sends four messages to the printer. (1) The F specifies that a real number is to be printed. (2) The 7 describes the *width* of the *field,* which in this case is the first seven available print positions. (3) The 3 indicates that the number is to be printed with three decimal places. (4) The printing is to be right justified on the field, which explains the one blank space to the left of the 1 in the actual printout.

The F format code is used for printing without exponents and the E format code uses exponents. Consequently, the E format is usually used when the numbers are very large, very small, or of unknown size. The code

```
E12.5
```

would cause the number −12.345 to be printed as

```
-0.12345E 02
```

You can see that the E format code requires seven positions beyond the decimal digits, namely, −0. for three and E*s*02 for four more, where *s* is either a minus sign or a blank. Thus for the E format $w \geqslant d + 7$.

We give two sets of examples and a brief discussion of each.

EXAMPLE 1 For the program

```
     A=3.95
     B=-46.36947
     C=-123.4567
     D=123456.789
     PRINT 39,A,B,C,D
39   FORMAT ('  ',F7.4,F10.4,F8.4,F7.3)
```

the next printed line would be

```
  3.9500   -46.3695 ******** *******
  F7.4     F10.4     F8.4     F7.3
```

Notice that the decimal point takes one of the field positions, and that B has been rounded off to fit into the number of decimal spaces specified. The last two numbers were not printed because the field specified in the format code is not wide enough. Since C = −123.4567 takes nine spaces, and only eight were specified and D = 123456.789 cannot fit into a 7-space field, the computer notes the error by printing asterisks the appropriate number of times.

EXAMPLE 2 The program

```
     A=-45.699
     B=.00009876
     C=-195.36E5
     D=.00004567
     PRINT  83,A,B,C,D
83   FORMAT('-',E14.3,E9.4,E16.5,E12.4)
```

would cause to be printed, after triple spacing, the following line:

```
     -0.457E  02 *********      -0.19536E  03   0.4567E-04
       E14.3       E9.4            E16.5           E12.4
```

Rounding occurs in the printing of A since only three significant figures are specified. The asterisks are printed instead of B because B cannot fit into the nine-digit field specified. C and D are printed as expected.

The I format code is used to print integers much as F and E are used to print real numbers. For example,

```
      J=65432
      K=23456
      L=-3999
      M=6543
      PRINT 12,J,K,L,M
   12 FORMAT('  ',I6,I4,2I7)
```

would cause the following to be printed on the next line:

```
 65432 ****    -3999      6543
 \____/\__/ \_____/ \_____/
   I6   I4     I7       I7
```

Since K is too long for the four print positions provided by the format code, four asterisks are printed to show this error. The last two variables, L and M, are both formated in fields of width 7 by the format code 2I7.

Spacing can be provided by specifying wide fields, but a separate format code can also be used. The format code nX causes n blanks to be printed (or left). Hence

```
      I=123
      J=456
      PRINT 7,I,J
    7 FORMAT ('O',I6,4X,I4)
```

causes

```
   123        456
 \____/\____/\___/
   I6    4X   I4
```

to be printed on the second line down.

Finally, literal strings can be printed in format codes. Either apostrophes or the format code *n*H can be used and interspersed between spacing codes where desirable. Thus the program

```
     PRINT 100
100  FORMAT('1',9H HARDWARE)
     PRINT 101
101  FORMAT('-',1X,'NUTS',2X,'BOLTS')
```

would cause the printer to skip to a new page and print the following lines:

```
 HARDWARE



 NUTS  BOLTS
```

Because the FORMAT statements can become quite long, it is sometimes impossible to get the full statement punched onto a single card. The statement can be continued onto the next card by placing any character except blank or zero in column 6 of the next, or continuation, card.

Naturally in most situations the user will need a mixture of these format codes. For example,

```
     IRAD=36
     CIRC=2 * 3.1416 * IRAD
     PRINT 988,IRAD,CIRC
988  FORMAT('O','THE CIRCUMFERENCE OF A CIRCLE OF
    1RADIUS',13,  'IS'F8.4)
```

would print the following after doublespacing.

```
THE CIRCUMFERENCE OF A CIRCLE OF RADIUS 36 IS 226.1952
```

We list a few rules for emphasis.

(a) FORMAT statements may appear anywhere in the program (except of course after END.).
(b) Do not exceed the maximum number of spaces on your printer.
(c) The format code must be appropriate for the variable to which it corresponds.
(d) Commas may be omitted unless ambiguity would result.
(e) Blanks may be placed anywhere in the format list to improve readability.

Finally, a few words about READ and WRITE.

```
     READ 27,X,Y,K
27   FORMAT (F9.4,6X,E12.3,2X,I4)
```

would read the data card

```
1234.56789136              .603E-04AX54  2176.
  F9.4        6X          E12.3      2X   I4
```

and assign X, Y, and K the values 1234.5678, .603E−04, and 5402, respectively. Notice that spacing in the FORMAT statement causes certain positions on the data card to be ignored. Also notice that blanks are read as zeros.

The general form of the READ statement is as follows.

READ(*a*,*b*,END=*c*) *list*

where

a is an unsigned integer constant or variable which represents a data set reference or unit number.

b is optional and is the statement number of the FORMAT statement describing the record being read.

END=*c* is optional and *c* is the number of the statement to which transfer is made when the end of the data set is reached.

list is optional and is the list of variable names to which the data are being assigned.

The general form of the WRITE statement is as follows.

WRITE(*a*,*b*) *list*

where *a*, *b* and *list* are as in the READ statement.

In either READ or WRITE the omission of *b* causes the data to be read from or written into the data set according to a format specified by the installation.

The unit numbers usually used for the card reader, the printer, and the punch are 5, 6, and 7, respectively. You should check your facility before running a program.

An example will show the usefulness of the END = feature of the READ statement.

Suppose an unspecified number of cards has one number per card on it. We wish to know the number of cards, the sum of the numbers and the average. This program would read the data from the cards and have the printer give the output.

```
      N=0
      SUM=0
   1  READ (5,99,END=36)  X
  99  FORMAT (F12.3)
      SUM=SUM+X
      N= N+1
      GO TO 1
  36  AVE=SUM / N
      WRITE (6,98)  N,SUM,AVE
  98  FORMAT ('O','THE',I4,'NUMBERS HAD SUM' F12.3'AND AVERAGE' F12.3)
```

Looping and Branching

In addition to the examples and comments given in Chapter 2, we mention a little more about the IF and DO statements.

The IF statement is of two kinds. There are logical IF and arithmetic IF statements. The general form of the logical IF statement is

IF(*e*)*fs*

The *e* stands for any logical-valued expression and *fs* stands for any executable FORTRAN statement except a DO statement or another logical IF statement. When the expression e is true, the FORTRAN statement fs is executed and then the next instruction is executed (unless fs is a transfer statement), when e is false, the next instruction is executed.

The logical IF allows the program to branch to one of two possible branches. The arithmetic IF allows a three-way branching. Its general form is as follows.

$$\text{IF}(e) \quad a_1,a_2,a_3$$

The e now stands for any arithmetic variable or expression, and a_1, a_2, a_3 are all executable statement numbers. The control passes to the statement with number a_1, a_2, or a_3 according as the arithmetic expression e is negative, zero, or positive. Notice that the statement following an arithmetic IF must be numbered, otherwise it can never be reached and executed.

The general form of the DO statement is

$$\text{DO } x \; i{=}m_1,m_2,m_3$$

where x is an executable statement number, i is a simple integer variable, and m_1, m_2 and m_3 are positive unsigned integer constants or simple integer variables whose value is greater than zero. If m_3 is omitted, it is assumed to have value 1.

The DO statement is a command to execute at least once the statements that follow the DO statement up to and including the statement numbered x. These statements are called the range of the DO. The first time through the range the variable i has value m_1 and each succeeding time i is increased by the value m_3. If i would have value larger than m_2 after adding m_3 to the current value of i, then control passes to the statement following the statement numbered x.

In our last remark, we note that an implied DO can be performed in READ and PRINT statements. Thus

```
READ, A,I,(B(J),J=1,25)
PRINT,X,K,(S(L),L=3,5,7,9)
```

are both allowable statements.

For those interested in learning more about FORTRAN IV, we suggest you consult one of the many textbooks and guides now available. We especially recommend *FORTRAN IV with WATFOR and WATFIV*, by P. Cress, P. Dirksen, and J. W. Graham (Prentice-Hall, Englewood Cliffs, N.J., 1970).

TABLE I SQUARES AND SQUARE ROOTS

n	n^2	$\sqrt{n}$	$\sqrt{10n}$	n	n^2	$\sqrt{n}$	$\sqrt{10n}$
1	1	1.000	3.162	51	2601	7.141	22.583
2	4	1.414	4.472	52	2704	7.211	22.804
3	9	1.732	5.477	53	2809	7.280	23.022
4	16	2.000	6.325	54	2916	7.348	23.238
5	25	2.236	7.071	55	3025	7.416	23.452
6	36	2.449	7.746	56	3136	7.483	23.664
7	49	2.646	8.367	57	3249	7.550	23.875
8	64	2.828	8.944	58	3364	7.616	24.083
9	81	3.000	9.487	59	3481	7.681	24.290
10	100	3.162	10.000	60	3600	7.746	24.495
11	121	3.317	10.488	61	3721	7.810	24.698
12	144	3.464	10.954	62	3844	7.874	24.900
13	169	3.606	11.402	63	3969	7.937	25.100
14	196	3.742	11.832	64	4096	8.000	25.298
15	225	3.873	12.247	65	4225	8.062	25.495
16	256	4.000	12.649	66	4356	8.124	25.690
17	289	4.123	13.038	67	4489	8.185	25.884
18	324	4.243	13.416	68	4624	8.246	26.077
19	361	4.359	13.784	69	4761	8.307	26.268
20	400	4.472	14.142	70	4900	8.367	26.458
21	441	4.583	14.491	71	5041	8.426	26.646
22	484	4.690	14.832	72	5184	8.485	26.833
23	529	4.796	15.166	73	5329	8.544	27.019
24	576	4.899	15.492	74	5476	8.602	27.203
25	625	5.000	15.811	75	5625	8.660	27.386
26	676	5.099	16.125	76	5776	8.718	27.568
27	729	5.196	16.432	77	5929	8.775	27.749
28	784	5.292	16.733	78	6084	8.832	27.928
29	841	5.385	17.129	79	6241	8.888	28.107
30	900	5.477	17.321	80	6400	8.944	28.284
31	961	5.568	17.607	81	6561	9.000	28.460
32	1024	5.657	17.889	82	6724	9.055	28.636
33	1089	5.745	18.166	83	6889	9.110	28.810
34	1156	5.831	18.439	84	7056	9.165	28.983
35	1225	5.916	18.708	85	7225	9.220	29.155
36	1296	6.000	18.974	86	7396	9.274	29.326
37	1369	6.083	19.235	87	7569	9.327	29.496
38	1444	6.164	19.494	88	7744	9.381	29.665
39	1521	6.245	19.748	89	7921	9.434	29.833
40	1600	6.325	20.000	90	8100	9.487	30.000
41	1681	6.403	20.248	91	8281	9.539	30.166
42	1764	6.481	20.494	92	8464	9.592	30.332
43	1849	6.557	20.736	93	8649	9.644	30.496
44	1936	6.633	20.976	94	8836	9.695	30.659
45	2025	6.708	21.213	95	9025	9.747	30.822
46	2116	6.782	21.448	96	9216	9.798	30.984
47	2209	6.856	21.679	97	9409	9.849	31.145
48	2304	6.928	21.909	98	9604	9.899	31.305
49	2401	7.000	22.136	99	9801	9.950	31.464
50	2500	7.071	22.361	100	10000	10.000	31.623

TABLE II AREAS UNDER THE NORMAL DISTRIBUTION

	.00	.01	.02	.03	.04	.05	.06	.07	.08	.09
0.0	.0000	.0040	.0080	.0120	.0160	.0199	.0239	.0279	.0319	.0359
0.1	.0398	.0438	.0478	.0517	.0557	.0596	.0636	.0675	.0714	.0753
0.2	.0793	.0832	.0871	.0910	.0948	.0987	.1026	.1064	.1103	.1141
0.3	.1179	.1217	.1255	.1293	.1331	.1368	.1406	.1443	.1480	.1517
0.4	.1554	.1591	.1628	.1664	.1700	.1736	.1772	.1808	.1844	.1879
0.5	.1915	.1950	.1985	.2019	.2054	.2088	.2123	.2157	.2190	.2224
0.6	.2257	.2291	.2324	.2357	.2389	.2422	.2454	.2486	.2517	.2549
0.7	.2580	.2611	.2642	.2673	.2704	.2734	.2764	.2794	.2823	.2852
0.8	.2881	.2910	.2939	.2967	.2995	.3023	.3051	.3078	.3106	.3133
0.9	.3159	.3186	.3212	.3238	.3264	.3289	.3315	.3340	.3365	.3389
1.0	.3413	.3438	.3461	.3485	.3508	.3531	.3554	.3577	.3599	.3621
1.1	.3643	.3664	.3686	.3708	.3729	.3749	.3770	.3790	.3810	.3830
1.2	.3849	.3869	.3888	.3907	.3925	.3944	.3962	.3980	.3997	.4015
1.3	.4032	.4049	.4066	.4082	.4099	.4115	.4131	.4147	.4162	.4177
1.4	.4192	.4207	.4222	.4236	.4251	.4265	.4279	.4292	.4306	.4319
1.5	.4332	.4345	.4357	.4370	.4382	.4394	.4406	.4418	.4429	.4441
1.6	.4452	.4463	.4474	.4484	.4495	.4505	.4515	.4525	.4535	.4545
1.7	.4554	.4564	.4573	.4582	.4591	.4599	.4608	.4616	.4625	.4633
1.8	.4641	.4649	.4656	.4664	.4671	.4678	.4686	.4693	.4699	.4706
1.9	.4713	.4719	.4726	.4732	.4738	.4744	.4750	.4756	.4761	.4767
2.0	.4772	.4778	.4783	.4788	.4793	.4798	.4803	.4808	.4812	.4817
2.1	.4821	.4826	.4830	.4834	.4838	.4842	.4846	.4850	.4854	.4857
2.2	.4861	.4864	.4868	.4871	.4875	.4878	.4881	.4884	.4887	.4890
2.3	.4893	.4896	.4898	.4901	.4904	.4906	.4909	.4911	.4913	.4916
2.4	.4918	.4920	.4922	.4925	.4927	.4929	.4931	.4932	.4934	.4936
2.5	.4938	.4940	.4941	.4943	.4945	.4946	.4948	.4949	.4951	.4952
2.6	.4953	.4955	.4956	.4957	.4959	.4960	.4961	.4962	.4963	.4964
2.7	.4965	.4966	.4967	.4968	.4969	.4970	.4971	.4972	.4973	.4974
2.8	.4974	.4975	.4976	.4977	.4977	.4978	.4979	.4979	.4980	.4981
2.9	.4981	.4982	.4982	.4983	.4984	.4984	.4985	.4985	.4986	.4986
3.0	.4987	.4987	.4987	.4988	.4988	.4989	.4989	.4989	.4990	.4990

TABLE III COMPOUND INTEREST $(1 + i)^n$

n	i = 1%	$1\frac{1}{2}$%	2%	$2\frac{1}{2}$%	3%	4%	5%	6%
1	1.0100	1.0150	1.0200	1.0250	1.0300	1.0400	1.0500	1.0600
2	1.0201	1.0302	1.0404	1.0506	1.0609	1.0816	1.1025	1.1236
3	1.0303	1.0457	1.0612	1.0769	1.0927	1.1249	1.1576	1.1910
4	1.0406	1.0614	1.0824	1.1038	1.1255	1.1699	1.2155	1.2625
5	1.0510	1.0773	1.1041	1.1314	1.1593	1.2167	1.2763	1.3382
6	1.0615	1.0934	1.1262	1.1597	1.1941	1.2653	1.3401	1.4185
7	1.0721	1.1098	1.1487	1.1887	1.2299	1.3159	1.4071	1.5036
8	1.0829	1.1265	1.1717	1.2184	1.2668	1.3686	1.4775	1.5938
9	1.0937	1.1434	1.1951	1.2489	1.3048	1.4233	1.5513	1.6895
10	1.1046	1.1605	1.2190	1.2801	1.3439	1.4802	1.6289	1.7908
11	1.1157	1.1779	1.2434	1.3121	1.3842	1.5395	1.7103	1.8983
12	1.1268	1.1956	1.2682	1.3449	1.4258	1.6010	1.7959	2.0122
13	1.1381	1.2136	1.2936	1.3785	1.4685	1.6651	1.8856	2.1329
14	1.1495	1.2318	1.3195	1.4130	1.5126	1.7317	1.9799	2.2609
15	1.1610	1.2502	1.3459	1.4483	1.5580	1.8009	2.0789	2.3966
16	1.1726	1.2690	1.3728	1.4845	1.6047	1.8730	2.1829	2.5404
17	1.1843	1.2880	1.4002	1.5216	1.6528	1.9479	2.2920	2.6928
18	1.1961	1.3073	1.4282	1.5597	1.7024	2.0258	2.4066	2.8543
19	1.2081	1.3270	1.4568	1.5987	1.7535	2.1068	2.5270	3.0256
20	1.2202	1.3469	1.4859	1.6386	1.8061	2.1911	2.6533	3.2071
21	1.2324	1.3671	1.5157	1.6796	1.8603	2.2788	2.7860	3.3996
22	1.2447	1.3876	1.5460	1.7216	1.9161	2.3699	2.9253	3.6035
23	1.2572	1.4084	1.5769	1.7646	1.9736	2.4647	3.0715	3.8197
24	1.2697	1.4295	1.6084	1.8087	2.0328	2.5633	3.2251	4.0489
25	1.2824	1.4509	1.6406	1.8539	2.0938	2.6658	3.3864	4.2919
26	1.2953	1.4727	1.6734	1.9003	2.1566	2.7725	3.5557	4.5494
27	1.3082	1.4948	1.7069	1.9478	2.2213	2.8834	3.7335	4.8223
28	1.3213	1.5172	1.7410	1.9965	2.2879	2.9987	3.9201	5.1117
29	1.3345	1.5400	1.7758	2.0464	2.3566	3.1187	4.1161	5.4184
30	1.3478	1.5631	1.8114	2.0976	2.4273	3.2434	4.3219	5.7435
31	1.3613	1.5865	1.8476	2.1500	2.5001	3.3731	4.5380	6.0881
32	1.3749	1.6103	1.8845	2.2038	2.5751	3.5081	4.7649	6.4534
33	1.3887	1.6345	1.9222	2.2589	2.6523	3.6484	5.0032	6.8406
34	1.4026	1.6590	1.9607	2.3153	2.7319	3.7943	5.2533	7.2510
35	1.4166	1.6839	1.9999	2.3732	2.8139	3.9461	5.5160	7.6861
36	1.4308	1.7091	2.0399	2.4325	2.8983	4.1039	5.7918	8.1473
37	1.4451	1.7348	2.0807	2.4933	2.9852	4.2681	6.0814	8.6361
38	1.4595	1.7608	2.1223	2.5557	3.0748	4.4388	6.3855	9.1543
39	1.4741	1.7872	2.1647	2.6196	3.1670	4.6164	6.7048	9.7035
40	1.4889	1.8140	2.2080	2.6851	3.2620	4.8010	7.0400	10.2857
41	1.5038	1.8412	2.2522	2.7522	3.3599	4.9931	7.3920	10.9029
42	1.5188	1.8688	2.2972	2.8210	3.4607	5.1928	7.7616	11.5570
43	1.5340	1.8969	2.3432	2.8915	3.5645	5.4005	8.1497	12.2505
44	1.5493	1.9253	2.3901	2.9638	3.6715	5.6165	8.5572	12.9855
45	1.5648	1.9542	2.4379	3.0379	3.7816	5.8412	8.9850	13.7646

TABLE IV PRESENT VALUE OF A DOLLAR $(1 + i)^{-n}$

				i				
n	1%	1½%	2%	2½%	3%	4%	5%	6%
1	.99010	.98522	.98039	.97561	.97087	.96154	.95238	.94340
2	.98030	.97066	.96117	.95181	.94260	.92456	.90703	.89000
3	.97059	.95632	.94232	.92860	.91514	.88900	.86384	.83962
4	.96098	.94218	.92385	.90595	.88849	.85480	.82270	.79209
5	.95147	.92826	.90573	.88385	.86261	.82193	.78353	.74726
6	.94205	.91454	.88797	.86230	.83748	.79031	.74622	.70496
7	.92372	.90103	.87056	.84127	.81309	.75992	.71068	.66506
8	.92348	.88771	.85349	.82075	.78941	.73069	.67684	.62741
9	.91434	.87459	.83676	.80073	.76642	.70259	.64461	.59190
10	.90529	.86167	.82035	.78120	.74409	.67556	.61391	.55839
11	.89632	.84893	.80426	.76214	.72242	.64958	.58468	.52679
12	.88745	.83639	.78849	.74356	.70138	.62460	.55684	.49697
13	.87866	.82403	.77303	.72542	.68095	.60057	.53032	.46884
14	.86996	.81185	.75788	.70773	.66112	.57748	.50507	.44230
15	.86135	.79985	.74301	.69047	.64186	.55526	.48102	.41727
16	.85282	.78803	.72845	.67362	.62317	.53391	.45811	.39365
17	.84438	.77639	.71416	.65720	.60502	.51337	.43630	.37136
18	.83602	.76491	.70016	.64117	.58739	.49363	.41552	.35034
19	.82774	.75361	.68643	.62553	.57029	.47464	.39573	.33051
20	.81954	.74247	.67297	.61027	.55368	.45639	.37689	.31180
21	.81143	.73150	.65978	.59539	.53755	.43883	.35894	.29416
22	.80340	.72069	.64684	.58086	.52189	.42196	.34185	.27751
23	.79544	.71004	.63416	.56670	.50669	.40573	.32557	.26180
24	.78757	.69954	.62172	.55288	.49193	.39012	.31007	.24698
25	.77977	.68921	.60953	.53939	.47761	.37512	.29530	.23300
26	.77205	.67902	.59758	.52623	.46369	.36069	.28124	.21981
27	.76440	.66899	.58586	.51340	.45019	.34682	.26785	.20737
28	.75684	.65910	.57437	.50088	.43708	.33348	.25509	.19563
29	.74934	.64936	.56311	.48866	.42435	.32065	.24295	.18456
30	.74192	.63976	.55207	.47674	.41199	.30832	.23138	.17411
31	.73458	.63031	.54125	.46511	.39999	.29646	.22036	.16425
32	.72730	.62099	.53063	.45377	.38834	.28506	.20987	.15496
33	.72010	.61182	.52023	.44270	.37703	.27409	.19987	.14619
34	.71297	.60277	.51003	.43191	.36604	.26355	.19035	.13791
35	.70591	.59387	.50003	.42137	.35538	.25342	.18129	.13011
36	.69892	.58509	.49022	.41109	.34503	.24367	.17266	.12274
37	.69200	.57644	.48061	.40107	.33498	.23430	.16444	.11579
38	.68515	.56792	.47119	.39128	.32523	.22529	.15661	.10924
39	.67837	.55953	.46195	.38174	.31575	.21662	.14915	.10306
40	.67165	.55126	.45289	.37243	.30656	.20829	.14205	.09722
41	.66500	.54312	.44401	.36335	.29763	.20028	.13528	.09172
42	.65842	.53509	.43530	.35448	.28896	.19257	.12884	.08653
43	.65190	.52718	.42677	.34584	.28054	.18517	.12270	.08163
44	.64545	.51939	.41840	.33740	.27237	.17805	.11686	.07701
45	.63905	.51171	.41020	.32917	.26444	.17120	.11130	.07265

TABLE V ACCUMULATED AMOUNTS $\frac{(1+i)^n - 1}{i}$

n	1%	1.5%	2%	2.5%	3%	3.5%	4%	5%	6%
1	1.00000	1.00000	1.00000	1.00000	1.00000	1.00000	1.00000	1.00000	1.00000
2	2.01000	2.01500	2.02000	2.02500	2.03000	2.03500	2.04000	2.05000	2.06000
3	3.03010	3.04522	3.06040	3.07562	3.09090	3.10622	3.12160	3.15250	3.18360
4	4.06040	4.09090	4.12161	4.15252	4.18363	4.21494	4.24646	4.31012	4.37462
5	5.10101	5.15227	5.20404	5.25633	5.30914	5.36247	5.41632	5.52563	5.63709
6	6.15202	6.22955	6.30812	6.38774	6.46841	6.55015	6.63298	6.80191	6.97532
7	7.21354	7.32299	7.43428	7.54743	7.66246	7.77941	7.89829	8.14201	8.39384
8	8.23567	8.43284	8.58297	8.73612	8.89234	9.05169	9.21423	9.54911	9.89747
9	9.36853	9.55933	9.75463	9.95452	10.1591	10.3685	10.5828	11.0266	11.4913
10	10.4622	10.7027	10.9497	11.2034	11.4639	11.7314	12.0061	12.5779	13.1808
11	11.5668	11.8633	12.1687	12.4835	12.8078	13.1420	13.4864	14.2068	14.9716
12	12.6825	13.0412	13.4121	13.7956	14.1920	14.6020	15.0258	15.9171	16.8699
13	13.8093	14.2368	14.6803	15.1404	15.6178	16.1130	16.6268	17.7130	18.8821
14	14.9474	15.4504	15.9739	16.5190	17.0863	17.6770	18.2919	19.5986	21.0151
15	16.0969	16.6821	17.2934	17.9319	18.5989	19.2957	20.0236	21.5786	23.2760
16	17.2579	17.9324	18.6393	19.3802	20.1569	20.9710	21.8245	23.6575	25.6725
17	18.4304	19.2014	20.0121	20.8647	21.7616	22.7050	23.6975	25.8404	28.2129
18	19.6147	20.4894	21.4123	22.3863	23.4144	24.4997	25.6454	28.1324	30.9075
19	20.8109	21.7967	22.8406	23.9460	25.1169	26.3572	27.6712	30.5390	33.7600
20	22.0190	23.1237	24.2974	25.5447	26.8704	28.2797	29.7781	33.0660	36.7856
21	23.2392	24.4705	25.7833	27.1833	28.6765	30.2695	31.9692	37.7193	39.9927
22	24.4716	25.8376	27.2990	28.8629	30.5368	32.3289	34.2480	38.5052	43.3923
23	25.7163	27.2251	28.8450	30.5844	32.4529	34.4604	36.6179	41.4305	46.9958
24	26.9735	28.6335	30.4219	32.3490	34.4265	36.6665	39.0826	44.5020	50.8156
25	28.2432	30.0630	32.0303	34.1578	36.4593	38.9499	41.6459	47.7271	54.8645
26	29.5256	31.5140	33.6709	36.0117	38.5530	41.3131	44.3117	51.1135	59.1564
27	30.8209	32.9867	35.3443	37.9120	40.7096	43.7591	47.0842	54.6691	63.7058
28	32.1291	34.4815	37.0512	39.8598	42.9309	46.2906	49.9676	58.4026	68.5281
29	33.4504	35.9987	38.7922	41.8563	45.2189	48.9108	52.9663	62.3227	73.6398
30	34.7849	37.5387	40.5681	43.9027	47.5754	51.6227	56.0849	66.4388	79.0582
31	36.1327	39.1018	42.3794	46.0003	50.0027	54.4295	59.3283	70.7608	84.8017
32	37.4941	40.6883	44.2270	48.1503	52.5028	57.3345	62.7015	75.2988	90.8898
33	38.8690	42.2986	46.1116	50.3540	55.0778	60.3412	66.2095	80.0638	97.3432
34	40.2577	43.9331	48.0338	52.6129	57.7302	63.4532	69.8579	85.0670	104.184
35	41.6603	45.5921	49.9945	54.9282	60.4621	66.6740	73.6522	90.3203	111.435
36	43.0769	47.2760	51.9944	57.3014	63.2759	70.0076	77.5983	95.8363	119.121
37	44.5076	48.9851	54.0343	59.7339	66.1742	73.4579	81.7022	101.628	127.268
38	45.9527	50.7199	56.1149	62.2273	69.1594	77.0289	85.9703	107.710	135.904
39	47.4123	52.4807	58.2372	64.7830	72.2342	80.7249	90.4091	114.095	145.058
40	48.8864	54.2679	60.4020	67.4026	75.4013	84.5503	95.0255	120.800	154.762
41	50.3752	56.0819	62.6100	70.0876	78.6633	88.5095	99.8265	127.840	165.048
42	51.8790	57.9231	64.8622	72.8398	82.0232	92.6074	104.820	135.232	175.951
43	53.3978	59.7920	67.1595	75.6608	85.4839	96.8486	110.012	142.993	187.508
44	54.9318	61.6889	69.5027	78.5523	89.0484	101.238	115.413	151.143	199.758
45	56.4811	63.6142	71.8927	81.5161	92.7199	105.782	121.029	159.700	212.744
46	58.0459	65.5684	74.3306	84.5540	96.5015	110.484	126.871	168.685	226.508
47	59.6263	67.5519	76.8172	87.6679	100.397	115.351	132.945	178.119	241.099
48	61.2226	69.5652	79.3535	90.8596	104.408	120.388	139.263	188.025	256.565
49	62.8348	71.6087	81.9406	94.1311	108.541	125.602	145.834	198.427	272.958
50	64.4632	73.6828	84.5794	97.4843	112.797	130.998	152.667	209.348	290.336

TABLE VI PRESENT VALUE OF AN ANNUITY $\frac{1-(1+i)^{-n}}{i}$

n	½%	1%	1.5%	2%	i 2.5%	3%	3.5%	4%	5%
1	.99502	.990099	.985222	.980392	.975610	.970874	.966184	.961538	.952381
2	1.98510	1.97040	1.95588	1.94156	1.92742	1.91347	1.89969	1.88609	1.85941
3	2.97025	2.94099	2.91220	2.88388	2.85602	2.82861	2.80164	2.77509	2.72325
4	3.95050	3.90197	3.85438	3.80773	3.76197	3.71710	3.67308	3.62990	3.54595
5	4.92587	4.85343	4.78264	4.71346	4.64583	4.57971	4.51505	4.45182	4.32948
6	5.89638	5.79548	5.69719	5.60143	5.50813	5.41719	5.32855	5.24214	5.07569
7	6.86207	6.72819	6.59821	6.47199	6.34939	6.23028	6.11454	6.00205	5.78637
8	7.82296	7.65168	7.48593	7.32548	7.17014	7.01969	6.87396	6.73274	6.46321
9	8.77906	8.56602	8.36052	8.16224	7.97087	7.78611	7.60769	7.43533	7.10782
10	9.73041	9.47130	9.22218	8.98259	8.75206	8.53020	8.31661	8.11090	7.72173
11	10.6770	10.3676	10.0711	9.78685	9.51421	9.25262	9.00155	8.76048	8.30641
12	11.6189	11.2551	10.9075	10.5753	10.2578	9.95400	9.66333	9.38507	8.86325
13	12.5562	12.1337	11.7315	11.3484	10.9832	10.6350	10.3027	9.98565	9.39357
14	13.4887	13.0037	12.5434	12.1062	11.6909	11.2961	10.9205	10.5631	9.89864
15	14.4166	13.8651	13.3432	12.8493	12.3814	11.9379	11.5174	11.1184	10.3797
16	15.3399	14.7179	14.1313	13.5777	13.0550	12.5611	12.0941	11.5623	10.8378
17	16.2586	15.5623	14.9076	14.2919	13.7122	13.1661	12.6513	12.1657	11.2741
18	17.1728	16.3983	15.6726	14.9920	14.3534	13.7535	13.1897	12.6593	11.6896
19	18.0824	17.2260	16.4262	15.6785	14.9789	14.3238	13.7098	13.1339	12.0853
20	18.9874	18.0456	17.1686	16.3514	15.5892	14.8775	14.2124	13.5903	12.4622
21	19.8880	18.8570	17.9001	17.0112	16.1845	15.4150	14.6980	14.0292	12.8212
22	20.7841	19.6604	18.6208	17.6580	16.7654	15.9369	15.1671	14.4511	13.1630
23	21.6757	20.4558	19.3309	18.2922	17.3321	16.4436	15.6204	14.8568	13.4886
24	22.5629	21.2434	20.0304	18.9139	17.8850	16.9355	16.0584	15.2470	13.7986
25	23.4456	22.0232	20.7196	19.5235	18.4244	17.4131	16.4815	15.6221	14.0939
26	24.3240	22.7952	21.3986	20.1210	18.9506	17.8768	16.8904	15.9828	14.3752
27	25.1980	23.5596	22.0676	20.7069	19.4640	18.3270	17.2854	16.3296	14.6430
28	26.0677	24.3164	22.7267	21.2813	19.9649	18.7641	17.6670	16.6631	14.8981
29	26.9330	25.0658	23.3761	21.8444	20.4535	19.1885	18.0358	16.9837	15.1411
30	27.7941	25.8077	24.0158	22.3965	20.9303	19.6004	18.3920	17.2920	15.3725
31	28.6508	26.5423	24.6461	22.9377	21.3954	20.0004	18.7363	17.5885	15.5928
32	29.5033	27.2696	25.2671	23.4683	21.8492	20.3888	19.0689	17.8736	15.8027
33	30.3515	27.9897	25.8790	23.9886	22.2919	20.7658	19.3902	18.1476	16.0025
34	31.1955	28.7027	26.4817	24.4986	22.7238	21.1318	19.7007	18.4112	16.1929
35	32.0354	29.4086	27.0756	24.9986	23.1452	21.4872	20.0007	18.6646	16.3742
36	32.8710	30.1075	27.6607	25.4888	23.5563	21.8323	20.2905	18.9083	16.5469
37	33.7025	30.7995	28.2371	25.9695	23.9573	22.1672	20.5705	19.1426	16.7113
38	34.5299	31.4847	28.8051	26.4406	24.3486	22.4925	20.8411	19.3679	16.8679
39	35.3531	32.1630	29.3646	26.9026	24.7303	22.8082	21.1025	19.5845	17.0170
40	36.1722	32.8347	29.9158	27.3555	25.1028	23.1148	21.3551	19.7928	17.1591
41	36.9873	33.4997	30.4590	27.7995	25.4661	23.4124	21.5991	19.9931	17.2944
42	37.7983	34.1581	30.9941	28.2348	25.8206	23.7014	21.8349	20.1856	17.4232
43	38.6053	34.8100	31.5212	28.6616	26.1664	23.9819	22.0627	20.3708	17.5459
44	39.4082	35.4555	32.0406	29.0800	26.5038	24.2543	22.2828	20.5488	17.6628
45	40.2072	36.0945	32.5523	29.4902	26.8330	24.5187	22.4955	20.7200	17.7741
46	41.0022	36.7272	33.0565	29.8923	27.1542	24.7754	22.7009	20.8847	17.8801
47	41.7932	37.3537	33.5532	30.2866	27.4675	25.0247	22.8994	21.0429	17.9810
48	42.5803	37.9740	34.0426	30.6731	27.7732	25.2667	23.0912	21.1951	18.0772
49	43.3635	38.5881	34.5247	31.0521	28.0714	25.5017	23.2766	21.3415	18.1687
50	44.1428	39.1961	34.9997	31.4236	28.3623	25.7298	23.4556	21.4822	18.2559

ANSWERS

TO ODD-NUMBERED EXERCISES

SECTION 1.1

1. $F_3 = 12¢$, $F_2 = 12¢$
3. $M = 97{,}945{,}000$, $W = 101{,}173{,}000$, $A_M = 26.4$, $A_W = 29.0$
5. $D_J = 11D_E$, $M_J > 300M_E$
7. $M = 16{,}710$, $16{,}710 > F + 1000$
9. Let A be the population of Asunción in 1962 and let P be the population of Paraguay in 1967. Then $A = 305{,}160$ and $P = 2{,}161{,}000$.
11. Let W and D be the weight and diameter of the average adult human eyeball. Let D_M and D_F be the diameters of the average male and female eye. Let W_B be the weight of an average baby's eye. Let T_A and T_B be the total weight of the average adult and baby. Then $W = \frac{1}{4}$ ounce, $D = 1$ inch, $D_M = D_F + \frac{1}{50}$ inch, $W/T_A = 1/4000$, and $W_B/T_B = 1/400$.
13. A is greater than B, A is larger than B, A exceeds B, A is more than B, B is less than A. (Many other answers are possible.)

SECTION 1.2

1. $T = .14(3500 - 3200)$, $T = \$42$
3. $1000 = 620 + .19(I - 7200)$
5. $.85 = (t/60)(3.24)$
7. $G > 2.5(1595)$, $G > 3987.5$
9. $Y < .406 - .100$, $Y < .306$
11. $18{,}752 = R - .228R$
13. $L > 90$, $H > 1500/2000 = 0.75$
15. $5.98x + 6.98y \leqslant 120$
17. $20CAB + 30COB < 200$
19. $6F_s + 9F_l \leqslant 15 \cdot 16 = 240$
21. $A + B + C \geqslant 100$
23. $30d + 12b + 14v > 500$
25. $.55S + .85C \geqslant 500$
27. $3400F + 2200C = 28{,}000$
29. $(120)(.15)A + (90)(.20)B > 400$

SECTION 1.3

1. $5.88R + 8.95 = 32.47$, $R = 4$
3. $12(130) + 9D = 2100$, $D = 60$
5. $20M = 300$, $M = 15$
7. $10{,}000B = 850(100)$, $B = 8.5$
9. $4(C + 2) + 6C = 88$, $C = 8$
11. $500 = 450 + 0.17(I - 6200)$, $I = 6494$
13. $.73 = (t/60)(4.38)$, $t = 10$
15. $1.50N - .50N - 10{,}000 > 0$, $N > 10{,}000$

17. $200(20) + P(25) \leqslant 10{,}000,\ P \leqslant 240$

19. $76 + 2C < 125,\ C < 24.5$

SECTION 1.4

1. 7.6 slices, 6.8 ounces

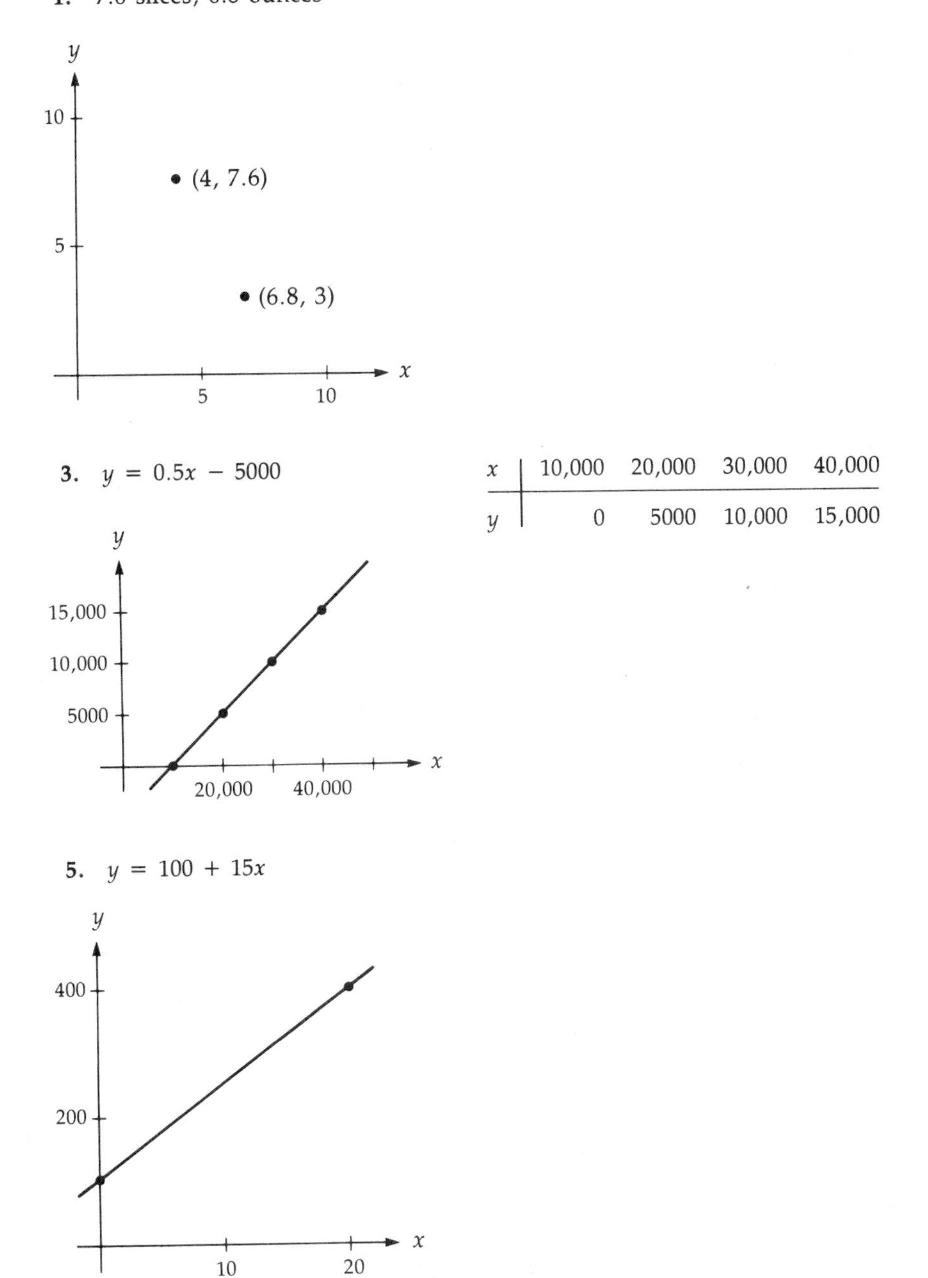

3. $y = 0.5x - 5000$

x	10,000	20,000	30,000	40,000
y	0	5000	10,000	15,000

5. $y = 100 + 15x$

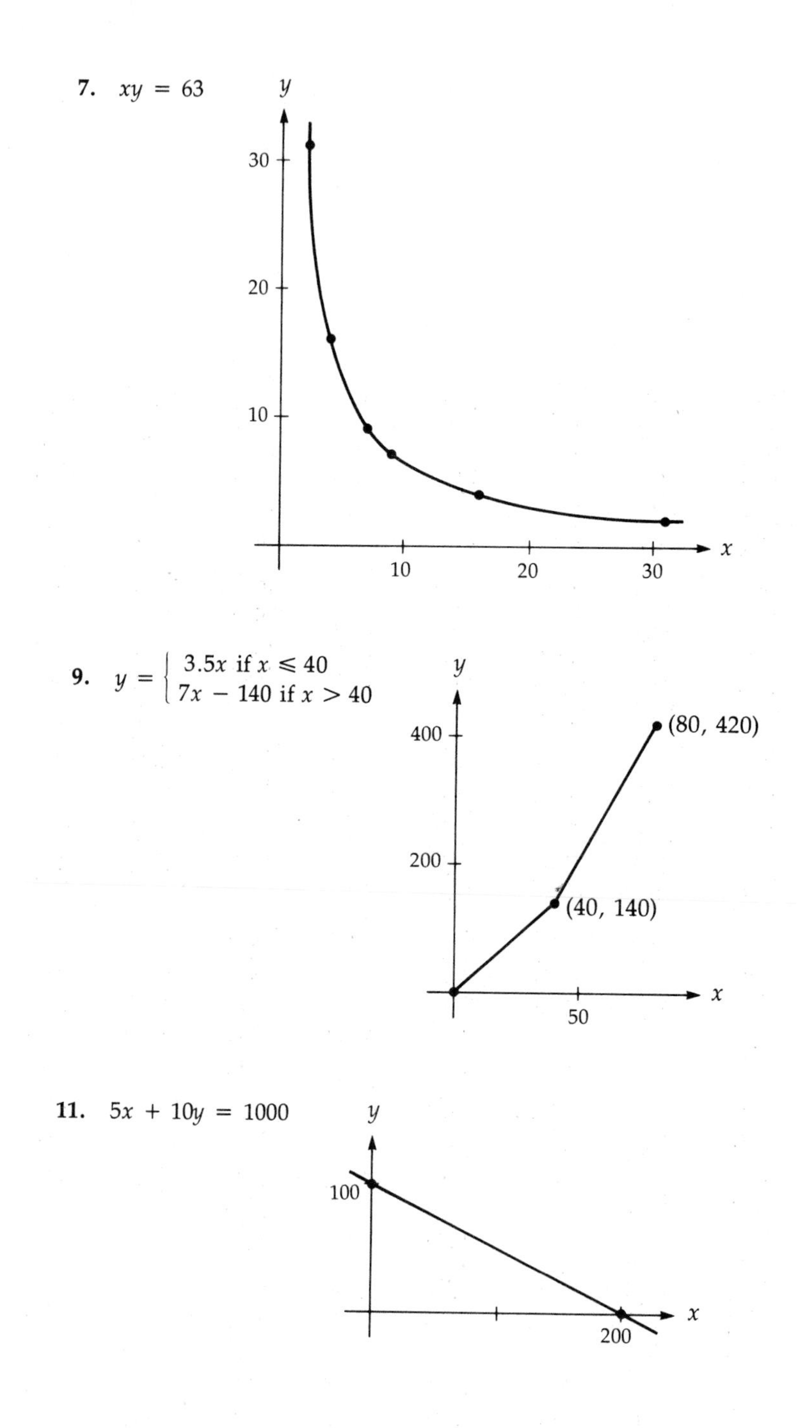

13. $25x + 10y \leq 500$

15. $150x + 60y \leq 1000$

SECTION 1.5

1. $218x + 113y \geqslant 1000$

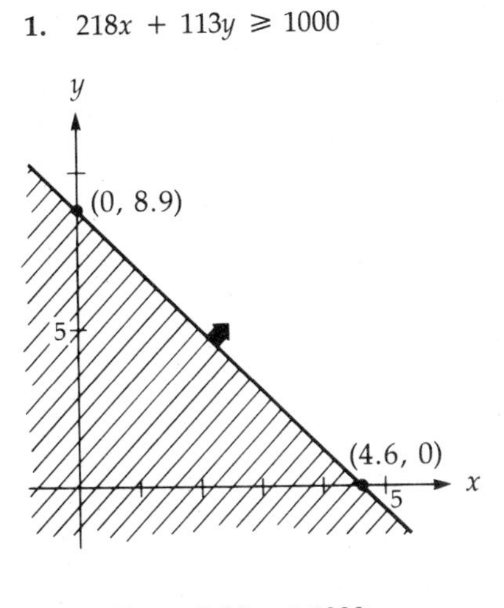

3. $3.5x + 2y \leqslant 10{,}000$

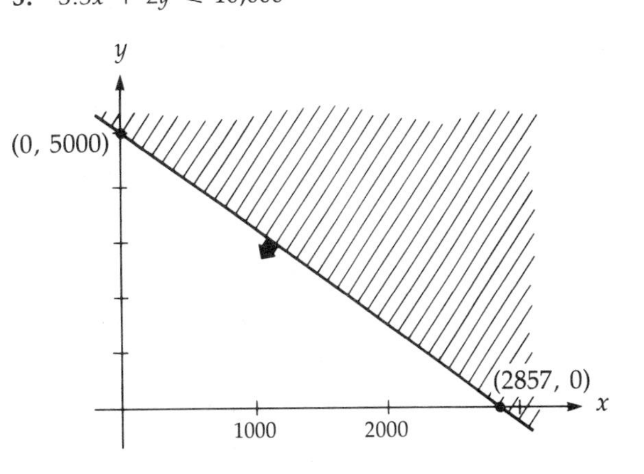

5. $4.98x + 5.98y \leqslant 1000$

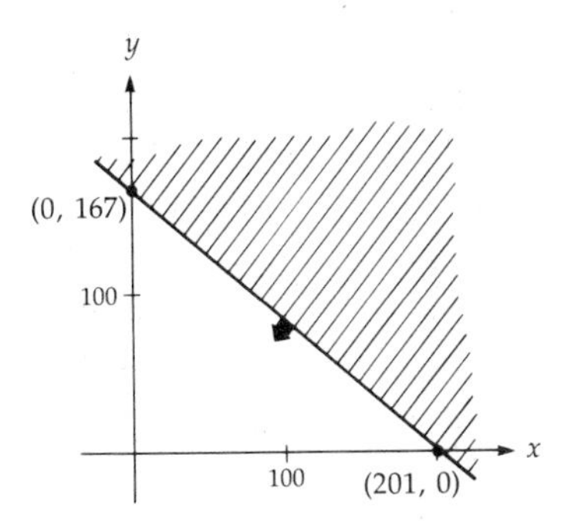

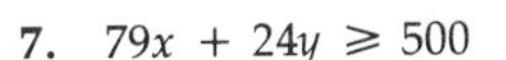

7. $79x + 24y \geqslant 500$

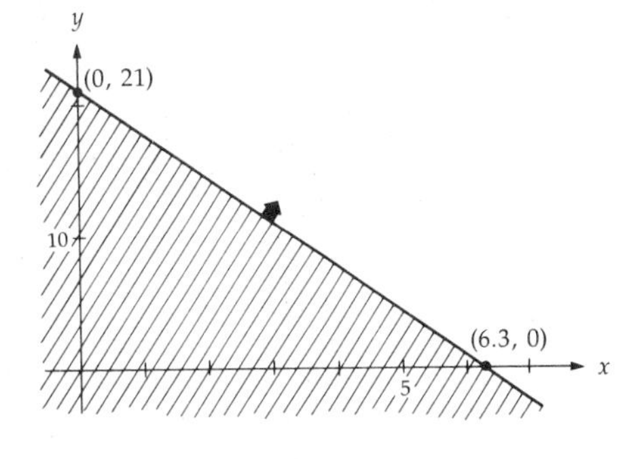

9. (1) $x + y \leqslant 4000$
(2) $.03x + .04y \geqslant 150$

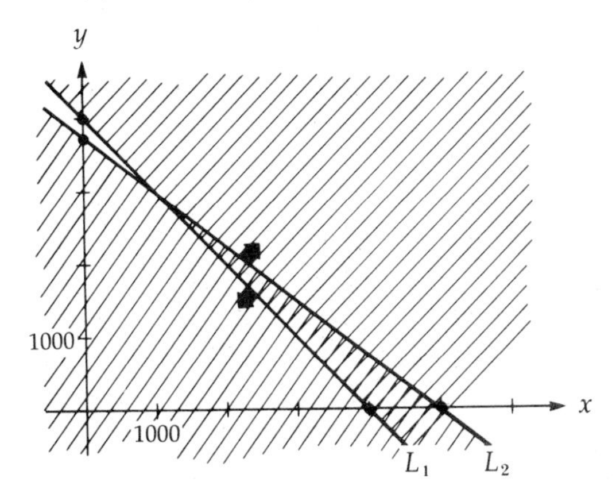

11. (1) $x \geqslant 4y$
(2) $x \geqslant 3000$
(3) $y \leqslant 1000$

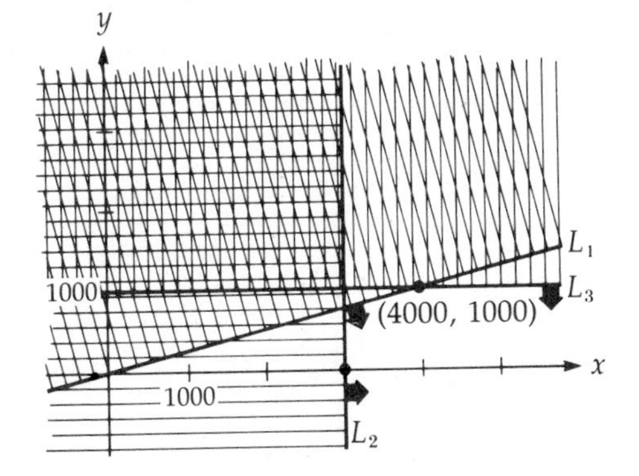

13. $40x + 30y \geq 2000$

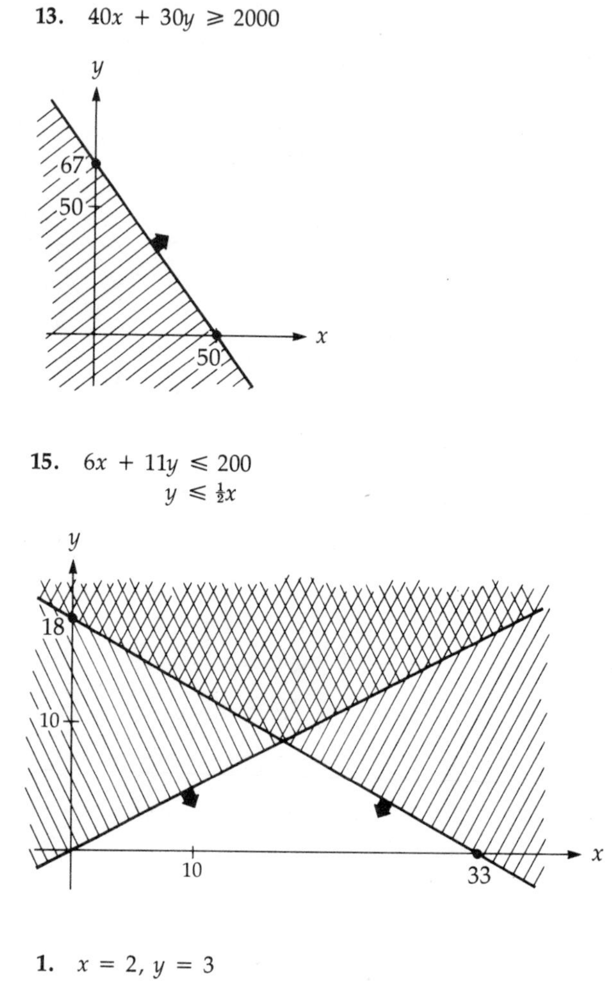

15. $6x + 11y \leq 200$
$y \leq \frac{1}{2}x$

SECTION 1.6

1. $x = 2, y = 3$

3. $x = 2.5, y = 3$

5. $x = 1.85, y = 1.46$

7. $x = .10, y = -.02$

9. $x = 3, y = 4$

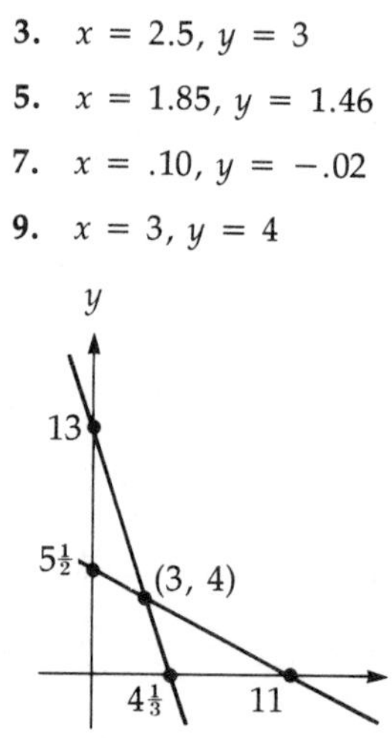

11. $x = 3, y = 2.5$

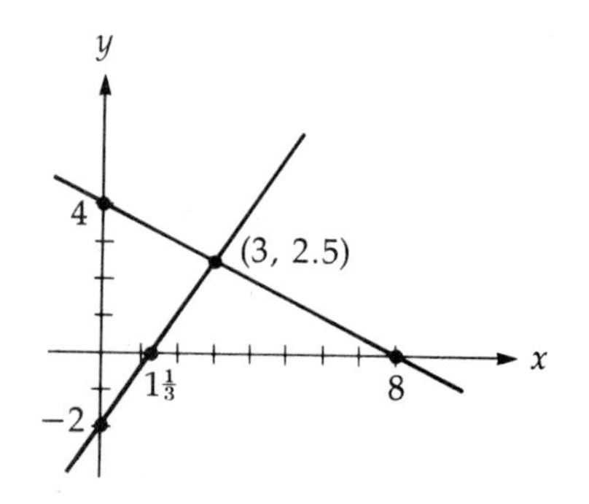

13. (1) $x + y = 1$
(2) $50x + 35y = 40$
$x = \frac{1}{3},\ y = \frac{2}{3}$

15. (1) $x + y = 1$
(2) $45x + 50y = 47$
$x = .6,\ y = .4$

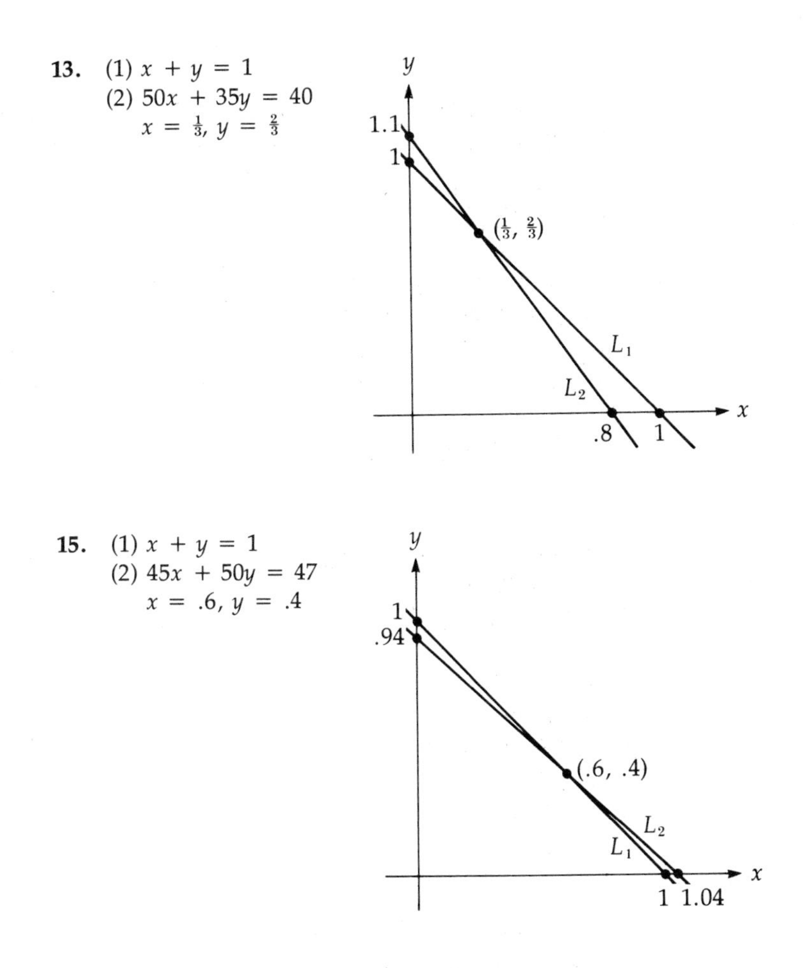

17. Let the trucker take x changers and y dust covers.
(1) $20x + 5y = 20000$
(2) $x + y = 2000$
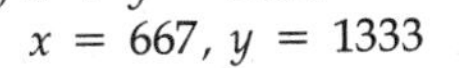
$x = 667,\ y = 1333$

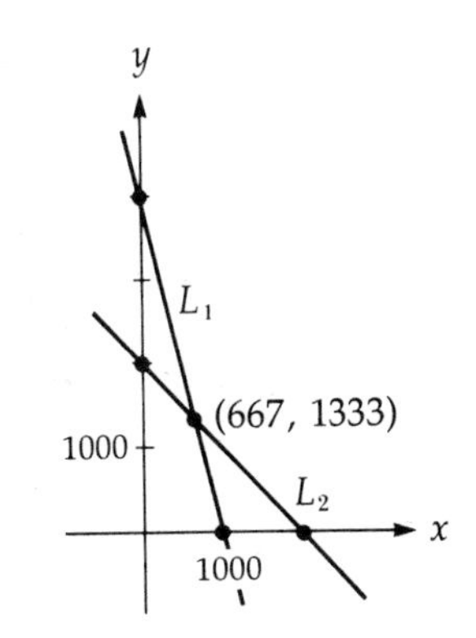

19. Let x cups of evaporated milk and y cups of skim milk be used.
(1) $346x + 87y = 166$
(2) $17.6x + 8.6y = 2(8.5)$
$x = -.04,\ y = 2.1$

Of course, the answer is not acceptable since x is negative; the nutritionist has set an impossible task.

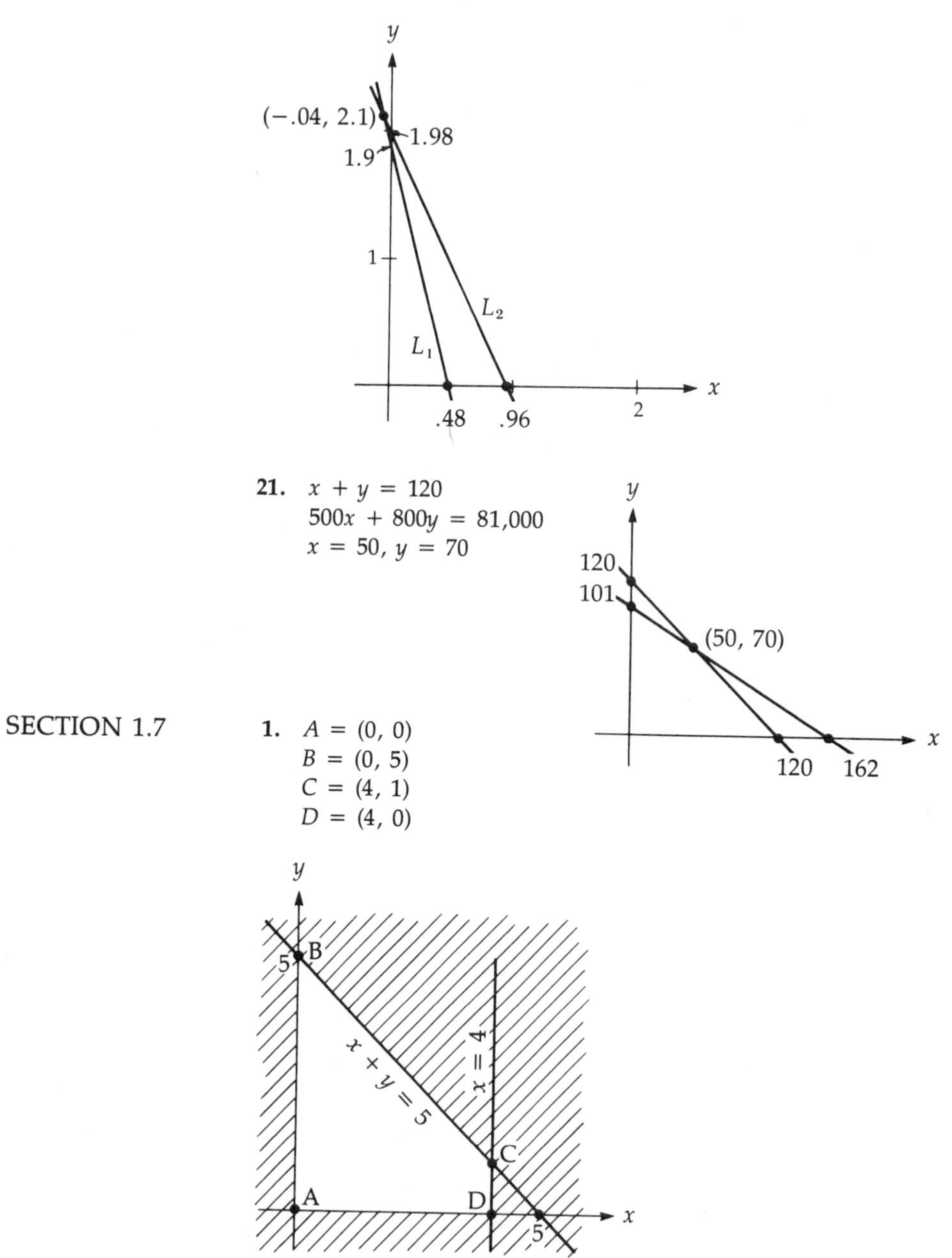

21. $x + y = 120$
$500x + 800y = 81{,}000$
$x = 50,\ y = 70$

SECTION 1.7

1. $A = (0, 0)$
$B = (0, 5)$
$C = (4, 1)$
$D = (4, 0)$

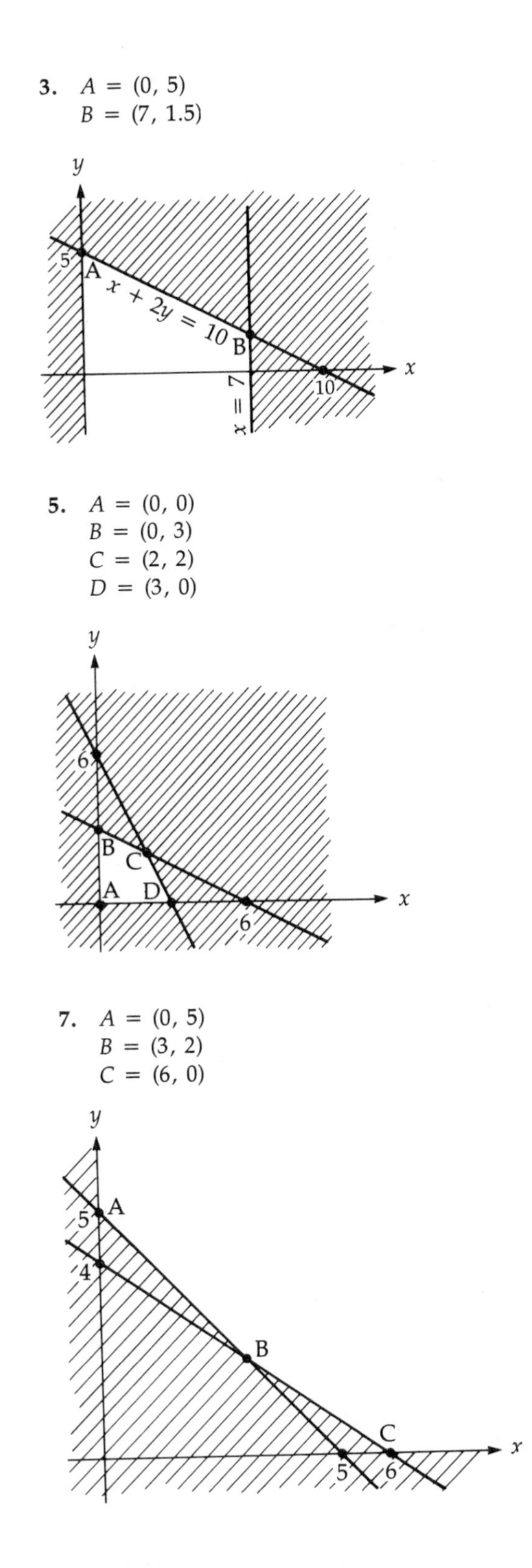

3. $A = (0, 5)$
$B = (7, 1.5)$

5. $A = (0, 0)$
$B = (0, 3)$
$C = (2, 2)$
$D = (3, 0)$

7. $A = (0, 5)$
$B = (3, 2)$
$C = (6, 0)$

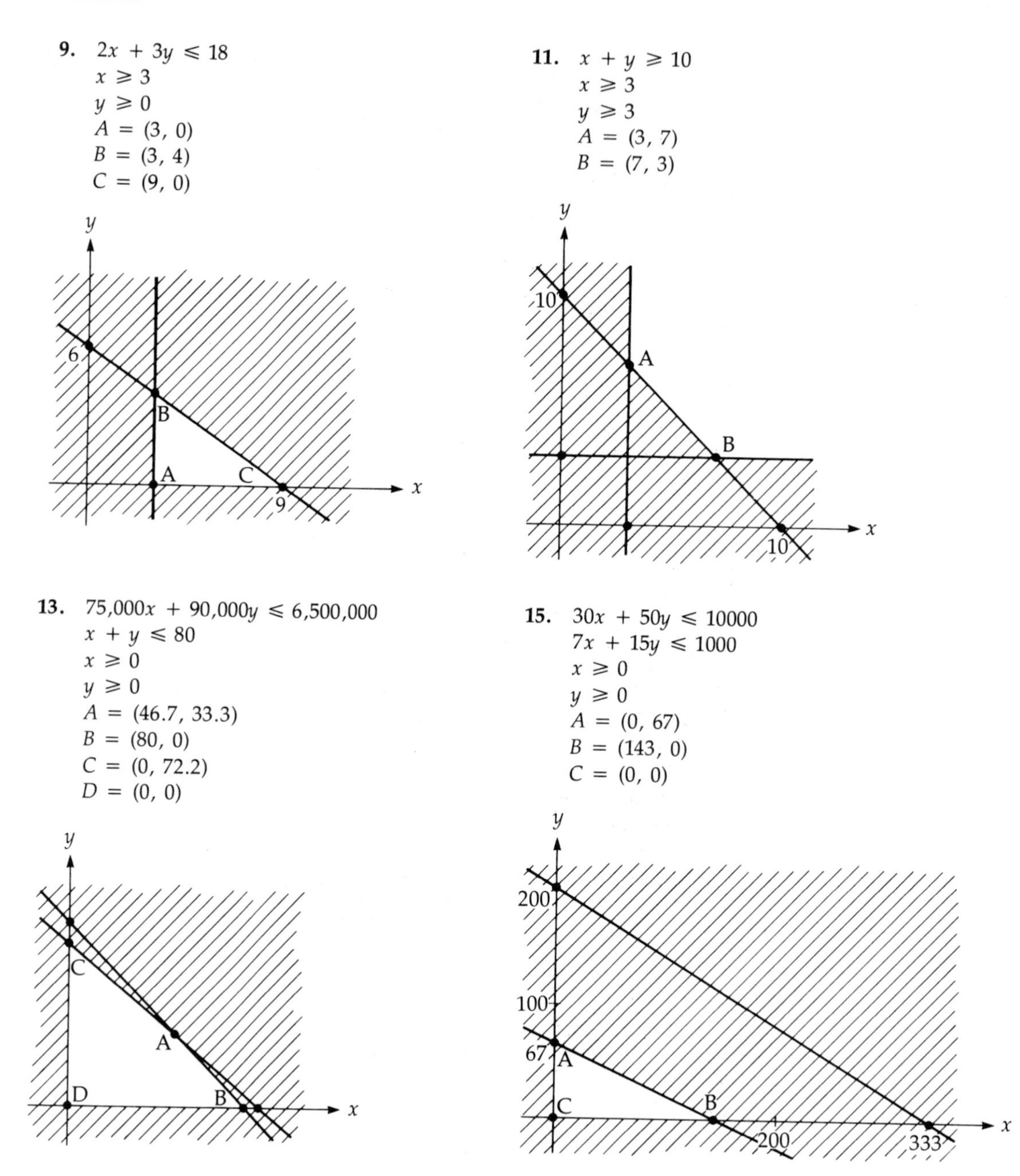

9. $2x + 3y \leqslant 18$
$x \geqslant 3$
$y \geqslant 0$
$A = (3, 0)$
$B = (3, 4)$
$C = (9, 0)$

11. $x + y \geqslant 10$
$x \geqslant 3$
$y \geqslant 3$
$A = (3, 7)$
$B = (7, 3)$

13. $75{,}000x + 90{,}000y \leqslant 6{,}500{,}000$
$x + y \leqslant 80$
$x \geqslant 0$
$y \geqslant 0$
$A = (46.7, 33.3)$
$B = (80, 0)$
$C = (0, 72.2)$
$D = (0, 0)$

15. $30x + 50y \leqslant 10000$
$7x + 15y \leqslant 1000$
$x \geqslant 0$
$y \geqslant 0$
$A = (0, 67)$
$B = (143, 0)$
$C = (0, 0)$

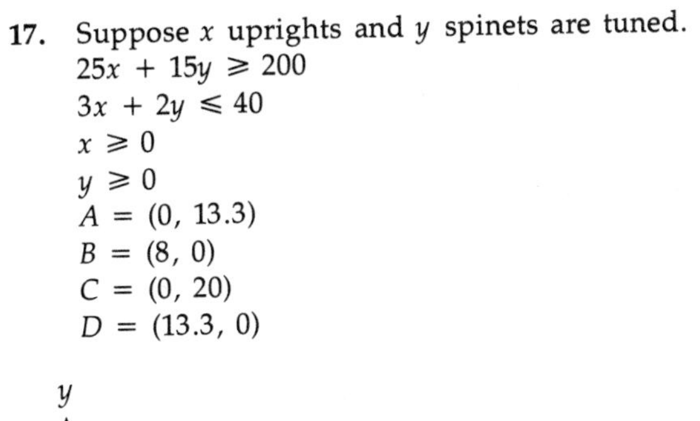

17. Suppose x uprights and y spinets are tuned.

$25x + 15y \geqslant 200$
$3x + 2y \leqslant 40$
$x \geqslant 0$
$y \geqslant 0$
$A = (0, 13.3)$
$B = (8, 0)$
$C = (0, 20)$
$D = (13.3, 0)$

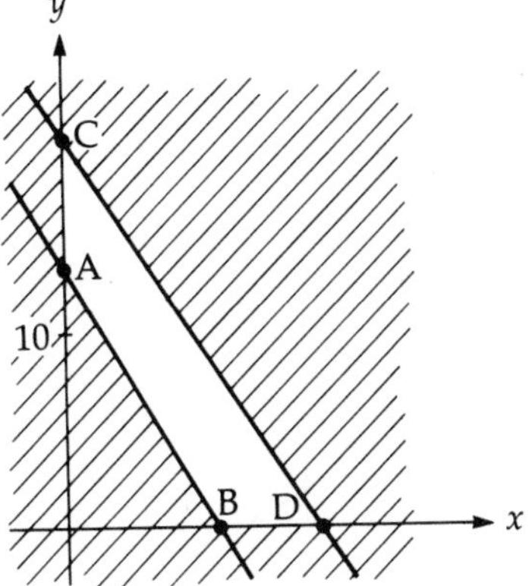

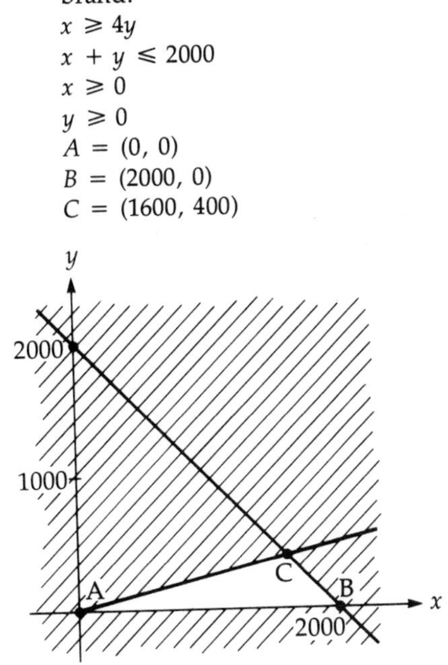

19. Suppose there are x cans of the well-known brand and y cans of the chain brand.

$x \geqslant 4y$
$x + y \leqslant 2000$
$x \geqslant 0$
$y \geqslant 0$
$A = (0, 0)$
$B = (2000, 0)$
$C = (1600, 400)$

SECTION 1.8

1. Maximum 9 at (0, 3)
   Minimum 2 at (1, 0)

3. Maximum 52 at (8, −2)
   Minimum −25 at (1, 5)

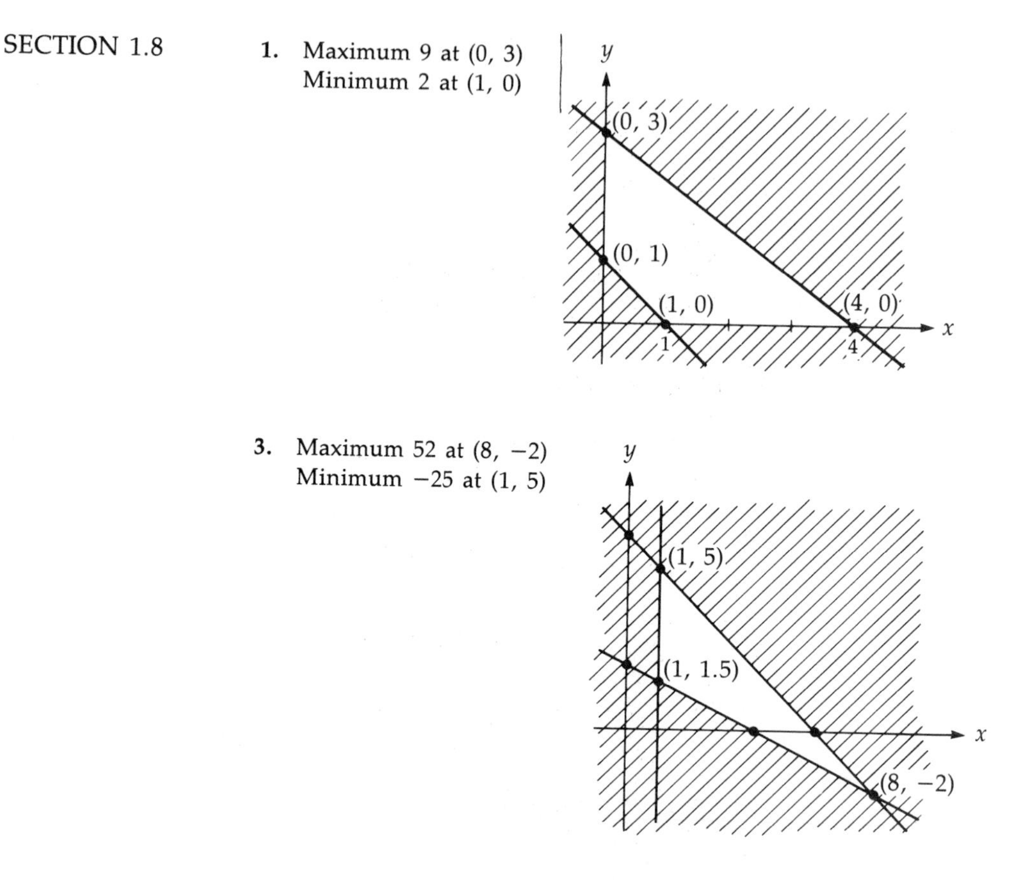

5. If he takes x men's and y women's brushes, then $x + \frac{1}{2}y \leqslant 40$, $x + y \leqslant 50$, $x \geqslant 0$, and $y \geqslant 0$. The profit $x + .8y$ is most for $x = 30$, $y = 20$.

7. If x Type A and y Type B are made, then $10x + 8y \leqslant 8(3600)$, $20x + 25y \leqslant 10(3600)$, $x \geqslant 0$, and $y \geqslant 0$. The number of brushes $x + y$ is most for $x = 1800$, $y = 0$.

9. If x Crustygood and y Ovenfresh are made, then $3x + 2y \leqslant 12$, $2x + 2.5y \leqslant 10$, $x \geqslant 0$, and $y \geqslant 0$. The profit $10x + 14y$ is most for $x = 0$, $y = 4$.

11. If x Fastrunners and y Zips are sold, then $0 \leqslant x \leqslant 80$, $0 \leqslant y \leqslant 50$, $3x + 5y \leqslant 300$. The profit is $x + 1.2y$ and is most for $x = 80$, $y = 12$.

13. If x cups of oatmeal and y cups of wheat germ are used, then $x \geqslant y$, $x + y \leqslant 2$, $x \geqslant 0$, and $y \geqslant 0$. The grams of protein $11x + 17y$ is most for $x = 1$, $y = 1$.

SECTION 1.9

1. $x = 0$ toasters and $y = 6667$ drills minimizes $12x + 8y$, where $20x + 15y \geq 100{,}000$, $x + y \geq 6000$, $x \geq 0$, and $y \geq 0$.

3. $x = 20$ Bibles and $y = 10$ dictionaries maximizes $1.4x + 1.8y$, where $2x + 3y \leq 70$, $x + y \leq 30$, $x \geq 0$, and $y \geq 0$.

5. $x = 86087$ blue spruce and $y = 28261$ Norway spruce maximizes $1.35x + 1.41y$, where $.13x + .11y \leq 14300$, $10x + 12y \leq 1{,}200{,}000$, $x \geq 0$, and $y \geq 0$.

7. $x = 667$ 9-by-12's and $y = 1333$ 9-by-9's maximizes $40x + 35y$, where $x + y \leq 2000$, $12x + 9y \leq 20{,}000$, $x \geq 0$, and $y \geq 0$.

9. The vertices are (0, 0), (0, 100), and (100, 0), and the value of the prize corresponding to each of these is \$0. Yet (50, 50) is within the region of possibilities, and there the prize is worth \$2500. The reason this does not contradict the linear programming theorem is that the prize value V is given by

$$V = xy,$$

and *this is not a linear function of x and y.*

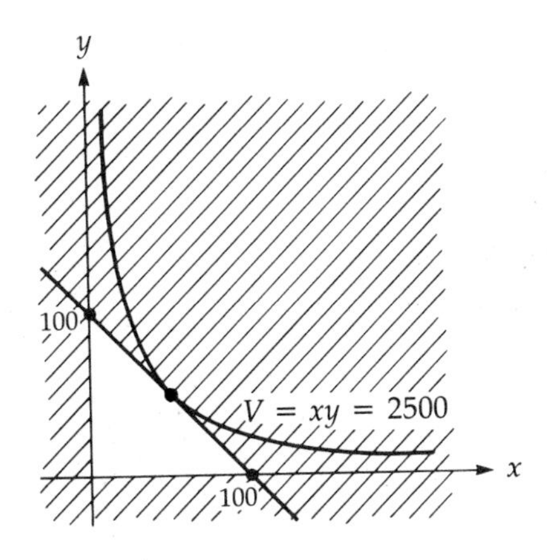

SECTION 2.2

1.

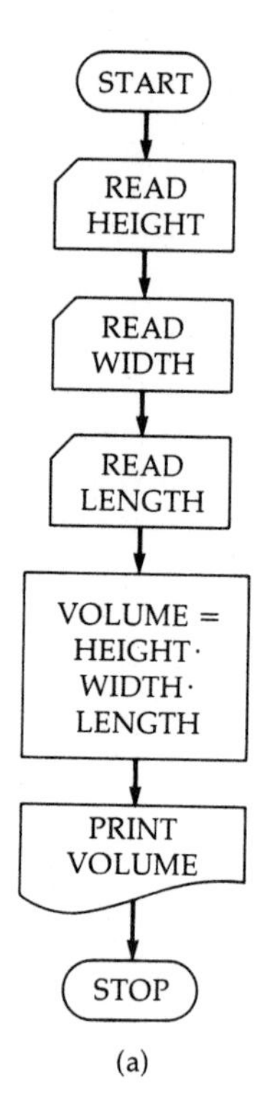

(a)

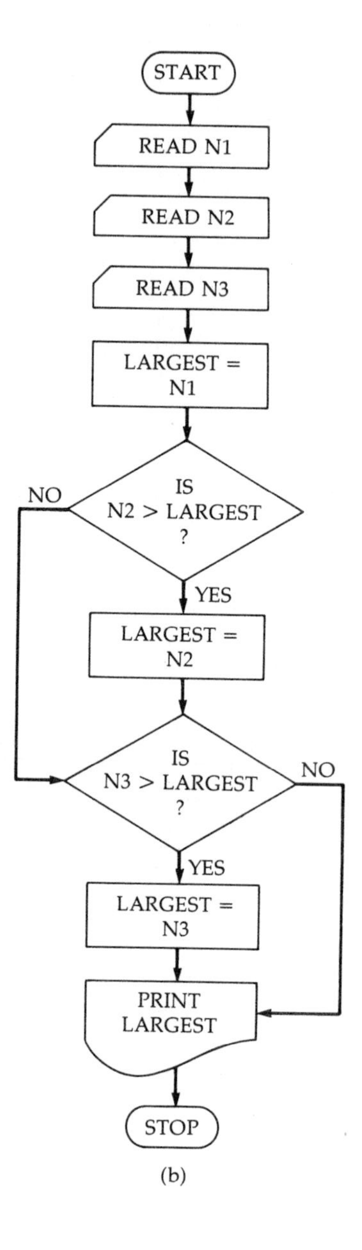

(b)

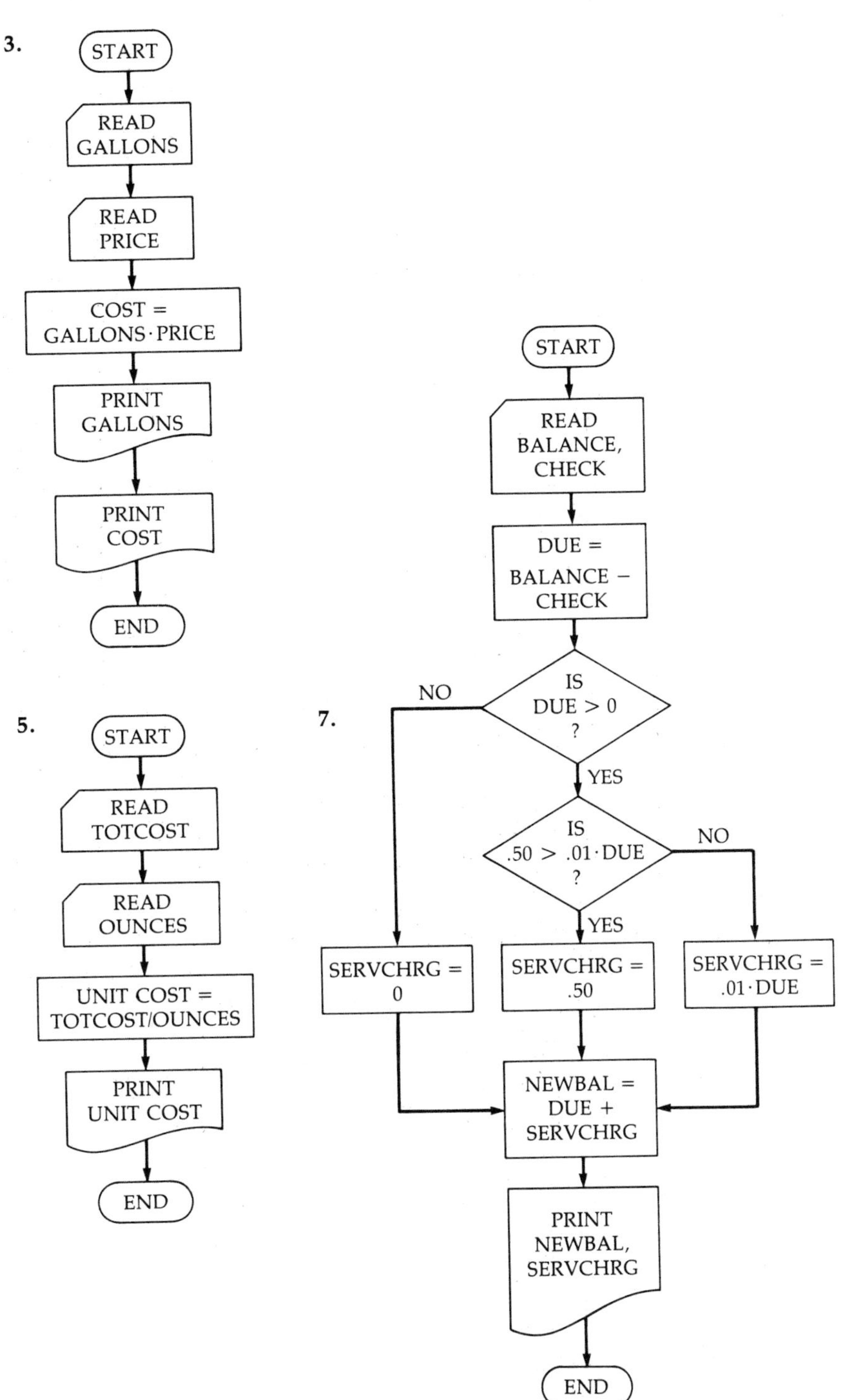
3.
START
READ GALLONS
READ PRICE
COST = GALLONS·PRICE
PRINT GALLONS
PRINT COST
END
5.
START
READ TOTCOST
READ OUNCES
UNIT COST = TOTCOST/OUNCES
PRINT UNIT COST
END
7.
START
READ BALANCE, CHECK
DUE = BALANCE − CHECK
IS DUE > 0 ?
NO
YES
IS .50 > .01·DUE ?
NO
YES
SERVCHRG = 0
SERVCHRG = .50
SERVCHRG = .01·DUE
NEWBAL = DUE + SERVCHRG
PRINT NEWBAL, SERVCHRG
END

9.

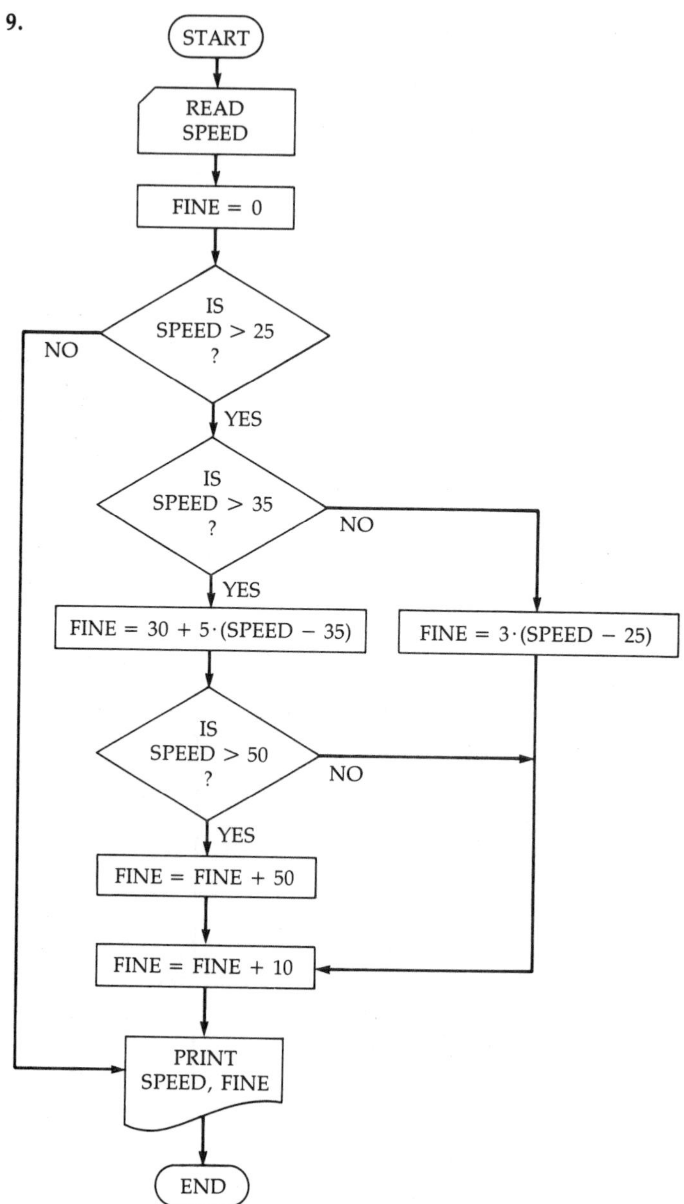
START
READ SPEED
FINE = 0
IS SPEED > 25 ?
NO
YES
IS SPEED > 35 ?
NO
YES
FINE = 30 + 5·(SPEED − 35)
FINE = 3·(SPEED − 25)
IS SPEED > 50 ?
NO
YES
FINE = FINE + 50
FINE = FINE + 10
PRINT SPEED, FINE
END

SECTION 2.3

1. (a) The procedure reads, consecutively, seven numbers while keeping a running total. The output is the seven numbers and their sum. 28.
   (b) The procedure reads three numbers, sets P equal to the third plus twice the sum of the first two and prints the 3 given numbers and P. 12, 10, 27, 71.
   (c) The procedure reads N, finds the product of N numbers read consecutively, and prints N and the Nth root of the product. In other words, it outputs N and the geometric mean of N given numbers. 3, 30.

3.

START

SET SUM to 0

SET counter to 0

Read a number

Add it to SUM

Add 1 to counter

If counter $<$ 10 go back to read another number

Otherwise divide sum by 10

Output the result

STOP

START
SUM = 0
C = 0
READ NUM
SUM = SUM + NUM
C = C + 1
IS $C \geqslant 10$?
NO
YES
AVE = SUM/10
PRINT AVE
END

5.

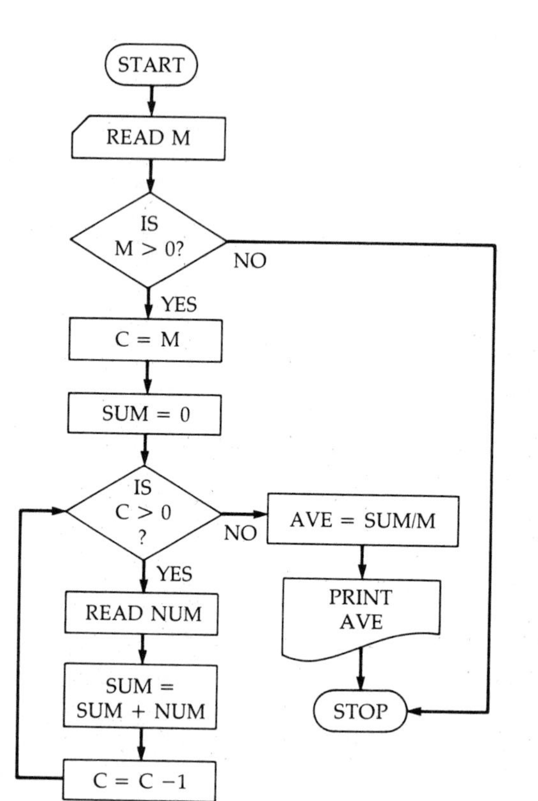
START
READ M
IS
M > 0?
NO
YES
C = M
SUM = 0
IS
C > 0
?
NO
AVE = SUM/M
YES
READ NUM
PRINT
AVE
SUM =
SUM + NUM
STOP
C = C −1

7.

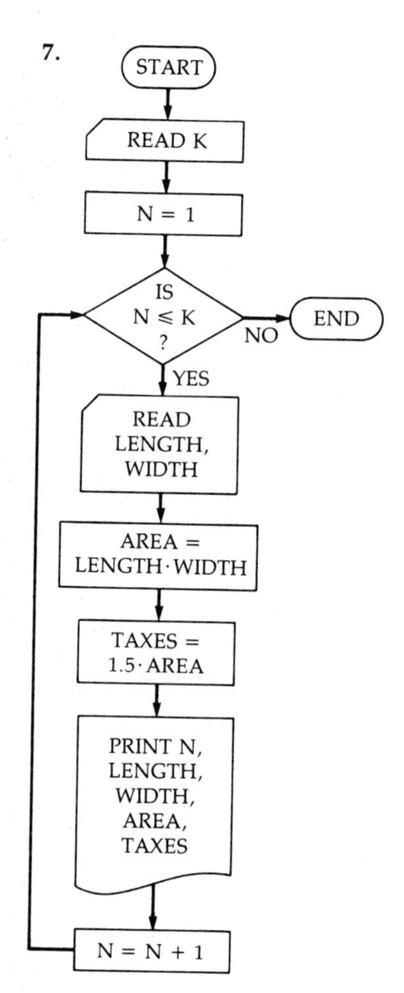
START
READ K
N = 1
IS
N ≤ K
?
NO
END
YES
READ
LENGTH,
WIDTH
AREA =
LENGTH · WIDTH
TAXES =
1.5 · AREA
PRINT N,
LENGTH,
WIDTH,
AREA,
TAXES
N = N + 1

9.

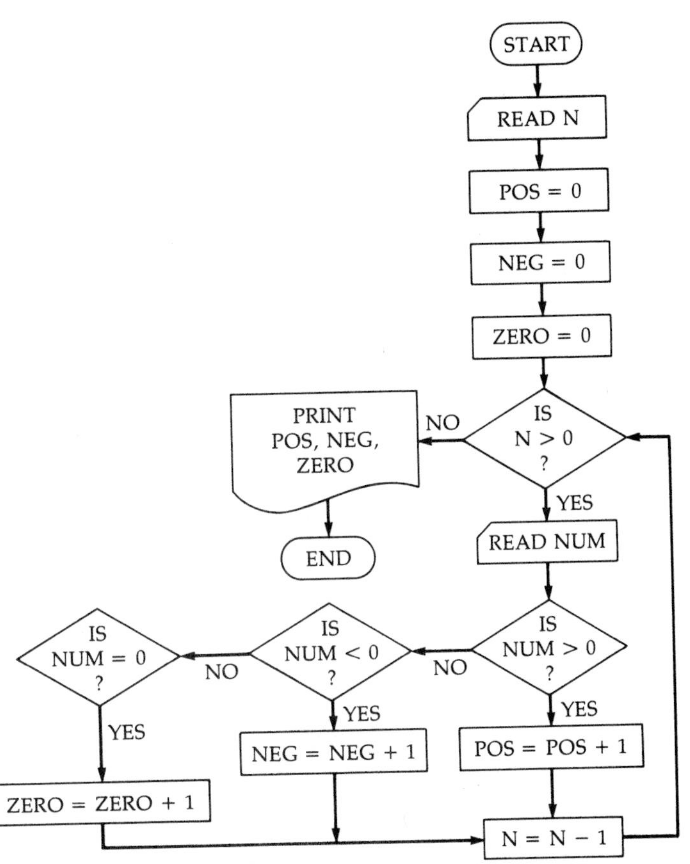

SECTION 2.4

1. (a) Integer (b) None
(c) Exponential (d) Decimal
(e) Decimal (f) Integer
(g) Integer (h) Decimal
(i) Exponential (j) None
(k) Decimal (l) Integer

3. (a) 32, integer (b) 15, integer
(c) 51.528, decimal (d) 17.23, decimal
(e) 15, decimal (f) 3.6, decimal

5. (a) $3a^2 + 15a + 43$ (b) $5x^2y^3 - 14.6xy^2 + 5.93y$
(c) $\dfrac{a^3b^4}{c} + \dfrac{2ba^{(2+c)}(-2)}{c}$ (d) $\dfrac{b^3(2^3)(b^{-2})}{a} = \dfrac{2^3b^3b^{-2}}{a} = \dfrac{8b}{a}$
(e) $3a^2 + 4a - 17$ (f) $x^2y + 3xy^2 - 17x^3y^2$

7. (a) X < = 3Y
(b) K > = 5
(c) X ↑ 2 = Y ↑ 3 + 16 or X**2 = Y**3 + 16
(d) X > Y

9. (a) (X ↑ 2 + Y ↑ 2) ↑ (1 / 2) or (X**2 + Y**2)**.5
(b) H * 3.14 * R ↑ 2

11. (29 − X) * M + (33 − Y) * N

13. 130 * (C ↑ 2 * 6.7) + (15 * T) * (B ↑ 3 * 78.9)

SECTION 2.5

1.
```
10 READ X, Y
20 LET A = (X + Y) / 2
30 PRINT X, Y, A
40 DATA 7, 11
50 END
```

3.
```
10 READ A, B
20 LET C = (A ↑ 2 + B ↑ 2) ↑ (1 / 2)
30 PRINT A, B, C
40 DATA 5, 12
50 END
```

5.
```
10 LET L = 6.3
20 LET A = L * L
30 LET V = A * L
40 PRINT L, A, V
50 END
```

7.
```
10 READ C, W
20 LET A = C / W
30 DATA 3.89,16
40 PRINT A
50 END
```

9. (a)
```
10 READ X, Y
20 LET H = (X**2 + Y**2)**.5
30 PRINT X, Y, H
40 DATA
50 END
```

(b)
```
10 READ H, R
20 LET V = H * 3.14 * R ↑ 2
30 PRINT H, R, V
40 DATA
50 END
```

11.
```
10 READ X, Y
20 READ M, N
30 LET P = (29 − X) * M + (33 − Y) * N
40 PRINT P
50 DATA
60 END
```

13.
```
10 READ C
20 LET S = C ↑ 2 * 6.7
30 READ B
40 LET I = B ↑ 3 * 78.9
50 READ T
60 LET W = 130 * S + 15 * T * I
70 DATA
80 PRINT W
90 END
```

SECTION 2.7

1. (a)
```
03 REM THIS PROGRAM ADDS 7 NUMBERS
10 LET S = 0
20 LET K = 1
```

```
30  PRINT "WHAT IS THE NEXT ADDEND"
40  INPUT A
50  LET S = S + A
60  IF K = 7 THEN 90
70  LET K = K + 1
80  GO TO 30
90  PRINT "THE SUM IS"; S
99  END
```

(b)

```
05  PRINT "WHAT ARE THE THREE NUMBERS"
10  READ A, B, C
20  LET P = 2 * (A + B) + C
30  PRINT A, B, C, P
40  DATA
50  END
```

(c)

```
05  REM THIS PROGRAM FINDS A GEOMETRIC MEAN
10  PRINT "HOW MANY NUMBERS WILL BE INPUT?"
15  INPUT N
20  LET K = N
25  LET G = 1
30  PRINT "WHAT IS THE NEXT NUMBER?"
35  INPUT F
40  LET G = G * F
45  LET K = K - 1
50  IF K > 0 THEN 30
55  LET G  =  G**(1 / N)
60  PRINT "THE PROGRAM READ"; N; "NUMBERS"
65  PRINT "THEIR GEOMETRIC MEAN IS"; G
70  END
```

3.

```
05  REM THIS PROGRAM AVERAGES 10 NUMBERS
10  LET S = 0
15  LET K = 0
20  PRINT "WHAT IS THE NEXT NUMBER?"
25  INPUT N
30  LET S = S + N
35  LET K = K + 1
40  IF K < 10 THEN 20
45  LET A = S / 10
50  PRINT "THE AVERAGE IS"; A
55  END
```

5.

```
05  REM THIS PROGRAM AVERAGES M NUMBERS
10  PRINT " HOW MANY NUMBERS"
12  INPUT M
15  IF M < = 0 THEN 70
20  LET C = M
25  LET S = 0
30  IF C < = 0 THEN 60
```

```
35  PRINT "NEXT NUMBER"
40  INPUT N
45  LET S = S + N
50  LET C = C - 1
55  GO TO 30
60  LET A = S / M
65  PRINT A
70  END
```

7.

```
02  REM THIS PROGRAM COMPUTES THE
03  REM TAXES ON K LOTS GIVEN
04  REM LENGTH, WIDTH, AND RATE.
10  PRINT "HOW MANY LOTS"
15  INPUT K
20  LET N = 1
25  IF N > K THEN 60
30  PRINT "TYPE THE LENGTH AND WIDTH"
35  INPUT L, W
40  LET A = L * W
45  LET T = A * 1.5
50  PRINT N, L, W, A, T
55  LET N = N + 1
58  GO TO 25
60  END
```

9.

```
02  REM THIS PROGRAM FINDS THE NUMBER
03  REM OF POSITIVE, NEGATIVE, AND ZERO
04  REM VALUES IN N NUMBERS
05  PRINT "HOW MANY NUMBERS"
10  INPUT K
15  LET P = 0
20  LET N = 0
25  LET Z = 0
30  IF K < = 0 THEN 90
35  PRINT "WHAT IS THE NEXT NUMBER"
40  INPUT M
45  IF M < = 0 THEN 60
50  LET P = P + 1
55  GO TO 80
60  IF M = 0 THEN 75
65  LET N = N + 1
70  GO TO 80
75  LET Z = Z + 1
80  LET K = K - 1
85  GO TO 30
90  PRINT P, N, Z
95  END
```

11.

```
02  REM THIS PROGRAM FINDS THE NEW
03  REM BALANCE FOR A CUSTOMER BY
```

```
04  REM DEDUCTING HIS PAYMENT FROM THE
05  REM OLD BALANCE AND ADDING
06  REM THE LARGER OF $.50 and 1% OF
07  REM REMAINING BALANCE.
10  PRINT "TYPE OLD BALANCE AND CHECK AMOUNT"
15  INPUT B, C
20  LET D = B − C
25  IF D < = 0 THEN 55
30  IF .50 < = .01 * D THEN 45
35  LET S = .50
40  GO TO 60
45  LET S = .01 * D
50  GO TO 60
55  LET S = 0
60  N = D + S
65  PRINT N, S
70  END
```

13.

```
02  REM THIS PROGRAM COMPUTES THE FINE
03  REM FOR A SPEEDING TICKET. THE
04  REM FINE IS $3/MILE OVER 25
05  REM AND UP TO 35, $5/MILE FOR
06  REM EACH MILE OVER 35, AND $50
07  REM ADDITIONAL FINE IF THE SPEED
08  REM IS OVER 50. THE FINE ALSO
09  REM INCLUDES A COURT FEE OF $10.
15  PRINT "TYPE SPEED"
20  INPUT S
25  LET F = 0
30  IF S < = 25 THEN 70
35  IF S < = 35 THEN 60
40  LET F = 30 + 5 * (S − 35)
45  IF S < = 50 THEN 65
50  LET F = F + 50
55  GO TO 65
60  LET F = 3 * (S − 25)
65  LET F = F + 10
70  PRINT "SPEED = "; S, "FINE = "; F
75  END
```

SECTION 2.8

1.

```
10  PRINT "TYPE YOUR SOCIAL SECURITY NUMBER"
15  PRINT "(SEPARATE THE DIGITS WITH COMMAS)"
20  INPUT A, B, C, D, E, F, G, H, I, J
25  PRINT "IF C IS EVEN, TYPE 1"
26  PRINT "IF C IS ODD, TYPE −1"
30  INPUT M
35  PRINT "IF F IS EVEN, TYPE 1"
36  PRINT "IF F IS ODD, TYPE −1"
```

```
 40  INPUT L
 45  LET N = L * M
 50  LET Y = M * (G * 10 + H + I * .1)**(N * (A + .1 * B))
 55  LET X = (100 * A + 10 * B + C + .1 * D + .01 * E)**F * Y**3
 60  LET P =100 * A + 10 * B + C
 65  LET Q = 10 * D + E
 70  LET R = 1000 * F + 100 * G + 10 * H + I
 75  PRINT "MY SOCIAL SECURITY NUMBER IS"
 80  PRINT P; "-"; Q; "-"; R
 85  PRINT "M ="; M
 90  PRINT "N ="; N
 95  PRINT "Y ="; Y
100  PRINT "X ="; X
110  END
```

3.

```
03  REM THIS PROGRAM COMPUTES THE SUM OF
06  REM THE CUBES OF THE FIRST N INTEGERS.
12  PRINT "TYPE N"
15  INPUT N
18  IF N < = 0 THEN 39
21  LET S = 0
24  FOR I = 1 TO N
27  LET S = S + I**3
30  NEXT I
33  PRINT "THE SUM OF THE CUBES OF THE FIRST";
36  PRINT N; "NUMBERS IS"; S
39  END
```

5.

```
03  REM THIS PROGRAM COMPUTES THE SUM OF
06  REM THE CUBES OF THE FIRST N INTEGERS
09  REM AND THE SQUARE OF THEIR SUM.
12  PRINT "TYPE N"
15  INPUT N
18  IF N < = 0 THEN 54
21  LET S = 0
24  LET C = 0
27  FOR I = 1 TO N
30  LET S = S + I
33  LET C = C + I**3
36  NEXT I
39  LET S = S**2
42  PRINT "THE SUM OF THE CUBES OF THE FIRST"; N;
45  PRINT "NUMBERS IS"; G
46  PRINT
48  PRINT "THE SQUARE OF THE SUM OF THE FIRST"; N;
51  PRINT "NUMBERS IS"; S
54  END
```

7.

```
03  REM THIS PROGRAM COMPUTES THE QUOTIENTS
04  REM WHEN A = 7**(2 * N) + 16 * N + 63 IS DIVIDED BY
```

```
05  REM 64 AND 128 FOR VALUES OF N FROM 0 TO 70
10  PRINT "N", "A/64 =", "A/128 ="
15  PRINT
20  FOR N = 0 TO 70
30  LET A = 7**(2 * N) + 16 * N + 63
40  LET B = A / 64
50  LET C = A / 128
60  PRINT N, B, C
70  NEXT N
80  END
```

11.

```
05  REM THIS PROGRAM PRINTS A TABLE
06  REM CONVERTING MILES TO KILOMETERS.
10  PRINT "MILES", "KILOMETERS"
11  PRINT
15  FOR M = 1 TO 10
20  LET K = 1.6093 * M
25  PRINT M, K
30  NEXT M
40  END
```

13.

```
02  REM THIS PROGRAM INPUTS THE NUMBER OF
03  REM HOUSES AND THE AREA OF EACH
04  REM ROOF IN SQUARE FEET. THE COST
05  REM FOR EACH ROOF IS COMPUTED ON
06  REM THE BASIS OF $22.97 FOR EACH
07  REM 100 SQUARE FEET. THE NUMBER,
08  REM AREA AND COST OF EACH ROOF IS
09  REM PRINTED AS WELL AS TOTAL AND
10  REM AVERAGE COST.
11  PRINT "HOW MANY HOUSES?"
12  INPUT M
13  IF M < = 0 THEN 85
14  IF M > 10 THEN 85
15  LET T = 0
20  FOR I = 1 TO M
25  PRINT "WHAT IS THE AREA?"
30  INPUT A(I)
35  LET C(I) = A(I) * .2297
40  LET T = T + C(I)
45  NEXT I
50  PRINT "HOUSE", "AREA", "COST"
55  PRINT
60  FOR I = 1 TO M
65  PRINT I, A(I), C(I)
70  NEXT I
71  PRINT
75  PRINT "TOTAL COST ="; T
80  PRINT "AVERAGE COST ="; T / M
85  END
```

SECTION 2.9

1. **(a)** Integer **(b)** Neither
(c) Floating point **(d)** Floating point
(e) Floating point **(f)** Integer
(g) Integer **(h)** Neither
(i) Floating point **(j)** Neither
(k) Floating point **(l)** Integer

3. **(a)** 32, integer **(b)** 15, integer
(c) 51.528, floating **(d)** 17.23, floating
(e) 15., floating **(f)** 3.6, floating

5. **(a)** $3a^2 + 15a + 43$ **(b)** $5x^2y^3 - 14.6xy^2 + 5.93y$

(c) $\dfrac{a^3b^4}{c} - \dfrac{4a^{(2+c)}\,b}{c}$ **(d)** $\dfrac{b^3}{a} \cdot 2^3b^{-2}$

(e) $3a^2 + 4a - 17$ **(f)** $x^2y + 3xy^2 - 17x^3y^2$

7. **(a)** (X**2 + Y**2)**(1/2)
or SQRT (X**2 + Y**2) **(b)** H * 3.14 * R**2

9. M * (29 − X) + N * (33 − Y)

11. 130 * C**2 * 6.7 + 15 * T * B**3 * 78.9

SECTION 2.10

1.
```
READ, A, B
AVG = (A + B) / 2
PRINT, A, B, AVG
STOP
END
```

3.
```
READ, A, B
H = (A**2 + B**2)**(1/2)
PRINT, A, B, H
STOP
END
```

5.
```
EDGE = 6.3
FACE = EDGE**2
VOL = EDGE**3
PRINT, EDGE, FACE, VOL
STOP
END
```

7.
```
COST = 3.89
NUMOZ = 16
CPOZ = COST / NUMOZ
PRINT, CPOZ
STOP
END
```

9. **(a)**
```
READ, X, Y
Z = (X**2 + Y**2)**.5
PRINT, X, Y, Z
STOP
END
```

(b)
```
READ, R, H
VOL = H * 3.14 * R**2
PRINT, R, H, VOL
STOP
END
```

11.
```
READ, X, Y
READ, M, N
P = M * (29 - X) + N * (33 - Y)
PRINT, X, Y, M, N, P
STOP
END
```

13.
```
READ, C, B, T
WEIGHT = 130 * C**2 * 6.7 + 15 * T * B**3 * 78.9
PRINT, WEIGHT
STOP
END
```

SECTION 2.11

1.

```
      K = 1
      READ, X
      Y = X
    3 K = K + 1
      READ, X
      IF (Y .GE. X) GO TO 5
      Y = X
    5 IF (K .LT. 5) GO TO 3
      PRINT, Y
      STOP
      END
```

3.

```
      SUM = 0
      N = 0
    5 READ, X
      SUM = SUM + X
      N = N + 1
      IF (N .LT. 10) GO TO 5
      AVG = SUM / 10
      PRINT, AVG
      STOP
      END
```

5.

```
      READ, M
      IF (M .LE. 0) GO TO 70
      K = M
      SUM = 0
   30 IF (K .LE. 0) GO TO 60
      READ, NUM
      SUM = SUM + NUM
      K = K - 1
      GO TO 30
   60 AVG = SUM / M
      PRINT, AVG
   70 END
```

7.

```
      READ, K
      N = 1
   25 IF (N .GT. K) GO TO 60
      READ, H, W
      A = H * W
      T = A * 1.5
      PRINT, N, H, W, A, T
      N = N + 1
      GO TO 25
   60 END
```

9.

```
      READ, N
      IPOS = 0
      NEG = 0
      IZERO = 0
    3 IF (N .LE. 0) GO TO 9
      READ, NUM
      IF (NUM .LE. 0) GO TO 6
      IPOS = IPOS + 1
      GO TO 8
    6 IF (NUM .EQ. 0) GO TO 7
      NEG = NEG + 1
      GO TO 8
    7 IZERO = IZERO + 1
    8 N = N - 1
      GO TO 3
    9 PRINT, IPOS, NEG, IZERO
      END
```

11. **(a)** X .LE. 3 * Y **(b)** K .GE. 5
(c) X**2 = Y**3 + 16 **(d)** X .GT. Y

13. Given 3 numbers x, y, and z, the program computes $(x + y + z) - 20 + y$ and prints y and the calculation result.

15. The program prints the number 16 and the sum of the cubes of the first 16 positive integers.

17. **(a)** Need a comma after READ
(b) N. E. should be .NE.
(c) Cannot have 2 assignment statements on one card
(d) Cannot print 21 since it is undefined

19. **(a)** The value of ISM will not change but it will contain a decimal.
(b) ISM would be the sum of the squares of the first 14 integers rather than the first 14 odd integers.
(c) It is printed inside the loop and outside.

SECTION 2.12

3.

```
      .
      .
      .
    2 YGROS = 52 * G - 775 * D
      IF (YGROS .LE. 9500) GO TO 10
      IF (YGROS .LE. 11500) GO TO 20
      TAX = .2 * G
      GO TO 4
   10 TAX = .16 * G
      GO TO 4
   20 TAX = .18 * G
    4 DED = .01 * (G - TAX)
      .
      .
      .
```

5.

```
      DO 5 I = 1, 9
      READ, K (I)
    5 CONTINUE
      KNUM = K(3) / 2
      IF (KNUM * 2 = K(3)) GO TO 10
      M = -1
      GO TO 15
   10 M = 1
   15 KNUM = K(6) / 2
      IF (KNUM * 2 = K(6)) GO TO 20
      N = -1 * M
      GO TO 25
   20 N = M
   25 Y = M * (10 * K(7) + K(8) + .1 * K(9))**(N * (K(1) + .1 * K(2)))
      X = (100 * K(1) + 10 * K(2) + K(3) + .1 * K(4) + .01 * K(5))**K(6) * Y**3
      I = 100 * K(1) + 10 * K(2) + K(3)
      J = 10 * K(4) + K(5)
      L = 1000 * K(6) + 100 * K(7) + 10 * K(8) + K(9)
      PRINT, I, "-",J,"-",L
      PRINT,"M=",M,"N=",N
      PRINT,"X=",X,"Y=",Y
      STOP
      END
```

7.

```
      KSUM = 0
      READ, N
      DO 5 I = 1, N
      KSUM = KSUM + I**3
    5 CONTINUE
      PRINT, N, KSUM
      STOP
      END
```

9.

```
      NSUM = 0
      NCUBES = 0
      READ, N
      DO 5 I = 1, N
      NSUM = NSUM + I
      NCUBES = NCUBES + I**3
    5 CONTINUE
      NSUM = NSUM**2
      PRINT, N, NSUM, NCUBES
      STOP
      END
```

11.

```
      PRINT, "N", "M / 64", "M / 128"
      PRINT
      DO 1 N = 0, 70
      M = 7**(2 * N) + 16 * N + 63
      B = M / 64
      C = M / 128
      PRINT, N, B, C
    1 CONTINUE
      STOP
      END
```

15.

```
      PRINT, "MILES", "KILOMETERS"
      DO 1 I = 1, 10
      C = I * 1.6093
      PRINT, I, C
    1 CONTINUE
      STOP
      END
```

17.

```
      PRINT, "HOUSE", "AREA", "COST"
      READ, M
      IF (M .LE. 0) THEN 25
      IF (M .GT. 10) THEN 25
      TOTAL = 0
      DO 5 I = 1, M
      READ, AREA (I)
      COST (I) = AREA (I) * .2297
      PRINT, I, AREA (I), COST (I)
      TOTAL = TOTAL + COST (I)
    5 CONTINUE
      AVGCST = TOTAL / M
      PRINT, "TOTAL COST =", TOTAL
      PRINT, "AVERAGE COST =", AVGCST
   25 STOP
      END
```

SECTION 3.1

1. 4/54 = 2/27

3. 27/54 = 1/2

5. 0

7. 3/8

9. 1/16

11. 8/52 = 2/13

13. 8/47

SECTION 3.2

1. −4/9 = −44¢

3. $-6\frac{1}{4}$¢

5. 5/8¢

7. −1/19 = −5¢

9. −10¢

11. 42¢

13. A, 2.6¢; B, 2.3¢; the game of problem 12, 2.5¢; game A is best.

SECTION 3.3

1. 36, 25/36

3. **(a)** 64/216 **(b)** 16/216 **(c)** 4/216 **(d)** 4/216 **(e)** 1/216 **(f)** 1/216

5. −16/216 dollars = −7.4¢

7. 136

9. −5/16¢

11. 6,760,000

13. 7! = 5040

15. \$100/5! = 83¢

SECTION 3.4

1. (2/31)(1/7) = 2/217

3. .3885

5. −19¢

7. $(.8)^5$ = .32768

9. **(a)** 4/52 **(b)** 1/2 **(c)** 2/52 **(d)** 28/52

11. (.9)(.6)(.5) = .27

13. −21¢

15. .2401

17. 4

19. (.05)(.2)(.1) = .001

21. (.7)(.8) = .56

SECTION 3.5

1. 1/40

3. 1/2000

5. (1/100)(19/20) = 19/2000

7. 1/2

9. 6/25

11. 21/38

13. **(a)** 4/52 + 6/52 = 5/26
(b) 4/52 + 6/52 − 2/52 = 2/13
(c) 4/52 + 6/52 − 4/52 = 3/26
(d) 1/2 + 6/52 − 3/52 = 29/52

15. 1/2 + 1/10 − 1/15 = 8/15

17. 1/4

19. .0014995

21. 1/4 + 1/4 − 1/16 = 7/16

SECTION 3.6

1. 24, 720 **3.** 210 **5.** 210 **7.** 35 **9.** 120

11. $10! / 7!3! = 120$ **13.** 36

15. Counting A-2-3-4-5 as a straight makes 5148 flushes and 10240 straights. Subtracting off the 40 straight flushes in each category gives the *Almanac*'s figures.

17. $52! / 47! = 311{,}875{,}200$ **19.** $40! / 30!10!$ **21.** $\binom{11}{5} = 462$

23. $12! / 9! = 1320$ **25.** $8! / 5! = 336$ **27.** $\binom{8}{3} = 56$

29. $5! = 120$ **31.** $30! / 27! = 24{,}360$ **33.** $\binom{21}{6} = 54{,}264$

SECTION 3.7

1. 1/36, 2/36, 4/36, 5/36, 6/36, 5/36, 3/36, 2/36, 1/36

3. 1/9 **5.** 15/216 **7.** 1/2 **9.** −41¢ **11.** $(.6)^5 = .078$

13. Six teams, −80¢ **15.** $(11/47)(10/46)(9/45) = .01$

17. $1 - (45/47)(44/46)(43/45) = .125$ **19.** 5/21 **21.** $52!/39!13!$

23. $26!/(13!)^2$ **25.** $6!20!(13!)^2/(3!)^2(10!)^2 26!$ **27.** 54912

29. 1,098,240 **31.** $\binom{39}{13} = 8{,}122{,}425{,}444$

SECTION 4.1

1. 75.4″, 76″ **3.** 16, 15.5 **5.** 3.48, 1.51

7. 7000

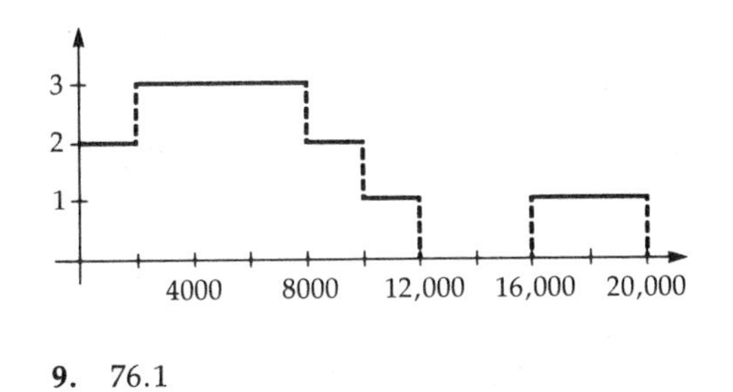

9. 76.1 **11.** 16,000

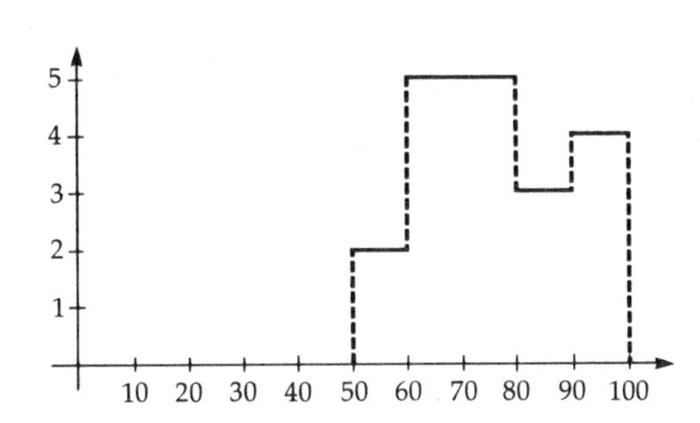

SECTION 4.2

1. $m = 8.4$, $s = 1.6$
3. Yes. $m = 1.002$, $s = .0098$
5. $m = 2.25$, $s = 5.5$
7. $m = \$38{,}000$, $s = \$4900$
9. .85 for A, .27 for B
11. 16.9 for first section, 19.7 for second section

First section

Second section

SECTION 4.3

1. .3643, .3708, .4979, .0279
3. .4955, .4713, .4989, .4726
5. .4452, .4982, .4987, .0239
7. .9312, .4379, .9681
9. .0069, .0244, .2514, .1357
11. .5596, .9945, .9868, .9987
13. .1071, .9842, .5197, .3808
15. **(a)** $.4 + .3 = .7$
    (b) $.88 - .4 - .3 = .18$
    (c) $1 - .88 = .12$
17. **(a)** .1915 **(b)** .1915 **(c)** .5
    (d) $.3413 + .5 = .8413$
    (e) $.5 - .3413 = .1587$
19. **(a)** .4686 **(b)** .9332
    (c) $.5 - .4332 = .0668$

SECTION 4.4

1. 2, 8, $-.4$
3. 4, -4, 6
5. .4772, .1915, .6687
7. .1587, .4013, .6915
9. .2620, .3811
11. .3085
13. .0228
15. .2033
17. 1.63
19. 2.12
21. 72.6
23. 73.2 inches

SECTION 4.5

1. .276
3. .098
5. .375
7. .941
9. 29.6%
11. .104
13. .309

SECTION 4.6

1. $m = 3200$, $s = 40$
3. $m = 42$, $s = 5.42$
5. $.5 + .4875 = .9875$

7. **(a)** .0062 **(b)** $P(x > 5)$, where x has the standard normal distribution. This is not in the table but is less than .001.

9. .0013 **11.** .0125

SECTION 5.1

1.

	SIMPLE	COMPOUND
(a)	$40	$48.89
(b)	$18	$21.54
(c)	$1600	$3801
(d)	$2400	$2690

3. 3% semiannually for 15 years, by $1.02.

5. **(a)** $340.33 **(b)** $3.71 **(c)** $610.27

7. $28,370 **9.** $23,433.82

11. Ed will have $6894.21.
Anja needs to buy a $4202.79 savings certificate.

13. 5.92¢

15. 6.98% compounded daily, 7.08% by the 360/365 method

17. $5.63

SECTION 5.2

1. **(a)** $2764.56 **(b)** $2867.80

3. $1256.70

5. $2449.26

7. $12,522, $399.30

9. $8061.93

11. $7800.08, $16,862.88

13. $117.18, $332.00, $625.51

15. $11.73

17. $250.40

SECTION 5.3

1. $6074.35

3. $19,541.30 − 900.45 = $18,640.85

5. $33,701.30 − $973.90 = $32,727.40, $31,913.55

7. $8126.33, $8126.33 − $79.81 − $68.75 = $7977.77

9. The first year Roberto earns no interest, so $n = 30$, $i = 4\%$. $66,101.88

SECTION 5.4

1. **(a)** $7090.58, $500
(b) $29,915.80, $4000

3. $113,922

5. $12,774.09

7. $1159.97, $13,919.64, $9598.20

9. $481,907.40

11. $451.07, $5412.84, $157,183.91

SECTION 5.5

1. \$175.53, \$1053.18, \$53.18
3. \$1604.85 per year, \$37,097, \$12,097
5. \$5827.71, \$63,277.10, \$13,277.10
9. \$2516.07
11. Rebecca should borrow from the savings and loan and put the \$1080 she would have had to pay in points (assuming a maximum mortgage of \$54,000) into a savings account for 30 years. The interest the \$1080 will earn in 30 years will be more than double the amount of extra interest Rebecca will have to pay to the savings and loan, even if she only earns at an annual rate of 6%.
13. \$274.83 15. \$238.03 17. \$438.01

SECTION 5.6

1. 24, \$5067.20, 14 3. 7 to $7\frac{1}{2}$, 16289. D.M.
5. 15 oz, 18 oz, more expensive by .03¢/oz
7. Larger can 9. Brand name 11. Half-gallon
13. 448 km 15. 606 km 17. 34 mi/gal
19. 30%, \$87.74, \$152.88
21. The credit union (by less than a dollar a month), about .775%
23. \$377.92 25. \$317.28, \$308 27. \$48,185.64, \$42,529.81

SECTION 6.1

1.

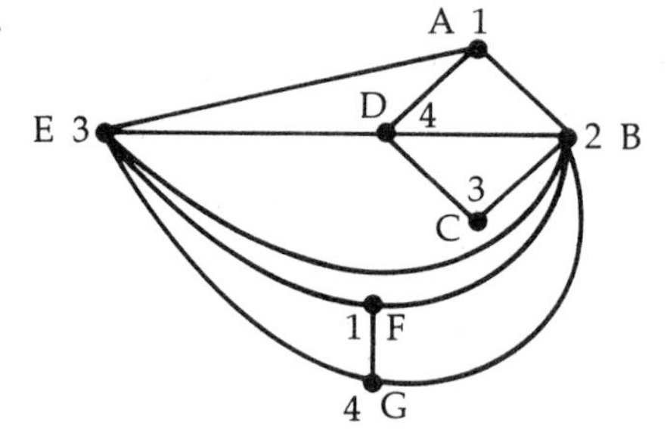

3.

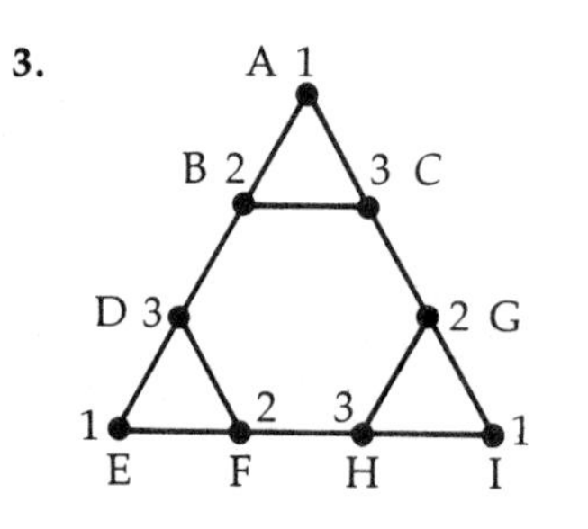

5.

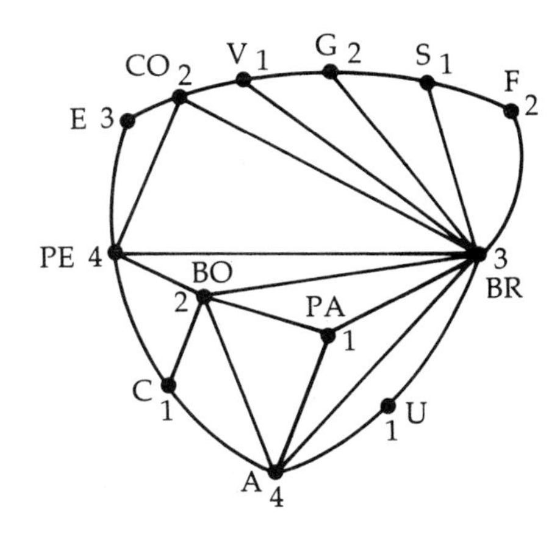

7.

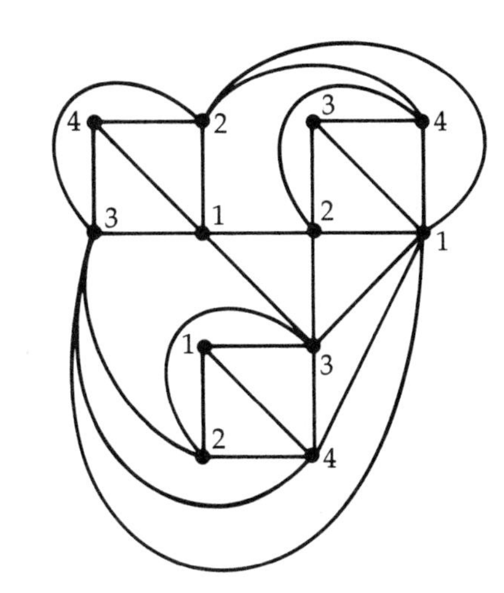

9. (a), (c), and (f); (b), (e), and (g)

11.

FIRST SLOT	SECOND SLOT	THIRD SLOT
Tabby	Alice	Spooksie
Tiger	Winky	Mildred
Foo Foo	Seymore	Thomson

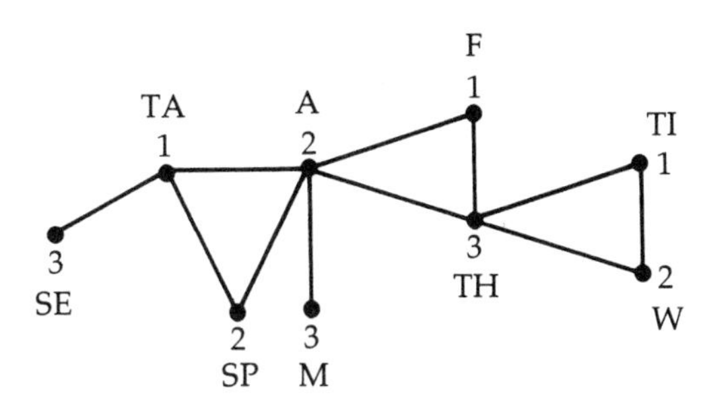

13. The schedule is impossible; there are 5 talks on computers.

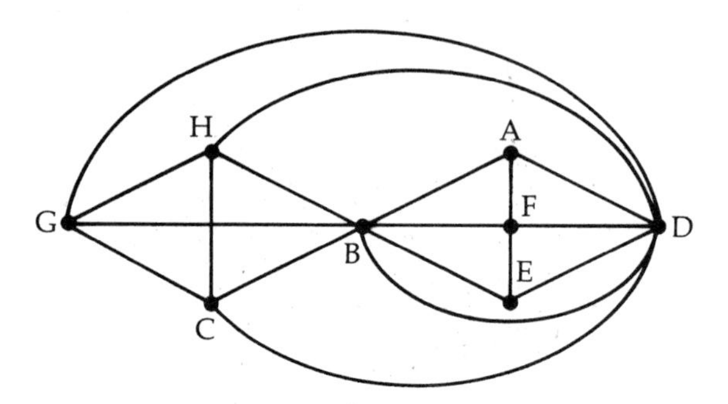

SECTION 6.2

1. A D C B E F A or A F E B C D A
3. A B D E G F C A or the reverse

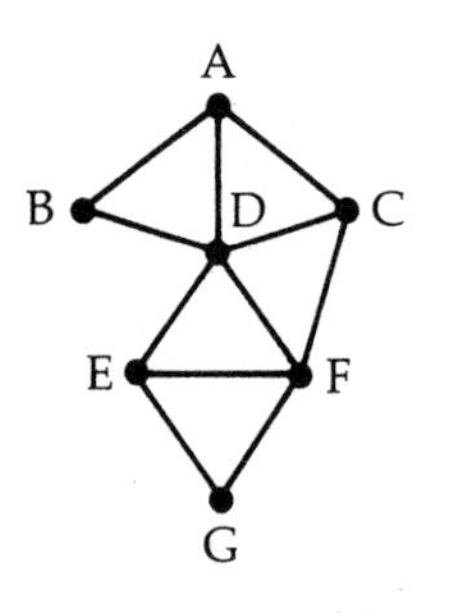

5. One way is A B C I D H G F E A
7. C E B A D F G or the reverse

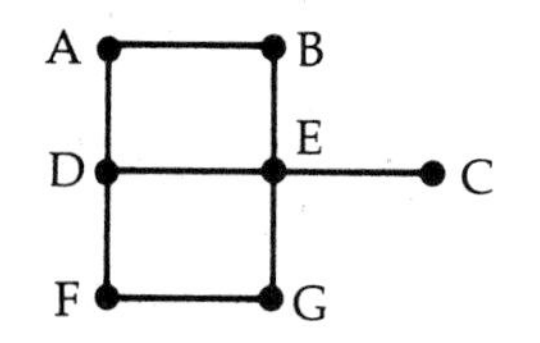

9. Impossible
11.

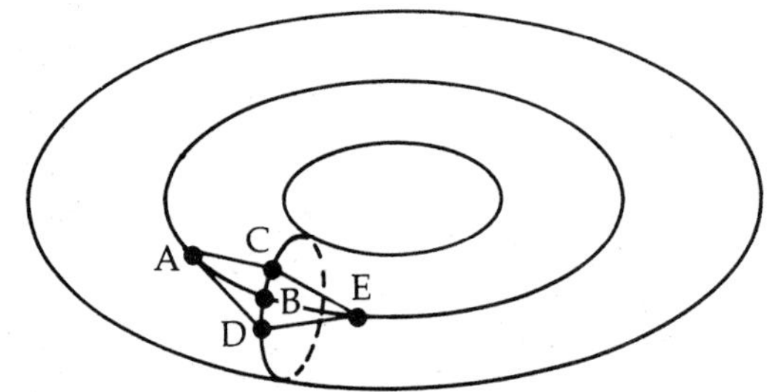

13.

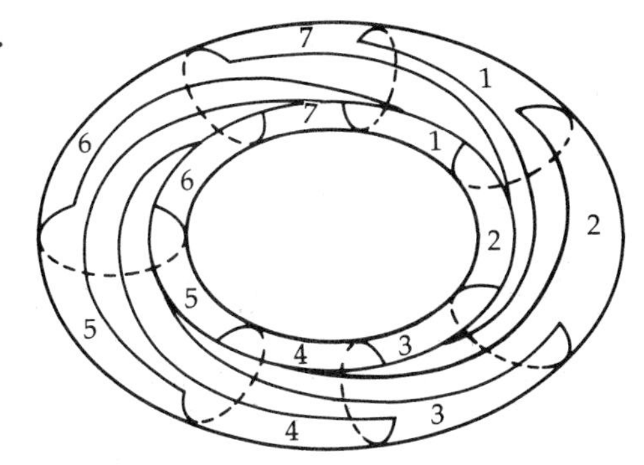

SECTION 6.3

1. Impossible

3. No. No.

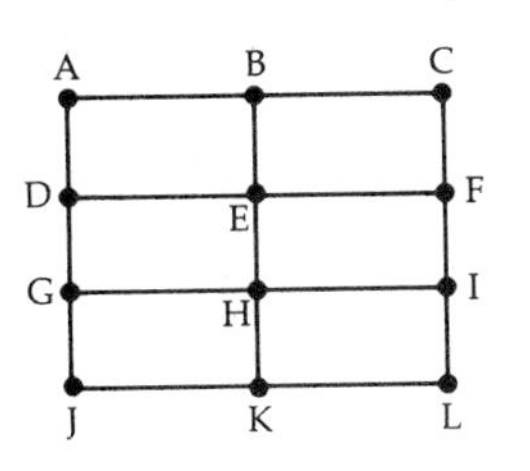

5. A B D B G F E C E D C A D G

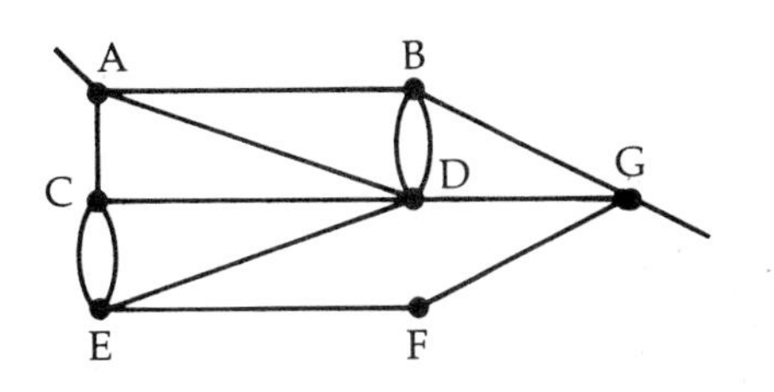

7. No.

9. Garrison, Drummond, Butte, Garrison, Helena, Great Falls, Townsend, Helena, Butte, Three Forks, Townsend

11. One way is M P O L M J I H L K N K H E A B E F B C F I G D C

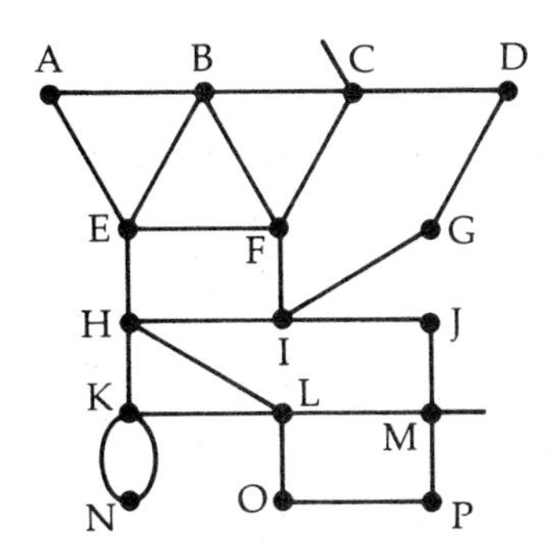

SECTION 6.4

1. A B C F E B E D E F I H E H G H I L K H K J G D A

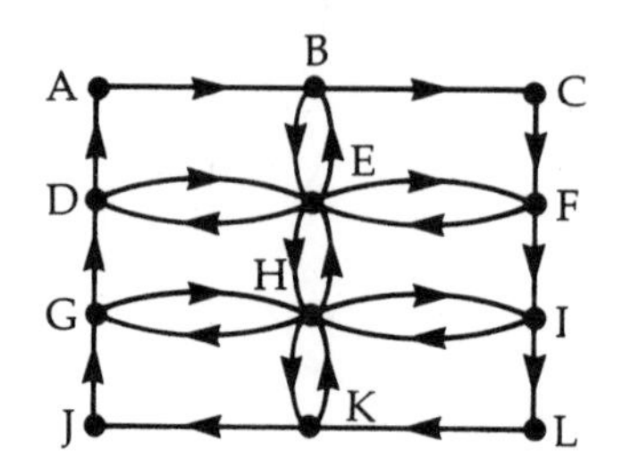

3. Yes 5 6 5 2 5 3 1 2 3 6 4 2

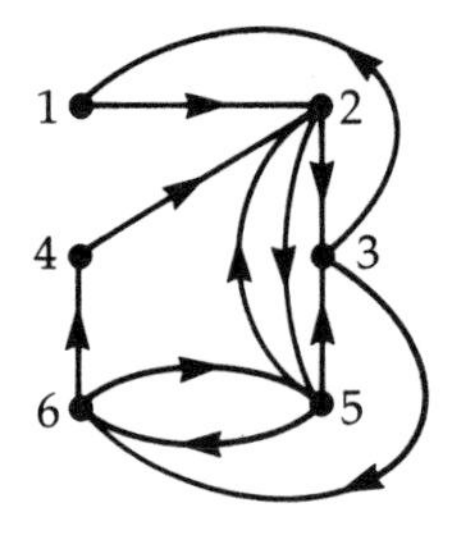

5. Yes. Al beat Bob beat Emil beat Gus beat Fred.

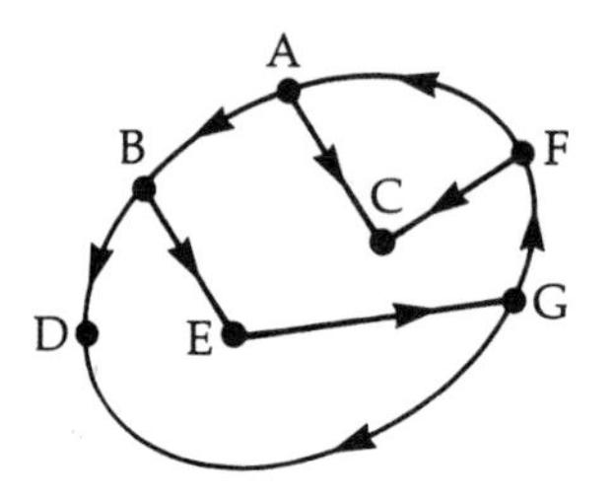

7. The murderer is still in the mansion. Two more sets of tracks lead toward the mansion than away from it. One set belongs to the victim; the other to someone still hiding there—the killer.

9. Yes. One way is A B C D C B M N O J P I B N P O M K J K F E I D E J F G H L G K L H

11.

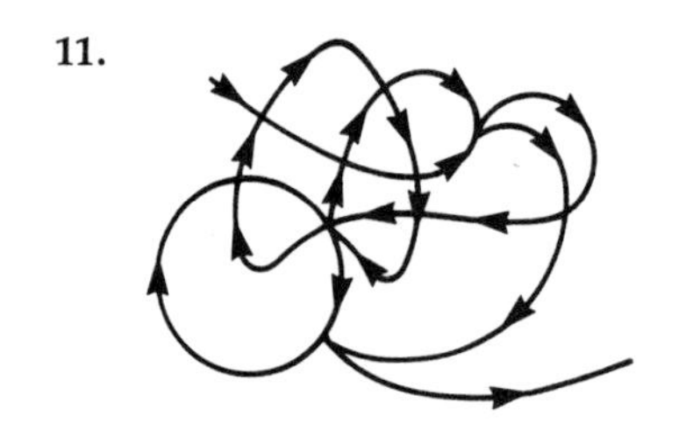

SECTION 6.5

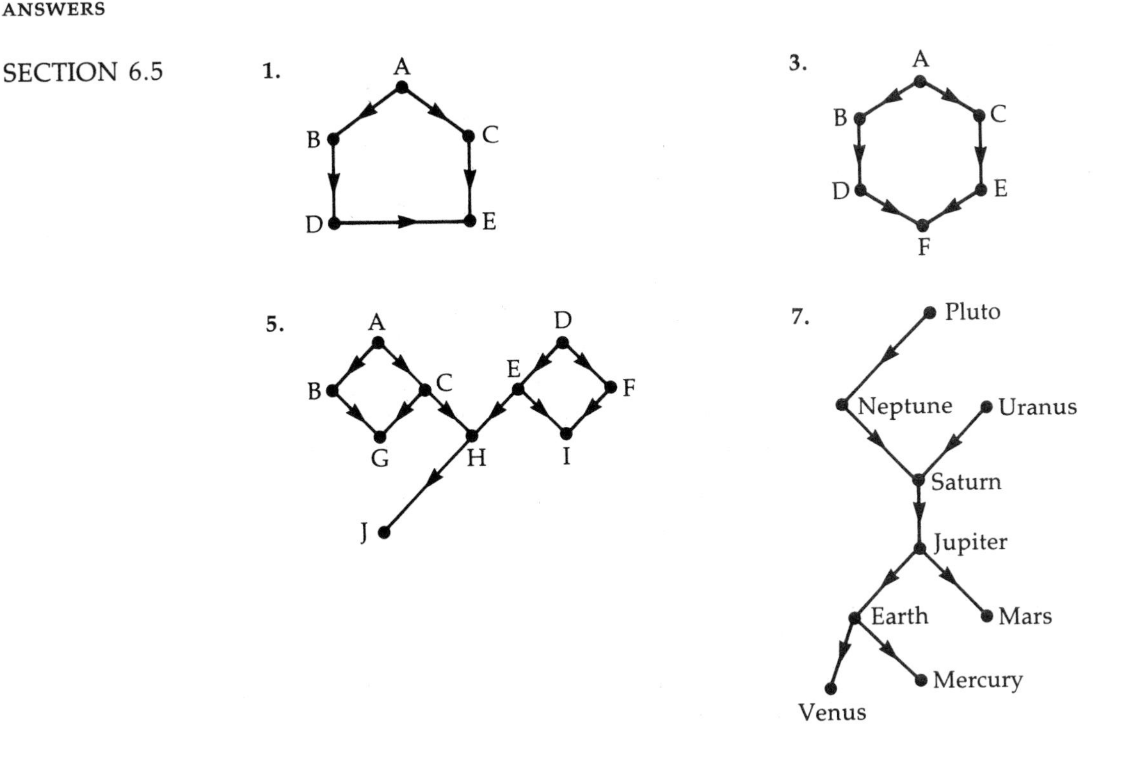

9. Actually these states may be arranged in any order around the circle. Transitivity is not a reasonable condition, if A → B is to mean that state A is larger than state B in either population or area. For example Arkansas has more people than Alaska, and Alaska has more area than Arizona. Yet Arkansas does not exceed Arizona in either population or area.

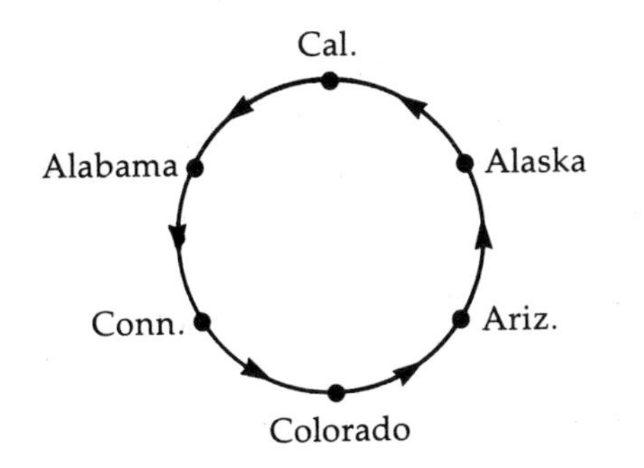

11.

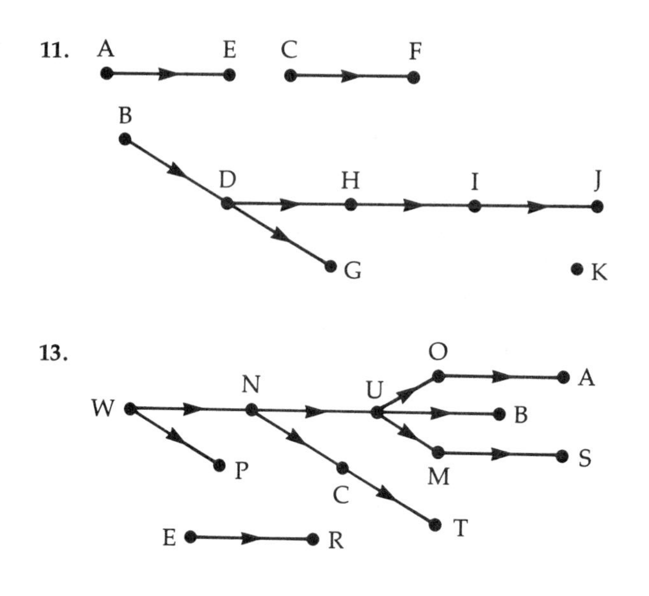

13.

SECTION 6.6

1. 28 days; A C F G H

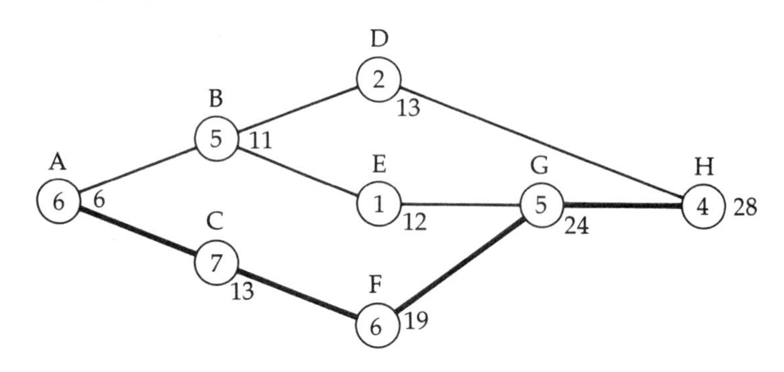

3. 60 days; A B D H I

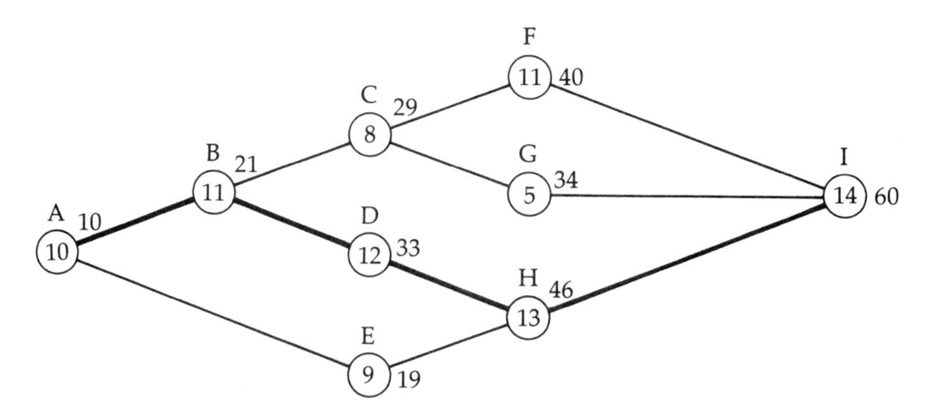

5. 290 days; A C D G H

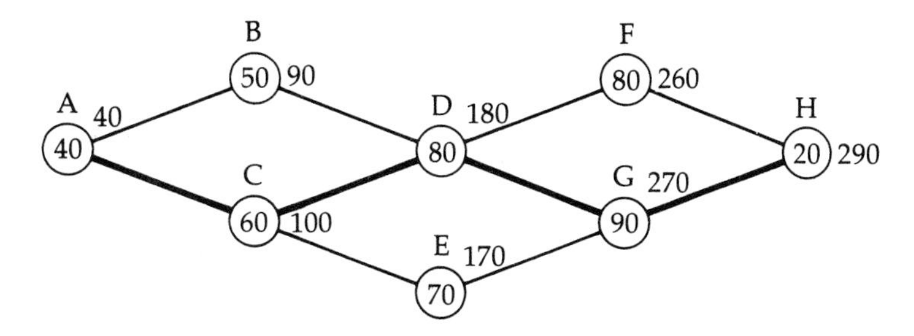

7. 27 days; A C F H I

9. 12 days; A B F I K

11. 143 minutes; A B E F G

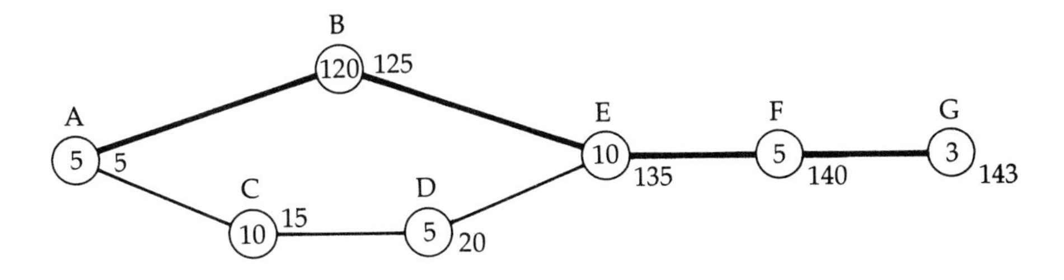

13. 104 minutes; A B C D F H J K L

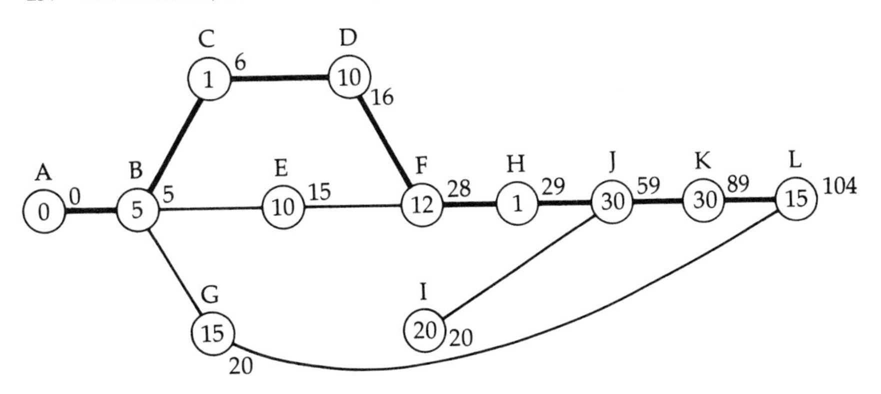

SECTION 7.1

1. Leonard Euler (1707–1783) (see Chapter 6 for more on Euler) and Augustin-Louis Cauchy (1789–1857). (An excellent account of the lives of these and other famous mathematicians is in *Men of Mathematics* by E. T. Bell, available in paperback.)

5.

		INCHES							CENTIMETERS												
		1/4	3/8	1/2	5/8	3/4	7/8	1	.5	1	1.5	2	2.5	3	1	2	3	4	English	Metric	Super
INCHES	1/4	5	0	0	0	0	0	0	0	0	0	0	0	0	0	0	0	0	4	0	3
	3/8	0	5	0	0	0	0	0	0	0	0	0	0	0	0	0	0	0	4	0	3
	1/2	0	0	5	0	0	0	0	0	0	0	0	0	0	0	0	0	0	4	0	3
	5/8	0	0	0	5	0	0	0	0	0	0	0	0	0	0	0	0	0	4	0	3
	3/4	0	0	0	0	3	0	0	0	0	0	0	0	0	0	0	0	0	4	0	1
	7/8	0	0	0	0	0	3	0	0	0	0	0	0	0	0	0	0	0	4	0	1
	1	0	0	0	0	0	0	3	0	0	0	0	0	0	0	0	0	0	4	0	1
CENTIMETERS ROUNDHEAD	.5	0	0	0	0	0	0	0	5	0	0	0	0	0	0	0	0	0	0	3	3
	1	0	0	0	0	0	0	0	0	5	0	0	0	0	0	0	0	0	0	3	3
	1.5	0	0	0	0	0	0	0	0	0	5	0	0	0	0	0	0	0	0	3	3
	2	0	0	0	0	0	0	0	0	0	0	3	0	0	0	0	0	0	0	3	1
	2.5	0	0	0	0	0	0	0	0	0	0	0	3	0	0	0	0	0	0	2	1
	3	0	0	0	0	0	0	0	0	0	0	0	0	3	0	0	0	0	0	2	1
CENTIMETERS FLATHEAD	1	0	0	0	0	0	0	0	0	0	0	0	0	0	5	0	0	0	0	3	3
	2	0	0	0	0	0	0	0	0	0	0	0	0	0	0	3	0	0	0	3	1
	3	0	0	0	0	0	0	0	0	0	0	0	0	0	0	0	3	0	0	2	1
	4	0	0	0	0	0	0	0	0	0	0	0	0	0	0	0	0	3	0	2	1

SECTION 7.2

1. (a) $\begin{pmatrix} 12 \\ 14 \end{pmatrix}$ (b) $\begin{pmatrix} 97 \\ 95 \\ 57 \end{pmatrix}$ (c) $\begin{pmatrix} 29 \\ 118 \\ 119 \end{pmatrix}$ (d) $\begin{pmatrix} 22 \\ 31 \\ 77 \\ 9 \end{pmatrix}$

3. (i)

	LARGE	SMALL
RACQUETS	4	2
BIRDIES	5	3

(ii) $\begin{pmatrix} 100 \\ 135 \end{pmatrix}$ RACQUETS, BIRDIES

5. (i)

	SMALLER	LARGER
RACQUETS	2	4
NETS	1	1
SHUTTLE-COCKS	3	5

(ii) $\begin{pmatrix} 210 \\ 66 \\ 276 \end{pmatrix}$ RACQUETS, NETS, SHUTTLECOCKS

7. (i)

	HOME	BAR	OFFICE
HIGHBALL	4	8	10
WATER	6	4	2
WINE	4	6	8
WHISKEY	0	2	4

(ii) $\begin{pmatrix} 150 \\ 122 \\ 128 \\ 28 \end{pmatrix}$ HIGHBALL, WATER, WINE, WHISKEY

9.
$$\begin{array}{l} \text{No. of } \tfrac{1}{4}'' \\ \text{No. of } \tfrac{3}{8}'' \\ \text{No. of } \tfrac{1}{2}'' \\ \text{No. of } \tfrac{5}{8}'' \\ \text{No. of } \tfrac{3}{4}'' \\ \text{No. of } \tfrac{7}{8}'' \\ \text{No. of } 1'' \\ \text{No. of 0.5 cm} \\ \text{No. of 1.0 cm} \\ \text{No. of 1.5 cm} \\ \text{No. of 2.0 cm} \\ \text{No. of 2.5 cm} \\ \text{No. of 3.0 cm} \\ \text{No. of 1 cm flatheads} \\ \text{No. of 2 cm flatheads} \\ \text{No. of 3 cm flatheads} \\ \text{No. of 4 cm flatheads} \end{array} \quad \begin{pmatrix} 15{,}500 \\ 15{,}500 \\ 15{,}500 \\ 15{,}500 \\ 9{,}000 \\ 9{,}000 \\ 9{,}000 \\ 17{,}500 \\ 17{,}500 \\ 17{,}500 \\ 9{,}000 \\ 9{,}000 \\ 9{,}000 \\ 17{,}500 \\ 9{,}000 \\ 9{,}000 \\ 9{,}000 \end{pmatrix}$$

SECTION 7.3

1. **(a)** $\begin{pmatrix} 3 & 7 \\ 11 & 3 \\ 3 & 8 \end{pmatrix}, \begin{pmatrix} 0 & 5 \\ 7 & -6 \\ 3 & -2 \end{pmatrix} \begin{pmatrix} 4 & 6 \\ 10 & 8 \\ 2 & 12 \end{pmatrix}, \begin{pmatrix} -2 & -8 \\ -12 & 2 \\ -4 & -4 \end{pmatrix}$

(b) $\begin{pmatrix} 3 & 2 \\ 4 & 9 \\ 0 & 10 \end{pmatrix}, \begin{pmatrix} -2 & -3 \\ -5 & -4 \\ -1 & -6 \end{pmatrix}, \begin{pmatrix} 2 & 3 \\ 5 & 4 \\ 1 & 6 \end{pmatrix}$

3. **(a)** Yes. They all have the same dimension.
(b) $A + E$ is not defined since A is 3×2 and E is 2×2.

5. **(a)** $A = \begin{pmatrix} 2 & 3 \\ 5 & 4 \\ 1 & 6 \end{pmatrix}, G = \begin{pmatrix} 1 & 1 \\ 1 & 1 \end{pmatrix}, H = \begin{pmatrix} -2 & 1 \\ -4 & -5 \end{pmatrix}, H = \begin{pmatrix} -2 & 1 \\ -4 & -5 \end{pmatrix}$

(b) $\begin{pmatrix} 0 & 0 \\ 4 & 4 \\ 3 & 3 \end{pmatrix}, \begin{pmatrix} -8 & -10 \\ 18 & 12 \end{pmatrix}, \begin{pmatrix} -12 & -10 \\ 3 & 15 \\ -6 & -2 \end{pmatrix}$

7. **(a)** No, since dimensions do not match up.
(b) Yes, since both equal $\begin{pmatrix} 6 & 8 \\ -12 & -6 \end{pmatrix}$

9. **(a)** $J \cdot M = \begin{pmatrix} -4 & 19 & -12 \\ -4 & 21 & -14 \\ 1 & 0 & 1 \end{pmatrix}, L \cdot M = \begin{pmatrix} 1 & 0 & 0 \\ 0 & 1 & 0 \\ 0 & 0 & 1 \end{pmatrix}$

$$M \cdot L = \begin{pmatrix} 1 & 0 & 0 \\ 0 & 1 & 0 \\ 0 & 0 & 1 \end{pmatrix}$$

9. (b)

$$L^2 = \begin{pmatrix} -5 & 9 & 0 \\ -3 & 4 & -3 \\ -3 & 3 & -5 \end{pmatrix}, \quad L^3 = \begin{pmatrix} -14 & 18 & -15 \\ -1 & -1 & -6 \\ 4 & -9 & -4 \end{pmatrix}$$

11. (a)

$$J + K = \begin{pmatrix} 2 & 1 & 1 \\ 3 & 2 & 5 \\ -1 & 3 & 5 \end{pmatrix}, \quad (J + K) \cdot L = \begin{pmatrix} -1 & 5 & 5 \\ -9 & 19 & 4 \\ -14 & 21 & -8 \end{pmatrix} \quad \text{yes}$$

(b) See 7(b) and 9(a) for examples where matrices commute. See 8(a) or use F and H for examples where two matrices do not commute.

(c) No. If I is the $n \times n$ identity matrix that has ones on the main diagonal (that is, in the iith place for every i) and zeroes elsewhere, then I commutes with every $n \times n$ matrix A, since $A \cdot I = I \cdot A = A$ in this case.

SECTION 7.4

1. (a)

$$A^2 = \begin{pmatrix} 4 & 0 & 0 \\ 3 & 0 & 2 \\ -1 & -2 & 0 \end{pmatrix}, \quad A^3 = \begin{pmatrix} 8 & 0 & 0 \\ 6 & -2 & 2 \\ -4 & -2 & -2 \end{pmatrix}$$

$$A^4 = \begin{pmatrix} 16 & 0 & 0 \\ 10 & -4 & 0 \\ -10 & 0 & -4 \end{pmatrix} \quad A^5 = \begin{pmatrix} 32 & 0 & 0 \\ 16 & -4 & -4 \\ -20 & 4 & -4 \end{pmatrix}$$

$$A^6 = \begin{pmatrix} 64 & 0 & 0 \\ 28 & 0 & -8 \\ -36 & 8 & 0 \end{pmatrix}$$

Several entries are predictable, but the first row in A^n will surely be $(2^n \quad 0 \quad 0)$.

(b) $B^2 = \begin{pmatrix} \frac{1}{4} & \frac{3}{4} \\ 0 & 1 \end{pmatrix}, \quad B^4 = \begin{pmatrix} \frac{1}{16} & \frac{15}{16} \\ 0 & 1 \end{pmatrix}, \quad B^8 = \begin{pmatrix} \frac{1}{256} & \frac{255}{256} \\ 0 & 1 \end{pmatrix}$

The entry in row 1 and column 1 is approaching zero.

(c) $C = C^2 = C^3$. All the same.

3. (a) $wH = (4 \quad -6 \quad 2)$, $Jx = \begin{pmatrix} \frac{3}{2} \\ \frac{3}{2} \\ 1 \end{pmatrix}$, $Hy = \begin{pmatrix} 1 \\ 2 \\ -1 \\ -2 \end{pmatrix}$, $zJ = z = (\frac{2}{9} \quad \frac{4}{9} \quad \frac{3}{9})$

(b) $yz = \begin{pmatrix} 0 & 0 & 0 \\ 0 & 0 & 0 \\ \frac{2}{9} & \frac{4}{9} & \frac{3}{9} \end{pmatrix}$, $\quad zy = \frac{3}{9}$

(c) $uH = (3 \quad 1 \quad -1)$, $\quad vH = (-1 \quad -2 \quad 2)$

5. (a) $A \cdot B = \begin{pmatrix} -12 & -8 \\ 3 & 7 \end{pmatrix}$, $B \cdot A = \begin{pmatrix} -8 & 19 & -5 \\ 2 & 5 & -2 \\ 0 & 6 & -2 \end{pmatrix}$

(b) Because they are not the same dimension.

(c) For $m \neq n$, if A is $m \times n$ and B is $n \times m$, then $A \cdot B$ and $B \cdot A$ are always defined. The matrices $A \cdot B$ and $B \cdot A$ are never equal since each is square, but of different dimensions.

7. (a) $\begin{pmatrix} 3 & -1 \\ 7 & -2 \end{pmatrix}$ (b) $\begin{pmatrix} \frac{7}{2} & -\frac{3}{2} \\ -\frac{1}{2} & \frac{1}{2} \end{pmatrix}$ (c) $\begin{pmatrix} 5 & 7 \\ 11 & 15 \end{pmatrix}$

9. $C = A \div B$ if and only if $A = C \cdot B$. Thus multiplying on the right by D, we get $A \cdot D = (C \cdot B) \cdot D = C\ (B \cdot D) = C \cdot I = C$. In the example,

$$C = A \cdot D = \begin{pmatrix} -1 & 1 & 0 \\ 40 & -12 & 10 \\ -19 & 6 & -5 \end{pmatrix}$$

SECTION 7.5

1. $\begin{pmatrix} 0 & 0 & 1 \\ \frac{1}{2} & 0 & \frac{1}{2} \\ \frac{3}{4} & 0 & \frac{1}{4} \end{pmatrix}$ $\begin{pmatrix} 0 & 1 & 0 \\ 0 & 0 & 1 \\ 1 & 0 & 0 \end{pmatrix}$ $\begin{pmatrix} \frac{1}{2} & \frac{1}{4} & \frac{1}{4} \\ \frac{1}{2} & 0 & \frac{1}{2} \\ 0 & \frac{2}{3} & \frac{1}{3} \end{pmatrix}$

3.

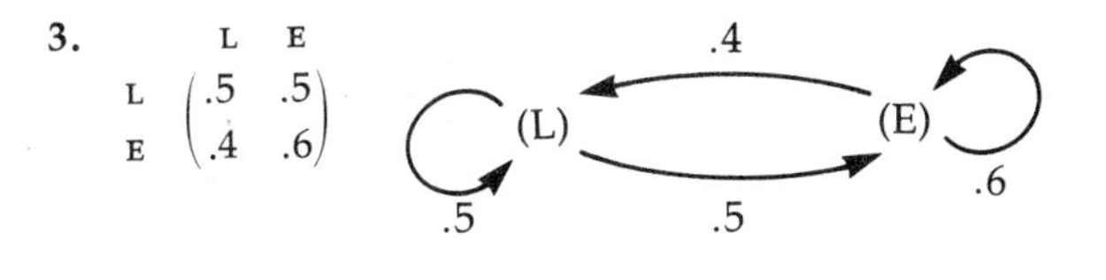

5.

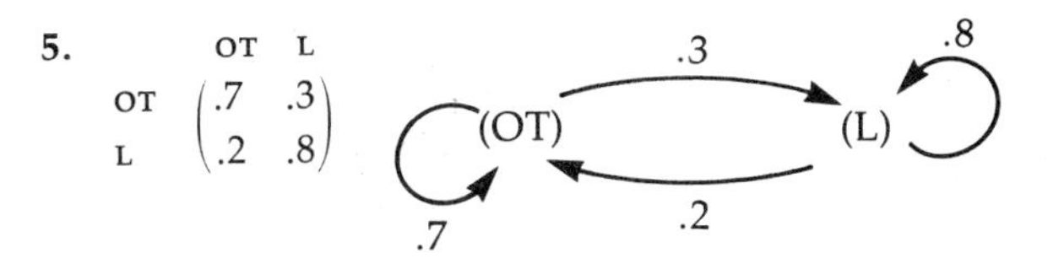

7.

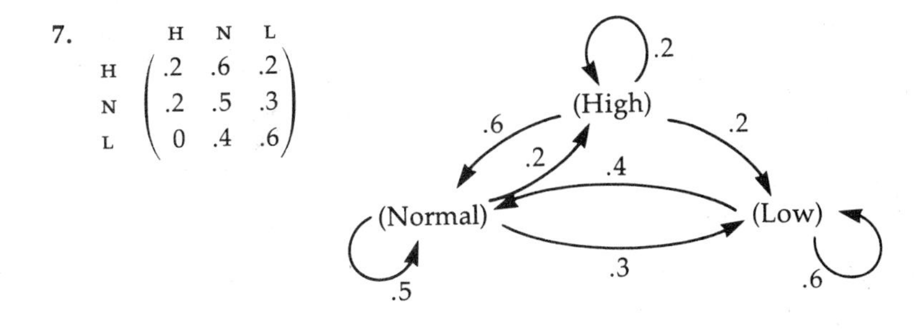

9.

$$\begin{array}{c} \\ \text{RAIN} \\ \text{CLEAR} \\ \text{CLOUDY} \end{array} \begin{array}{c} \begin{array}{ccc} \text{RAIN} & \text{CLEAR} & \text{CLOUDY} \end{array} \\ \begin{pmatrix} 0 & .5 & .5 \\ .25 & .5 & .25 \\ .25 & .25 & .5 \end{pmatrix} \end{array}$$

(Rain)
(Clear)
(Cloudy)
.5
.5
.25
.25
.25
.25
.5
.5

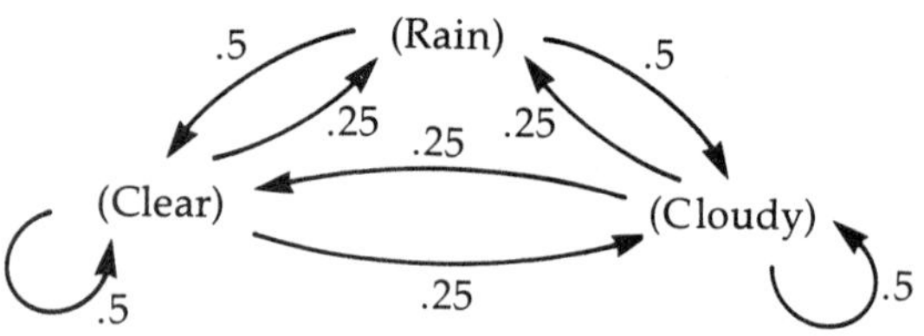

SECTION 7.6

1. The situation is unchanged. $(.4 \quad .2 \quad .4)A = (.4 \quad .2 \quad .4)$.

3. .4445 **5.** .375 **7.** .14 **9.** .375, .40

SECTION 7.7

1. $w = (\frac{1}{2}, \frac{1}{2})$ **3.** $w = (\frac{1}{3}, \frac{2}{3})$

5. $w = (\frac{2}{9}, \frac{7}{9})$ **7.** $w = (\frac{3}{11}, \frac{8}{11})$

9. $\frac{5}{11} = .45454545$ **11.** 80

SECTION 7.8

1. $(\frac{4}{9}, \frac{1}{9}, \frac{4}{9})$ **3.** 56, $(\frac{4}{9}, \frac{5}{9})$ **5.** 60%

7. $w = (\frac{2}{17}, \frac{8}{17}, \frac{7}{17})$ so the patient will be normal about 14 days and low about 12 days during a 30-day month.

9. 20% of the time **11.** 40% of the time

13. (.143, .214, .286, .214, .143)

15. (.125, .25, .25, .25, .125)

17. 10, (.188, .267, .337, .208)

SECTION 7.9

1. **(a)** $x = 2$, $y = -2$, $z = 1$ **(b)** $x = 5$, $y = 2$, $z = 2$ **(c)** $x = \frac{26}{17}$, $y = \frac{26}{17}$, $z = \frac{11}{17}$ **(d)** $x = \frac{3}{2}$, $y = \frac{1}{2}$, $z = 2$

3. $(\frac{2}{5}, \frac{1}{5}, \frac{2}{5})$

5. Mönch: 4100 meters
Gross-Wannenhorn: 3900 meters
Weisshorn: 4500 meters

7. $(\frac{2}{17}, \frac{8}{17}, \frac{7}{17})$

9. $(\frac{1}{5}, \frac{2}{5}, \frac{2}{5})$

11. $(\frac{1}{7}, \frac{3}{14}, \frac{2}{7}, \frac{3}{14}, \frac{1}{7})$, $(\frac{1}{63}, \frac{2}{63}, \frac{4}{63}, \frac{8}{63}, \frac{16}{63}, \frac{32}{63})$, $(\frac{1}{8}, \frac{1}{4}, \frac{1}{4}, \frac{1}{4}, \frac{1}{8})$

13. $(\frac{35}{124}, \frac{30}{124}, \frac{33}{124}, \frac{26}{124})$

SECTION 7.10

1. **(a)** (Tory, Labour) = $(\frac{154}{357}, \frac{203}{357})$
(b) (TT, TL, LT, LL) = $(\frac{112}{671}, \frac{182}{671}, \frac{182}{671}, \frac{195}{671})$

3. .123077

5. The gambler has a 60% chance of stopping with $500 and a 40% chance of going broke. Notice the gambler's expected gain is zero.

7. It would need to be 100001×100001. It is not feasible with current computers. For more on absorbing Markov chains including more on this problem of Gambler's Ruin, see Kemeny, Snell, and Thompson, *Introduction to Finite Mathematics*.

INDEX

Absorbing Markov chain, 439–441
Adding and subtracting equations, 42–46
Addition principle, refined, 198
Addition rule for probabilities, 196-202
Addressing, 159-160
Adjacent vertices of a graph, 327
Alphabetic numeration (Greek), 82-83
Amortization, 295-296
Annual yield, 278
Annuity, 291-294
  formula, 291
Appel, Kenneth, 337
Assignment statement, 97, 123-124, 156

Babbage, Charles, 85-86, 88, 89
Bar graph, 223
BASE, 445
BASIC, 82, 96, 111-143
  branching, 118, 133, 447
  commands, 121-138, 444-453
    BASE, 445
    BYE, 132
    CATLIST, 138
    DATA, 122, 447
    DIM, 444-445
    END, 127
    FILENAME, 129
    FOR, 138, 140
    GOTO, 133
    IF. . .THEN, 133
    INPUT, 133-134, 448
    LET, 123
    LIST, 132
    LNH, 132
    MAT, 451-453
    NEXT, 138, 140
    NODATA, 447
    PRINT, 124, 449
    READ, 122, 448
    RECOVER/SYSTEM, 129
    REM, 130, 444
    RESEQ, 136
    RESTORE, 447
    RNH, 130, 132
    RUN, 130, 132
    STOP, 135
  constants
    decimal, 113
    exponential, 114
    integral, 113
    string, 444
  diagnostics, 454
  literal string, 125, 444
  log-on, 111
  log-off, 111
  matrix, 450-453
    arithmetic, 451
    input, output, 451
  special functions, 446, 453
  string, 443-444
    constants, 444
    variables, 444
  transfers, 132-133, 447
  variable
    simple numeric, 114-115
    string, 444
    subscript, 115
Big six, 174, 179
Bimonthly, 277
Binomial distribution, 258-264
  mean, 261
  standard deviation, 261
Bridge, 217-218
Buying a car, 314-317
  depreciation, 314
  insurance, 316-317
Buying a house, 319--323
  closing costs, 320
  monthly payment, 319
  points, 320
BYE, 132

Calculator, 83, 276
Cards
  charge, 310-313
  deck of, 178
CATLIST, 138
Cayley, Arthur, 380
Charge cards, 310-313
Chuck-a-luck, 185

Colmar, Thomas de, 84
Combinations, 208-211
Common ratio, 280
Complement of an event, 192
Compound event, 199-200
Compound interest formula, 271, 273
Computer
  compiler, 93
  components, 89
  CPU, 91-93
  hardware, 91
  history, 82-88
  software, 91, 93
  tasks, 89
Connected graph, 347
Consumer Reports, 316
CONTINUE, 168
Continuous probability
  curve, 236
  distribution, 237
Core, 87
Craps, 214
Critical path, 370
Current situation vector, 411, 412

DATA, 122, 447
Depreciation, car, 314
DIM, 444
DIMENSION, 457
Dimension of a matrix, 390
Directed graph, 351-354
  transitive, 359-363
Discrete probability distribution, 236
DO, 167-168, 467
Dominant characteristic, 382
Draw poker, 216

Edge of a graph, 327
Elementary row operations, 431
END, 127, 156-157
Entry of a matrix, 391
Equation(s), linear, 25
  graph of, 26
  systems of, 37-46
  substitution, 37
  adding and subtracting, 42-46
Equations, solving
  single, 15
  systems of
    addition and subtraction, 42-46
    matrix, 429-435
    substitution, 37
Equilibrium, 421
Euler, Leonhard, 347
Euler's theorem for tracing graphs, 347
Event(s), 176
  complement of, 192
  compound, 199
  independent, 191-192, 199
  mutually exclusive, 196, 199
  simple, 199
Expectation, 180
Experiment, 176
Explosion, simple, 384

Factorials, 189
Fair game, 180
FILENAME, 129
Finance charge, 312
Fixed point, 420
Flow chart(s), 94-110
  counter, 105
  decision symbol, 98
  looping, 98, 105
  start, 95
  stop, 96
  symbols, 97
Flush, in poker, 204
FOR, 138, 140
FORMAT, 460
FORTRAN IV, 82, 145-169, 454-467
  addressing, 160
  branching, 160-162, 466-467
  built-in functions, 459
  commands
    CONTINUE, 168
    DIMENSION, 457
    DO, 167-168, 467
    END, 156
    FORMAT, 460
    GO TO, 160
    IF, 466-467
    IF (logical), 161
    IMPLICIT, 458
    INTEGER, 458
    LOGICAL, 458
    PRINT, 155, 459
    READ, 153, 465

FORTRAN IV commands (continued)
    REAL, 458
    STOP, 156
    WRITE, 466
  constants
    fixed point, 145-146
    floating point, 145-146
    Hollerith, 456
    integer, 145
    literal, 456
    logical, 455
    real, 145-146
  JCL, 152
  logical
    constants, 455
    expressions, 160-161
    operations, 455
    variables, 455
  numbers, 145-148
  relational operators, 160-161
  source deck, 152
  transfers, 160, 161
  variables
    fixed point, 147-148
    floating point, 147-148
    integer, 147-148
    logical, 455
    real, 147-148
    subscripted, 457
Four-color theorem, 336-339
Frequency, 224
Full house, in poker, 219

Geometric series, 280
GIGO, 88
GO TO, 133, 160
Graph(s), 19-78, 327-328
  adjacent vertices of, 327
  bar, 223
  choosing units, 24
  connected, 347
  directed, 351-354
  edges, 327
  of a linear inequality, 30
  map-coloring, 327
  scheduling, 326
  of a statement, 23
  of a system of inequalities, 33
  vertices, 327
Graphing grouped data, 223

Haken, Wolfgang, 337
Hieroglyphic symbols, 82
Hollerith, Herman, 86
Hybrid, 382

IF, 466-467
IF (logical), 161
IF. . .THEN, 133
IMPLICIT, 458
Implicit conditions, 51
Income tax table, 7
Independent events, 192
Inequalities
  in English, 4
  graphing, 29-37
  linear, 26
    systems of, 48-57
  solving, 16
Inflation, 301-302
  table of U.S., 278
INPUT, 133-134, 448
Installment buying, 295-297
Insurance
  automobile, 316-317
  interest adjusted price, 318
  life, 318
    term, 318
    whole life, 318
INTEGER, 458
Interest, 268-324
  add-on, 306-307
  compound, 271
  compounded, 273
  periods, 274
  simple, 270
  360/365 method, 279
  true annual, 307
Inventory-file matrix, 385

JCL, 152

Kemeny, John, 383
Kemeny tree, 410
Key punch, 157
Königsberg, seven bridges of, 346
Kronecker, Leopold, 389

Leibniz, Gottfried Wilhelm, 84
LET, 123
Linear
  equation, 25
  function, 62
  inequality, 26
  programming, 60-78
    graphical method, 69, 78
    maximum and minimum, 63
  theorem, 63
LIST, 132
LNH, 132
LOGICAL, 458

Mailman problem, 344-347
Markov chain(s), 406-441
  absorbing, 439-441
  defined, 419
  main theorem, 423
  state, 419, 437
MAT, 451-453
Matrix, 383, 390-441, 450-453
  addition, 391-394
  basic, 450-453
  dimension, 390
  inventory-file, 385
  multiplication, 386, 394-398
  negative, 393
  square, 400
  terminology, 390
  transition, 407, 420
    regular, 422
  zero, 393
Mean, 222-225
  of binomial distribution, 261
  of grouped data, 224
Median, 226
Metric units, 304
Multiplication principle, 185-189
  for counting, 186
Multiplication rule for probabilities, 191
Mutually exclusive events, 196

$\binom{n}{r}$ notation, 210
Napier, John, 83
NEXT, 138, 140
NO DATA, 447
Normal curve, 238-250
  standard, 245
Normal distribution
  general, 244-250
  standard, 235-242
Number lines, 21
Numeration, Greek alphabetic, 82

Outcome, 176

Parlays, 215
Pascal, Blaise, 84
Percent, 2
Permutations, 206-208
PERT, 366-372
  critical path, 370
Poker, 204, 212, 216, 219
Precedence of operations, 116-117, 148-149
Present value formula
  of an amount, 274-275
  of an annuity, 291
PRINT, 124, 155, 449, 459
Probability
  addition rule, 196
  of an event, 176
  of independent events, 191
  multiplication rule, 191
  vector, 420
Probability curve, continuous, 236, 237
Probability distribution
  continuous, 237
  discrete, 236
Production vector, 385

READ, 122, 153, 448, 465
REAL, 458
Recessive characteristic, 382
RECOVER/SYSTEM, 129
Refined addition principle, 198
Regular transition matrix, 422
REM, 130, 444
Repeated trials, 252-258
  formula, 254-255
RESEQ, 136
RESTORE, 447
Result vector, 385
Retirement planning, 297
RNH, 130-132
Roman numerals, 82-83
Roulette, 178, 196

Row operations, elementary, 431
RUN, 130-132

Σ (sigma) notation, 229
Salesman problem, 339
Savings plan formula, 113-114, 283
Scientific notation, 113, 146
Series, geometric, 280
Seven bridges of Königsberg, 346
Simple event, 199
Simple interest, 270
Source deck, 152
Square root table, explanation of, 231
Standard deviation, 231-234
  alternative formula, 233
  of binomial distribution, 261
Standard normal curve, 245
Standard normal distribution, 235-242
Statistics, 222-264
STOP, 135, 156
Straight, in poker, 204
Straight-flush, in poker, 212
Subscripts, 2
Substitution, 37
Sylvester, James Joseph, 84
Systems of equations
  adding and subtracting, 42-46
  linear, 37-46
  matrix method, 429-435
  substitution method, 37
Systems of linear inequalities, graphing, 48-57

Table II explanation, 239
Table V, extending, 289
Torus, 343, 344
Transistor, 87
Transition
  diagram, 407
  matrix, 407, 420
    regular, 422
Transitive directed graph, 360
Tree for solving probability problems, 200
True annual interest, 307
Truth in Lending Act, 310

Unit cost, 303
Units, 10
  metric, 304

Vector, 383, 390
  column, 390, 402
  current situation, 412
  probability, 420
  production, 385
  result, 385
  row, 391
Vertex
  of a graph, 327
  of a region, 50
Vonnegut, Kurt, 84

WATFIV, 82, 145
WRITE, 466

x-axis, *x*-coordinate, 22

y-axis, *y*-coordinate, 22
Yarborough, 218
Years to double, 302

Zero matrix, 393

1 2 3 4 5 6 7 8 9 10